Planning Curriculum in Mathematics

Jodean E. Grunow
Mathematics Consultant

John T. Benson, State Superintendent
Wisconsin Department of Public Instruction
Madison, Wisconsin

This publication is available from:

Publication Sales
Wisconsin Department of Public Instruction
Drawer 179
Milwaukee, WI 53293-0179
(800) 243-8782 (U.S. only)
(608) 266-2188
(608) 267-9110 Fax
www.dpi.state.wi.us/pubsales

Bulletin No. 2008

ISBN 1-57337-097-5

Printed on recycled paper

Foreword

The Wisconsin Department of Public Instruction's Strategic Plan identifies the following among its several goals:

Goal 1: Help all students reach challenging academic standards so they are prepared for further learning and productive employment.
Goal 2: Build a solid foundation for learning for all children.
Goal 3: Help all students become caring, contributing, responsible citizens.
Goal 4: Ensure talented, dedicated, and well-prepared educators are in every classroom and public school.

In a technological society that is continually requiring a better and deeper understanding of mathematics, meeting the cited goals becomes insistent and compelling. Because knowledge is increasing so rapidly, this document contends that we have all become learners of mathematics. And because we cannot begin to predict where it will all lead, it is necessary that we teach our children how to think, how to learn, and how to be critical consumers of knowledge. Teaching and learning mathematics with understanding is the goal of the curriculum, instruction, and assessment that is evolving. Professional development assumes a major role because teachers are learners, too.

Putting the standards into action requires addressing the content strands identified in *Wisconsin's Model Academic Standards for Mathematics* as well as the processes identified in Standard A: application of a variety of mathematical skills and strategies—including reasoning, oral and written communication, and the use of appropriate technology—when solving mathematical, real-world, and nonroutine problems. Implementation of the standards also requires an understanding of current educational research regarding the teaching and learning of mathematics.

There are certain foundations the task force deemed necessary in the curriculum design process, many of which have been foci of agency efforts as well: equity, technology, and service-learning. Additionally, the task force identified mathematics-specific foundations: openness, time, commitment, and adaptability. All of these factors often weigh heavily in the success of viable efforts.

This document has been prepared to help school districts with curriculum planning as they build programs to address these goals. Many thoughtful educators contributed to these efforts. The department extends its sincere gratitude to these people and thanks them for their genuine interest in the welfare of Wisconsin's mathematics students.

John T. Benson
State Superintendent of Public Instruction

Task Force

The department wishes to express appreciation and gratitude to the task force who contributed expertise and time to the development of this document. Devoting countless hours and much research to making sure this document reflects wise and responsible thinking regarding mathematics program development, the dedication of the task force members to their profession and to mathematics education is commendable. Wisconsin's students are fortunate to have such caring and concerned people working for them.

Task Force Advisors

Mary M. Lindquist
Mathematics Education Department
Columbus State College
Columbus, Georgia

Thomas A. Romberg
Department of Curriculum and Instruction
University of Wisconsin–Madison
Madison, Wisconsin

Task Force Co-Chairs

J. Marshall Osborn
Mathematics Department (Retired)
University of Wisconsin–Madison
Madison, Wisconsin
(Also a member of High School Task Force)

John Janty
Mathematics Department Chair
Waunakee Community School District
Waunakee, Wisconsin
(Also a member of High School Task Force)

Writing Groups

Elementary

DeAnn Huinker, Chair
Mathematics Education Department
University of Wisconsin–Milwaukee
Milwaukee, Wisconsin

Janet Alekna
Mixed Age Third/Fourth Teacher
Grove Elementary School
Wisconsin Rapids School District
Wisconsin Rapids, Wisconsin

William Breisch
Director of Curriculum and Instruction
Monona Grove School District
Monona , Wisconsin

Diana Kasbaum
Title I/Mathematics Consultant
Wisconsin Department of Public Instruction
Madison, Wisconsin

Phil Makurat
Mathematics Education Department
University of Wisconsin–Whitewater
Whitewater, Wisconsin

Lynne Miller
Early Education Administrator
Sheboygan Area School District
Sheboygan, Wisconsin

Middle School

Jane Howell, Chair
Mathematics Teacher (Retired)
Platteville School District
Platteville, Wisconsin

Faye Hilgart
Middle School Projects Coordinator
Madison Metropolitan School District
Madison, Wisconsin

High School

James F. Marty, Chair
Mathematics Instructor
Waukesha School District/Carroll College
Waukesha, Wisconsin

Ann Krause
Dean of General Education
Blackhawk Technical College
Janesville, Wisconsin

Jennifer Thayer
Director of Curriculum and Instruction
Monroe School District
Monroe, Wisconsin

Chapter Authors

James M. Moser
Mathematics Consultant (Retired)
Wisconsin Department of Public Instruction
Madison, Wisconsin

Billie Earl Sparks
Mathematics Education Department
University of Wisconsin–Eau Claire
Eau Claire, Wisconsin

John C. Moyer
Department of Mathematics, Statistics, and Computer Science
Marquette University
Milwaukee, Wisconsin
(Also a member of Middle School Task Force)

Steve Reinhart
Mathematics Teacher (Retired)
Chippewa Falls School District
Chippewa Falls, Wisconsin
(Also a member of Middle School Task Force)

Glen Miller
Director of Curriculum and Instruction
Kiel Area School District
Kiel, Wisconsin

Contributing Authors

Thomas P. Carpenter
Department of Curriculum and Instruction
University of Wisconsin–Madison
Madison, Wisconsin

Richard Lehrer
Educational Psychology Department
University of Wisconsin–Madison
Madison, Wisconsin

Contributors

Carol J. Apuli
Director of Curriculum and Instruction
Muskego-Norway School District
Muskego, Wisconsin

Eileen P. Reider
Principal
Muskego Elementary School
Muskego-Norway School District
Muskego, Wisconsin

Henry Kepner
Mathematics Education Department
University of Wisconsin–Milwaukee
Milwaukee, Wisconsin

Julie Stafford
Project Director
Wisconsin Academies Staff Development Institutes
Chippewa Falls, Wisconsin

Harlan Weber
Mathematics Department Chair
Sheboygan Area School District
Sheboygan, Wisconsin

Marge Wilsman
Director WECB On-Line Professional Development Services
Educational Communications Broadcasting
Madison, Wisconsin

Reactors

Joan Grampp
Mathematics Teacher
Milwaukee Public School District
Milwaukee, Wisconsin

Linda McQuillen
Teacher Leader
Madison Metropolitan School District
Madison, Wisconsin

Acknowledgments

Special thanks to:

Office of the State Superintendent
Steven Dold, Deputy State Superintendent

Division for Learning Support: Instructional Services
John Fortier, Assistant State Superintendent
Sue Grady, Director, Content and Learning Team
Gerhard Fischer, Education Program Coordinator
Kathryn Lind, Teacher Education and Licensing Consultant
Theresa Hall, Office of Educational Accountability
Beverly Kniess, Program Assistant

Division for Learning Support: Equity and Advocacy
Barbara Bitters, Director, Equity Mission

Division for Libraries, Technology, and Community Learning
Calvin J. Potter, Assistant Superintendent
Greg Doyle, Director, Education Information Services
Sandi Ness, Publications Director
Victoria Horn, Graphic Designer
Mark Ibach, Publications Editor
Mary Kathleen Boguszewski, Technology Education Consultant
Jeffery Miller, Service Learning

Wisconsin Mathematics Council Executive Board

University of Wisconsin–Madison Systems Administration
Fran Garb, Office of Academic Affairs

Very special thanks to:

The chapter authors for writing, rewriting, editing, supplying sources and resources, checking and rechecking, coming to DPI to refine yet another piece, and remaining smiling and accommodating throughout the process.

Jane Howell, who has been invaluable, researching when needed, proofing when under a deadline, finding specific pieces when needed, and providing inspiration, ideas, and guidance when called upon.

Beverly Kniess, who has dealt with all of the input for the document, scheduled meetings and sent letters, copied and mailed, produced charts, graphs, and forms, and has continued to be positive and upbeat.

An undertaking of this nature is huge in scope. These people have made the task so much easier.

Copyrighted Materials

Contents

From the National Perspective

Mary M. Lindquist
Chair, Commission on the Future of the Standards
Past President, National Council of Teachers of Mathematics

Often those of us who have worked on standards at the national level are asked why we have developed national standards when educational decisions are the prerogative of states. The mission of the National Council of Teachers of Mathematics (NCTM), the world's largest professional organization of mathematics teachers, is to improve mathematics education for all students. To make improvement, the council needed a vision that included what mathematics students should know and be able to do. Its first major effort in enunciating a vision was the publication of the *Curriculum and Evaluation Standards for School Mathematics* (NCTM 1989).

The *Standards* set broad goals that give direction but not prescription. Most states took the challenge of moving toward this vision as they created their own documents. Wisconsin was a leader because it had already created a document that was visionary for that time. Those of us developing the NCTM document often turned to *A Guide to Curriculum Planning in Mathematics* (Wisconsin Department of Public Instruction 1986) for guidance. This symbiotic relationship between state and national standards continues to develop. The national group provides a valuable resource for states and the states inform future documents of the council. For example, the writers of the *Principles and Standards for School Mathematics* (NCTM 2000) studied many state documents to ascertain the status of the nation's thinking and progress to date.

Wisconsin has again built on its own documents and on the national efforts in producing *Planning Curriculum in Mathematics.* This document complements *Wisconsin's Model Academic Standards for Mathematics* (Wisconsin Department of Public Instruction 1998), which sets forth what mathematics students should know and be able to do at the end of grade levels 4, 8, and 12. Setting content standards is a necessary step, but not a sufficient action. This document will help those involved in the mathematics education of Wisconsin students to move forward. In particular, the following aspects of this document are noteworthy:

Focus on teaching and learning. Teachers shape what students learn and their disposition toward learning mathematics. This document takes the position that learning mathematics with understanding is at the heart of making improvement. This position itself needs to be understood and teachers need support as they strive to teach for understanding. Not only of this charge were NCTM's three standards documents published in 1989, 1991, and 1995.

The central tenet underlying this charge is that students need to become mathematically literate.

The central tenet underlying this charge is that students need to become mathematically literate. Note that the term *literacy* refers to the human use of language. In fact, one's ability to read, write, listen, and speak a language is the most important tool through which human social activity is mediated. Each human language and each human use of language has an intricate design and a variety of functions to which it is tied in complex ways. For a person to be literate in a language implies that the person knows many of the design resources of the language and is able to use those resources for several different social functions. Analogously considering mathematics as a language implies that students must learn not only the concepts and procedures of mathematics (its design features), but also to use such ideas to solve nonroutine problems and to mathematize in a variety of situations (its social functions). The epistemological shift involves moving from judging student learning in terms of mastery of concepts and procedures to making judgments about student understanding of the concepts and procedures and their ability to mathematize problem situations. In the past, too little instructional emphasis was placed on understanding and the tests used to judge learning failed to adequately provide evidence about understanding or the ability to solve nonroutine problems.

Reform Schooling

All students can and must learn more and somewhat different mathematics than has been expected in the past in order to be productive citizens in tomorrow's world.

A set of assumptions about instruction and schooling practices has been associated with this vision of mathematical literacy. First, all students can and must learn more and somewhat different mathematics than has been expected in the past in order to be productive citizens in tomorrow's world. In particular, all students must have the opportunity to learn important mathematics regardless of socioeconomic class, gender, and ethnicity. Second, we have long underestimated the capability of all students to learn mathematics. Third, some of the important notions we expect students to learn have changed due to changes in technology and new applications. Thus, at every stage in the design of instructional settings we must continually ask, Are these important ideas in mathematics for students to understand? Wisconsin's *Model Academic Standards for Mathematics* were developed in light of these assumptions. Fourth, technological tools increasingly make it possible to create new, different, and engaging instructional environments. Fifth, the critical learning of mathematics by students occurs as a consequence of building on prior knowledge via purposeful engagement in activities and by discourse with other students and teachers in classrooms. The new NSF-supported mathematics curricula have been developed with these notions in mind.

This last assumption about the learning of mathematics is based on research carried out in the last decade showing that, in classrooms where the emphasis of instruction has shifted from mastery of facts and skills to understanding, students have become motivated to learn and achievement at all levels has increased. Carpenter and Lehrer (1999) have characterized understanding in terms of five interrelated forms of mental activity from which mathematical and scientific understanding emerges:

1. Constructing relationships
2. Extending and applying mathematical and scientific knowledge
3. Reflecting about mathematical and scientific experiences
4. Articulating what one knows
5. Making mathematical and scientific knowledge one's own

Since all learning occurs as a consequence of experiences and all humans have a variety of experiences, virtually all complex ideas in mathematics are understood by a student at a number of different levels in quite different ways. Furthermore, a student's level of understanding will change as a consequence of instructional experiences. Thus, the challenge is to create classroom experiences so that a student's understanding grows over time. As recently stated in *How People Learn*:

The challenge is to create classroom experiences so that a student's understanding grows over time.

> Students come to the classroom with preconceptions about how the world works. If their initial understanding is not engaged, they may fail to grasp the new concepts and information that are taught, or they may learn them for purposes of a test but revert to their preconceptions outside the classrooms.
>
> *(Donovan, Bransford, and Pellegrino 1999, 10)*

Research in mathematics instruction has identified a series of steps that lead students to understanding. Students should begin their investigations with a problem that needs to be addressed and that makes sense to them. The initial instructional activity should be experientially real to students so they are motivated to engage in personally meaningful mathematical work. This step involves raising questions about the problem situation. Hypothesis generation is a critical aspect of mathematical and scientific reasoning that has rarely been taught. Any instructional sequence assumes that each activity is justifiable in terms of some potential end points in a learning sequence. Paul Cobb states:

Students should begin their investigations with a problem that needs to be addressed and that makes sense to them.

> This implies that students' initially informal mathematical activity should constitute a basis from which they can abstract and construct increasingly sophisticated mathematical conceptions. At the same time, the starting point situations should continue to function as paradigm cases that involve rich imagery and thus anchor students' increasingly abstract mathematical activity.
>
> *(1994, 23–24)*

Next, students need to identify information and procedures they could use to answer their questions. Cobb goes on to argue that "Instructional sequences should involve activities in which students create and elaborate symbolic models of their informal mathematical activity. This modeling activity might involve making drawings, diagrams, or tables, or it could involve developing informal notations or using conventional mathematical notations" (p. 24).

Students need to identify information and procedures they could use to answer their questions.

to identify a vision, in setting the international and state mathematics scene, and in providing moral support as this daunting task was pursued.

We are ALL learners of mathematics.

Chapter 1, "We Are All Learners of Mathematics: Mathematics in Wisconsin," paints a picture of where mathematics as a field is today and briefly provides a history of mathematical development in the United States. A synopsis of the national standards pieces is given and their relationship to the *Wisconsin's Model Academic Standards for Mathematics* (1998) is discussed. An overview of legislation affecting Wisconsin's mathematics programs is offered and a recounting of Wisconsin's assessment results is cited. Contemplation of future thrusts for mathematics in Wisconsin is conjectured and the implications of the standards and assessment movement for the state are pondered. Concluding with possible mathematical missions for Wisconsin, the original contention that, "We are all learners of mathematics!" is supported. The chapter provides a national and state background against which the rest of the guidelines can be posited. A special note of thanks goes to Ann Krause, Dean of General Education, Blackhawk Technical College, for information regarding technical college expectations and world of work requirements. Thanks also go to J. Marshall Osborn, Mathematics Department, University of Wisconsin–Madison, for insights into university expectations.

"Development of understanding . . . should pervade everything that happens in mathematics classrooms."

—Carpenter and Lehrer 1999, 15

Chapter 2, "Teaching and Learning Mathematics with Understanding," has been reprinted with permission of Thomas P. Carpenter and Richard Lehrer of the University of Wisconsin–Madison from the book *Mathematics Classrooms That Promote Understanding* (Fennema and Romberg 1999, 19–32). These outstanding researchers present compelling evidence for teaching for understanding to be the goal of instruction. Noting the availability of "an emerging research base about teaching and learning," the authors suggest that "the most important feature of learning with understanding is that such learning is generative" (19). Discussing what understanding is and how it is developed through construction of relationships, extending and applying mathematical knowledge, reflecting about experiences, articulating what one knows, and making mathematical knowledge one's own, the authors pose the question, "Is understanding the same for everyone?" (6). Critical dimensions of classrooms that promote understanding are then identified and discussed. Finally, conjecturing that "understanding is a goal not only for students but also for teachers" (13), the authors conclude that "development of understanding . . . should pervade everything that happens in mathematics classrooms" (15). Sincere gratitude is expressed to the authors of the chapter and to the publishers, Lawrence Erlbaum Associates, for use of the chapter. It states the understanding case in terminology that could not be duplicated nor recounted.

Chapter 3, "CIA—Curriculum, Instruction, and Assessment—An Integrated Whole" looks at three crucial components of a mathematics program designed to foster understanding. Characterized by Fortier and Moser (November 1992) as "CIA," the three pieces are offered as a package; one cannot function adequately without attention to each of the others. The heart of the mathematics program discussion, curriculum, instruction, and assessment are its most important foundations. The curriculum segment contends that an excellent curriculum cannot achieve what is intended unless it is well taught and appropriately assessed. The instruction piece suggests that a district can adopt the finest curriculum available, but, if those materials are not well taught, it will not reap the desired results. Additionally, the enacted curriculum is much more than curricular materials. It is the interaction between and among the materials, the teacher, the students, and the mathematical environment. Excellent instructors of mathematics listen to students' cognitions to inform curricular decision-making. Finally, an excellent curriculum can be in place. Wise and wonderful teaching can take place. Yet, it is assessment, both formal and informal, that will convince us that teaching and learning mathematics with understanding is occurring. "It is assessment that helps us distinguish between teaching and learning."

The heart of the mathematics program discussion, curriculum, instruction, and assessment are its most important foundations.

Chapter 4, "Designing Professional Development to Promote Understanding," is written by Jodean E. Grunow, mathematics consultant, Department of Public Instruction. Having been involved in mathematics professional development for Wisconsin teachers for the last several years and having recently completed research (Grunow 1999) that deals specifically with building conceptual understanding, teacher knowledge, and appropriate assessment of the growth of such knowledge, Dr. Grunow gives a synopsis of current research regarding effective professional development. The professional development chapter is placed at the beginning of this document because the guide team felt that if we are to accomplish the goal of truly improving mathematics education for the children of Wisconsin, professional development needs to come first, not last (or maybe never), the position to which such endeavors have often been assigned in the past. "The scary thing about these efforts to truly address mathematical understanding is that those who are charged with the reform, were the most successful in the traditional classroom and often have never been themselves taught in the manner in which we need to teach" (Sparks 1998). Fortunately, Wisconsin has had several very successful professional development efforts that have advanced the cause considerably. Examples of those outstanding professional development programs are given at the end of the chapter as models of professional development programs that can be emulated.

"Understanding is a goal not only for students, but also for teachers."

—Carpenter and Lehrer 1999, 13

Having laid the groundwork for the document in the first five chapters, chapter 5, "Putting the Standards into Action," is the heart and soul of the document. In discussing district requests for help in implementing standards-based mathematics education, the guide team felt that several components were essential:

The heart and soul of the document is "Putting the Standards into Action."

- Rationale for process and procedures
- Discussion of sequencing and timing of mathematics concepts reflective of *Wisconsin's Model Academic Standards* (WMAS) (1998) and *Principles and Standards for School Mathematics* (PSSM) (2000)
- Presentation of challenging mathematics investigations complete with hoped-for understandings, appropriate assessment suggestions, and extension questions

The introduction to this section, written by Billie Earl Sparks, University of Wisconsin–Eau Claire, and James M. Moser, mathematics consultant, De-

Mathematics investigations that transcend grade levels are presented.

An approach believed to be replicable in classrooms is carefully outlined in each of the content strands.

partment of Public Instruction (retired), gives an "umbrella" view of essentials of rich school mathematics programs as explicated through each of the WMAS strands: mathematical processes, number and number relationships, geometry, measurement, statistics and probability, and algebraic relationships. Drs. Moser and Sparks present mathematical investigations that are broad enough to transcend grade levels as examples of the kinds of questions that can be used to build understandings that are developed over the course of time. The introduction is followed by grade-level sections very carefully assembled, negotiated, edited, and reedited by outstanding Wisconsin mathematics educators at the various grade levels. "Early Beginnings in Mathematics" was prepared by Dr. DeAnn Huinker, University of Wisconsin–Milwaukee; Lynne M. Miller, Early Learning Center, Sheboygan Area Schools; and Harlan Weber, mathematics coordinator, Sheboygan Area Schools. It focuses on young children's learning, mathematical programs for young learners, important ideas for early mathematics, and looks at the early mathematical beginnings represented in each of the content strands. Prepared by Dr. DeAnn Huinker, University of Wisconsin–Milwaukee; Dr. Phil Makurat, University of Wisconsin–Whitewater; Diana Kasbaum, Title I/mathematics, Wisconsin Department of Public Instruction; Janet Alekna, Grove Elementary School, Wisconsin Rapids School District; and Bill Breisch, Curriculum Director, Monona Grove School District, "Learning Mathematics in the Elementary Grades" gives an overview of ideas important in the development of elementary mathematics programs, comments on each of the standards for elementary grades, provides an illustrative task for each standard, discusses the task, looks at student work, considers the standards addressed, cites connections to other standards, asks extension questions, and notes other sample tasks. Believing this approach to be replicable in classrooms, it is carefully outlined in each of the content strands. Pursuing exactly the same course, "Learning Mathematics in the Middle School" was prepared by Jane Howell, Platteville Middle School (retired); Dr. John M. Moyer, Marquette University; Steve Reinhart, Chippewa Falls Middle School (retired); and Faye Hilgart, Madison Metropolitan School District. An interesting article by Steve Reinhart that appeared in the NCTM journal *Mathematics Teaching in the Middle School,* "Never Say Anything a Kid Can Say," appears in Appendix E, "Readings," and is offered as a springboard for additional contemplation. "Learning Mathematics in the High School," prepared by Dr. James F. Marty, Waukesha North High School and Carroll College; Dr. J. Marshall Osborn, University of Wisconsin–Madison; John Janty, Waunakee High School; Jennifer Thayer, Director of Curriculum and Instruction, Monona School District; and Ann Krause, Dean of General Education, Blackhawk Technical College, also worked with this model to present examples of mathematical investigations at the high school level.

Chapter 6, "Using Learning Research to Guide Mathematics Program Development," was written by John M. Moyer, Marquette University. This outstanding task force member has done an analysis of current mathematics research to identify some important research elements that school districts must consider in designing their mathematics programs. Dr. Moyer has identified the principles for teaching, designing curriculum, and conducting assessment that have grown out of cognitive science research and that focus on mathematical understanding. Noting the difference between learning research and instructional research, Dr. Moyer presents some principles of instruction based on a growing body of effective instruction research. Organizing the research findings around five dimensions of instruction that should be considered in designing mathematics instruction that promotes understanding (Hiebert et al. 1997), Moyer explores the nature of classroom tasks, the role of the teacher and the social culture of the classroom, mathematical tools as learning supports, and, because these guidelines discuss equity and access, he left discussion of the fifth dimension, equity and accessibility, to that piece of the guide. In an era of research quoting, Dr. Moyer has done a learned analysis of the research that exists, has pinpointed the research that needs to be considered in developing school mathematics programs, and has provided an understandable synopsis for users of this guide. Director of several outstanding middle school mathematics projects in the Milwaukee area, Dr. Moyer has also made valuable contributions to the middle school section and to the outstanding professional development programs pieces.

Research elements that school districts must consider in designing their mathematics programs are discussed.

Chapter 7, "Foundations for Mathematics Program Development," is a discussion of the assumptions that undergird this guide. Those foundations are openness; time; equity; technology; and service-learning. Another foundation piece, commitment and adaptability, is the final chapter of this document because it lays the groundwork for where we need to go from here. Openness suggests that there are several pieces presently in place for mathematics education—standards, assessments, and curricula—and that we have much knowledge about how children (and teachers) learn mathematics. The challenge is then presented to "use appropriately, in an unbiased manner, the knowledge regarding teaching and learning mathematics for understanding that exists." Often identified as a major deterrent, time is a component of learning for understanding; we must, as a nation, look at alternative methods for gaining the time needed to facilitate mathematical learning. Equity contends that, in a mathematized society, we cannot afford to slight one student mathematically; *all* students can learn substantive mathematics, and *all* students must be afforded the opportunity to learn the mathematics necessary to be productive citizens. Technology works hand-in-hand with mathematics, one providing seed for the other; employment of technological tools in an information age needs to be an underlying tenet for mathematics program development. Service-learning, a thrust of the Wisconsin DPI, provides rich, compelling problems for mathematical investigation as well as opportunities to apply mathematical skills in real-world situations.

The assumptions that undergird this document, the foundations, are: openness, time, equity, technology, and service-learning.

Chapter 8, "How Does a School District Look at Mathematics Program Development?" helps school districts look at steps that can be taken in contemplating their districts' mathematics programs. Offered only as suggestions, the ideas presented mirror many of the steps taken very effectively by school districts throughout the state as they work with standards-led mathematics programs. Many resources are cited in this offering to which districts can refer as they develop their own processes for mathematics program assessment. Several districts in Wisconsin have developed outstanding processes for examining their programs and for developing "next steps." A reflective piece on the

Resources to which districts can refer as they develop standards-led programs are offered.

process, "Getting Started at Improving Your Math Education Program: One District's Ongoing Story," by Glen Miller, Director of Curriculum, Kiel School District, is eloquently offered in Appendix E.

Chapter 9, "Commitment and Adaptability," provides a look backward, a view of today, and a vision for tomorrow in mathematics education. Suggesting that we have many important components in place, that we have many models to help us advance in our efforts, and that we still have challenges, it is suggested that we need two ideas as we carry forward: commitment and adaptability. Once we have selected an avenue of pursuit, we need to stay the course. We need to commit to the effort, to provide the necessary resources to bring the task to fruition, and to assess continually to see what we have accomplished. Then we need to be adaptable, carefully weighing changes to be made, making those changes in a timely manner, and assessing what has transpired to inform our next decisions. To quote a wonderful book called *Who Moved My Cheese?* (Johnson 1998): "They keep moving our cheese [our goals]. . . . Today we need flexible people . . . because we want not only to survive, but to stay competitive." Newman and Wehlage (1995) put it this way: "Disequilibrium is everywhere, the brain is equipped to deal with a turbulent world, the change process is intrinsically transformational, and to function best in a new environment, we need to embrace a fundamentally different world view or perceptual orientation" (10–11).

Commitment and adaptability are the name of the game.

There are seven appendixes in the document. Appendix A offers a listing of the National Science Foundation-funded mathematics curricula, contacts for those programs, and impact studies that can be referenced to assess effectiveness of the programs. Content specifications for the newly adopted Chapter PI 34: Teacher Education Program Approval and Licenses (July 1, 2000) are given in Appendix B. Specific performance standards for each of the grade levels, compiled by James M. Moser, Ph.D., DPI Mathematics Consultant (retired), in conjunction with his work with school districts on alignment, are offered in Appendix C. Several resources for classroom assessment are offered in Appendix D. Appendix E offers readings designed to stimulate contemplation. Thumbnail sketches of several of Wisconsin's exemplary professional development programs are given in Appendix F and can provide assistance to districts as they plan their professional development experiences. Finally, Appendix G recounts NCTM Position Papers. Also available on the NCTM Web site (www.nctm.org), these papers are offered here as readily reproducible synopses of thought on a host of issues.

The appendices offer specifics.

This document is designed to assist developers of Wisconsin's school mathematics programs to make wise decisions about improving their mathematics programs in an era of great change. It is the continuation of effort by the Wisconsin Department of Public Instruction (DPI) to assist Wisconsin educators as they strive to maintain outstanding schools and excellent mathematics programs.

References

Fennema, E., and T. A. Romberg, eds. 1999. *Mathematics Classrooms That Promote Understanding.* Mahwah, NJ: Lawrence Erlbaum Associates.

Grunow, J.E. 1999. *Using Concept Maps in a Professional Development Program to Assess and Enhance Teachers' Understanding of Rational Number.* Ann Arbor, MI: UMI Company. (Ph.D. diss., UMI Microform 9910421).

Hiebert, J., T. P. Carpenter, E. Fennema, K. Fuson, D. Wearne, H. Murray, A. Olivier, and P. Human. 1997. *Making Sense: Teaching and Learning Mathematics with Understanding.* Portsmouth, NH: Heinemann.

Johnson, S. J. 1998. *Who Moved my Cheese?* New York: G. P. Putnam's Sons.

Newmann, F. M., and G. G. Wehlage. 1995. *Successful School Restructuring: A Report to the Public and Educators by the Center on Organization and Restructuring of Schools.* Madison: Wisconsin Center for Education Research.

Romberg, T. A. and D. M. Stewart. 1987. *The Monitoring of School Mathematics: Background Papers.* Madison: Wisconsin Center for Education Research.

Sparks, B. E. 1998. Keynote address to the *Mathematics for the New Millennium* participants, August 10–13, Stevens Point, Wisconsin.

Standards 2000 Project. 2000. *Principles and Standards for School Mathematics.* Reston, VA: National Council of Teachers of Mathematics.

Wisconsin Department of Public Instruction. 1967. *Guidelines to Mathematics, K–6.* Madison: Wisconsin Department of Public Instruction.

———. 1981. *Guidelines to Mathematics, K–8.* Madison: Wisconsin Department of Public Instruction.

———. 1981. *The Mathematics Curriculum: 9–12.* Madison: Wisconsin Department of Public Instruction.

———. 1986. *A Guide to Curriculum Planning in Mathematics.* Madison: Wisconsin Department of Public Instruction.

———. 1998. *Wisconsin's Model Academic Standards for Mathematics.* Madison: Wisconsin Department of Public Instruction.

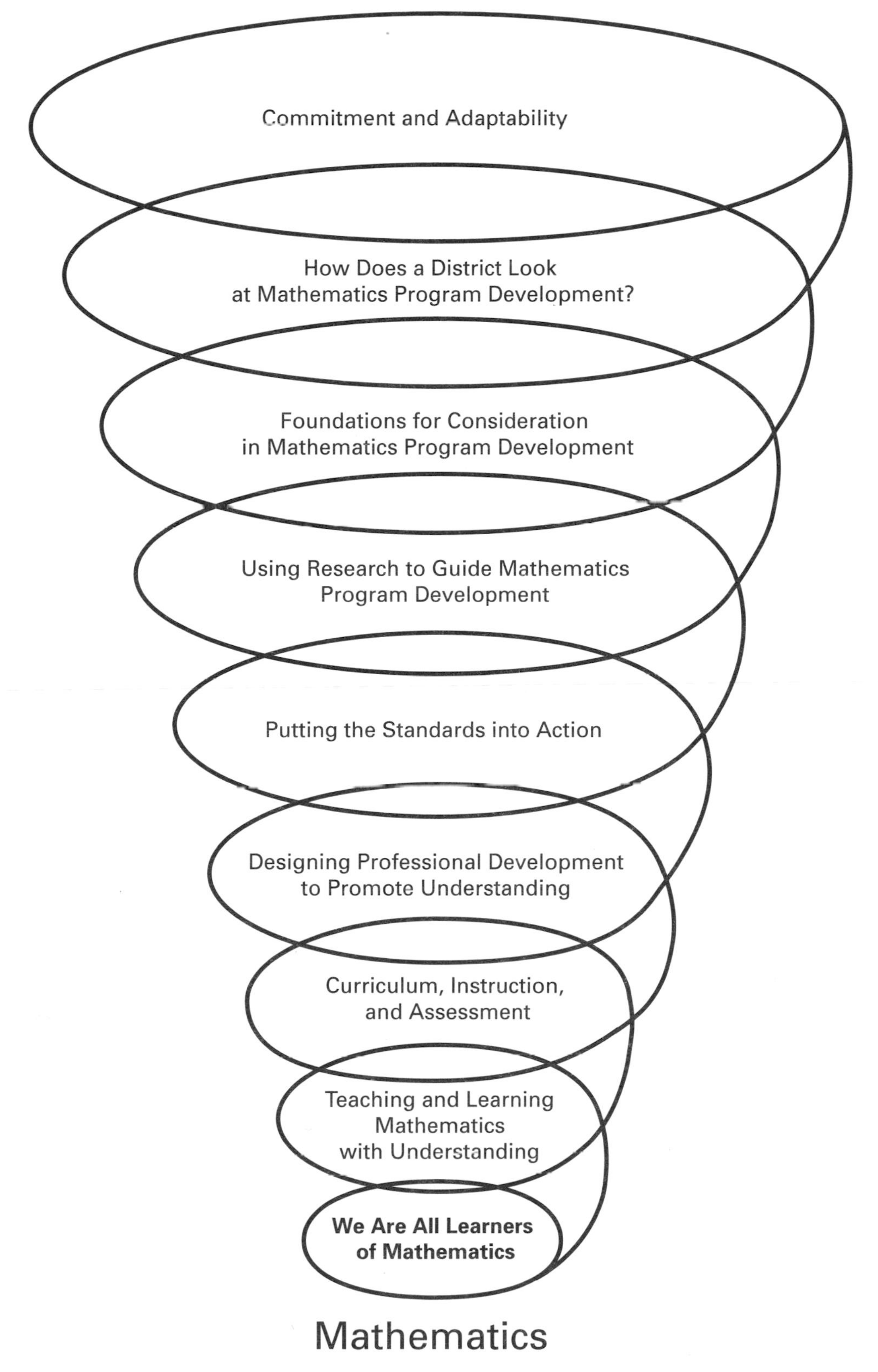

Spring (spring) *n.* 1. a source; origin; or beginning: *a spring into action.* 2. elasticity; resilience: *a spring forward with commitment.* (Morris, W., ed., 1971. *The American Heritage Dictionary of the English Language.* New York: American Heritage Publishing Company, Inc. and Houghton Mifflin Company, 1250, 2nd definition.)

We Are All Learners of Mathematics

Mathematics in Wisconsin

1

Where Is Mathematics Today?

We are living in an information age. Knowledge is increasing at an exponential rate. Mathematical knowledge that a few years ago fueled the fires of technological advances is now being augmented by those same advances; new and more complex mathematics is being generated. Ideas that existed only in the mind—fractals, genetic codes, the chaos of natural phenomena—can now be represented mathematically. The economics of existence have become more and more complicated. Stock market maneuvers and international monetary exchanges have become familiar entities of everyday life. We have learned much about how people learn; we can technologically map learning sequences, and we have research that helps us facilitate knowledge building.

"We live in an age of mathematics—the culture has been 'mathematized.'"

—National Academy of Science 1986

> In the past quarter century mathematics and mathematical techniques have become an integral, pervasive, and essential component of science, technology, and business. In our technically oriented society, "innumeracy" has replaced illiteracy as our principal educational gap. One could compare the contribution of mathematics to our society with needing air and food for life. In fact, we could say that we live in an age of *mathematics*—the culture has been *"mathematized."* (National Academy of Sciences 1986)

We Are *All* Learners of Mathematics

The task of mathematics education, then, becomes one of helping students and those who work with them to learn how to learn mathematics, how to solve problems, and how to acquire the automaticity with skills and procedures that is necessary for efficient and effective problem solving. Mathematics is best learned in meaningful and memorable contexts. Conceptual and procedural knowledge is best developed when a need for it has been established. Teachers of mathematics need to know the ever-changing mathematics content, effective methodologies, and the structure and pedagogy of mathematics. Others who work with students need to know what research says about learning and the importance of inquiry, and they need to be aware of the economic setting in which the mathematics of today and of the future is couched.

Mathematics is best learned in meaningful and memorable contexts.

Our students . . . must know how to learn the mathematics they will need to function in the future.

If today's mathematics students are to survive in an increasingly global economy, it is imperative that *all* of them know, understand, and are able to do complex mathematics. Likewise, with predictions that a majority of our students will pursue educational endeavors beyond high school, it is necessary that *all* graduating students have access to additional education; they must have mathematical backgrounds that will allow them to do this. Our students need to be confident in their mathematical abilities; they should be able to use problem solving techniques to tackle real-life dilemmas; they must know how to use the mathematics they presently understand; and they especially must know how to learn the mathematics they will need to function in the future.

It is an exciting time to be involved in mathematics education! Developing mathematics programs that capitalize on the strengths that now exist, that respond to arising needs, and that envision what can be becomes an exciting, compelling, and demanding challenge.

How Did Mathematics Education Get Where It Is Today?

World War II had a profound effect on people's ideas about the need for mathematics. War operations demanded much knowledge of mathematics; crash courses in mathematics and other subjects often had to be provided for recruits. The National Council of Teachers of Mathematics (NCTM) created the Commission on Post-War plans, which issued three reports (Commission on Post-War Plans of the National Council of Teachers of Mathematics 1944, 1945, 1947) recommending that schools ensure mathematical literacy for all who could achieve it, though, at the same time, "apparently believing that the majority of students ought not to be expected to study anything so advanced as first-year algebra" (Willoughby 2000, 3).

The value of educational research [was] established during the world wars.

Education in 1950 found itself breathing a sigh of relief after World War II and responding to the influx of postwar baby boomers. Though the war effort had called for mathematical innovation, postwar mathematics addressed the tasks at hand. It was standard practice for many students who were not college intending to study "practical" courses such as consumer mathematics, installment buying, principles of insurance, and so forth. Mothers who had worked during the war returned to homemaking, and daughters viewed themselves as eventually fulfilling that role, too; a college education for females was often considered optional as was higher-level mathematics.

The value of educational research had been established during the world wars, when academic research was used to address urgent national needs. Thus, in 1954 Congress passed the Cooperative Research Act, which authorized the United States Office of Education (USOE) to enter into cooperative arrangements with universities, colleges, and state education agencies to conduct educational research, surveys, and demonstrations, thus beginning "the modern era in educational research and development" (Klausmeier and Wisconsin Associates 1990, xiii). In 1954 the National Science Foundation was authorized to support course content improvement in the sciences, and in

1956 Congress appropriated funds to support educational research and other activities in 30 colleges and universities and in six state education agencies.

The launching of *Sputnik* by the Soviets in 1957 caught much of the United States unawares and heightened anxieties concerning the preparation of U.S. students in mathematics, science, and technology. At that time, the National Defense Education Act was passed, and the U.S. government committed to investing millions of dollars, under the auspices of the National Science Foundation (NSF, established in 1950 with the primary purpose of enhancing scientific research), in mathematics and science research in an effort to determine how both are best learned. In 1963, the USOE initiated a Research and Development Center program and founded four R & D centers in 1964 and six in 1965. The Wisconsin Center, the third center established in 1964, was charged with improving education "through programs of research, development, and dissemination, including demonstrations in the schools. Human and financial resources were to be concentrated over an extended period of time to solve particular educational problems" (Klausmeier and Wisconsin Associates 1990, xiii). The establishment of the Wisconsin Center offered education a valuable partner that helped shape mathematics education in the state and led it to renown in the nation.

Establishment of the Wisconsin Center for Educational Research in 1964 offered education a valuable partner that helped shape mathematics education in the state and led it to renown in the nation.

In 1959, the College Entrance Examination Board (CEEB) published the *Report of the Commission on Mathematics* (Commission on Mathematics 1959), which called for much more mathematics for the college-capable student. Calling the traditional one year of algebra and one year of geometry "a shocking inadequacy," a minimum recommendation of three years of mathematics was offered, and a strong call was issued for four years of high school mathematics, with the "most talented" needing to attempt an advanced placement program. Those four years should include algebra, geometry, trigonometry, an introduction to statistical thinking, and advanced mathematics including elementary functions, introductory probability with statistical applications, and introduction to modern algebra to prepare students for entry into college at the calculus level. Students able to participate in accelerated programs could pursue calculus.

The decade following *Sputnik* and governmental partnership in educational research saw the development of "new math," a concept that focused on bringing meaning to the discipline. Though "new math" had been conceived in 1951 with the creation of the University of Illinois Committee on School Mathematics (UICSM), the federal support following *Sputnik* gave it impetus, and the 1960s saw incorporation of its concepts in school materials. Backed with federal funds for support and innovation, new ideas were introduced into traditional classrooms. Federally funded teacher training programs proliferated, especially those that addressed advancing technologies. The effort was noble, but teachers were not specifically trained in how to use the mathematical materials that had been developed, parents were not given explanations regarding the shift in mathematical content, and students were caught between a taught mathematics and an experienced mathematics. Yet students who experienced well-taught "new math" and capitalized on the knowledge that it brought them found themselves very well prepared for

ESEA (1965) brought into focus attention:

To equality of opportunity for all

To addressing the specific needs of each student

To the allocation of resources to the benefit of each child.

mathematics- and science-oriented fields. Increasing numbers of doctors, research fellows, and so forth emanated from this era.

A parallel movement, begun with the Elementary and Secondary Education Act (1965) that launched the Title I program for disadvantaged students, brought into focus an attention to the equality of opportunity for all, to addressing the specific needs of each student, and to the allocation of resources to the benefit of each child. The need to serve all students mathematically, coupled with a public that on the whole did not understand "new math," led to calls for "back to the basics." The materials that were produced in the 1970s focused on practice, basic facts, and development of skills and procedures. But the retrenching of the 1970s reaped unwanted results. Mathematical skills, as measured on the National Assessment of Educational Progress (NAEP), declined (National Center for Educational Statistics 1998).

An Agenda for Action (National Council of Teachers of Mathematics, 1980)

In 1980, the NCTM—prompted by national test results; an extensive survey, Priorities in School Mathematics (PRISM), in which the profession and the public voiced their opinions on what was needed in mathematics programs; and emerging research on the building of mathematical understanding—issued *An Agenda for Action: Recommendations for School Mathematics of the 1980s.* It called for

- a problem solving focus,
- a definition of basic skills to encompass more than computational facility,
- taking full advantage of power of calculators and computers,
- standards for effectiveness and efficiency in the teaching of mathematics,
- the evaluation of mathematics programs on a "wider range of measures than conventional testing" (NCTM 1980, 1),
- mathematics study for more students with a greater range of options to accommodate diverse needs of the student population,
- a high level of professionalism for mathematics teachers, and
- public support commensurate with the importance of mathematical understanding to individuals and society.

Calls for change were issued:

NCTM, An Agenda for Action

NCEE, A Nation at Risk

Paulos, Innumeracy

NRC, Everybody Counts.

Other calls for change were heard. In an open letter to the American people, *A Nation at Risk: The Imperative for Educational Reform* (National Commission on Excellence in Education 1983, p. 4), a call was issued to maintain the economic viability of the nation and to afford each child an opportunity to access "the tools for developing their individual powers of mind and spirit to the utmost." John Allen Paulos's *Innumeracy: Mathematical Illiteracy and its Consequences*, became a national bestseller after its publication in 1988, calling attention to a "disease that has ravaged our technological society" (Hofstadter review, cover, Paulos, 1988). In *Everybody Counts: A Report to the Nation on the Future of Mathematics Education,* the National Research Council, the National Academy of Sciences, the National Academy of Engineering, and the Institute of Medicine committed to active participation "in the long-term work of rebuilding mathematics education in the United States" (National Research Council 1989, foreword).

Curriculum and Evaluation Standards for School Mathematics (Commission on Standards for School Mathematics, March 1989)

Thus, in 1989, the NCTM issued its *Curriculum and Evaluation Standards for School Mathematics*. After the fluctuations of the preceding decades—in light of the rapid advances that technology was making, the research on mathematics teaching and learning, and the uncertainty within the mathematics curriculum precipitated by changes in mathematics itself, the standards offered a vision of mathematics for the 1990s and appropriate curricular emphases for the era. The first document of its kind in a discipline, the standards were a landmark publication developed to "ensure quality, indicate goals, and promote change" (Commission on Standards for School Mathematics 1989, 2). The standards set forth societal goals of mathematically literate workers, lifelong learning, opportunity for all, and an informed electorate; the goals for students asked that they value mathematics, become confident in their abilities to do mathematics, become problem solvers, communicate mathematically, and reason mathematically. Urging mathematical empowerment for *all* students, the standards called for a paradigm shift in mathematical thinking—away from mathematics as "increasingly constricted sections of pipe through which all students must pass," toward "a pipeline for human resources that flows from childhood experiences to scientific careers" (Steen 1990).

The first document of its kind in a discipline, the **Curriculum and Evaluation Standards for School Mathematics** ***(Commission on Standards for School Mathematics 1989) were a landmark publication developed to ensure quality, indicate goals, and promote change.***

Professional Standards for Teaching Mathematics (Commission on Teaching Standards for School Mathematics, March 1991)

Recognizing the implications of research on mathematics teaching and learning, the NCTM next emphasized the importance of instruction in the process of building mathematical understanding. *Professional Standards for Teaching Mathematics* (Commission on Teaching Standards for School Mathematics 1991) called for teachers who are proficient in: (1) selecting mathematical tasks to engage students' interests and intellect, (2) providing opportunities to deepen their own understanding of the mathematics being studied and its applications, (3) orchestrating classroom discourse to promote the investigation and growth of mathematical ideas, (4) using, and helping students use, technology and other tools to pursue mathematical investigations, (5) seeking, and helping students seek, connections to previous and developing knowledge, and (6) guiding individual, small-group, and whole-class work.

The **Professional Standards for Teaching Mathematics** ***(Commission on Teaching Standards for School Mathematics, March 1991) emphasized the importance of instruction in the process of building mathematical understanding.***

Citing commonly held beliefs and dispositions regarding mathematics learning as major detractors to progress, the standards called for shifts toward classrooms as mathematical communities; logic and evidence as verification; reasoning, conjecturing, inventing, and problem-solving; and connecting mathematics, its ideas, and its applications. Pulling teachers from textbook-driven activities to decision-making roles in the setting of goals and the selection of tasks—as managers of classroom discourse, creators of environments to support teaching and learning mathematics, and analyzers of student learning—the standards called for the empowerment of teachers for the orchestration of learning opportunities. Continual and varied assessment was urged to inform teacher decision-making and program goal-setting. To help move teachers toward both goals, professional development for both pre-service and in-service teachers was urged as a primary thrust, and a call was made to policy makers, schools and school systems, colleges and universities, and professional organizations to support teacher development and worthy mathematics programs. The tenets espoused in the professional teaching standards served as a beacon in a new decade.

Assessment Standards for School Mathematics (Assessment Standards Working Groups, May 1995)

The third set of standards of the trilogy, Assessment Standards for School Mathematics *(Assessment Standards Working Groups 1995), suggested that "assessment should be a method of fostering growth toward high expectations" (p. 1).*

We teach what we assess. The third set of standards of the trilogy, *Assessment Standards for School Mathematics,* was published in 1995. The standards were designed to "expand on and complement" (Assessment Standards Working Groups 1995, 1) the NCTM's *Curriculum and Evaluation Standards for School Math.* Believing teachers to be the best judges of student progress and perceiving all students as capable of learning mathematics and of having that learning assessed, the working group concluded that "assessment should be a means of fostering growth toward high expectations" (1). The standards called for a shift to:

- a rich variety of mathematical topics and problem situations, away from [an] "arithmetic" [emphasis]
- investigating problems, away from memorizing and repeating
- questioning and listening in teaching, away from telling
- the use of evidence from several sources judged by teachers, away from a single externally judged test

The identified assessment standards were:

- Assessment should reflect the mathematics that all students need to know and be able to do.
- Assessment should enhance mathematics learning.
- Assessment should promote equity.
- Assessment should be an open process.
- Assessment should promote valid inferences about mathematics learning.
- Assessment should be a coherent process.

Citing the different purposes of assessment—monitoring student progress, making instructional decisions, evaluating students' achievement, and evaluating programs—the standards proposed that in order to develop mathematical power in *all* students, assessment "needs to support the continued mathematics learning of *each* student . . . Assessment occurs at the intersection of important mathematics content, teaching practices, and student learning" (6). *Assessment Standards for School Math* was a revolutionary document, offering a vision for a new decade.

National Science Foundation-Funded Curricula

Much has ensued in the decade since the release of the "Curriculum and Evaluation Standards." The standards movement brought focus to educational reform and prompted other disciplines to develop parallel standards. Educators, finding that the standards answered many of their needs, next inquired about curricular materials to address standards-led learning. The National Science Foundation, finding this to be a worthwhile focus, offered funding to groups willing to develop total curricula at the elementary, middle, and high school levels that reflected knowledge and research about the teaching and learning of mathematics and that incorporated the vision of the standards. Several consortia of interested mathematics researchers, professors, teachers, and others took up the challenge and developed curricula to reflect the intent of the standards. Those curricula had to be conceived using student research and had to show that they did, indeed, positively impact mathematics teaching and learning. Publishers for the materials were then sought. Additionally, support centers for the implementation of the materials were established to provide professional development for teachers in the use of the materials, to connect users of the materials for collegial dialogue, to further research the use of the materials, and to answer questions regarding the materials. (See Appendix A for a listing of the materials, supporting research pertaining to the materials, and support site centers' addresses, Web sites, etc.) Following the lead of the National Science Foundation, textbook publishers began to address standards-led education, revamping and augmenting their series to reflect those visions. States that had lists of recommended materials and texts demanded that the materials or texts to be considered were in sync with the intent of the standards.

The National Science Foundation . . . offered funding to groups willing to develop total curricula at three levels: elementary, middle, and high school.

Systemic Reform

Concurrent with the introduction of the standards and with the curricula development, the National Science Foundation initiated systemic reforms in states, cities, rural areas, and other local regions. The NCTM Standards documents proved to be useful frameworks for mathematics in those efforts. It was the intent of the developers of the national standards that states take those standards and "fit" them to their specific needs. States began the process of developing standards, adopting them, and then measuring attainment of them. Throughout the decade following the introduction of the standards, the states did this in various forms. (See the Mathematics in Wisconsin section in this chapter.)

Another goal of the original standards developers was that following the development of state standards, districts within the state would study the implications of the state standards for their programs and align their efforts appropriately. This part of the process has, likewise, often come to fruition.

Concurrent with the introduction of the standards . . . the National Science Foundation initiated systemic reforms in states, cities, rural areas, and other local regions.

Ultimately, however, it was the intent of the standards that they "influence the classroom practice of teachers directly" (Ferrini-Mundy 2000, 39). The adoption of the standards by the policy community through systemic efforts and curricula development brought an unintended focus: "The [influence of the standards] on classroom practice was, by an large, mediated through the translation of standards into other forms" (39).

Principles and Standards for School Mathematics (National Council of Teachers of Mathematics, 2000)

The PSSM reaffirmed the process already begun, added the principles of equity, curriculum, teaching, learning, assessment, and technology, and called for a highly ambitious, cohesive plan committed to excellence.

A decade after the release of the first national standards for school mathematics, the NCTM stepped back to evaluate the results of the standards movement, assess what still needed to be done, and consider foci for the next decade. In April 2000 at the NCTM Conference in Chicago, the *Principles and Standards for School Mathematics* were unveiled. Reaffirming the process already begun, reflecting requests for specific emphases, and planning for the future, the PSSM narrowed the content focus to five strands—number and operations, algebra, geometry, measurement, and data analysis and probability; this validated *Wisconsin's Model Academic Standards for Mathematics* (WMAS), which also listed those five content strands. Five process strands were identified: problem solving, reasoning and proof, communication, connections, and representations (WMAS lists all process standards as Standard A, the all-encompassing Mathematical Processes). To the standards of the PSSM were added six principles for school mathematics: equity, curriculum, teaching, learning, assessment, and technology; this reflects Wisconsin's concentration on CIA (curriculum, instruction, and assessment), openness, time, equity, technology, service-learning, and commitment and adaptability. In *A Vision for School Mathematics,* PSSM calls for a plan that "is highly ambitious. Achieving it requires solid mathematics curricula, competent and knowledgeable teachers who can integrate instruction with assessment, education policies that enhance and support learning, classrooms with ready access to technology, and a commitment to both equity and excellence" (NCTM 2000, ix).

Mathematics in Wisconsin

How Has Wisconsin Addressed the Standards and Assessment Movement?

The WMAS were the result of efforts over several years to ascertain what the people of Wisconsin wanted from their schools and from specific disciplines.

In 1998, *Wisconsin's Model Academic Standards for Mathematics* (Wisconsin Department of Public Instruction 1998) were released. In a joint letter from the lieutenant governor and state superintendent to the citizens of Wisconsin, the standards were characterized as guidelines for "greater academic achievement" for meeting "the challenges of tomorrow" and "significant because they herald the dramatically different way in which student achievement will be judged." The standards were the result of efforts over several years to ascertain what the people of Wisconsin wanted from their schools and from specific disciplines. Focus groups worked on identifying those desired outcomes.

Task forces made up of people from diverse backgrounds were appointed to identify the content and performance standards. The task forces grounded themselves in mathematical learning research, familiarized themselves with national and other state standards, and investigated international standards and assessments. The results of their efforts were offered three times for public hearing and feedback. Revisions were made in accordance with what was heard. The resulting document was published and widely disseminated. For instance, in the spring following the release of the document, each teacher who attended the Wisconsin Mathematics Council Green Lake Conference received a copy of the document.

Required by legislation [Wisconsin School Law 118.30 Pupil assessment (1g)(a)] to have a set of standards in place by August 1, 1988, for at least reading and writing, geography and history, mathematics, and science, school districts were offered the option of adopting the state standards or developing their own. Many districts, finding that the standards gave them good direction for their mathematics programs, began a process of examining their enacted curricula in light of the standards. This alignment process proved to be extremely valuable to districts. It afforded teachers within a discipline an opportunity to meet with the entire discipline staff, K–12, to examine the curriculum across grade levels and to identify gaps and overlaps. Focusing on developing a total K–12 mathematics experience and assuring that all pieces were in place for district students led districts to further identify important concepts to be developed at each grade level.

The process of aligning district mathematics programs with standards proved to be very valuable to districts.

With these processes complete, many districts began reexamining their curricular materials to see if they were adequately addressing the identified sequence and the focus on the building of understanding called for in the standards. Finding on the market for the first time materials that had been developed to reflect the intent of the standards and student effectiveness research, many school districts began a pilot process to evaluate materials before adopting them. Considering that in many cases school districts use their materials for 7 to 10 years, this review process has strengthened the basis of an important decision. Many Wisconsin districts are, at present, in this phase of standards implementation.

A piloting process to evaluate materials before adopting them has strengthened the basis of an important decision.

In conjunction with a call for standards, Wisconsin School Law 118.30 Pupil assessment called for examinations to: (1)(a) . . . "measure pupil attainment of knowledge and concepts in the fourth, eighth, and tenth grades"; and to (1)(b) ". . . measure whether pupils meet the pupil academic standards issued by the governor as executive order no. 326, dated January 13, 1998." These criteria are applicable to charter schools as well as public schools. In addition to spelling out graduation criteria that include proficiency demonstration in specified coursework (118.33 High school graduation standards; criteria for promotion), by September 1, 2002, each school board "shall develop a written policy specifying criteria for granting a high school diploma that, in addition to the proficiency criteria, shall include the pupil's academic performance and the recommendations of teachers" (Wisconsin Department of Public Instruction 2000b). Likewise, each school board must adopt a written policy specifying the criteria for promoting a pupil from fourth grade to fifth grade and from eighth grade to ninth grade, including a proficiency measure and "the pupil's academic performance; the recommendations of teachers, . . . and any other academic criteria specified by the school board" (6)(a)1. This measure also pertains to charter schools. Districts may adopt the state-developed proficiency examinations or they may develop their own. The state-developed examinations reflect the content and performance standards and the thrust for the development of student understanding.

What Is Wisconsin's Mathematics "of Record"?

Wisconsin has always been a national leader in the field of mathematics.

Wisconsin has always been a national leader in the field of mathematics. The recent *National Education Goals Report: Building a Nation of Learners* (National Education Goals Panel 1999) identified Wisconsin as one of 12 states that achieved "outstanding performance" in public school education improvement and cited six areas of "highest performing" ratings—including Wisconsin fourth graders, 27 percent of whom performed at or above proficient in mathematics, and Wisconsin eighth graders, 32 percent of whom performed at or above proficient. Additionally, Wisconsin eighth graders achieved "most-improved" status, with a nine-point gain in the percentage of public school eighth graders scoring at or above proficient in mathematics (34–35). "Improvement over time" also cited Wisconsin eighth graders for the significant increase in the percentages of public school eighth graders who scored at or above proficient in mathematics.

In 2000, 69% of Wisconsin high school students took the ACT test.

On an international basis, a study (Johnson and Owen 1998) done for the U.S. Department of Education statistically linked state results from the 1996 NAEP with country results from the 1995 Third International Mathematics and Science Study (TIMSS). The linking study investigated how each state would have performed on TIMSS, relative to the 41 countries that actually participated in the international assessment. Wisconsin achieved "highest-performing" status, along with Iowa, Maine, Minnesota (based on actual scores, not estimated scores), Montana, Nebraska, and North Dakota. Only Belgium (Flemish educational system), the Czech Republic, Hong Kong, Japan, Korea, and Singapore would be expected to outperform these seven states in eighth grade mathematics.

The University of Wisconsin (UW) System and Wisconsin Technical College System (WTCS) seek, as part of their admission requirements, results from ACT testing. (If WTCS applicants do not have ACT or SAT scores, they are asked to perform on alternative measures such as ASSET, TABE, and COMPASS.) Because the ACT Assessment (http://www.act.org/news/data/00/00states.html) is designed for students who plan to attend college, its focus is on students who have completed the recommended college preparatory courses as defined by ACT. This includes three years or more of mathematics, including one year's credit each for algebra I, algebra II, and geometry and one half-year's credit for trigonometry, calculus (not pre-calculus), other math courses beyond algebra II, and computer math or computer science.

In 2000, 69 percent of Wisconsin high school students took the ACT test. The average composite state score was 22.2. On a national level, 38 percent of graduates were tested and the average composite score was 21.0. Although the assessment companies strongly discourage comparison of state-by-state scores, Wisconsin has consistently placed among the top states in the United States in average composite scores.

Some students intending to attend out-of-state schools take the SAT (http://www.sat.org/press/senior00/html/table4.html). Seven percent of Wisconsin students took the SAT in 2000. They received an average verbal score of 584 and an average mathematics score of 597, in comparison with a national verbal score of 505 and mathematics score of 514. In the last 10 years, both the verbal and mathematics scores have seen small but steady increases.

Wisconsin's students participation in Advanced Placement programs has expanded 73 percent over the past five years.

Additionally, Wisconsin students' participation in Advanced Placement (AP) programs has expanded 73 percent over the past five years (Wisconsin Department of Public Instruction 2000a). Sixty-five percent of high schools in Wisconsin participated in the AP program in the 1999–2000 school year, as compared to a national rate of 57 percent. Nationally, the passing rate on the AP exams that year was 63.7 percent; Wisconsin had a passing rate of 68.8 percent.

Where Does Wisconsin Need to Focus Its Efforts in Mathematics?

Further analysis of the NAEP results shows that, while 3 percent of Wisconsin's fourth graders are advanced and 24 percent are at or above proficient, 47 percent are at or above basic and 26 percent are below basic. At the eighth grade level, while 5 percent are advanced and 27 percent are at or above proficient, 43 percent are at or above basic and 25 percent are below basic. When the data are disaggregated, other findings surface. Although statistically male and female performances do not differ significantly in Wisconsin, performances of white students differ from those of Black and Hispanic students; improvements that were noted from 1992 to 1996 for white students in Wisconsin were not noted for Black or Hispanic students in Wisconsin. Other factors in these results besides ethnicity were the educational level of one or both parents, Title I participation, free/reduced-price lunch program eligibility, and the type of location of the school district.

Consistency in performance across grade levels is a Wisconsin goal.

Analysis of the Wisconsin Knowledge and Concepts Examination (WKCE) (www.dpi.state.wi.us) results show small but steady growth from 1997–1998 to 1998–1999 to 1999–2000. Yet, consideration needs to be given to the fact that a 75 percent proficiency rate for fourth grade means that 25 percent of fourth graders are at basic or minimal; a 42 percent proficiency rate for eighth grade means that 58 percent of eighth graders are at basic or minimal; and a 39 percent proficiency rate for tenth grade means that 61 percent of tenth graders are at basic or minimal. There is a decline in performance from fourth to eighth to tenth grades. Disaggregated data show little difference in male and female performances, but percentage performances of nonwhite, limited English-proficient, disabled, and economically disadvantaged populations are below the average state performance. In all instances there is a decline in performance from fourth to eighth to tenth grade. Although this phenomenon mirrors findings on the NAEP, it is not an international finding. TIMSS results for other countries show much more consistency in performance across the school years.

What Are the Implications of the Standards and Assessment Movement for Wisconsin?

Students need to become mathematical thinkers and builders of knowledge.

Mathematics is changing. As technology mushrooms, new mathematics is developed. Increasingly sophisticated mathematics is demanded of each person for mere economic survival. Consequently, we cannot have some of our students at basic or minimal. We need to make sure that each student is equipped to handle the mathematical thinking that will be necessary to succeed in the twenty-first century. With a huge growth in knowledge, the mere grasping of skills and procedures is inadequate. Students need to learn how to learn mathematics, to know how to use mathematics to solve problems, and to have enough confidence in themselves to tackle complex mathematical tasks. They need to be able to break a complex problem into "do-able" pieces so they can analyze the situation bit by bit. Then, they need to be able to synthesize their several solutions into a cohesive, whole solution. Students need to become mathematical thinkers and builders of knowledge.

Educators are finding that they need to reevaluate and readjust their instructional practices and expectations as well as their curricula in order to engage today's students in meaningful and relevant learning experiences.

As mathematical endeavors are contemplated for today's students, educators need to recognize that most have life experiences quite different from those of educators themselves. Many of today's students have spent countless hours having their world defined for them on television screens, computer displays, and video games. Their three-dimensional world has often been experienced through two-dimensional displays. Although children in the past spent much time in their developmental years engaging in hands-on activities—such as playing with tinker toys, erector sets, dollhouses, toy trucks and cars, sandboxes, and toy hammers and nails and interacting with their world by touching, smelling, and tasting—many of today's students have had their individual activities carefully staged and orchestrated and their interactive activities formalized and organized by adults, in stark contrast to the "pick-up" games of the past. Clearly, students are arriving at the steps of the schoolhouses with much different experiences and expectations than students of the past. Educators are finding that they need to reevaluate and readjust their instructional practices and expectations as well as their curricula in order to engage today's students in meaningful and relevant learning experiences.

Education and Employability

Educational attainment affects employability. A recent report from the National Center for Education Statistics (Ibach 2000) noted:

> The employment rate of male and female 25- to 34-year-olds was generally higher among those individuals with a higher level of education; in 1998, males and females ages 25–34 with a bachelor's degree or higher were more likely to be employed than their peers who had lower levels of educational attainment.
>
> Between 1971 and 1998, the employment rate of males ages 25–34 decreased for those who had not finished high school and those with a high school diploma or GED, and remained relatively constant for those with some college and those with a bachelor's degree or higher.

Between 1971 and 1998, the employment rate of females ages 25–34 increased across all education levels. The rate of increase for females who did not complete high school was lower than the rate of increase for females who attained higher levels of education.

Mathematics Education and Post-High School Educational Experiences

Demographics are showing us that growing numbers of students will be seeking post-high school educational experiences. State requirements (PI 18.03 High school graduation standards) that have called for two credits of high school mathematics including (3.) "instruction in the properties, processes, and symbols of arithmetic and elements of algebra, geometry, and statistics" are undergoing scrutiny. There are many calls for additional credit requirements and for a change to "coursework, instruction, and assessment that reflects the Wisconsin Model Academic Standards and Wisconsin state assessments" (Proposed PI 18.03 Changes for High School Graduation in Mathematics Discussion Issues). The University of Wisconsin–Madison requires three years of high school mathematics and notes that most students seeking admission have an average of four years of high school mathematics (http://www.uwsa.edu/opar/oparpub.htm). Indeed, a call for four years of challenging mathematics for all high school students was issued at the press conference held during the April 2000 National Council of Teachers of Mathematics Annual Meeting in Chicago, when the *Principles and Standards for School Mathematics* were released.

Growing numbers of students will be seeking post-high school educational experiences.

Mathematics and Economic Viability

Mathematical ability is increasingly key in maintaining economic viability in the changing world of work. Employers in the state of Wisconsin and across the nation contend that creative thinking and problem solving are key skills for their employees. The SCANS (Secretary's Commission on Achieving Necessary Skills) 2000 report defined Workplace Know-How as a three-part foundation of skills and personal qualities that are needed for solid job performance (http://www.scans.jhu.edu/workreq.html). Competent workers in the high performance workplace need:

Creative thinking and problem solving are key skills for employees.

Basic skills—reading, writing, **arithmetic and mathematics,** speaking, and listening

Thinking skills—the ability to learn, to reason, to think creatively, to make decisions, and to **solve problems**

Personal qualities—individual responsibility, self-esteem and self-management, sociability, and integrity

In Wisconsin, the technical colleges have defined these foundational skills as core abilities. Business and industry representatives from across the state have validated these core abilities, transferable skills, and attitudes as essential to an individual's success regardless of occupation or community setting. Though these core abilities are loosely based on the SCANS report, each technical college in the state has developed its own set of core abilities. The core abilities for each of the 16 technical colleges in Wisconsin are available through the Wisconsin Instructional Design System (WIDS) and the Wisconsin Technical System Foundation. The general themes represented by these core abilities are:

Many different success stories testify to the crucial value of mathematics in important real-world problems.

Communicate clearly
Work effectively in teams
Use appropriate technology
Solve problems and demonstrate critical thinking
Apply basic skills effectively
Demonstrate appropriate work behavior
Show respect for diversity
Value self positively

Consequently, Wisconsin graduates must possess these foundational skills in order to be attractive to employers.

In addition, the SIAM (Society for Industrial and Applied Mathematics 2000) Report on Mathematics in Industry (http://www.siam.org/mii/node1.html) confirms the remarkable range and variety of the applications of mathematics in industry and government. Many different success stories testify to the crucial value of mathematics in important real-world problems, including materials processing, automobile design, medical diagnosis, development of financial products, network management, and weather prediction. Some of the most important traits identified include: **skill in formulating, modeling, and solving problems from diverse and changing areas;** interest in, knowledge of, and flexibility across disciplines; knowledge of and experience with computation; communication skills, spoken and written; and adeptness at working with colleagues.

Mathematics Education and Higher-Education Admission Requirements

A recent landmark agreement between the UW-System and the WTCS regarding transfer of credits has made the mathematical choices students make even more important.

Academic requirements for entrance to the UW System and the WTCS are becoming more aligned. The Wisconsin Alignment Project was undertaken in October 1999 to address issues of continuity between what students are expected to learn in high school and what they must know to be prepared to succeed in college, and to strengthen the articulation between K–12 and higher education (http://www.uwsa.edu/acadeaff/align/). In math and science, a curriculum based on WMAS would prepare a student to enter the WTCS without remediation. However, additional preparation would be required beyond that described by WMAS for students applying to the UW System. Due to a recent landmark agreement between the UW System and the WTCS regarding transfer of credit (Wisconsin Technical College System Board and Board of Regents of the University of Wisconsin System 2000), the mathematical choices students make are even more important. Higher education, business, and industry are demanding that mathematics education be encouraged for *all,* not only a select few.

The U.S. Department of Labor's Bureau of Labor Statistics focuses on the high-tech era into which students will be entering. Citing the 30 occupations having the largest projected growth in the period from 1996 to 2006, they note that collectively, the 30 occupations will account for more than 8.6 million new jobs (Linn 1999). Three involve large amounts of computer knowledge and will account for one million new jobs, 12 percent of the new jobs projects. Eight occupations involve considerable understanding of computer applications and account for 2.6 million new jobs, 30 percent of the jobs projected. The 19 remaining occupations will involve minimal computer skills, but of those 5 million jobs, 3 million will pay low wages. That leaves 20 percent of the new jobs not requiring much in the way of computer skills, but paying better than low wages. Mathematics and computer technology are inextricably wed. These figures speak of the era in which our students will live and function. We have to fit our students for the mathematical thinking that such technological expertise will demand.

We have to fit our students for the mathematical thinking that a high-tech world demands.

What, Then, Is Wisconsin's Mathematical Mission?

With the mathematical vision that stands on the horizon of a new millennium, with WMAS as a guideline, with examinations that reflect more and more the intent of the standards in operation, and with curricular materials available that have been developed to reflect knowledge of how children learn and the intent of the standards, the focus must now shift to developing instructional philosophies and methodologies that will help students build conceptual understanding, develop mathematical facility, and bolster confidence in their abilities to "do math." This focus must be on helping *all* students build mathematical power and on working with *all* mathematics teachers to capitalize on their considerable skills to facilitate the development of such knowledge.

Wisconsin's mathematical mission must focus on helping ALL students build mathematical power and on working with all mathematics teachers to capitalize on their considerable skills to facilitate the development of conceptual understanding, mathematical facility, and mathematical confidence.

References

Assessment Standards Working Groups. 1995. *Assessment Standards for School Mathematics.* Reston, VA: National Council of Teachers of Mathematics.

Commission on Mathematics. 1959. *Report of the Commission on Mathematics: Program for College Preparatory Mathematics.* New York: College Entrance Examination Board.

Commission on Standards for School Mathematics. 1989. *Curriculum and Evaluation Standards for School Mathematics.* Reston, VA: National Council of Teachers of Mathematics.

Commission on Teaching Standards for School Mathematics. 1991. *Professional Standards for Teaching Mathematics.* Reston, VA: National Council of Teachers of Mathematics.

Ferrini-Mundy, J. 2000. "The Standards Movement in Mathematics Education." In *Learning Mathematics for a New Century,* edited by M. J. Burke and F. R. Curcio. Reston, VA: National Council of Teachers of Mathematics.

Ibach, M., ed. 2000. "NCES Data Describes Employment of Young Adults by Educational Attainment." In *Education Forum,* 3 (35) (April / May). Madison: Wisconsin Department of Public Instruction. (http://www.dpi. state.wi.us/dpi/edforum/ef0335_7.html)

Johnson, E. G., and E. Owen. August 1998. *Linking the National Assessment of Educational Progress (NAEP) and the Third International Mathematics and Science Study (TIMSS): A Technical Report.* Washington, D.C.: U.S. Department of Education, National Center for Education Statistics Research and Development Report.

Klausmeier, J. H., and Wisconsin Associates. 1990. *The Wisconsin Center for Education Research: Twenty-Five Years of Knowledge Generation and Educational Improvement.* Madison: Wisconsin Center for Educational Research, School of Education, University of Wisconsin–Madison.

Linn, E. 1999. "Tomorrow's Jobs: How High-Tech Are They?" In *Equity Coalition, V* (Fall). University of Michigan School of Education: Programs for Educational Opportunity.

National Academy of Sciences. 1986. *Renewing U.S. Mathematics.* Overhead package for *Curriculum and Evaluation Standards for School Mathematics 1989.* Reston, VA: National Council of Teachers of Mathematics.

National Center for Education Statistics. 1998. *NAEP Facts (3)* 2 (September): 1. Washington D.C.: U.S. Department of Education.

National Council of Teachers of Mathematics. 1980. *An Agenda for Action: Recommendations for School Mathematics of the 1980s.* Reston, VA: National Council of Teachers of Mathematics.

———. 2000. *Principles and Standards for School Mathematics.* Reston, VA: National Council of Teachers of Mathematics.

National Education Goals Panel. 1999. *The National Education Goals Report: Building a Nation of Learners.* Washington, D.C.: U.S. Government Printing Office.

National Research Council. 1989. *Everybody Counts: A Report to the Nation on the Future of Mathematics Education.* Washington, D.C.: National Academy Press.

Office of Educational Research and Improvement. 1998. *National Center for Education Statistics NAEP Facts, 3* (2) (September). Washington, D.C.: U.S. Department of Education.

Paulos, J. A. 1988. *Innumeracy: Mathematical Illiteracy and Its Consequences.* New York: Vintage Books, Random House, Inc.

Society for Industrial and Applied Mathematics (SIAM). February 11, 2000. *The SIAM Report on Mathematics in Industry*. (www.siam.org/mii/node1.html).

Steen, L. A. 1990. *On the Shoulders of Giants: New Approaches to Numeracy.* Washington, D.C.: National Academy Press.

Willoughby, S.S. 2000. Perspectives on Mathematics Education. In *Learning Mathematics for a New Century*, edited by M.J. Burke and F.R. Curcio. Reston, VA: National Council of Teachers of Mathematics.

Wisconsin Department of Public Instruction. 1998. *Wisconsin's Model Academic Standards for Mathematics.* Madison: Wisconsin Department of Public Instruction.

———. 2000a. *Suggestions for Local School Boards in Approaching the Development of High School Graduation and Fourth/Eighth Grade Advancement Policies: Implementing the Provisions of 1999 Wisconsin Act 9.* Madison: Wisconsin Department of Public Instruction.

———. 2000b. *DPI Information* (DPI 2000, 55). Madison: Wisconsin Department of Public Instruction.

Wisconsin Technical College System Board and Board of Regents of the University of Wisconsin System. 2000. *Resolution on Student Transfer Collaboration*. Madison: University of Wisconsin.

Planning Curriculum in

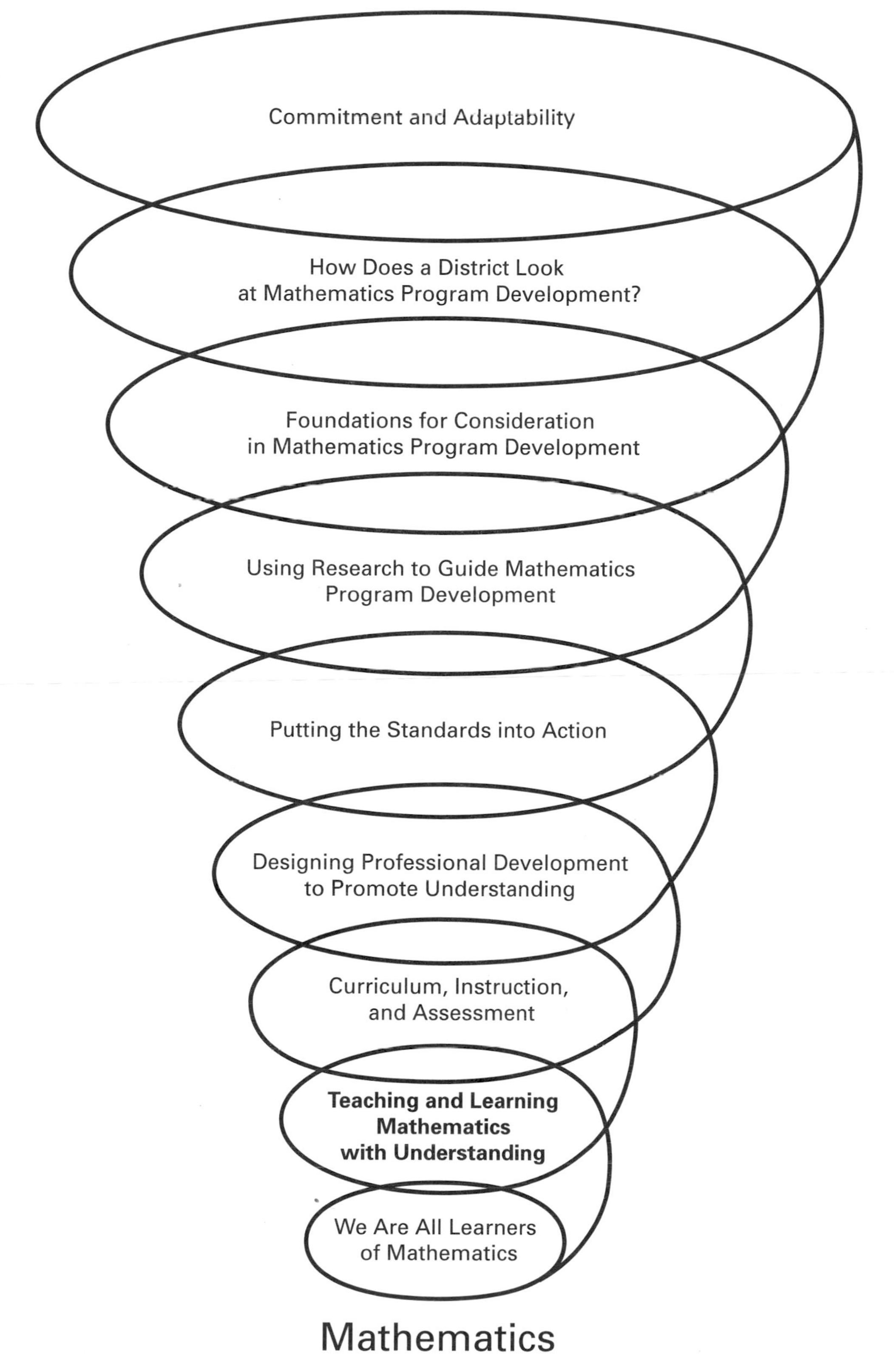

Mathematics

Spring (spring) *n.* 1. a source; origin; or beginning: *a spring into action.* 2. elasticity; resilience: *a spring forward with commitment.* (Morris, W., ed., 1971. *The American Heritage Dictionary of the English Language.* New York: American Heritage Publishing Company, Inc. and Houghton Mifflin Company, 1250, 2nd definition.)

Teaching and Learning Mathematics With Understanding

Thomas P. Carpenter
Richard Lehrer
University of Wisconsin–Madison

2

This article is reprinted with permission of the authors, Thomas P. Carpenter and Richard Lehrer of the University of Wisconsin–Madison, and the publisher, Lawrence Erlbaum Associates. It appears in E. Fennema and T. A. Romberg, eds., *Mathematics Classrooms That Promote Understanding*, (Mahwah, NJ: Lawrence Erlbaum Associates 1999), 19–32. References in the article that address other chapters refer to chapters in that book. Sincere thanks is extended to the authors and the publisher for authorizing reprinting of the article. One of the books in the Studies in Mathematical Thinking and Learning Series (Alan Schoenfeld, advisory editor), the entire book addresses the building of understanding in classrooms and is recommended reading.

To prepare mathematically literate citizens for the twenty-first century, classrooms need to be restructured so that mathematics can be learned with understanding. This is not a new goal of instruction: school reform efforts since the turn of the twentieth century have focused on ways to create learning environments so that students learn with understanding. In earlier reform movements, notions of understanding were often derived from ways that mathematicians understood and taught mathematics. What is different now is the availability of an emerging research base about teaching and learning that can be used to decide what it means to learn with and to teach for understanding. This research base describes how students themselves construct meaning for mathematical concepts and processes and how classrooms support that kind of learning.

The availability of an emerging research base about teaching and learning can be used to decide what it means to learn with and to teach for understanding.

Why Understanding?

Perhaps the most important feature of learning with understanding is that such learning is generative. When students acquire knowledge with understanding, they can apply that knowledge to learn new topics and solve new and unfamiliar problems. When students do not understand, they perceive each topic as an isolated skill. They cannot apply their skills to solve problems not explicitly covered by instruction, and neither can they extend their learning to new topics. In this day of rapidly changing technologies, we cannot anticipate all the skills that students will need over their lifetimes or the problems they will encounter. We need to prepare students to learn new skills and knowledge and to adapt their knowledge to solve new problems. Unless students learn with understanding, whatever knowledge they acquire is likely to be of little use to them outside the school.

What Is Understanding?

Learning with understanding is generative.

Understanding is not an all-or-none phenomenon. Virtually all complex ideas or processes can be understood at a number of levels and in quite different ways. Therefore, it is more appropriate to think of understanding as emerging or developing rather than presuming that someone either does or does not understand a given topic, idea, or process. As a consequence, we characterize understanding in terms of mental activity that contributes to the development of understanding rather than as a static attribute of an individual's knowledge.

How Understanding Is Developed

We propose five forms of mental activity from which mathematical understanding emerges: (1) constructing relationships, (2) extending and applying mathematical knowledge, (3) reflecting about experiences, (4) articulating what one knows, and (5) making mathematical knowledge one's own. Although these various forms of mental activity are highly interrelated, for the sake of clarity we discuss each one separately.

Constructing Relationships

It is appropriate to think of understanding as emerging or developing rather than presuming someone either does or does not understanding a given topic, idea, or process.

Things take meaning from the ways in which they are related to other things. People construct meaning for a new idea or process by relating it to ideas or processes that they already understand. Children begin to construct mathematical relations long before coming to school, and these early forms of knowledge can be used as a base to further expand their understanding of mathematics. Formal mathematical concepts, operations, and symbols, which form the basis of the school mathematics curriculum, can be given meaning by relating them to these earlier intuitions and ideas. For example, children as early along in the educational process as kindergarten and first grade intuitively solve a variety of problems involving joining, separating, or comparing quantities by acting out the problems with collections of objects. Extensions of these early forms of problem-solving strategies can be used as a basis to develop the mathematical concepts of addition, subtraction, multiplication, and division (see chapter 4, Carpenter et al., in Fennema and Romberg 1999).

Unless instruction helps children build on their informal knowledge and relate the mathematics they learn in school to it, they are likely to develop two separate systems of mathematical knowledge: one they use in school and one they use outside of school. For example, children often are not bothered by the fact that they get one answer when they calculate with paper and pencil and another when they figure out the same problem using counters or some other material. They do not see that the answers they get with procedures they learn in school should be the same as the answers they get when they solve problems in ways that make sense to them.

Extending and Applying Mathematical Knowledge

It is not sufficient, however, to think of the development of understanding simply as the appending of new concepts and processes to existing knowledge. Over the long run, developing understanding involves more than simply connecting new knowledge to prior knowledge; it also involves the creation of rich, integrated knowledge structures. This structuring of knowledge is one of the features that makes learning with understanding generative. When knowledge is highly structured, new knowledge can be related to and incorporated into existing networks of knowledge rather than connected on an element-by-element basis. When students see a number of critical relationships among concepts and processes, they are more likely to recognize how their existing knowledge might be related to new situations. Structured knowledge is less susceptible to forgetting. When knowledge is highly structured, there are multiple paths to retrieving it, whereas isolated bits of information are more difficult to remember.

Unless instruction helps children build on their informal knowledge and relate the mathematics they learn in school to it, they are likely to develop two separate systems of mathematical knowledge: one they use in school and one they use outside of school.

Although developing structure is a hallmark of learning with understanding, the nature of that structure is critical because not all relationships are mathematically fruitful. Learning with understanding involves developing relationships that reflect important mathematical principles. The examples in the chapters that follow (Fennema and Romberg 1999) illustrate how addition and subtraction of multidigit numbers is related to basic concepts of place value, how fractions are related to the concept of division, how knowledge of graphing is extended to more general forms of data representation and interpretation, and how informal ideas about space can be developed into the mathematical structures of geometry.

Over the long run, developing understanding involves more than simply connecting new knowledge to prior knowledge; it also involves the creation of rich, integrated knowledge structures.

One of the defining characteristics of learning with understanding is that knowledge is gained in ways that clarify how it can be used. It often has been assumed, however, that basic concepts and skills need to be learned before applications are introduced. This is a faulty assumption: Children use their intuitively acquired knowledge to solve problems long before they have been taught basic skills.

We have come to understand that applications provide a context for developing skills, so in teaching for understanding, skills are linked to their application from the beginning. For example, in the projects described in Carpenter et al. (chapter 4, Fennema and Romberg 1999), children start out solving problems involving joining, separating, and comparing—before they have learned about addition and subtraction. The operations of addition and subtraction are presented as ways of representing these problem situations. In the classroom episode described in Lehrer, Jacobson, Kemeny, and Strom (chapter 5, Fennema and Romberg 1999), children's natural language about shape and form eventually is transformed into mathematical propositions and definitions.

Reflecting About Experiences

Reflection involves the conscious examination of one's own actions and thoughts. Routine application of skills requires little reflection: One just follows a set of familiar procedures. Reflection, however, plays an important role in solving unfamiliar problems. Problem solving often involves consciously examining the relation between one's existing knowledge and the conditions of a problem situation. Students stand a better chance of acquiring this ability if reflection is a part of the knowledge-acquisition process.

It has often been assumed that basic concepts and skills need to be learned before applications are introduced. This is a faulty assumption.

To be reflective in their learning means that students consciously examine the knowledge they are acquiring and, in particular, the way it is related both to what they already know and to whatever other knowledge they are acquiring. But learning does not only occur with the addition of new concepts or skills: It also comes about through the reorganization of what one already knows. Reflecting about what one knows and how one knows can lead to this sort of reorganization.

Our notion of the emerging nature of understanding is seen in students' developing ability to reflect on their knowledge. Initially students have limited ability to reflect on their thinking. One characteristic of students' developing understanding is that they become increasingly able to do so.

Articulating What One Knows

The ability to communicate or articulate one's ideas is an important goal of education, and it also is a benchmark of understanding. Articulation involves the communication of one's knowledge, either verbally, in writing, or through some other means such as pictures, diagrams, or models. Articulation requires reflection in that it involves lifting out the critical ideas of an activity so that the essence of the activity can be communicated. In the process, the activity becomes an object of thought. In other words, to articulate our ideas, we must reflect on them to identify and describe critical elements. Articulation requires reflection, and, in fact, articulation can be thought of as a public form of reflection.

One characteristic of students' developing understanding is that they become increasingly able to do so.

As with reflection, students initially have difficulty articulating their ideas about an unfamiliar topic or task, but by struggling to articulate their ideas, especially with means such as mathematical symbols or models, students develop the ability to reflect on and articulate their thinking.

Making Mathematical Knowledge One's Own

Understanding involves the construction of knowledge by individuals through their own activities so that they develop a personal investment in building knowledge. They cannot merely perceive their knowledge simply as something that someone else has told them or explained to them; they need to adopt a stance that knowledge is evolving and provisional. They will not view knowledge in this way, however, if they see it as someone else's knowledge, which they simply assimilate through listening, watching, and practicing.

When students struggle to articulate their ideas, they develop the ability to reflect on and articulate their thinking.

This does not mean that students cannot learn by listening to teachers or other students, but they have to adopt what they hear to their own ends, not simply accept the reasoning because it is clearly articulated by an authority figure. Neither does this mean that understanding is entirely private. The development of students' personal involvement in learning with understanding is tied to classroom practices in which communication and negotiation of meanings are important facets.

In this more general sense, students author their own learning. They develop their own stances about different forms and practices of mathematics. For example, some students are fascinated by number; others by space; still others, by questions of chance and uncertainty. Students who understand mathematics often define interests that guide their activity. Ideally, learning is guided by personal histories of aptitude and interest, not simply by curricular sequences.

Understanding involves the construction of knowledge by individuals through their own activities so that they develop a personal investment in building knowledge.

An overarching goal of instruction is that students develop a predisposition to understand and that they strive to understand because understanding becomes important to them. This means that students themselves become reflective about the activities they engage in while learning or solving problems. They develop relationships that may give meaning to a new idea, and they critically examine their existing knowledge by looking for new and more productive relationships. They come to view learning as problem solving in which the goal is to extend their knowledge.

Is Understanding the Same for Everyone?

In proposing the five forms of mental activity from which mathematical understanding emerges, we are not suggesting that all students learn in exactly the same way or that understanding always looks the same in all individuals. What we are proposing is that the development of understanding involves these forms of mental activity in some form. For an idea to be understood, it must be related to other ideas, but there are many ways that ideas might be related. We are not suggesting that relations must be formed in the same way or through the same activities, only that understanding depends on ideas being organized in some productive way that makes them accessible for solving problems. The ability to extend and apply knowledge is a hallmark of understanding, but this does not imply that all people extend and apply their knowledge in the same way. By the same token, reflection and articulation can take on a variety of forms, but we cannot conceive of understanding developing without some sort of reflection and articulation. Personal histories of developing understanding will vary. Such variation is often an important catalyst for conceptual change as students reconcile their own views with those of others.

Critical Dimensions of Classrooms That Promote Understanding

What does all this mean for instruction? As previously discussed, for learning with understanding to occur on a widespread basis, classrooms need to provide students with opportunities to (1) develop appropriate relationships, (2) extend and apply their mathematical knowledge, (3) reflect about their own mathematical experiences, (4) articulate what they know, and (5) make mathematical knowledge their own. To organize a classroom that enables students to engage in these activities, there are at least three dimensions of instruction that need to be considered: (1) tasks or activities in which students engage and the problems that they solve, (2) tools that represent mathematical ideas and problem situations, and (3) normative practices, which are the standards regulating mathematical activity, agreed on by the students and teacher.

All students [do not] learn in exactly the same way; understanding [does not] always look the same in all individuals.

Tasks

Mathematics lessons frequently are planned and described in terms of the tasks in which students engage. Tasks can range from simple drill-and-practice exercises to complex problem solving tasks set in rich contexts. Almost any task can promote understanding. It is not the tasks themselves that determine whether students learn with understanding: The most challenging tasks can be taught so that students simply follow routines, and the most basic computational skills can be taught to foster understanding of fundamental mathematical concepts. For understanding to develop on a widespread basis, tasks must be engaged in for the purpose of fostering understanding, not simply for the purpose of completing the task. (For an example of how learning to use multidigit numbers within the context of computational activities becomes a task in which students' understanding of place value grows, see Carpenter et al., chapter 4, Fennema and Romberg 1999.)

Tools

Tools are used to represent mathematical ideas and problem situations. They include such things as paper and pencil, manipulative materials, calculators and computers, and symbols. Problems are solved by manipulating these tools in ways that follow certain rules or principles (see Lajoie, chapter 7, Fennema and Romberg 1999), on tools for data representation and visualization). Computational algorithms, for example, involve the manipulation of symbols to perform various arithmetic calculations. These same operations can be performed by representing the numbers with counters or base-10 blocks and combining, grouping, or partitioning the counters or blocks in appropriate ways. Connections with representational forms that have intuitive meaning for students can greatly help students give meaning to symbolic procedures. In Carpenter et al. and Sowder and Philipp (chapters 4 and 6, Fennema and Romberg 1999), examples involving adding and subtracting whole numbers and dividing fractions illustrate how such connections can be developed.

For understanding to develop on a widespread basis, tasks must be engaged in for the purpose of fostering understanding, not simply for the purpose of completing the task.

Standard mathematical representations and procedures involve symbols and operations on those symbols that have been adopted over centuries and have been constructed for the purposes of efficiency and accuracy. The connections between symbols and symbolic procedures and the underlying mathematical concepts that they represent are not always apparent. As a consequence, practicing formal procedures involving abstract symbols does little to help students connect the symbols or procedures to anything that would give them meaning. One of the ways to resolve this dilemma is for students to link the critical steps in procedures with abstract symbols to representations that give them meaning (see Kaput, chapter 8, Fennema and Romberg 1999).

Representations may be introduced by the teacher or constructed by students. In Lehrer et al. (chapter 5, Fennema and Romberg 1999) classroom episodes are described in which students invented representations for quantities, forms, measures, and large-scale space. Each form of representation provided opportunities for developing new mathematical knowledge. As noted in

Romberg and Kaput (chapter 1, Fennema and Romberg 1999), the syntactically guided manipulation of formal representations is an important goal of instruction and, if we want students to understand the representations they use, we should encourage them to reflect explicitly on the characteristics of those representations used for understanding and communicating about mathematical ideas and for solving problems.

Normative Practices (Norms)

The norms in a particular class determine how students and the teacher are expected to act or respond to a particular situation. Normative practices form the basis for the way tasks and tools are used for learning, and they govern the nature of the arguments that students and teachers use to justify mathematical conjectures and conclusions. These norms can be manifest through overt expectations or through more subtle messages that permeate the classroom environments.

Mathematical tools include such things as paper and pencil, manipulative materials, calculators and computers, and symbols.

Although the selection of appropriate tasks and tools can facilitate the development of understanding, the normative practices of a class determine whether they will be used for that purpose. In classrooms that promote understanding, the norms indicate that tasks are viewed as problems to be solved, not exercises to be completed using specific procedures. Learning is viewed as problem solving rather than drill and practice. Students apply existing knowledge to generate new knowledge rather than assimilate facts and procedures. Tools are not used in a specific way to get answers: They are perceived as a means to solve problems with understanding and as a way to communicate problem-solving strategies. The classrooms are discourse communities in which all students discuss alternative strategies or different ways of viewing important mathematical ideas such as what is a triangle (see Lehrer et al. and Sowder and Philipp, chapters 5 and 6, Fennema and Romberg 1999). Students expect that the teacher and their peers will want explanations as to why their conjectures and conclusions make sense and why a procedure they have used is valid for the given problem. In this way, mathematics becomes a language for thought rather than merely a collection of ways to get answers.

Structuring and Applying Knowledge

For students to learn with understanding, they must have opportunities to relate what they are learning to their existing knowledge in ways that support the extension and application of that knowledge. In classrooms where students learn with understanding, there are a number of ways that instruction can provide them the opportunities to structure their knowledge. For example, students may be asked specifically to identify relevant relationships. Students may be expected to specify explicit links between symbolic procedures and manipulations of physical materials, as in the Conceptually Based Instruction classes described in Carpenter et al. (chapter 4, Fennema and Romberg 1999). The relationships may also be drawn in less direct ways, as

In classrooms that promote understanding, the norms indicate that tasks are viewed as problems to be solved, not exercises to be completed using specific procedures.

when students compare and contrast alternative strategies that they have generated to solve a problem, as in the Cognitively Guided Instruction classes described in the same chapter.

For students to learn with understanding, they must have opportunities to relate what they are learning to their existing knowledge in ways that support the extension and application of that knowledge.

It is critical that providing opportunities for students to develop structured knowledge is a major and continuing focus of instruction. Students cannot be expected to develop critical knowledge structures by practicing procedures. Watching a demonstration, listening to an explanation of how things are related, or even engaging in a few teacher-directed hands-on tasks is not enough. Students need time to develop knowledge structures, and instruction should offer students extended opportunities to develop relationships through the tasks in which they engage.

The selection and sequencing of tasks and tools is critical. They should not be selected exclusively on mathematical structure. We must take into account children's thinking, the knowledge they bring to a situation, and the way their thinking typically develops. A tacit assumption underlying much of the traditional mathematics curriculum has been that problem solving involves the application of skills and, consequently, that skills must be learned before students can profitably engage in problem solving. The examples in the chapters that follow document that this is not the case. In these examples, problem solving and the learning of basic concepts and skills are integrated; in fact, problems and applications provide the context for learning fundamental mathematical concepts and skills.

Students need time to develop knowledge structures.

As noted earlier, there is an extensive body of research documenting that children acquire a great deal of intuitive or informal knowledge of mathematics and begin developing problem-solving abilities outside of school. The formal concepts and skills of the mathematics curriculum need to be related to these informal concepts and problem-solving skills, or students will not see how the mathematics they learn in school applies to solving problems in the world. Furthermore, this informal knowledge can provide a solid foundation for giving meaning to the abstract mathematical symbols, concepts, and skills that students learn in school.

Tasks and tools need to be selected such that mathematics instruction builds on children's informal mathematical knowledge and that problems and applications, and the related mathematical concepts and skills, are connected from the beginning.

Reflection and Articulation

Children acquire a great deal of intuitive or informal knowledge of mathematics and begin developing problem-solving abilities outside of school.

One of the primary ways that learning with understanding occurs is through reflection. Initially, students generally use concrete tools as implements to solve a given task. As students reflect on the use of the tools, the manipulations of the physical materials become abstracted. Eventually students no longer have to actually manipulate the physical tools themselves; they can think directly about more-abstract symbolic representations of the tools (see Carpenter et al. and Lehrer et al., chapters 4 and 5, Fennema and Romberg 1999). The process is recursive. The more abstract representations become themselves objects of reflection, leading to an awareness of the underlying mathematical concepts that the tools, and their symbolic abstractions, embody. As the concepts and principles embodied in a given tool become objects of reflection, higher-level mathematical principles emerge and so on. For example, students start out solving addition and subtraction problems by modeling the joining and separating action using counters. By reflecting on their procedures and their emerging knowledge of groupings of 10, they come to use more efficient procedures that involve the use of some sort of 10-structured material, such as base-10 blocks. As students describe and reflect about the solutions using materials grouped by 10, they become increasingly less dependent on the base-10 materials themselves; they start to use abstract representations of the base-10 materials (see Carpenter et al., chapter 4, Fennema and Romberg 1999). As students compare different abstract strategies and reflect on these differences, they begin to see that certain procedures have advantages over others, and they begin to see explicitly how properties like commutativity and associativity are involved in their procedures.

The process [of building understanding] is recursive.

Encouraging Reflection

The question is: How do we encourage this type of reflection? Providing explicit guidelines for encouraging reflection is difficult, but a critical factor is that teachers recognize and value reflection. When that is the case, teachers establish classroom norms that support reflection. A specific norm that plays a critical role in supporting reflection in the descriptions of classrooms that follow is the expectation that students articulate their thinking. Asking students why their solutions work, why a given solution is like another solution, how they decided to solve the problem as they did, and the like not only helps to develop students' ability to articulate their thinking, it encourages them to reflect.

A critical factor is that teachers recognize and value reflection.

At this point, we should distinguish between two types of reflection, both of which are important: reflection by students about what they are doing and why, as tasks are being carried out, and reflection about tasks and their solutions after the tasks have been completed. Discussion of alternative strategies that students have used to solve a given problem involves reflection about a task after it has been completed. The probing questions that a teacher might ask at this point address this type of reflection.

Discussing alternative strategies, however, addresses more than the issue of reflection on completed tasks. When students know that they are expected to explain their responses, they are more likely to reflect on a task as they are carrying it out. Reflection on the task while carrying it out can also be encouraged directly by asking students to articulate what they are doing during the process of solving a problem. One possibility is for teachers to talk to students as they are solving a problem, asking the students to explain assumptions and why they are pursuing the strategy that they have chosen. Questions like "What are you doing?" "Why are you doing that?" and "How will that help you to solve the problem?" encourage reflection. Being asked such questions on a regular basis helps students internalize them so that they will ask themselves the same questions as they think about a given task.

Another way to encourage reflection on tasks in progress is to have students work in small cooperative groups. When students are actually solving a problem together, they must articulate their assumptions, conjectures, and plans to one another. For this kind of reflection and articulation to occur, however, classroom norms must establish that these kinds of interactions are what cooperative group work is all about.

A Basis for Articulation

Two types of reflection, both of which are important, are: reflection by students about what they are doing and why, as tasks are being carried out, and reflection about tasks and their solutions after the tasks have been completed.

For articulation to be meaningful to all the participants in a class, there must be a common basis for communication. The selection of appropriate tools can fill this role, but teachers must ensure that everyone has a consistent interpretation of the tools and their use. Manipulative materials can provide common referents for discussion, but students do not always impart the same meanings to manipulations of physical materials that knowledgeable adults do. It is important that discussions include opportunities for students to articulate how they are thinking about and using tools.

Notations (e.g., those developed for representing quality, two-dimensional graphs, etc.) can provide a common basis for discussion, and they can help students to clarify their thinking. Notations thus play a dual role: first, as a window (for teachers and others) into the evolution of student thinking and, second, as a tool for thought. Notations are records that communicate about thinking. Appropriate notational systems allow students to articulate their thinking in very precise ways, and the precision demanded by the notational system can make students sharpen their thinking so that it can be articulated.

Classroom Norms

For articulation to be meaningful to all participants in a class, there must be a common basis for communication.

One norm that underlies teaching for understanding is that students apply existing knowledge in the generation of new knowledge. Learning is not perceived by either the teacher or students as assimilation and practice. Learning is viewed as problem solving, and students are expected to actively work to relate new concepts and procedures to their existing knowledge.

A specific class norm that supports this conception of learning is that students regularly discuss alternative strategies (which they have generated to solve a given problem) with the teacher, with other students, and within the context of whole-class discussion. It is not enough to have an answer to a problem; students are expected to be able to articulate the strategy they used to solve the problem and explain why it works. This means discussing how the solution is related to the parameters of the problem and how the procedures used in the solution are related to underlying mathematical concepts or some external representation that has established meaning.

In discussing alternative strategies, students not only explain their own solutions and their own thinking but also discuss how strategies used by different students are alike and different. In other words, they consider the connections between alternative solutions. This is one of the important ways in which relationships are made explicit. As students report and discuss solutions representing different levels of abstraction and understanding, they have the opportunity to link more abstract strategies with more basic strategies. For example, when some students solve a problem using manipulative materials and other students solve the problem with symbolic representations, the discussion of the relationship between the two strategies draws attention to connections that give the abstract symbolic procedures meaning (see the class interaction at the beginning of Carpenter et al., chapter 4, Fennema and Romberg 1999).

Making Knowledge One's Own

As with reflection and articulation, classroom norms play a central role in helping students develop a personal sense of ownership of their knowledge. Again, specifying guidelines is difficult, but it is critical that teachers place a high value on the individual student's involvement and autonomy. All students should have opportunity to discuss their ideas, and each student's ideas should be taken seriously by everyone else in the class. The overriding goal of the classroom should be the development of understanding.

It is not enough to have answers to a problem; students are expected to be able to articulate the strategy they used to solve the problem and explain why it works.

Reflection is inherently personal, and encouraging reflection is critical in helping students develop a sense of ownership of their knowledge. Students need to be given some control over the tasks in which they engage and the tools they use to solve them so that they believe they have control over their own learning.

Teachers and Understanding

Inherent in much of the previous discussion is the assumption that understanding is a goal not only for students but also for teachers. Understanding plays a critical role in the solution of any complex problem, and teaching certainly involves solving complex problems. Our conception of teacher understanding is based on the same principles as our conception of student understanding.

We focus on two components of teachers' understanding and the relations between them: understanding of mathematics and understanding students' thinking. To provide instruction of the kind envisioned in this volume, teachers need to understand the mathematics they are teaching, and they need to understand their own students' thinking. The mathematics to be taught and the tasks and tools to be used might be specified by an instructional program, but without requisite understanding of mathematics and students, teachers will be relegated to the routine presentation of (someone else's) ideas neither written nor adapted explicitly for their own students. In short, their teaching will be dominated by curriculum scripts, and they will not be able to establish the classroom norms necessary for learning with understanding to occur. They will not be able to engage students in productive discussion of alternative strategies because they will not understand the students' responses; neither will they be able readily to recognize student understanding when it occurs.

It is critical that teachers place a high value on the individual student's involvement and autonomy.

Teachers need to understand the mathematics they are teaching, and they need to understand their own students' thinking.

Understanding mathematics for instruction involves more than understanding mathematics taught in university mathematics content courses. It entails understanding how mathematics is reflected in the goals of instruction and in different instructional practices. Knowledge of mathematics must also be linked to knowledge of students' thinking so that teachers have conceptions of typical trajectories of student learning and can use this knowledge to recognize landmarks of understanding in individuals.

Understanding is a goal not only for students but also for teachers.

Teachers need to reflect on their practices and on ways to structure their classroom environment so that it supports students' learning with understanding. They need to recognize that their own knowledge of mathematics and of students' thinking, as well as any student's understanding, is not static.

Teachers must also take responsibility for their own continuing learning about mathematics and students. Class norms and instructional practices should be designed to further not only students' learning with understanding but also teachers' knowledge of mathematics and of students' thinking. Tasks and tools should be selected to provide a window on students' thinking, not just so that the teacher can provide more appropriate instruction for specific students but also so the teacher can construct better models for understanding students' thinking in general.

Conclusion

For students and teachers, the development of understanding is an ongoing and continuous process and one that should pervade everything that happens in mathematics classrooms. For many years there has been a debate on whether individuals should learn skills with understanding from the outset or whether they should acquire a certain level of skill mastery first and then develop an understanding of why skills work the way they do. A mounting body of evidence supports the importance of learning with understanding from the beginning. When students learn skills without understanding, the rote application of the skills often interferes with students' subsequent attempts to develop understanding. When students learn skills in relation to developing an understanding, however, not only does understanding develop, but mastery of skills is also facilitated.

Teachers must also take responsibility for their own continuing learning about mathematics and students.

The development of understanding takes time and requires effort by both teachers and students.

If we have learned one thing through our studies, it is that the development of understanding takes time and requires effort by both teachers and students. Learning with understanding will occur on a widespread basis only when it becomes the ongoing focus of instruction, when students are given time to develop relationships and learn to use their knowledge, when students reflect about their own thinking and articulate their own ideas, and when students make mathematical knowledge their own.

We do not have precise prescriptions for how classrooms should be organized to accomplish these goals. In this chapter we have provided some issues and components to consider when thinking about instruction, but ultimately responsibility for learning with understanding rests with the teachers and students themselves. Teachers must come to understand what it means for their students to learn with understanding and must appreciate and value learning with understanding. Like their teachers, students must come to value understanding and make understanding the goal of their learning. That is the ultimate goal of instruction.

In the chapters that follow (Fennema and Romberg 1999), we provide examples of instruction that offer opportunity for the development of understanding as we have characterized it. There are a number of similarities among the examples, but we are not suggesting that understanding will occur only in classes similar to the ones described. We do, however, argue that for learning with understanding to occur, instruction needs to provide students the opportunity to develop productive relationships, extend and apply their knowledge, reflect about their experiences, articulate what they know, and make knowledge their own.

Note

Authors contributed equally to the writing of this chapter.

For Further Reading

Anderson, J. 1990. *Cognitive Psychology and Its Implications.* Cambridge, MA: Harvard University Press.

Hiebert, J., T. P. Carpenter, E. Fennema, K. Fuson, P. Human, H. Murray, and A. Olivier. 1997. *Making Sense: Teaching and Learning Mathematics with Understanding.* Portsmouth, NH: Heinemann.

Kaput, J. 1992. Technology and Mathematics Education. In *Handbook of Research on Mathematics Teaching and Learning*, edited by D. Grouws. New York: Macmillan.

Lehrer, R., and D. Chazan. 1998. *Designing Learning Environments for Developing Understanding of Geometry and Space.* Mahwah, NJ: Lawrence Erlbaum Associates.

Schoenfeld, A. 1995. *Mathematical Problem Solving.* New York: Academic Press.

Schon, D. 1987. *Educating the Reflective Practitioner.* San Francisco: Jossey-Bass.

Wertsch, L. 1991. *Voices of the Mind.* Cambridge, MA: Harvard University Press.

Yakel, E., and P. Cobb. 1996. Sociomathematical Norms, Argumentation, and Autonomy in Mathematics. *Journal of Research in Mathematics Education,* 27: 458–77.

Planning Curriculum
in

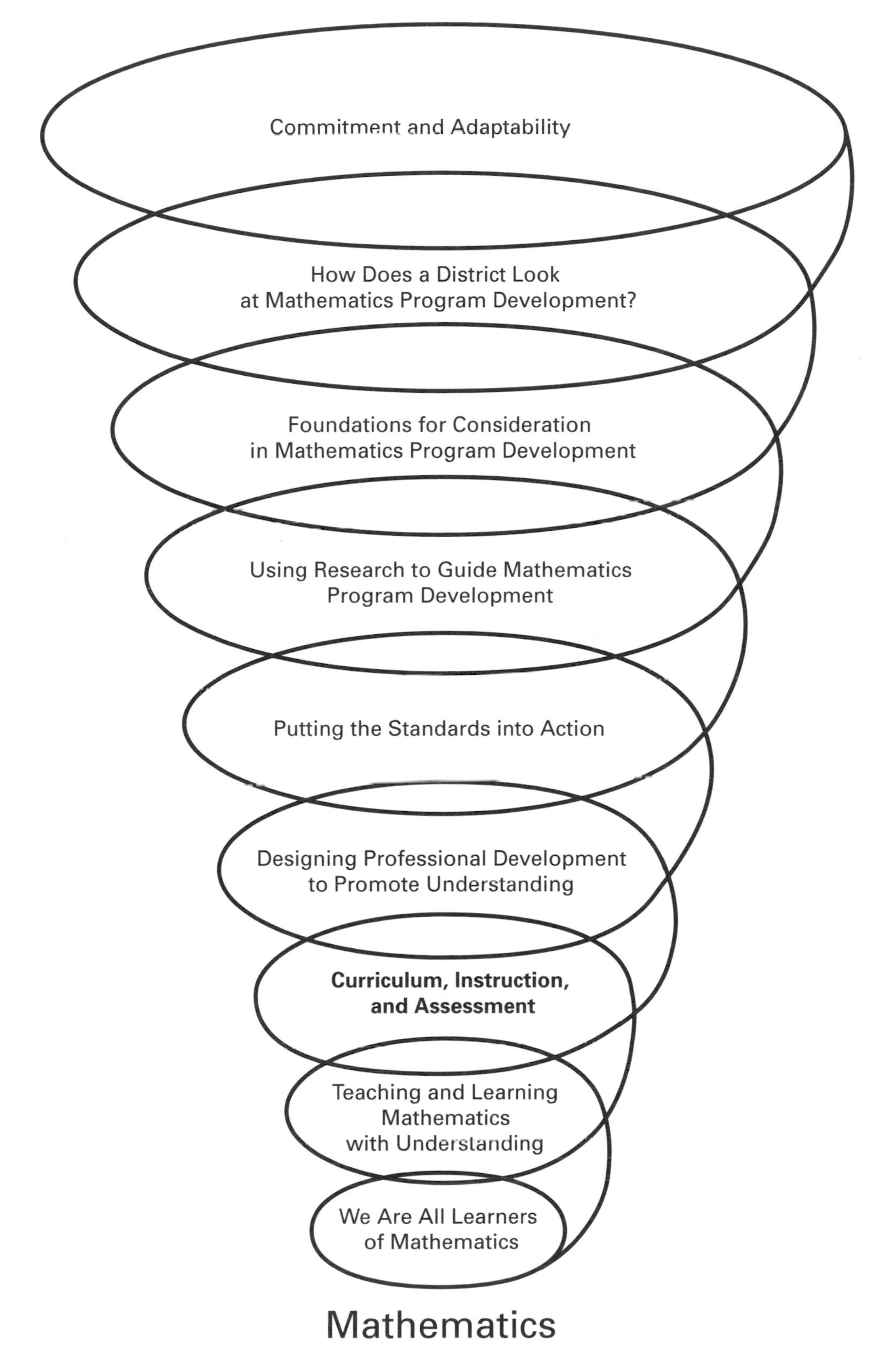

Spring (spring) *n.* 1. a source; origin; or beginning: *a spring into action.* 2. elasticity; resilience: *a spring forward with commitment.* (Morris, W., ed., 1971. *The American Heritage Dictionary of the English Language.* New York: American Heritage Publishing Company, Inc. and Houghton Mifflin Company, 1250, 2nd definition.)

Curriculum, Instruction, and Assessment

An Integrated Whole

3

Curriculum

Interestingly characterizing reform efforts prompted by the *Curriculum and Evaluation Standards* (Commission on Standards for School Mathematics 1989) as ***CIA,*** Fortier and Moser (1992) pointed out that curriculum, instruction, and assessment are a package; one component cannot function adequately without attention to each of the others. An excellent curriculum cannot achieve what is intended unless it is well taught and appropriately assessed. Realizing that the enacted curriculum is jointly constructed by the teacher and students as they interact with instructional materials, the selection of materials that will best facilitate construction of student knowledge is one of the teacher's most important tasks (Ball and Cohen 1996; Council of Chief State School Officers, Wisconsin Center for Educational Research, and Eleven State Collaborative 2000). A curriculum is not a textbook or a set of materials, but the selection of investigations, the sequencing of them, the assessment of understandings gleaned from pursuit of the tasks at hand, and the additional planning for capitalizing on the learning that takes place as a result of the explorations. Teachers continually make decisions regarding pacing, augmenting, deleting, and delivery. Students indicate understandings gleaned and lack of understandings. Continual assessment by the teacher and the students of knowledge gained is necessary to the construction of an ever-evolving curriculum.

A curriculum is more than a collection of activities.

What Is a Curriculum?

The curriculum principle from the National Council of Teachers of Mathematics (NCTM) *Principles and Standards for School Mathematics* (Standards 2000 Project 2000, 14–16) states, "A curriculum is more than a collection of activities; it must be **coherent, focused on important mathematics,** and **well articulated** across the grades." A school mathematics curriculum is a strong determinant of what students have an opportunity to learn and what they do learn. In a **coherent** curriculum, mathematical ideas are linked to and build on one another so that students' understanding and knowledge deepens and their ability to apply mathematics expands. An effective mathematics curriculum **focuses on important mathematics**—mathematics that will prepare students for continued study and for solving problems in a variety of school, home, and work settings. A **well-articulated** curriculum challenges students to learn increasingly more sophisticated mathematical ideas as they continue their studies and gives guidance about when closure is expected for particular skills or concepts.

The driving forces of curricula are sequencing, connectedness within a strand, interconnectedness among strands, development of thinking processes, and the relationship between procedural and conceptual knowledge. Curricula need to capitalize on the knowledge that we have about how understanding is built. Kieren's Model of Mathematical Knowledge Building and his Model of Recursive Understanding are instructive when considering these relationships.

How Is Mathematical Knowledge Built?

Mathematical knowledge building becomes increasingly complex. It moves from knowledge shared by people living in a particular society (ethnomathematical knowledge) to knowledge based on the use of thinking tools, imagery, and the informal use of mathematical language (intuitive knowledge) to knowledge acquired as a result of working with symbolic expressions (technical-symbolic knowledge) to knowledge which is derived through logically situating a statement in an axiomatic structure (axiomatic deductive knowledge) (Kieren 1993, 66–71). (See Figure 3.1.)

Each kind of knowing successively embeds the previous kind of knowing. The more complex outer layers, while dependent on inner ways, do not sequentially result from the inner, more intuitive ways of knowing. Actually, the outer ways "involve or organize the inner ways of knowing" (Kieren 1993, 69). Knowledge is recursive, the inner ways of knowing contributing to the large outer ways of knowing, and the outer ways of knowing checking the understandings gleaned from the inner ways of knowing.

In the inner less complex stages of knowing, validation is based on "real mathematics, the applicational mathematics of the 'real world.'" At the outer stages, validation is based strictly on structural, logical, or philosophical arguments. The line across the model, "the Hilbert Line," delineates the point at which validation moves from the "real" to the "philosophical." This model contributes much to the applications versus structure debate. It clearly points out that both are essential in the building of mathematical knowledge, that each makes a contribution to knowledge building, and that the two need to work together in a symbiotic relationship.

Knowledge is recursive, the inner ways of knowing contributing to the large outer ways of knowing, and the outer ways of knowing checking the understandings gleaned from the inner ways of knowing.

The model for the recursive theory of mathematical understanding (see Figure 3.2.) builds on the idea of mathematical knowledge building and sees mathematical understanding as "dynamic, nonlinear, and transcendently recursive" (Kieren 1993, 72). Understanding is a "whole dynamic process" (Pirie 1988) and "a human activity" (Romberg and Kaput 1999). As people develop understanding, they are capable, through language, of self-referencing, of making distinctions in their own behavior based on taking effective action,

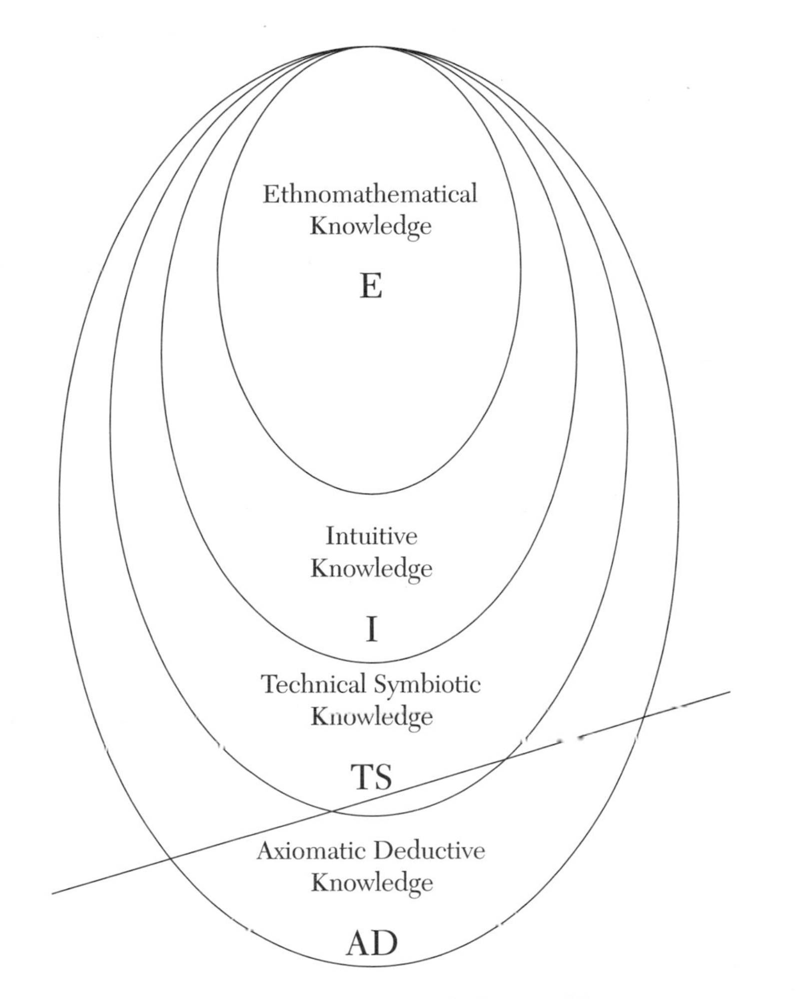

FIGURE 3.1 **A Model of Mathematical Knowledge Building**
From "Rational and Fractional Numbers: From Quotient Fields to Recursive Understanding," by T. E. Kieren, 1993, in T. P. Carpenter, E. Fennema, and T. A. Romberg, eds., *Rational Numbers: An Integration of Research,* pp. 197–218. Copyright © 1993 by Lawrence Erlbaum Associates, Publishers. Reprinted with permission of the author.

and on making distinctions on previous knowing. People are also able to grow and gain in understanding as a result of interactions with others. A gain in understanding is thus dynamic and nonlinear.

Knowing, then, grows recursively, making distinctions on previous knowing. Mathematical understanding is a dynamic process that folds back on itself for "growth, extension, and re-creation" (Kieren 1993, 73).

The model for the recursive theory of mathematical understanding recognizes the importance at rudimentary stages of "doing" so that an image of what happens is gained—image making. After several experiences with doing and image making, the knowledge gleaned becomes mental objects—image having. Property noticing calls into action recursion on the three previous stages so that activities undertaken, images languaged, and mental images interact to identify patterns. All of these stages are related to actions or images. When the thinking moves to generalizations and self-conscious thinking, formalization occurs. With these mental structures in place, observation can take

A gain in understanding is dynamic and nonlinear.

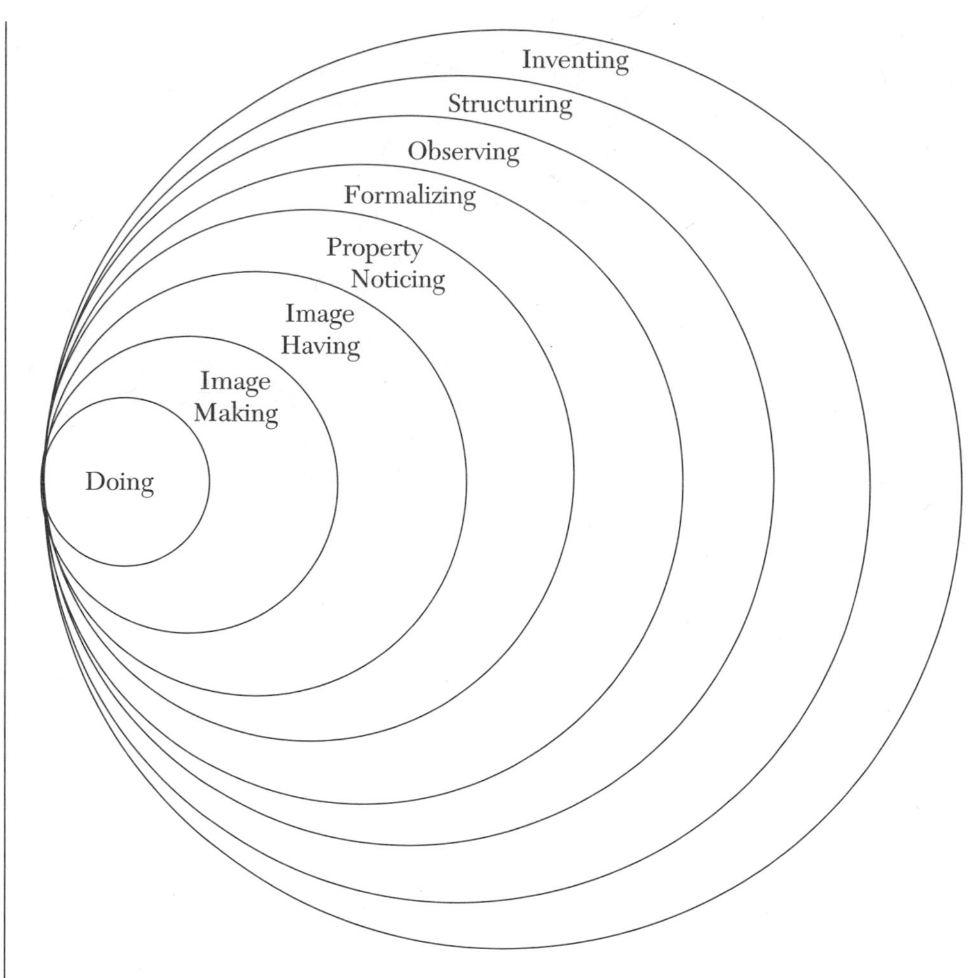

FIGURE 3.2 **A Model for the Recursive Theory of Mathematical Understanding**

From "Rational and Fractional Numbers: From Quotient Fields to Recursive Understanding," by T. E. Kieren, 1993, in T. P. Carpenter, E. Fennema, and T. A. Romberg, Eds., *Rational Numbers: An Integration of Research,* pp. 197–218. Copyright © 1993 by Lawrence Erlbaum Associates, Publishers. Reprinted with permission of the author.

place regarding previous ideas. Finally, structuring of observations is done in terms of a systematic set of assumptions. The outer, creative level of understanding—inventing—occurs when one can alter thinking about a topic without destroying the structure and understanding already built.

What Are the Implications for Curricula?

The model of recursive understanding has great pertinence for curriculum construction.

This model of recursive understanding (Kieren 1993, 72) has great pertinence for curriculum construction. Students need to move from informal to formal reasoning as a result of working with problems created to facilitate the process (Romberg 1998). Curricular investigations need to be rich in connections between and among the strands and must foster reasoning and communication through vocabulary, forms of representation, choice of materials, tools, techniques, and intellectual methods so problems are defined and solved with reason, insight, inventiveness, and technical proficiency (Lappan and Phillips

1998, 83). Problems initially addressed must be based in realism (Streefland 1993), reflecting the society from which students emanate (Kleiman, Tobin, and Isaacson 1998) and capitalizing on informal knowledge (Mack 1993). Appropriate manipulatives need to be used to develop transferable images (Mercer and Henningsen 1998). Since students are at different levels of understanding with different strands, multiple representations (Ball 1993), congruent contexts (Lampert 1991), and multiple approaches (Murphy 1998) are important. For self-referencing purposes, students need opportunities for metacognition (Lamon 1993) and reflection (Phillips and Lappan 1998). They also need opportunities for discourse (Laughlin and Moyer 1998), for sharing of solutions, and for opportunities to reference their knowledge against that of their peers. Pacing is important, with continual opportunity for growth (Billstein 1998), but with latitude to check understanding and to make connections.

Students need to move from informal to formal reasoning as a result of working with problems created to facilitate the process.

"To sequence instruction based on the tenets of mathematics as a human activity implies that tasks cannot be organized in a particular sequence to cover each detail; instead, instructional units or tasks that focus on investigation of problem situations need to be created. From such situations, the key ideas in each strand can be acquired or constructed and related to other ideas in other strands or substrands, and the uses of these ideas in other disciplines identified" (Romberg and Kaput 1999, 9). Continuing, the authors argue that investigations that promote mathematics worth teaching and worth understanding should consider the following questions:

> Do the tasks lead anywhere? (Do they begin with prior knowledge and extend the map of the domain . . . [realizing that] there is no one correct sequence of concepts and activities in any domain?)
>
> Do the tasks lead to model building? (Essential to professional practice in mathematics, model construction, model evaluation, and model revision, allow for further exploration and identification of important big ideas in a domain.)
>
> Do the tasks lead to inquiry and justification? (Students should learn to conjecture, investigate, conclude, and evaluate, as well as explain and defend findings. They need to learn to invent strategies and to discuss why such strategies are viable. "Genuine inquiry . . . provides them with practice in and eventual understanding of the criteria for evidence and explanation patterns characteristic of mathematics" (12).)
>
> Are the tasks relevant to students? (Are the tasks "intrinsically motivating?" Do they rest on what students know, "produce cognitive incongruity," and create solution curiosity, with appropriate challenge (neither too easy nor too hard)?) (Romberg and Kaput 1999, 9–12)

What Are Conceptual and Procedural Knowledge?

A major component of the curriculum question in the United States is the conceptual/procedural knowledge question. Conceptual knowledge is "rich in relationships" (Hiebert and Lefevre 1986, 3), a connected web of knowledge, a network in which both the links and clusters of knowledge are important. "A

Conceptual knowledge must be learned meaningfully and cannot be generated directly by rote learning.

unit of conceptual knowledge cannot be an isolated piece of information" (4). Conceptual knowledge is developed by the construction of relationships between pieces of information. The pieces of information can be stored in memory, or one can be stored and one can be newly learned. There are levels of relationships. The primary level simply connects the two pieces. The higher level, the reflective level (5), ties together more abstract components, less tied to specific contexts, by recognizing similar core features in the pieces of information. Conceptual knowledge must be learned meaningfully and cannot be generated directly by rote learning.

Procedural knowledge may or may not be learned with meaning.

Procedural knowledge has two parts as described by Hiebert and Lefevre (1986, 6)—formal language, the symbol representation system, and algorithms, rules for completing mathematical tasks. The procedural system is "structured" (7), procedures are hierarchically arranged, and subprocedures contribute to superprocedures. Procedural knowledge may or may not be learned with meaning. Rote learning produces knowledge that is "notably absent in relationships" and is "closely tied to the context in which it is learned" (8); it does not generalize to other situations. Facts, propositions, and procedures can be learned by rote, but are stored as isolated pieces of information until the learner later recognizes or constructs relationships among the pieces and then creates conceptual knowledge. There are benefits derived from procedural knowledge for procedural knowledge: developing meaning for symbols (connecting symbols with referents, (Schoenfeld 1986)), recalling procedures, and using procedures. There are benefits of procedural knowledge for conceptual knowledge: symbols enhance concepts, procedures apply concepts to solve problems, and procedures promote concepts.

Rote learning produces knowledge that is "notably absent in relationships," is "closely tied to the context in which it is learned," and does not generalize to other situations.

—Hiebert and Lefeure 1986, 8

There are reasons for algorithms (Usiskin 1998): they are powerful, they are reliable, they are accurate, they are fast, they furnish a written record, they establish a mental image, they are instructive, they can be used in other algorithms, and they can be objects of study. There are dangers inherent in all algorithms: blind acceptance of results, overzealous application, a belief that algorithms train the mind, and helplessness if the technology (including paper and pencil) is not available. In a recent work, *The Advent of the Algorithm: The Idea that Rules the World,* Berlinski (2000), heralds the power of the algorithm, the set of instructions that drives computers, and enthuses that it is the algorithm "that has made possible the modern world" (xv). He also speculates on where use of the algorithm might lead us.

Mathematical knowledge, in its fullest sense, includes significant, fundamental relationships between conceptual and procedural knowledge.

The NCTM devoted an entire yearbook to algorithms: *The Teaching and Learning of Algorithms in School Mathematics: 1998 Yearbook.* In an analysis of algorithmic thinking and recursive thinking, Mingus and Grassl (1998, 34–36) compare and contrast the two. They characterize algorithmic thinking as "a method of thinking and guiding thought processes that uses step-by-step procedures, requires inputs and produces outputs, requires decisions about the quality and appropriateness of information coming in and going out, and monitors the thought processes as a means of controlling and directing the thinking process." Recursive thinking is algorithmic in the sense that it proceeds in a step-by-step manner. However, it is iterative and self-referential. It builds on itself. "Algorithmic thinking has an end, a stopping point, while recursive theory never ends" (35). The process relies on both inputs and outputs while the algorithm relies only on the initial input value. Kieren and Pirie (1991) used recursion as a metaphor in describing how children understand, (35).

What Is the Relationship Between Conceptual and Procedural Knowledge?

Hieber and Lefeure (1986, 9) explain, "Mathematical knowledge, in its fullest sense, includes significant, fundamental relationships between conceptual and procedural knowledge. Students are not fully competent in mathematics if either kind of knowledge is deficient or if they both have been acquired but remain separate entities." Both contribute to and acquire meaning in the process of doing mathematics and reflecting on the gains. Realizing the importance of this question of conceptual and/or procedural knowledge, the *NCTM Curriculum and Evaluation Standards for School Mathematics* (Commission on Standards for School Mathematics 1989) clearly states (contrary to popular interpretation) the relationship between conceptual and procedural knowledge production. "Knowing mathematics is doing mathematics" (7). Informational knowledge is valuable "in the extent to which it is useful in the course of some purposeful activity . . . It is clear that the fundamental concepts and procedures from some branches of mathematics should be known by all students" (7). "The curriculum for all students must provide opportunities to develop an understanding of mathematical models, structures, and simulations applicable to many disciplines" (7). "The availability of calculators does not eliminate the need for students to learn algorithms" (8).

Knowing mathematics is doing mathematics.

—Commission on Standards for School Mathematics 1989, 7

Erickson (1998), called conceptual knowledge the "missing link in performance-based theory." "The idea that teachers can develop performances that demonstrate deep understanding assumes that they have consciously identified the kinds of deep understandings that the performance should demonstrate" (4). Asking for "thinking beyond the facts," a classroom-based structure of knowledge model is offered (See Figure 3.3).

Noting that the traditional curriculum design "emphasizes the lower cognitive levels, centering around topics and related facts" and only addresses the key concepts of the disciplines "offhandedly within topics of study," Erickson cites the Third International Mathematics and Science Study (TIMSS) findings regarding these issues.

What Has the TIMSS Study Found in Curricular Comparisons?

The TIMSS, one of the most encompassing and thorough international studies of curricula ever done, found some major flaws in the way the United States has traditionally addressed curriculum development (McNeely 1997). Curricula in both mathematics and science in U.S. schools are unfocused, but the lack of curricular focus is more true in mathematics than in science. Mathematics curricula in the United States consistently cover far more topics than is typical in other countries.

Mathematics curricula in the United States consistently cover far more topics than is typical in other countries.

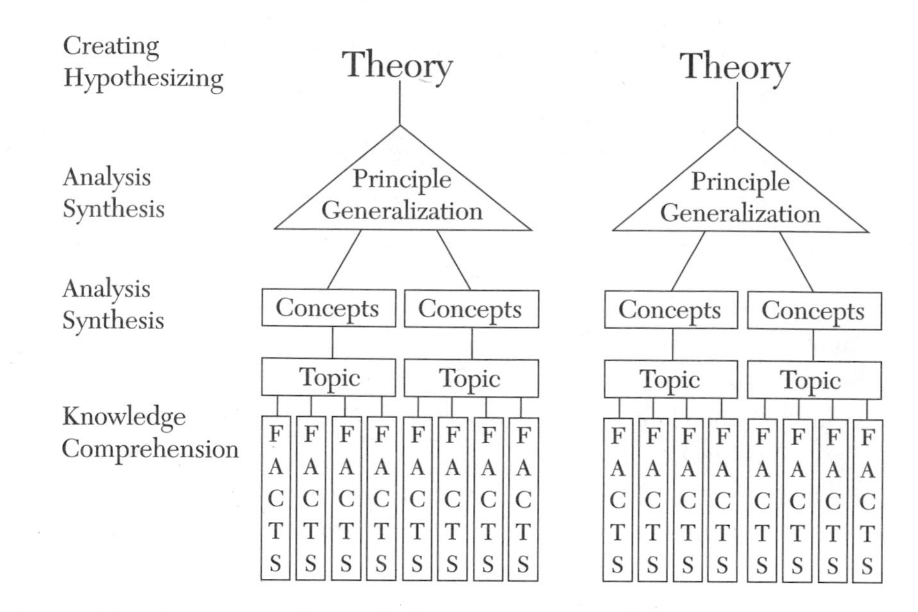

FIGURE 3.3 **Structure of Knowledge**
From H. L. Erickson, 1998, *Concept-Based Curriculum and Instruction: Teaching Beyond the Facts.* Copyright © 1998 by Corwin Press, Inc. Reprinted with permission of the author.

In both mathematics and science, topics remained in composite U.S. curricula for more grades than all but a few other TIMSS countries. The practice is to add far more topics earlier than other countries do and repeat them. Mathematics topics remain in the U.S. composite curriculum for two years longer than the international median. Mastering pieces is emphasized to the exclusion of focusing on a few key goals, linking content together, and setting higher demands on students. U.S. curricula seek to cover something of everything. Composite textbooks are unfocused, include far more topics, and look quite different from other nations' texts.

Mathematics topics remain in the U.S. composite curriculum for two years longer than the international median.

United States texts emphasize using routine procedures and the less complex, more easily taught expectations for student performance. Most U.S. schools and teachers make selective use of textbook contents and rarely cover all of a textbook's content. Teachers spend little time on concepts, trying to cover as many of the topics as possible. U.S. teachers engage in more teaching activities per lesson than their counterparts in other countries.

Most U.S. schools and teachers make selective use of textbook contents and rarely cover all of a textbook's content.

The pervasive influence of industrial and assembly-line production, incremental assembly, behavioral psychology, behavioral objects, and programmed instruction have evidenced themselves in a "curricula of many small topics, frequent low demands, and interchangeable pieces of learning to be assembled later" (McNeely 1997, 171).

What Efforts Have Been Made to Address the Curriculum Question?

The *Principles and Standards for School Mathematics* (Standards 2000 Project 2000) focus squarely on the development of curricula that reflect the five content strands; that address the processes of problem solving, reasoning and proof, connections, communication, and representation; that highlight the "increasing sophistication of ideas across the grades" (7); that link the content and process standards; and that urge a focused curriculum that "moves on." Curricula that have been developed to reflect the intent of the standards movement have attempted to embody the vision of the standards philosophy.

Wisconsin's Model Academic Standards for Mathematics (WMAS) (Wisconsin Department of Public Instruction 1998), taking thrust from the national standards, other state standards, international standards, and input from the people of Wisconsin regarding their expectations for mathematics in Wisconsin schools, were established for grades 4, 8, and 12. This was a conscious decision of the authors so that individual school districts could exercise autonomy in their establishment of curriculum choices within each of these grade level bands. Many districts found, after contemplation of the grade band expectations, that specific grade-level expectations would be helpful. Those districts have since written benchmarks for each grade level. An example of how a district might assign benchmarks based on *WMAS* for each grade level is offered in Appendix C. It was prepared by James M. Moser, Wisconsin Mathematics Consultant, Retired, as a guide for helping districts facilitate the process. It is not offered as a recipe because each district needs to contemplate its needs and design its own plan to fit district goals. It also needs to remain cognizant of the "big picture" so that overall goals are not lost in minutiae.

In all discussions of curriculum, it must be remembered that the actual curriculum of the classroom is "an enacted curriculum [that is] jointly constructed by teachers, students, and materials in particular contexts" (Ball and Cohen 1996, 7). Excellent curricula—good mathematics—still have to be "well taught" (Even and Lappan 1994) and appropriately assessed (Grunow 1999).

The actual curriculum of the classroom is "an enacted curriculum [that is] jointly constructed by teachers, students, and materials in particular contexts."

—Ball and Cohen 1996, 7

Instruction

The heart of the CIA (curriculum, instruction, and assessment) trilogy is, of course, teaching. A district can adopt the finest curriculum available, but, if those materials are not well taught, it will not reap the desired results. Excellent assessment procedures and instruments can be available, but, if student understanding is not facilitated by knowledgeable and astute teachers, the results will be dismal. Darling-Hammond, Wise, and Klein (1999), suggested, "Efforts to restructure America's schools to meet the demands of a knowledge-based economy are redefining the mission of schooling and the job of teaching . . . This new mission for education requires substantially more knowledge and radically different skills for teachers" (2).

What Is Good Mathematics Teaching?

The NCTM *Principles and Standards for School Mathematics,* Teaching Principle (Standards 2000 Project 2000, 16–18), states, "Effective mathematics teaching requires understanding what students know and need to learn and then challenging and supporting them to learn it well." It continues, "Students learn mathematics through the experiences that teachers provide.

Thus, students' understanding of mathematics, their ability to use it to solve problems, and their confidence in and disposition toward mathematics are all shaped by the teaching they encounter in school. The improvement of mathematics education for all students requires effective teaching in all classrooms."

The improvement of mathematics education for all students requires effective teaching in all classrooms.

—*Standards 2000 Project 2000, 16–18*

Recognizing the expanding role of teachers and the importance of the creation of a challenging curriculum and a stimulating environment in which student learning can thrive, the *Professional Standards for Teaching Mathematics* (Commission on Teaching Standards for School Mathematics 1991) called for teachers who are proficient in:

selecting engaging mathematical tasks;
providing opportunities for students to deepen their understanding of the mathematics being studied;
stimulating investigation and growth of mathematical ideas through the orchestration of classroom discourse;
using, and helping students use, technology in mathematical investigations;
helping students connect their learning to previous and developing knowledge;
and guiding individual, small-group, and whole-class work.

In essence, The Professional Standards for Teaching Mathematics emphasized the importance of teacher decision making, made teachers curriculum writers, and recognized the context of teaching and learning as paramount to accomplishment.

In essence, these standards emphasized the importance of teacher decision making, made teachers curriculum writers, and recognized the context of teaching and learning as paramount to accomplishment. With increased autonomy, of course, also came an implied increase in responsibility.

Characterizing the kind of teaching needed to help students succeed in building understanding, in pursuing disciplined inquiry, and in solving problems connected to the world with an audience beyond school, as "authentic instruction," Newmann, Secada, and Wehlage (1995) offer criteria of intellectual quality consistent with "authentic tasks." Support for all students to master challenging work calls for an "instructional climate [that] communicates high expectations for all and cultivates enough trust and respect to reward serious effort" (29). The standards of authentic instruction are listed as:

Support for all students to master challenging work calls for an instructional climate [that] communicates high expectations for all and cultivates enough trust and respect to reward serious effort.

—Newmann, Secada, Wehlage 1995, 29

Standard 1. Higher order thinking: Instruction involves students in manipulating information and ideas by synthesizing, generalizing, explaining, hypothesizing, or arriving at conclusions that produce new meaning and understandings for them.
Standard 2. Deep knowledge: Instruction addresses central ideas of a topic or discipline with enough thoroughness to explore connections and relationships and to produce relatively complex understandings.
Standard 3. Substantive conversations: Students engage in extended conversational exchanges with the teacher and/or their peers about subject matter in a way that builds an improved and shared understanding of ideas or topics.
Standard 4. Connections to the world beyond the classroom: Students make connections between substantive knowledge and either public problems or personal experiences. (See Service Learning in this document.)

What Kind of Expertise Is Required?

The improvement of education has long been viewed as a policy question or as a function of educational spending, but current studies (Darling-Hammond and Ball 1998; Cohen and Hill 1998) are bringing the influence of teacher expertise back squarely to the forefront. Accounting for variance of as much as 43 percent in mathematics test score gains and an effect five times as great as investments in lowering pupil–teacher ratios, teacher influence on learning reigns supreme.

Teacher expertise includes content knowledge, knowledge of pedagogy, and pedagogical content knowledge.

Contemplation of what teacher expertise encompasses leads to a realization of the complexity of teacher knowledge. Fennema and Franke (1992) offered a learned analysis of that knowledge and a model of teacher knowledge situated in the context of the classroom. (See model in the professional development chapter in this document.) "Examination of teachers is beginning to indicate that knowledge can be and is transformed through classroom interaction" (p. 162). Teacher expertise includes knowledge of mathematics (content knowledge), knowing how to teach (pedagogical knowledge), knowing how to teach mathematics (pedagogical content knowledge), and knowing how students think and learn—and, in particular, how this occurs within specific mathematics content.

Examination of teachers is beginning to indicate that knowledge can be and is transformed through classroom interaction.

—Fennema and Franke 1992, 162

How Is Teacher Expertise Achieved?

While political forces attempt to "standardize" teaching, "teachers continue to talk about their work as demanding and unpredictable rather than straightforward and routine" (Darling-Hammond 1997). Additionally, teachers view teaching as an art—"creativity and judgment are always needed" (71). Indeed, flexibility, adaptability, and creativity have been identified as some of the most important determinants of teachers' effectiveness (Schalock 1979; Darling-Hammond, Wise, and Pease 1983). Teachers who "connect" with students, who plan with regard to students' abilities and needs, and who capitalize on students' prior experiences, motivations, and interests "are more effective at stimulating higher-order thinking" (72).

Flexibility, adaptability, and creativity have been identified as some of the most important determinants of teachers' effectiveness.

But, contends Hargreaves (1994), teaching has changed. Citing inclusion, curricular change, diverse assessment strategies, a different kind of parent involvement, and increasing responsibilities, he presents the interesting contrast between the positive effects of *professionalization* and the negative effects of *intensification*. Professionalization implies an increase in the teacher's role. Intensification, on the other hand, asks teachers to respond to greater pressures, to comply with prescribed programs, mandated curricula, and step-by-step methods of instruction, and "to collaborate willingly in their own exploitation as more and more is extracted from them" (118). Recognizing this as an era of educational reform influenced heavily by intensification, Hargreaves notes that we can't make progress with our present educational bureaucracy, nor can we return to the past. "It is time for a change in the rules of teaching and teachers' work" (262). We need to increase professionalization and avoid intensification.

Professionalization implies an increase in the teacher's role and changes in the occupations to make them more like established professions.

In *The Teaching Gap,* Stigler and Hiebert (1999) state, "Our goal . . . is to convince our readers that improving the quality of teaching must be front and center in efforts to improve students' learning. Teaching is the one process in the educational system that is designed specifically to facilitate students' learning" (3). Stigler and Hiebert note the high quality of teaching that already exists, but recognize the inadequacies in student achievement that have been identified nationally through the National Assessment of Educational Progress (NEAP) and internationally through the TIMSS. Stigler, who coauthored *The Learning Gap: Why Our Schools Are Failing and What We Can Learn from Japanese and Chinese Education* (1992), and Hiebert identify three things that were learned from TIMSS: (1) teaching, not teachers, is the critical factor, (2) teaching is a cultural activity, and (3) there is a gap in methods for improving teaching.

Intensification asks teachers to respond to greater pressures.

Suggesting that "designing and building a research-and-development system that explicitly targets steady, gradual improvement of teaching and learning" (131) is needed, the authors identify six principles for gradual, measurable improvement:

Principle #1: Expect improvement to be continual, gradual, and incremental.
Principle #2: Maintain a constant focus on student learning goals.
Principle #3: Focus on teaching, not teachers.
Principle #4: Make improvements in context.
Principle #5: Make improvement the work of teachers.
Principle #6: Build a system that can learn from its own experience.

Concluding, the authors suggest that the popular solution of professionalizing teachers needs redefinition. "Blaming teachers for not improving teaching is unfair" (174). Instead, they offer, we can find a "lasting solution in professionalizing teaching" (175).

Professionalization or Professionalism?

Professionalism is adherence to a set of high standards internal to the practice under consideration.

In these discussions, it is essential to distinguish between professionalization and professionalism. According to Noddings (1992), professionalism is "adherence to a set of high standards internal to the practice under consideration," while professionalization is "changes in occupations that will make them more like established professions" (197). Noting the lack of professional status for teachers, Noddings suggests changes. She proposes cooperation among mathematicians, mathematics users, and educators in the design of courses "especially for teachers." She encourages a press for professional working conditions and an emphasis on "teacher–student relationships as central to the very concept of professionalism." Encouraging teachers and administrators to work together, she notes that "professionals depend on the trust of their clients . . . teachers need opportunities to earn that trust, and such opportunities can only arise as teachers are entrusted by those now controlling their work" (206).

Echoing the call for professionalization, the National Commission on Teaching and America's Future (1996; 1997) offer five major recommendations:

Get serious about standards for both students and teachers.
Reinvent teacher preparation and professional development.
Fix teacher recruitment, and put qualified teachers in every classroom.
Encourage and reward teacher knowledge and skill.
Create schools that are organized for student and teacher success. (Darling-Hammond and Ball 1998, 2)

What Steps Have Been Taken to Professionalize Teaching?

In *Teaching As the Learning Profession*, a compendium of suggestions for policy and practice, Darling-Hammond and Sykes (1999) explore the whole of teacher education: teaching as the learning profession, rethinking teacher education, rethinking professional development for teachers, rethinking organizations for teacher learning, and rethinking policy for teacher learning. An excellent resource, this book should be in every school's professional library.

There are national and state efforts to improve certification and licensing.

In response to the increased call for professionalization and "the growing realization that the intense national quest to improve education for students requires attention to the knowledge and skills of teachers" (Darling-Hammond, Wise, and Klein 1999, 1), there are national and state efforts to improve certification and licensing. In 1986, the Carnegie Forum on Education and the Economy's Task Force on Teaching as a Profession, in *A Nation Prepared: Teachers for the Twenty-First Century*, called for the formation of a national board to improve student performance by strengthening teaching.

In 1987, the National Board for Professional Teaching Standards was established with the mission of establishing high and vigorous standards for what teachers should know and be able to do, developing and operating a national voluntary system to assess and certify teachers who meet those standards, and advancing related education reforms for the purpose of improving student learning in America's schools. Thus, the National Board Certification process is a collaborative, standards-based, ongoing professional development opportunity. Aimed at keeping fine teachers in classrooms, the National Board believes that excellent teachers "should have incentives, rewards, and career paths" as they progress. With national criteria, the certification process is removed from local constraints and is not limited by numbers. The National Board Standards describe what teachers should know and be able to do:

National Board Standards describe what teachers should know and be able to do.

1. Teachers are committed to students and their learning.
2. Teachers know the subjects they teach and how to teach those subjects to students.
3. Teachers are responsible for managing and monitoring student learning.
4. Teachers think systematically about their practice and learn from experience.
5. Teachers are members of learning communities. (National Board for Professional Teaching Standards, 1993)

Certification is available in several fields (e.g., mathematics) and is granted in accordance with populations taught: early childhood (ages 3–8);

middle childhood (ages 7–12); early adolescence (ages 11–15); and adolescent young adulthood (ages 14–18+). Candidates for certification create a portfolio that includes samples of their students' work, videotapes of selected lessons with analytical commentary on those pieces of evidence, and documentation of work as a professional, in the community, and with families. They are tested on subject-matter knowledge, as well as their understanding of how to teach those subjects to their students. The portfolio and assessments are scored by trained assessors, weighted, and combined in an overall score. "This process is a demanding one for the educator, but it is worthwhile" (Kickbusch 1999, 3). Presently (1999–2001), financial incentives from the state (1997 Wisconsin Act 237, s. 115.42, Stats.) are given to teachers who become board certified and are certified and teaching in Wisconsin. It is hoped that school districts with certified teachers will recognize the professionalism of the certification and reward those teachers with pay incentives, increased responsibilities, and so forth.

The State of Wisconsin has changed its teacher licensing and certification requirements.

In line with National Board Certification and in an effort to reflect the standards effort and a move toward performance-based or competency-based teacher preparation, the State of Wisconsin (Chapter PI 34, Approved by the legislature 2/17/00 and Taking effect 7/1/00) has changed its teacher licensing and certification requirements. (See the DPI Web site: www.dpi.state.wi.us, Divisions and Teams, Teacher Certification and Licensing and/or Appendix B in this document.) After August 31, 2004, teachers graduating from education programs will receive an initial five-year, nonrenewable license—based on standards, including assessments of content knowledge and teaching skills, and requiring institutional endorsement, mentoring during the initial teaching period, orientation, and seminars based on teacher standards.

The 10 Wisconsin Standards for Teacher Development and Licensure are:

The requirements reflect the standards effort and a move toward performance based on competency-based evaluation.

- Teachers know the subjects they are teaching. (In accordance with this standard, the content for each subject area has been explicated in PI34 Content Guidelines. See Appendix B.)
- Teachers know how children grow.
- Teachers understand that children learn differently.
- Teachers know how to teach.
- Teachers know how to manage a classroom.
- Teachers communicate well.
- Teachers are able to plan different kinds of lessons.
- Teachers know how to test for student progress.
- Teachers are able to evaluate themselves.
- Teachers are connected with other teachers and the community.

There will be three stages of licensing:
- *Initial five-year nonrenewable*
- *Five-year professional license renewable*
- *Ten-year master educator license renewable.*

The licenses will be granted based on developmental levels—early childhood, middle childhood, early adolescence, and adolescence—and in discipline fields. A five-year professional license may be applied for after at least three years of teaching experience. The professional licensure calls for development of a professional improvement plan based on one or more of the national teaching standards. The plan must be approved by a panel of three, a peer, an administrator, and a teacher educator. Teachers then gather evidence of having met the plan requirements and the panel certifies successful completion of the plan.

This license may be renewed by completing and implementing a professional development plan and having the plan and successful implementation approved by a professional development team of three teachers. After five years of experience and successful completion of the professional plan, a teacher may apply for the 10-year master educator license. This requires a related master's degree, evidence of contributions to the profession, demonstration of exemplary classroom performance on video or on-site, and formal assessment by a professional development team. National Board Certification can serve in lieu of these requirements.

This new plan moves certification from a course and credit orientation to a performance and competency orientation. The three stages of licensing provide for self-directed, planned professional development guided by the 10 Wisconsin teacher standards. The mentoring component will allow excellent seasoned teachers an opportunity to work with incoming teachers to assist in their growth and to augment their own through the experience. Current teachers who hold renewable licenses have the option of taking university credits or of creating a professional development plan. Lifetime license holders will have no additional requirements, but will have the opportunity for professional growth in keeping with the new plan. It is hoped that the plan will move the profession toward increasing professionalization and, at the same time, address certification issues that arise because of increased demand for teachers, especially in certain disciplines.

The three stages of licensing provide for self-directed, planned professional development guided by the 10 Wisconsin teacher standards.

Citing the demands of today's classrooms—student bodies that are more diverse and have greater needs, parents and community groups that are asking for more influence in the school, the social, economic, and technological needs of the workplace that are placing greater demands on workers, the increasing federal and state mandates and the increase of special programs, and, yet, the love of the status quo—Rallis, Rossman, Phlegar, and Abeille (1995) discuss "dynamic teachers." Characterizing those teachers as moral stewards, constructors, philosophers, facilitators, inquirers, bridgers, and changemakers, the authors present the cases of dynamic teachers in innovative schools, all of whom succeeded in helping children think and learn and prepare for the unknown challenges of their futures.

In all instances, those teachers focused on continued learning for themselves and continued professional improvement. Teachers teach because they care about their students and their futures. They also teach because they love to learn. It is this spark of learning that must be kept alive and burning if we are to retain excellent teachers in classrooms. In his timeless manner, Confucius offered, "Never tire of learning; Never tire of teaching" (Levine 1999, 115).

Teachers teach because they care about their students and their futures. They also teach because they love to learn.

Assessment

Curriculum, instruction, and assessment have been included in this document as an integrated whole because one component cannot function adequately without attention to each of the others. An excellent curriculum can be in place. Wise and wonderful teaching can be taking place. But it is assessment, both formal and informal, that will convince us that teaching and learning mathematics with understanding is occurring. Indeed, "it is assessment that enables us to distinguish between teaching and learning. Many times because a topic has been taught, it is assumed it is learned. Only until it is appropriately assessed do we know" (Clarke, Clarke, and Lovitt 1990, 128–29). Romberg (see the preface in this document) contends that assessment consistent with the standards-based vision—assessment of "mathematical literacy"—needs to be developed to adequately measure attainment of understanding.

A comprehensive accounting includes knowledge of math and disposition toward math.

What Is Assessment?

A number of terms are used to refer to the collection of data and information for the purposes of describing an individual's or group's level of knowledge, performance, or achievement—for example, measurement, test, assessment, and evaluation. *Mathematical assessment,* according to Webb (1992, 662–63) is the "comprehensive accounting of an individual's or group's functioning with mathematics or in the application of mathematics." Wood (1987) included in that comprehensive accounting "a variety of evidence, qualitative as well as quantitative reflected in "a variety of contexts," and "including *knowledge of mathematics* and *disposition toward mathematics*" (2). The Joint Committee on Standards for Educational Evaluation (1981) defines evaluation as "the systematic collection of evidence to help make decisions regarding (1) students' learning, (2) materials development, and (3) programs to systematically investigate the worth or merit of some objective" (Webb, 663). Thus, assessment can be used to provide evidence of what students know and are able to do; to express what is valued regarding what students are to know, do, or believe; to provide information to decision makers; and to provide information on the effectiveness of the educational system as a whole.

Assessment can be used to provide evidence of what students know and are able to do; to express what is valued regarding what students are to know, do, or believe; to provide information to decision makers; and to provide information on the effectiveness of the educational system as a whole.

Cognizant of the many purposes of assessment, the *Assessment Standards for School Mathematics* (Assessment Standards Working Group 1995) offers an excellent listing of ideas to consider in planning assessments that help to clarify the principal points at which critical decisions need to be made, along with the decisions and actions that occur within each phase (4–5). The listing is given in its entirety because of its usefulness for assessment planners:

Planning the assessment
- What purpose does the assessment serve?
- What framework is used to give focus and balance to the activities?
- What methods are used for gathering and interpreting evidence?
- What criteria are used for judging performances on activities?
- What formats are used for summarizing judgments and reporting results?

Gathering evidence
- How are activities and tasks created or selected?
- How are procedures selected for engaging students in the activities?
- How are methods for creating and preserving evidence of the performances to be judged?

Interpreting the evidence
- How is the quality of the evidence determined?
- How is an understanding of the performances to be inferred from the evidence?
- What specific criteria are applied to judge the performances?
- Have the criteria been applied appropriately?
- How will the judgments be summarized as results?

Using the results
- How will the results be reported?
- How should inferences from the results be made?
- What action will be taken based on the inferences?
- How can it be ensured that these results will be incorporated in subsequent instruction and assessment?

A variety of assessment tools will be used to create a complete picture of what the student knows.

What Are the Elements of a Good Assessment?

"Fundamental to mathematical assessment is an explicit definition of the content to be assessed" (Webb 1992, 664). If knowledge of mathematics is integrated and determination of students' mathematical power requires assessment of all aspects of their mathematical knowledge, as well as the extent to which they have integrated this knowledge, then the approach will require that students "apply a variety of mathematical concepts and procedures and that they engage in reasoning and problem-solving" (667). Such an approach requires "ample time to define appropriate assessment situations and to allow students to work through them effectively." There is also an expectation that a variety of assessment tools will be used to create a complete picture of what the student knows. If disposition is important, then assessment of attitude, motivation, and individual connate characteristics also needs to be included.

A domain-based assessment program includes informal assessment, formal assessment, and reflections of growth over time.

In "Assessment in Classrooms That Promote Understanding," Shafer and Romberg (1999, 160) identify assumptions underlying the design of those assessments: (1) assessment should be viewed as an ongoing process that is integrated with instruction, (2) multiple sources of evidence are needed to assess students' developing knowledge in a domain, and (3) assessment should involve the deliberate documentation of information derived from classroom interaction as well as from written work. Giving attention to the mental activities described in Carpenter and Lehrer's "Teaching and Learning Mathematics with Understanding" (chapter 2 of this document), a domain-based assessment program was described that included informal assessment, formal assessment, and reflections of growth over time. Attention to increasingly complex levels of reasoning was also considered.

Assessments, then, need to check understanding of significant and correct mathematics, mathematical reasoning, connections among mathematical ideas, and communication about mathematics. Romberg, Zarinnia, and Collis (1990) suggest that two types of assessment items facilitate measuring these understandings: (1) an open response format where student work is evaluated on the basis of the selection and use of data given and the processes and concepts used and (2) the "superitem" (30), where the student responds to a series of questions and the response is taken to indicate methods of using the data and the employment of certain concepts and techniques.

How Should Assessments Be Evaluated?

Assessments that ask for conceptual understanding need to be scored on a different basis, such as rubrics.

Generally speaking, assessments that ask for conceptual understanding need to be scored on a different basis, such as rubrics. Rubrics describe the performance levels that might be expected in response to open-ended questions, performance items, and so forth. The Wisconsin Student Assessment System (WSAS), in moving to performance criteria, implemented rubric scoring. (See WSAS *General Proficiency Categories and Mathematics Proficiency Descriptors* rubrics in Appendix D.) Several school districts have designed benchmark assessments to accompany their grade-specific goals and are using rubric scoring to evaluate them. (See Sun Prairie's *Sample Scoring Process, General Rubric,* and *Student Rubric* and Milwaukee's *Rubric for Scoring Mathematics Performance Assessment* in Appendix D.) Many districts are finding the use of rubrics to be beneficial in the classroom as well as for assessment. Every student is given credit for attempting to solve the problem, which, of course, encourages students to try each problem. At the other end of the performance scale, there is no ceiling. Students are given the opportunity to construct an advanced response that can include explanatory charts, graphs, extension questions, and so forth. Outstanding students no longer need to settle only for a "right answer."

Many districts are finding the use of rubrics to be beneficial in the classroom as well as for assessment.

Taking scoring for developing mathematical reasoning a step farther, Peressini and Webb (1999) suggest that assessment needs to look at foundational knowledge, solution process, and communication. Students need to learn how to communicate their mathematical reasoning and leave "clear trails of their mathematical work" for assessors to get "more robust and complete understandings of students' mathematical power" (Peressini and Webb 1999, 173). Citing Moss's hermeneutic approach that called for "holistic, integrative interpretations of collected performances that seek to understand the whole in light of its parts," the authors suggest that such analyses "approach the interpretation of students' mathematical work from a different perspective" (173).

Wisconsin State Assessments

Two statutes have impacted assessment in Wisconsin. The first statute—Wisconsin School Law 118.30(1)(a)—called for examinations to measure pupil attainment of knowledge and concepts in the fourth, eighth, and tenth grades. Additionally, the statute [118.30(1)(b)] called for a high school graduation examination that is designed to measure whether pupils meet the pupil academic standards issued by the governor as executive order no. 326, dated January 13, 1998. Another component of the law [118.30(1m)(a) and 118.30(1m)(am)] stated that the school board shall provide a fourth or eighth grade pupil with at least two opportunities to take the test, and [118.30(1m)(d)] for high school students, must administer the examination at least twice each school year and shall administer the examination only to pupils enrolled in the eleventh and twelfth grades.

After much discussion regarding the validity of resting promotion and graduation on a single score, the law was modified to state that each school board shall adopt a written policy specifying the criteria for promoting a pupil from the fourth grade to the fifth grade and from the eighth grade to the ninth grade [118.33(6)(a)]; and requiring that by September 1, 2002, each school board operating high school grades shall develop a written policy specifying criteria for granting a high school diploma [118.33(1)(f)]. Effective September 1, 2002, the criteria for fourth and eighth grade promotion shall include the pupil's score on the knowledge and concepts examination, the pupil's academic performance, recommendations of teachers based on academic performance, and any other academic criteria specified by the school board [118.39(6)(a)]. Also effective September 1, 2002, the school board must adopt a written policy specifying the criteria it will use for granting a high school diploma. The criteria must include the number of credits required and other elements as identified in paragraph (a) of ss.118.33, the other elements being the pupil score on the high school graduation test, pupil academic performance, and recommendations of teachers. Guidelines for developing advancement policies in keeping with 1999 Wisconsin Act 9 have been developed by the Department of Public Instruction for this purpose (March 18, 2000).

The WKCE and the HSGT have been designed to measure proficiency achievement of Wisconsin's Model Academic Standards.

Designed to measure proficiency achievement of *Wisconsin's Model Academic Standards,* both the *Wisconsin Knowledge and Concepts Exam* (WKCE) and the forthcoming *High School Graduation Test* (HSGT) have been carefully developed to yield information about mathematical disposition; knowledge and understanding of the nature of mathematics including concepts, procedures, and skills; mathematical reasoning; the ability to use mathematical language in communication of mathematical ideas; and the ability to apply mathematical knowledge to solve problems in a variety of contexts and disciplines. Different types of items are used—multiple choice, constructed response, and extended constructed response. Using a cross-hatching of content strands based on the standards and process strands, items are often developed to measure multiple facets of knowledge in a single item (CTB McGraw-Hill 1997a).

The assessments are rubric scored.

The assessments are rubric scored (see Appendix D). Proficiency Score Standards have been set (Wisconsin Department of Public Instruction 1997). The Knowledge and Concepts examinations have been carefully aligned with *Wisconsin's Model Academic Standards for Mathematics* (Wisconsin Department of Public Instruction 1998). Assessment information is available on the Office of Educational Accountability Web site: http://www.dpi.state.wi.us/oea/assessmt.html. In-depth information about individual school districts and their performances is now available through the Wisconsin Information Network for Successful Schools (WINSS) project (www.dpi.state.wi.us/sig/index.html) to facilitate data-driven decision-making.

The *High School Graduation Test (HSGT)* is being subjected to scrutiny in its development. *Wisconsin's Model Academic Standards (WMAS)* have been examined to determine the content that could be assessed appropriately on a large-scale standardized test. For instance, there is domain knowledge at

The State of Wisconsin and its teachers are taking positive actions to gain the knowledge necessary for construction of the opportunities and assessments that will help move mathematics endeavors forward.

a reproduction level that can be tested very adequately with a good multiple-choice item. There is knowledge at a connections or analysis level that requires explanation and a constructed response item format. Knowledge at an analysis level often lends itself well to an extended constructed response item. Finally, there are standards such as "E.12.1 Work with data in the context of real world situations" that cannot be assessed other than with performance in the classroom. The *Wisconsin High School Graduation Test Educator's Guide* (Department of Public Instruction 2000) is a valuable tool for developing an understanding of how the assessment is being developed to measure specified components.

In the preface to this document, T. A. Romberg suggested that if there is improvement to be made in education, a major change needs to be made in assessment. The State of Wisconsin and its teachers are becoming more and more aware of this need and are taking positive actions to gain the knowledge necessary for construction of the opportunities and assessments that will help move mathematics endeavors forward.

There are many assessment tools and techniques that can be used in classrooms to facilitate the development and assessment of student learning. A detailed set of those aids is found in Appendix D.

References

Curriculum

Ball, D. L. 1993. "Halves, Pieces, and Twoths: Constructing and Using Representational Contexts in Teaching Fractions." In *Rational Numbers: An Integration of Research*, edited by T. P. Carpenter, E. Fennema, and T. A. Romberg. Hillsdale, NJ: Lawrence Erlbaum Associates, Publishers.

Ball, D. L., and D. K. Cohen. 1996. "Reform by the Book: What Is—or Might Be—the Role for Curriculum Materials in Teacher Learning and Instructional Reform?" *Educational Researcher*, 25 (9), 6–8, 14.

Berlinski, D. 2000. *The Advent of the Algorithm: The Idea That Rules the World.* New York: Harcourt, Inc.

Billstein, R. 1998. "Middle Grades Mathematics: The STEM Project—a Look at Developing a Middle School Mathematics Curriculum." In *Mathematics in the Middle*, edited by L. Leutzinger. Reston, VA: National Council of Teachers of Mathematics.

Commission on Standards for School Mathematics. 1989. *Curriculum and Evaluation Standards for School Mathematics.* Reston, VA: National Council of Teachers of Mathematics.

Council of Chief State School Officers, Wisconsin Center for Education Research, and Eleven State Collaborative. 2000. *Using Data on Enacted Curriculum in Mathematics and Science: Sample Results from a Study of Classroom Practices and Subject Content.* Washington, D.C.: Council of Chief State School Officers.

Erickson, H. L. 1998. *Concept-Based Curriculum and Instruction: Teaching Beyond the Facts.* Thousand Oaks, CA: Corwin Press, Inc.

Even, R. and G. Lappan. 1994. "Constructing Meaningful Understanding of Mathematics Content." In *Professional Development for Teachers of Mathematics: 1994 Yearbook*, edited by D. B. Aichele and A. F. Coxford. Reston, VA: National Council of Teachers of Mathematics.

Fortier, J., and J. M. Moser. 1992. *Wisconsin Criteria for Quality Tasks.* Madison, WI: Wisconsin Department of Public Instruction and Wisconsin Center for Educational Research.

Grunow, J. E. 1999. *Using Concept Maps in a Professional Development Program to Assess and Enhance Teachers' Understanding of Rational Number.* Ann Arbor, MI: UMI (Microform 9910421).

Hiebert, J., and P. Lefevre. 1986. "Conceptual and Procedural Knowledge in Mathematics: An Introductory Analysis." In *Conceptual and Procedural Knowledge: The Case of Mathematics,* edited by J. Hiebert. Hillsdale, NJ: Lawrence Erlbaum Associates, Publishers.

Kieren, T. E. 1988. "Personal Knowledge of Rational Numbers." In *Number Concepts and Operations in the Middle Grades,* edited by J. Hiebert and M. Behr. Reston, VA: National Council of Teachers of Mathematics.

———. 1993. "Rational and Fractional Numbers: From Quotient Fields to Recursive Understanding." In *Rational Numbers: An Integration of Research,* edited by T. P. Carpenter, E. Fennema, and T. A. Romberg. Hillsdale, NJ: Lawrence Erlbaum Associates, Publishers.

Kieren, T., and S. Pirie. 1991. "Recursion and the Mathematical Experience." In *Epistemological Foundations of Mathematical Experience,* edited by L. Steffe. New York: Springer-Verlag.

Kleiman, G. M., D. Tobin, and S. Isaacson. 1998. "What Should a Middle School Mathematics Classroom Look Like? Watching the 'Seeing and Thinking Mathematically' Curriculum in Action." In *Mathematics in the Middle,* edited by L. Leutzinger. Reston, VA: National Council of Teachers of Mathematics.

Lamon, S. J. 1993. "Ratio and Proportion: Children's Cognitive and Metacognitive Processes." In *Rational Numbers: An Integration of Research,* edited by T. P. Carpenter, E. Fennema, and T. A. Romberg. Hillsdale, NJ: Lawrence Erlbaum Associates, Publishers.

Lampert, M. 1991. "Connecting Mathematical Teaching and Learning." In *Integrating Research on Teaching and Learning Mathematics,* edited by E. Fennema, T. P. Carpenter, and S. J. Lamon. Albany, N.Y.: State University of New York Press.

Lappan, G., and E. Phillips. 1998. "Teaching and Learning in the Connected Mathematics Project." In *Mathematics in the Middle,* edited by L. Leutzinger. Reston, VA: National Council of Teachers of Mathematics.

Laughlin, C., and J. C. Moyer. 1998. "The Power of Discourse." In *Mathematics in the Middle,* edited by L. Leutzinger. Reston, VA: National Council of Teachers of Mathematics and National Middle School Association.

Mack, N. K. 1993. "Learning Rational Numbers with Understanding: The Case of Informal Knowledge." In *Rational Numbers: An Integration of Research,* edited by T. P. Carpenter, E. Fennema, and T. A. Romberg. Hillsdale, N. J.: Lawrence Erlbaum Associates, Publishers.

Maturana, H. R., and F. J. Varela. 1987. "Assessment of Rational Number Understanding: A Schema-Based Approach." In *Ratio: An Integration of Research,* edited by T. P. Carpenter, E. Fennema, and T. A. Romberg. Mahwah, NJ: Erlbaum.

McNeely, M.E., ed. 1997. *Attaining Excellence TIMSS as a Starting Point to Examine Curricula: Guidebook to Examine School Curricula.* Washington D.C.: U.S. Department of Education, Office of Educational Research and Improvement.

Mercer, S., and M. A. Henningsen. 1998. "The Pentomino Project: Moving Students from Manipulatives to Reasoning and Thinking about Mathematical Ideas." In *Mathematics in the Middle,* edited by L. Leutzinger. Reston, VA: National Council of Teachers of Mathematics and National Middle School Association.

Murphy, D. A. 1998. "A Multiple-Intelligence Approach to Middle School Mathematics." In *Mathematics in the Middle,* edited by L. Leutzinger. Reston, VA: National Council of Teachers of Mathematics.

Office of Educational Research and Improvement. 1997. *Attaining Excellence: TIMSS As a Starting Point to Examine Curricula: Guidebook to Examine School Curricula.* Washington, D.C.: U.S. Department of Education.

Phillips, E., and G. Lappan. 1998. "Algebra: The First Gate." In *Mathematics in the Middle,* edited by L. Leutzinger. Reston, VA: National Council of Teachers of Mathematics.

Pirie, S. E. B. 1988. "Understanding: Instrumental, Relational, Intuitive, Constructed, Formalized . . . How Can We Know?" *For the Learning of Mathematics,* 8 (3), 7–11.

Romberg, T. A. 1998. "Designing Middle School Mathematics Materials Using Problems Created to Help Students Progress from Informal to Formal Mathematical Reasoning."

In *Mathematics in the Middle,* edited by L. Leutzinger. Reston, VA: National Council of Teachers of Mathematics.

Romberg, T. A., and J. J. Kaput. 1999. "Mathematics Worth Teaching: Mathematics Worth Understanding." In *Mathematics Classrooms That Promote Understanding,* edited by E. Fennema and T. A. Romberg. Mahwah, NJ: Lawrence Erlbaum Associates, Publishers.

Schoenfeld, A. H. 1985. *Mathematical Problem Solving.* New York: Academic Press, Inc.

Standards 2000 Project. 2000. *Principles and Standards for School Mathematics.* Reston, VA: National Council of Mathematics.

Streefland, L. 1993. "Fractions: A Realistic Approach." In *Rational Numbers: An Integration of Research,* edited by T. P. Carpenter, E. Fennema, and T. A. Romberg. Hillsdale, NJ: Lawrence Erlbaum Associates, Publishers.

Tomm, K. January 1989. *Consciousness and Intentionality in the Work of Humberto Maturana.* Paper presented at the meeting of the faculty of education, University of Alberta, Edmonton.

Usiskin, Z. 1998. "Paper-and-Pencil Algorithms in a Calculator-and-Computer Age." In *The Teaching and Learning of Algorithms in School Mathematics: 1998 Yearbook,* edited by L. J. Morrow and M. J. Kenney. Reston, VA: National Council of Teachers of Mathematics.

Wisconsin Department of Public Instruction. 1998. *Wisconsin's Model Academic Standards.* Madison: Wisconsin Department of Public Instruction.

Instruction

Carnegie Forum on Education and the Economy. 1986. *A Nation Prepared: Teachers for the 21st Century.* Washington, D.C.: Carnegie Forum on Education and the Economy, Task Force on Teaching as a Profession.

Cohen, D. K., and H. C. Hill. 1998. "State Policy and Classroom Performance: Mathematics Reform in California." *Consortium for Policy Research in Education Policy Briefs, RB 23,* 1–16.

Commission on Teaching Standards for School Mathematics. 1991. *Professional Standards for Teaching Mathematics.* Reston, VA: National Council of Teachers of Mathematics.

Darling-Hammond, L. 1997. *The Right to Learn: A Blueprint for Creating Schools That Work.* San Francisco: Jossey-Bass Publishers.

———. 1998. "Teachers and Teaching: Testing Policy Hypotheses from a National Commission Report." *Educational Researcher, 27,* 5–15.

Darling-Hammond, L., and D. L. Ball. 1998. *Teaching for High Standards: What Policymakers Need to Know and Be Able to Do.* Philadelphia, PA: CPRE Publications.

Darling-Hammond, L., and G. Sykes. 1999. *Teaching As the Learning Profession: Handbook of Policy and Practice.* San Francisco: Jossey-Bass Publishers.

Darling-Hammond, L., A. E. Wise, and S. P. Klein. 1999. *A License to Teach: Raising Standards for Teaching.* San Francisco: Jossey-Bass Publishers.

Darling-Hammond, L., A. E. Wise, and S. R. Pease. 1983. "Teacher Evaluation in the Organizational Context: A Review of the Literature." *Review of Educational Research, 53,* 285–297.

Fennema, E., and M. L. Franke. 1992. "Teachers' Knowledge and Its Impact." In *Handbook of Research on Mathematics Teaching and Learning,* edited by D. A. Grouws. New York: Macmillan.

Hargreaves, A. 1994. *Changing Teachers, Changing Time: Teachers' Work and Culture in the Postmodern Age.* New York, NY: Columbia University Teachers College Press.

Kickbusch, K. 1999. National Board Certification. *Focus: Educational Issues Series,* 4 (October): 6. Madison: Wisconsin Education Association Council.

Levine, S. L., ed. 1999. *A Passion for Teaching.* Alexandria, VA: Association for Supervision and Curriculum Development. (www.ascd.org)

Mingus, T. T. Y, and Grassl, R. M. 1998. "Algorithmic and Recursive Thinking: Current Beliefs and Their Implications for the Future." In *The Teaching and Learning of Algorithms in School Mathematics: 1998 Yearbook,* edited by L. J. Morrow and M. J. Kenney. Reston, VA: National Council of Teachers of Mathematics.

National Board for Professional Teaching Standards (NBPTS). 1993. *What Teachers Should Know and Be Able to Do.* Detroit: NBPTS.

National Commission on Teaching and America's Future. 1996. *What Matters Most: Teaching for America's Future.* New York: National Commission on Teaching and America's Future.

———. 1997. *Doing What Matters Most: Investing in Quality Teaching.* New York: National Commission on Teaching and America's Future.

Newmann, F. M., W. G. Secada, and G. G. Wehlage. 1995. *A Guide to Authentic Instruction and Assessment: Vision, Standards and Scoring.* Madison: Wisconsin Center for Education Research.

Noddings, N. 1992. "Professionalization and Mathematics Teaching." In *Handbook of Research on Mathematics Teaching and Learning,* edited by D. A. Grouws. New York: Macmillan Publishing Company.

Rallis, S. F., G. B. Rossman, J. M. Phlegar, and A. Abeille. 1995. *Dynamic Teachers: Leaders of Change.* Thousand Oaks, CA: Corwin Press, Inc.

Schalock, D. 1979. "Research on Teacher Selection." In *Review of Research in Education* (Vol. 7), edited by D. C. Berliner. Washington, D.C.: American Educational Research Association.

Standards 2000 Project. 2000. *Principles and Standards for School Mathematics.* Reston, VA: National Council of Teachers of Mathematics.

Stigler, J. W., and J. Hiebert. 1999. *The Teaching Gap: Best Ideas from the World's Teachers for Improving Education in the Classroom.* New York: The Free Press.

Assessment

Assessment Standards Working Groups. 1995. *Assessment Standards for School Mathematics.* Reston, VA: National Council of Teachers of Mathematics.

Azzolino, A. 1990. "Writing As a Tool for Teaching Mathematics: The Silent Revolution." In *Teaching and Learning Mathematics in the 1990s: 1990 Yearbook,* edited by T. J. Cooney and C. R. Hirsch. Reston, VA: National Council of Teachers of Mathematics.

Bartels, B. 1995. "Promoting Mathematics Connections with Concept Mapping." *Mathematics Teaching in the Middle School, 1(7)* (November–December): 542–549.

Beyerback, B. A. 1986. *Concept Mapping in Assessing Prospective Teachers' Concept Development.* (ERIC Document Reproduction Service No. ED 291 800.)

Biggs, J. B., and K. F. Collis. 1989. *Educational Research: An Introduction, Fifth Ed.* New York: Longman.

Brown, M. 1988. *The Graded Assessment in Mathematics Project.* Unpublished manuscript, King's College, London, U.K.

Carpenter, T. P., and E. Fennema. 1988. "Research and Cognitively Guided Instruction." In *Integrating Research on Teaching and Learning Mathematics,* edited by E. Fennema, T. P. Carpenter, and S. J. Lamon. Madison: Wisconsin Center for Education Research.

Clarke, D. 1988. *Assessment Alternatives in Mathematics.* Melbourne, Australia: Jenkin Buxton Printers Pty. Ltd.

———. 1997. *Constructive Assessment in Mathematics: Practical Steps for Classroom Teachers.* Berkeley, CA: Key Curriculum Press.

Clarke, D. J., Clarke, D. M., and Lovitt, C. J. 1990. "Changes in Mathematics Teaching Call for Assessment Alternatives." In *Teaching and Learning Mathematics in the 1990s: 1990 Yearbook,* edited by T. J. Cooney and C. R. Hirsch. Reston, VA: National Council of Teachers of Mathematics.

Commission on Standards for School Mathematics. 1989. *Curriculum and Evaluation Standards for School Mathematics.* Reston, VA: National Council of Teachers of Mathematics.

Commission on Teaching Standards for School Mathematics. 1991. *Professional Standards for Teaching Mathematics.* Reston, VA: National Council of Teachers of Mathematics.

CTB/McGraw-Hill. 1997a. *Teacher's Guide to Terra Nova.* Monterey, CA: CTB/McGraw-Hill.

———. 1997b. "Wisconsin Student Assessment System Knowledge and Concepts Examinations Sample Items." From *Teacher's Guide to Terra Nova.* Monterey, CA: CTB/McGraw-Hill.

Danielson, C. 1997. *A Collection of Performance Tasks and Rubrics.* Larchmont, NY: Eye On Education.

Dold, S. B., J. D. Fortier, W. J. Erpenbach, M. Burke, and S. K. Ketchum. 1998. *Wisconsin Knowledge and Concepts Examinations: An Alignment Study (at Grade 4, Grade 8, Grade 12).* Madison: Wisconsin Department of Public Instruction.

Entrekin, V. S. 1992. "Sharing Teaching Ideas: Mathematical Mind Mapping." *Mathematics Teacher, 85,*(September): 444–445.

Fennema, E. 1998. Conversation with author. Madison, WI, UW-Madison.

Fortier, J., H. G. Cook, and M. Burke. 2000. *Wisconsin High School Graduation Test Educator's Guide.* Madison: Wisconsin Department of Public Instruction.

Glaser, R. 1986. "The Integration of Instruction and Testing." In *The Redesign of Testing for the 21st Century: Proceedings of the 1985 ETS Invitational Conference,* edited by E. E. Freeman. Princeton, NJ: Educational Testing Service.

Grunow, J. E. 1999. *Using Concept Maps in a Professional Development Program to Assess and Enhance Teachers' Understanding of Rational Number.* Ann Arbor, MI: UMI (UMI Microform 9910421).

Howell, J. A. 1999. Cray Academy WASDI Teacher Leader Workshop Aug. 2–6, 1999. Chippewa Falls, WI.

Joint Committee on Standards for Educational Evaluation. 1981. *Standards for Evaluations of Educational Programs, Projects and Materials.* New York: McGraw-Hill.

MacDonell, A., and C. Anderson (Eds.). 1999. *Balanced Assessment for the Mathematics Curriculum: Berkeley, Harvard, Michigan State, Shell Centre.* White Plains, N.Y.: Dale Seymour Publications. (Available for elementary, middle school, and high school, two levels.)

Mathematical Sciences Education Board. 1993. *Measuring Up: Prototypes for Mathematics Assessment.* Washington, D.C.: National Academy Press.

Moss, P. A. 1994. "Can There Be Validity without Reliability?" *Educational Researcher, 18* (9), 5–12.

New Standards. 1998. *Performance Standards.* Pittsburgh, PA: National Center on Education and the Economy and the University of Pittsburgh.

Office of Educational Accountability. 1997. *Final Summary Report of the Proficiency Score Standards for the Wisconsin Student Assessment System (WSAS) Knowledge and Concepts Examinations for Elementary, Middle, and High School at Grades 4, 8, and 10.* Madison: Wisconsin Department of Public Instruction.

———. 1995. *Wisconsin Student Assessment System Performance Assessment Sampler.* Madison: Wisconsin Department of Public Instruction.

Peressini, D., and N. Webb. 1999. "Analyzing Mathematical Reasoning in Students' Responses Across Multiple Performance Assessment Tasks." In *Developing Mathematical Reasoning in Grades K–12: 1999 Yearbook,* edited by L. V. Stiff and F. R. Curcio. Reston, VA: National Council of Teachers of Mathematics.

Romberg, T. A., E. A. Zarinnia, and K. F. Collis. 1990. "A New World View of Assessment in Mathematics." In *Assessing Higher Order Thinking in Mathematics,* edited by G. Kulm. Washington, D.C.: American Association for the Advancement of Science.

Ruiz-Primo, M. A., and R. J. Shavelson. 1996. "Problems and Issues in the Use of Concept Maps in Science Assessment." *Journal of Research in Science Teaching, 33* (6), 569–600. (Paper presented at NSF Assessment Conference, Washington, D.C., 11/30–12/2/95.)

Shafer, M. C., and T. A. Romberg. 1999. "Assessment in Classrooms That Promote Understanding." In *Mathematics Classrooms That Promote Understanding,* edited by E. Fennema and T. A. Romberg. Mahwah, NJ: Lawrence Erlbaum Associates, Publishers.

Shell Centre for Mathematical Education. 1984. *Problems with Patterns and Numbers.* University of Nottingham, U.K.: Joint Matriculation Board.

———. 1985. *The Language of Functions and Graphs.* University of Nottingham, U.K.: Joint Matriculation Board.

Silver, E. A., J. Kilpatrick, and B. Schlesinger. 1990. *Thinking through Mathematics.* New York: College Entrance Examination Board.

Standards 2000 Project. 2000. *Principles and Standards for School Mathematics.* Reston, VA: National Council of Teachers of Mathematics.

Stenmark, J. K., ed. 1991. *Mathematics Assessment: Myths, Models, Good Questions, and Practical Suggestions.* Reston, VA: National Council of Teachers of Mathematics.

Stenmark, J. K., EQUALS Staff, and Assessment Committee of the *Campaign for Mathematics.* 1989. *Assessment Alternatives in Mathematics.* Berkeley, CA: University of California.

Stine, M.A. 2000. "State Achievement Tests Can Be a Positive Force in Your Classroom." *ENC Focus: Assessment That Informs Practice* 7 (2). Columbus, OH: Eisenhower National Clearinghouse for Mathematics and Science Education.

Webb, N. L. 1992. "Assessment of Students' Knowledge of Mathematics: Steps Toward a Theory." In *Handbook of Research on Mathematics Teaching and Learning,* edited by D. A. Grouws. New York: Macmillan Publishing Company.

Webb, N. L., and D. Briars. 1990. "Assessment in Mathematics Classrooms, K–8." In *Teaching and Learning Mathematics in the 1990s: 1990 Yearbook,* edited by T. J. Cooney and C. R. Hirsch. Reston, VA: National Council of Teachers of Mathematics.

Wood, R. 1987. "The Agenda for Educational Measurement." In *Assessing Educational Achievement,* edited by D. L. Nuttall. London, U.K.: Falmer Press.

Wisconsin Department of Public Instruction. 1997. *Final Summary Report of the Proficiency Score Standards for the Wisconsin Student Assesment System (WSAS) Knowledge and Concepts Examinations for Elementary, Middle, and High School at Grades 4, 8, and 10.* Madison: Wisconsin Department of Public Instruction.

———. 1998. *Wisconsin's Model Academic Standards for Mathematics.* Madison: Wisconsin Department of Public Instruction.

———. March 18, 2000. *Suggestions for Local School Boards in Approaching the Development of High School Graduation and Fourth/Eighth Grade Advancement Policies: Implementing the Provisions of 1999 Wisconsin Act 9.* Madison: Wisconsin Department of Public Instruction.

Planning Curriculum
in

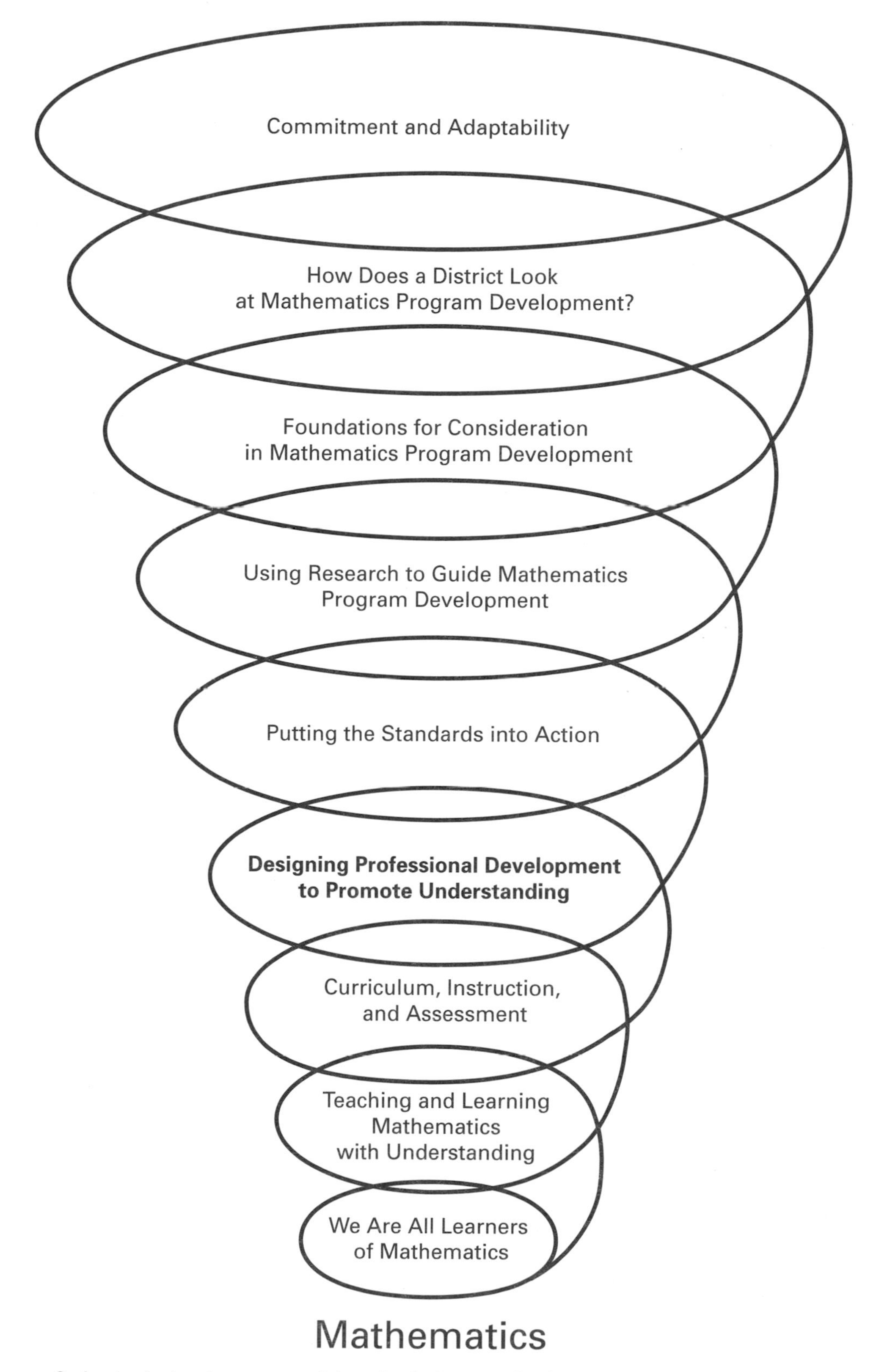

Spring (spring) *n.* 1. a source; origin; or beginning: *a spring into action.* 2. elasticity; resilience: *a spring forward with commitment.* (Morris, W., ed., 1971. *The American Heritage Dictionary of the English Language.* New York: American Heritage Publishing Company, Inc. and Houghton Mifflin Company, 1250, 2nd definition.)

Designing Professional Development to Promote Understanding

4

Jodean E. Grunow
Mathematics Consultant, Wisconsin Department of Public Instruction

Professional Development? The traditional placement of a professional development chapter is at the end . . . and now we "do" professional development . . . if there are funds, if there is time, if there is interest. Professional development in this document is placed "up front." We have worthy standards that have been developed nationally and on a state level through organizational and community consensus. Assessments, though there continues to be work to be done, are moving toward reflecting those standards and toward judging mathematical literacy (Romberg 2001, Preface of this document). Curricula have been developed and are being conceived to reflect the intent of the standards. The component that will pull all of these elements together is, of course, instruction. Instruction for the building of understanding is the vision of the standards. Teachers, both pre-service and in-service, will need to experience the excitement of learning for understanding themselves, will need to hone their content skills since the content knowledge ante has been "upped" with this vision, will need to learn how to read and how to build on learner cognitions, and will need to learn how to assess in a manner congruent with curriculum and instruction. There has never been a greater need for professional development in the field of mathematics. It is an insistent need because of the demands of an information and technological society.

There has never been a greater need for professional development in the field of mathematics.

Why Is There a Need for Professional Development?

Student Performance Is Closely Linked to Teacher Expertise

The goal of all of the standards, curriculum, instruction, and assessment movements is, of course, increase in student performance: what students know and are able to do. Just as there has been intensive research on teaching and learning to increase effectiveness, there has been in-depth research on professional development. Some of the findings from this research are eye opening.

There has been in-depth research on professional development.

Every dollar spent on more highly qualified teachers produced greater increases in student achievement than a dollar spent on any other single program."

—Sparks and Hirsh 2000, 2

Student learning, teacher learning, and professional development are inextricably interrelated (Cohen and Hill 1998). The National Commission on Teaching and America's Future (NCTAF), in its report *What Matters Most: Teaching for America's Future* (1996), as described by Sparks (1998), recounts a Texas study (Ferguson 1991) that found that teachers' expertise accounts for about 40 percent of the variance in students' reading and mathematics achievement at grades 1 through 11, "more than any other single factor" (2). Additionally, it was found that "every dollar spent on more highly qualified teachers produced greater increases in student achievement than a dollar spent on any other single program" (Sparks and Hirsh 2000, 2).

Darling-Hammond and Ball (1998) have summarized the research about the relationships between teacher knowledge and student performance, reporting a 43 percent variance in the mathematics test score gains of students in grades 3 through 6 as a result of teacher expertise. They cite a 1983 review of 65 studies that concluded that teacher effectiveness depends on the amount and kind of teacher education and training, including not only content knowledge but also preparation in teaching methods specific to mathematics itself. Another analysis of 60 studies was cited. These studies found that investments in increasing teacher education had five times the effect on student achievement gains as did investments in lowering pupil–teacher ratios.

Teacher Expertise Is an Outgrowth of Professional Development

Professional development was one of the goals identified by the National Education Goals Panel in 1995. Governors and business leaders called for world-class standards in critical content areas in March 1996, and President Clinton's January 1997 State of the Union address focused attention on upgrading teacher expertise for the twenty-first century. Most recently, the report to the nation from the National Commission on Mathematics and Science Teaching for the Twenty-First Century (2000, 7), stated:

> The primary message of this report holds that America's students must improve their performance in mathematics and science if they are to succeed in today's world and if the United States is to stay competitive in an integrated global economy. The Report's second message points in the direction of a solution: the most direct route to improving mathematics and science achievement for all students is better mathematics and science teaching.

The most direct route to improving mathematics and science achievement for all students is better mathematics and science teaching.

—National Commission on Mathematics and Science Teaching for the Twenty-First Century 2000, 7

Several challenges require more extensive and more effective approaches to professional development—diversity, enhanced goals, and new organizations (Hixson and Tinzmann 1990). Citing current inadequacies—"significant number of teachers who have few or no professional development opportunities"; workshops, courses, and institutes not congruent with learning goals and lacking in support mechanisms; a focus on individual development without attention to organization development; and "pockets of innovation" with little potential for wide impact—Loucks-Horsley, Hewson, Love, and Stiles (1998, xi) lament the gap between knowledge and practice. Recognizing the large knowledge base that does exist about professional development and the extensive research findings that are available regarding mathematics learning, they continue:

> There is a large body of literature on adult learning and staff development; it is not, however, connected to the discipline of mathematics . . . and many of the mathematics educators who design and conduct professional development programs for teachers do not know this literature. Second, although there is consensus about the general characteristics of effective professional development, there is much less known about how to put those principles into practice. (xii)

Mathematics educators who design and conduct professional development do not always know the literature on adult learning and staff development.

NCTAF (1996) makes five sets of policy recommendations:

- Link standards for teachers to standards for students.
- "Reinvent" teacher preparation and professional development.
- Overhaul teacher recruitment with the goal of putting a qualified teacher in every classroom.
- Encourage and reward teacher competence.
- Reorganize schools to ensure teacher and student success.

What Should This Professional Development Look Like?

Professional Development Needs to Undergo a Paradigm Shift

If there is indeed a greater need for effective professional development and past efforts have not always been productive, much less judged successful by teachers themselves, then it would appear that a change in approach and thought is necessary. Sparks and Hirsh (1997) have called for a paradigm shift in staff development. Reflecting on "the days when educators sat passively while an 'expert' exposed them to new ideas or 'trained' them in new practices," they recalled that "the success of the effort was judged by a 'happiness quotient' that measured participants' satisfaction with the experience and their recounted off-the-cuff assessment regarding its usefulness" (1). They also call for a move to staff development that "affects the knowledge, attitudes, and practices of individual teachers, administrators, and other school employees [and] alters the cultures and structures of the organizations in which those individuals work" (2). Major components of the paradigm shift were identified (12–16):

Past professional development efforts have not always been productive, much less judged successful by teachers themselves.

- From individual development to individual development AND organization development
- From fragmented, piecemeal improvement efforts to staff development driven by a clear, coherent strategic plan for the school district, each school, and the departments that serve schools

- From district-focused to school-focused approaches to staff development
- From a focus on adult needs and satisfaction to a focus on student needs and learning outcomes and changes in on-the-job behaviors
- From training conducted away from the job as the primary delivery system for staff development to multiple forms of job-embedded learning
- From an orientation toward the transmission of knowledge and skills to teachers by (so-called) experts to the study by teachers of the teaching and learning processes
- From a focus on generic instructional skills to a combination of generic and content-specific skills
- From staff developers who function primarily as trainers to those who provide consultation, planning, and facilitation services as well as training
- From staff development provided by one or two departments to staff development as a critical function and major responsibility performed by all administrators and teacher leaders
- From staff development directed toward teachers as the primary recipients to continuous improvement in performance for everyone who affects student learning
- From staff development as a frill that can be cut during difficult financial times to staff development as an indispensable process without which schools cannot hope to prepare young people for citizenship and productive employment

Staff development will be judged by whether it alters instructional behavior in a way that benefits students.

Noting three "powerful ideas" (Sparks and Hirsh 1997, 4–16) that have fueled the paradigm shift—results-driven education, systems thinking, and constructivism—Sparks and Hirsh suggest that "results-driven education for students requires results-driven staff development for educators" and note that "staff development's success will be judged not by how many teachers and administrators participate in staff development programs or how they perceive its value, but by whether it alters instructional behavior in a way that benefits students. The goal is improved performance—by students, staff, and the organization."

The goal is improved performance—by students, staff, and the organization.

—Sparks and Hirsh 1997, 5

Professional Development Experiences Need to Consider Systems Thinking and Learning Environments

Professional development then needs to look at a broader picture and a wider range of considerations. Systems thinking has been described by Senge (1990, 69) as "a discipline for seeing wholes, for seeing inter-relationships rather than things, for seeing patterns of change rather that static 'snapshots.'" He suggests that systems thinking can help lessen complexity. Applying "leverage," change can be introduced into certain areas—assessment strategies, for example—that can have a positive effect throughout the organization. Unfortunately, selection of those areas in which to leverage is not always abundantly clear and is hampered by damaging "nonsystemic ways of thinking." Professional development then needs to help build sufficient awareness of the organization so that teachers can appreciate the whole and be cognizant of the components to leverage to prompt organizational movement.

"Just as young people create their cognitive structures based on their interactions with the world, so, too, do adults construct reality based on 'schemes'—categories, theories, ways of knowing" (Clinchy 1995, 383). Learning occurs when events require adaptive changes in these schemes, when there is a discrepant event (Romberg 1991). Thus, it is important that teachers provide a learning environment where students search for meaning, appreciate uncertainty, and inquire responsibly (Jackson 1993). The implication for professional development is that transmittal forms of staff development are ineffectual, that staff development must model teaching for understanding for teachers "if those teachers are expected to be convinced on the validity of those practices and to understand them sufficiently well to make them an integrated part of their classroom repertoires" (Sparks and Hirsh 1997, 11).

The Focus of Professional Development Needs to Be "Understanding"

How Is Understanding Built?

Hiebert and Carpenter (1992, 66) have suggested that "to think about and communicate mathematical ideas, we need to represent them in some way." Communication is external and takes the form of written symbols, pictures, spoken symbols, real-world situations, and manipulative aids (Post et al. 1993, 352) (See Figure 4.1). All are dynamic processes and translation between and among them facilitates communication and knowledge construction.

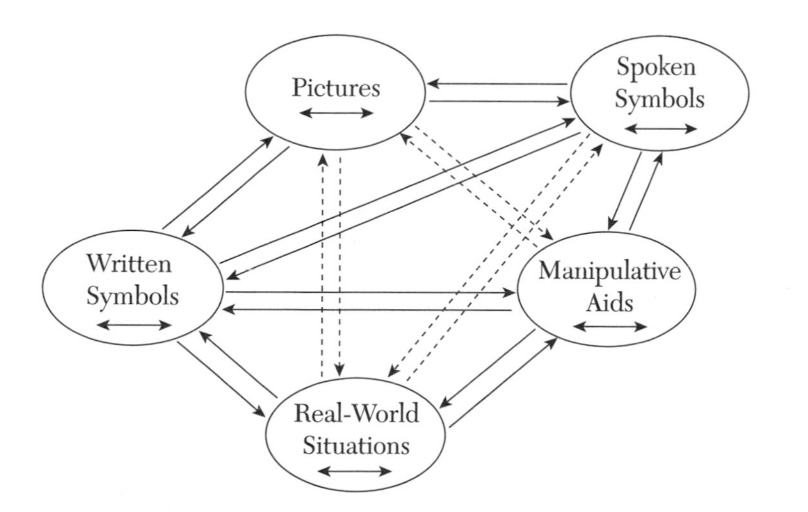

FIGURE 4.1 **Lesh Translation Model (1979)**
From T. R. Post, K. A. Cramer, M. Behr, R. Lesh, and G. Harel, "Curriculum Implications of Research on the Learning, Teaching, and Assessing of Rational Number Concepts," in *Rational Numbers: An Integration of Research*, T. P. Carpenter, E. Fennema, and T. A. Romberg (Eds.), pp. 327–62. Copyright © 1993 by Lawrence Erlbaum Associates, Publishers. Used with permission of the publisher.

It is in the connections of the components [external and internal] that understanding occurs.

Thinking is internal. Internal representations are visualized as components of networks. These networks are believed to be either hierarchically structured or weblike in nature; that is, they are semantic nets (Geeslin and Shavelson 1975; Greeno 1978; Leinhardt and Smith 1985). It is conceived that both types of nets can coexist and interconnect. The idea of internal representations remained a topic of debate for years because mental ideas cannot be seen. Cognitive science has, however, accepted mental representations as legitimate. The connection between external and internal representations also remains a topic of debate, but assumptions that external representations do contribute to internal representational schemes have been made, and "evidence from task situations suggests that this is a reasonable assumption" (Hiebert and Carpenter 1992, 66).

It is in the connections of the components that understanding occurs. The more connections that are made, the richer the understanding. The richer the network, the greater the learning. Complicated networks that develop at fairly high levels of abstraction and that are quite stable are called *schemata.* These networks seem to serve as references against which specific events can be interpreted; they are situational. It is theorized that connections are built when similarities and differences are considered. It would appear that similarities and differences between and among representational forms would be the basis for the development of connections, but similarities and differences can also occur within a representation. Connections can be built on inclusion. The consideration of knowledge building moving from hands-on to formal mathematical manipulation speaks to the connectedness of external and internal representations.

Growth in networks can occur through the addition of a new fact or procedure to already existing networks or the reorganization of networks with new connections being established or old connections being changed or deleted. Reorganization can be slight or massive. Some additions may require restructuring of old networks. Exhibited results can be revelations or confusions. In the last analysis, reorganization should result in more connections and strengthened networks.

What Implications Does the Building of Understanding Have for Classrooms and Professional Development Experiences?

If external representations do influence the development of internal networks, then choice of appropriate representations becomes even more important. (The importance of representation is reflected in the National Council of Teachers of Mathematics *Principles and Standards for School Mathematics* [2000]; representation has been included as a fifth process standard [67–71].) If explorations with concrete materials build to mental explorations and if knowledge is recursive, it would seem especially important to build "from the ground up" so that learners could rebuild if necessary. This is the powerful message conveyed by Kieren's model for the recursive theory of mathematical understanding (1993) (See chapter 3: *Curriculum, Instruction, and Assessment* in this document.). It would also indicate the importance of building understanding of situations before attaching symbols to them. It would call for attention to patterning, seeing similarities and differences in relationships, and for extending patterning from observed situations to mental manipulations through a series of building activities (Osborn 2001).

Because knowledge is built both externally and internally, it would seem that rich tasks, appropriate representations, and discourse are necessary on an external level and that metacognition and reflection are necessary in the move from the external to the internal. Realistic, situated contexts should move to challenging, thought-provoking stimuli. There should always be an extension question. Learners should be led to develop their own "what ifs?" (Grunow 1999).

Hiebert and Carpenter (1992) offer compelling reasons for developing understanding: (1) understanding is generative; (2) understanding promotes remembering; (3) understanding reduces the amount that must be remembered; (4) understanding enhances transfer; and (5) understanding influences beliefs (74–77).

Understanding is generative.

Understanding promotes remembering.

Understanding reduces the amount that must be remembered.

Understanding enhances transfer.

Understanding influences beliefs.

Professional Development Needs to Be Continuous and Ongoing

"What everyone appears to want for students—a wide array of learning opportunities that engage students—is for some reason denied to teachers when they are learners," Lieberman (1995, 591) contends. She suggests that teachers need to "involve themselves as learners—in much the same way as they wish their students would" (592). Staff development needs to move from "direct teaching" to opportunities to conduct research on practice (Richardson 1994) to a culture of inquiry "where professional learning becomes part of the expectations for teachers' roles and is an integral part of the culture of a school" (59). Darling-Hammond and McLaughlin (1995) have sought professional development that "prepares teachers to see complex subject matter from the perspectives of diverse students" and provides occasions for "teachers to reflect critically on their practice and to fashion new knowledge and beliefs about content, pedagogy, and learners" (597).

Teachers need to involve themselves as learners—in much the same way as they wish their students would.

—Lieberman 1995, 591

Knowledge Gain Needs to Be the Primary Focus of Professional Development Experiences

Characterizing "content" as "the missing paradigm" in research on teaching and educational policy decision, Shulman (1986, 7) has called for increased emphasis on subject-matter content knowledge, pedagogical content knowledge, and curricular knowledge. Content knowledge, he contends, needs to include not only knowledge of the concepts of a domain but also an understanding of the structures of the subject matter (Shulman 1987). Pedagogical content knowledge includes "ways of representing and formulating the subject [to] make it comprehensible to others" (9) as well as understanding what makes the learning of specific topics easy or difficult. Comparing the curriculum and its associated materials to a physician's full range of treatments, Shulman maintains that "we are delinquent in building awareness of the variety of materials available" and in knowledge of the characteristics that serve as "both the indications and contraindications for the use of particular curriculum or

Content is the missing paradigm in research on teaching and educational policy decision.

—Shulman 1986, 7

program materials in particular instances" (10). Ball (1991) adds that "teachers' subject-matter knowledge interacts with their assumptions and explicit beliefs about teaching and learning, about students, and about context to shape the ways in which they teach mathematics" (1–2).

Teachers have to "take their complex knowledge and somehow change it so that their students are able to interact with the materials and learn.

—Fennema and Franke 1992, 162

Exploring the many facets of teachers' knowledge, Fennema and Franke (1992) have identified several components of knowledge (see Figure 4.2.): knowledge of mathematics, pedagogical knowledge, knowledge of learners' cognitions in mathematics, and context-specific knowledge, all of which are impacted by beliefs (Peterson et al. 1989). They then situated those components in "in context." "Some components of teacher knowledge evolve through teaching." Teachers have to "take their complex knowledge and somehow change it so that their students are able to interact with the materials and learn" (162)

Professional Development Approached through Content Is Powerful

"Human learning and teaching is highly specific and situated," and although generic staff development has some value, staff development approached through the content is "richer and more textured" (Shulman, in Sparks 1992, 14). Continuing, he notes that "the most powerful form of staff development occurs at the intersection of content and pedagogy" (15).

Loucks-Horsley, Hewson, Love, and Stiles (1998) have stated that "traditional ways in which professional development has occurred are inadequate" (xi) and offer some professional development values:

The most powerful form of staff development occurs at the intersection of content and pedagogy.

—Shulman, in Sparks 1992, 74

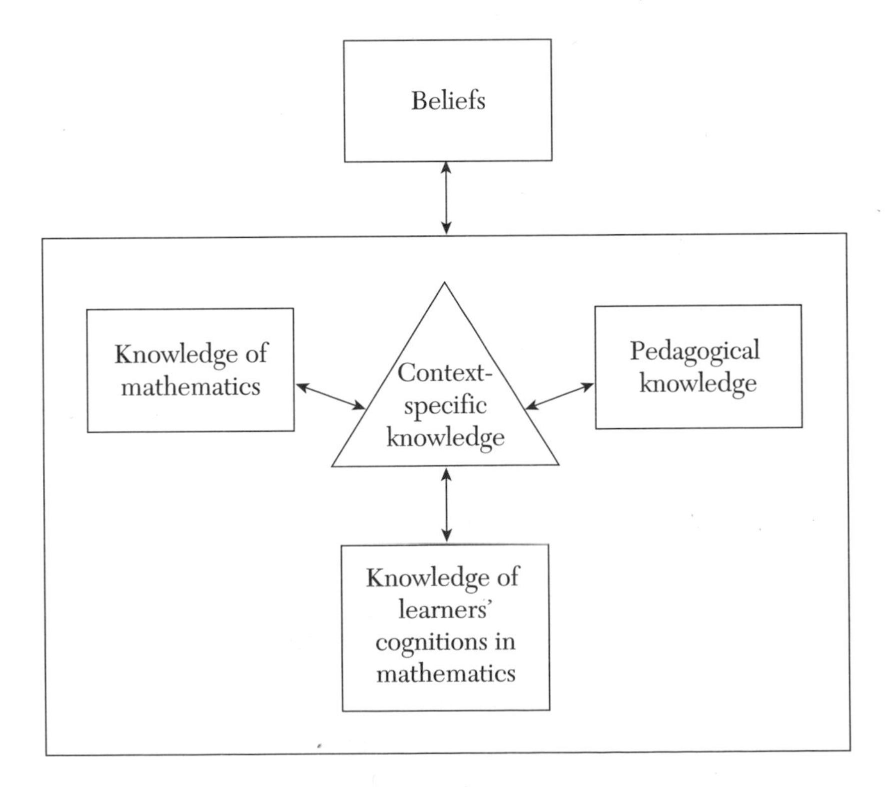

FIGURE 4.2 **Teachers' Knowledge: Developing in Context**
From E. Fennema and M. L. Franke, "Teachers' Knowledge and Its Impact," in D. A. Grouws (Ed.), *Handbook of Research on Mathematics Teaching and Learning*, pp. 147–64. Copyright © 1992 by National Council of Teachers of Mathematics. Used with permission of the author.

1. Professional development experiences must have *all* learners and their learning at their core. This implies a new perspective on content and on teaching and learning strategies.
2. Pedagogical content knowledge (knowing how to teach specific mathematical concepts and principles at different developmental levels) must be the focus of professional development.
3. Principles that guide the reform of student learning should also guide professional learning for educators.
4. The content of professional learning must come from both research and practice, from external research and consultants and from internal wisdom gained by the teacher over time.
5. Professional development must be aware of context. It must align with and bolster system-based changes that support learning such as standards, assessment, curriculum, culture, and capacity for continuous improvement xviii–xix).

Grunow (1999) has designed a professional development experience to enhance knowledge growth, specifically middle school teachers' understanding of rational number. (See Figure 4.3.) Basing the explorations on existing

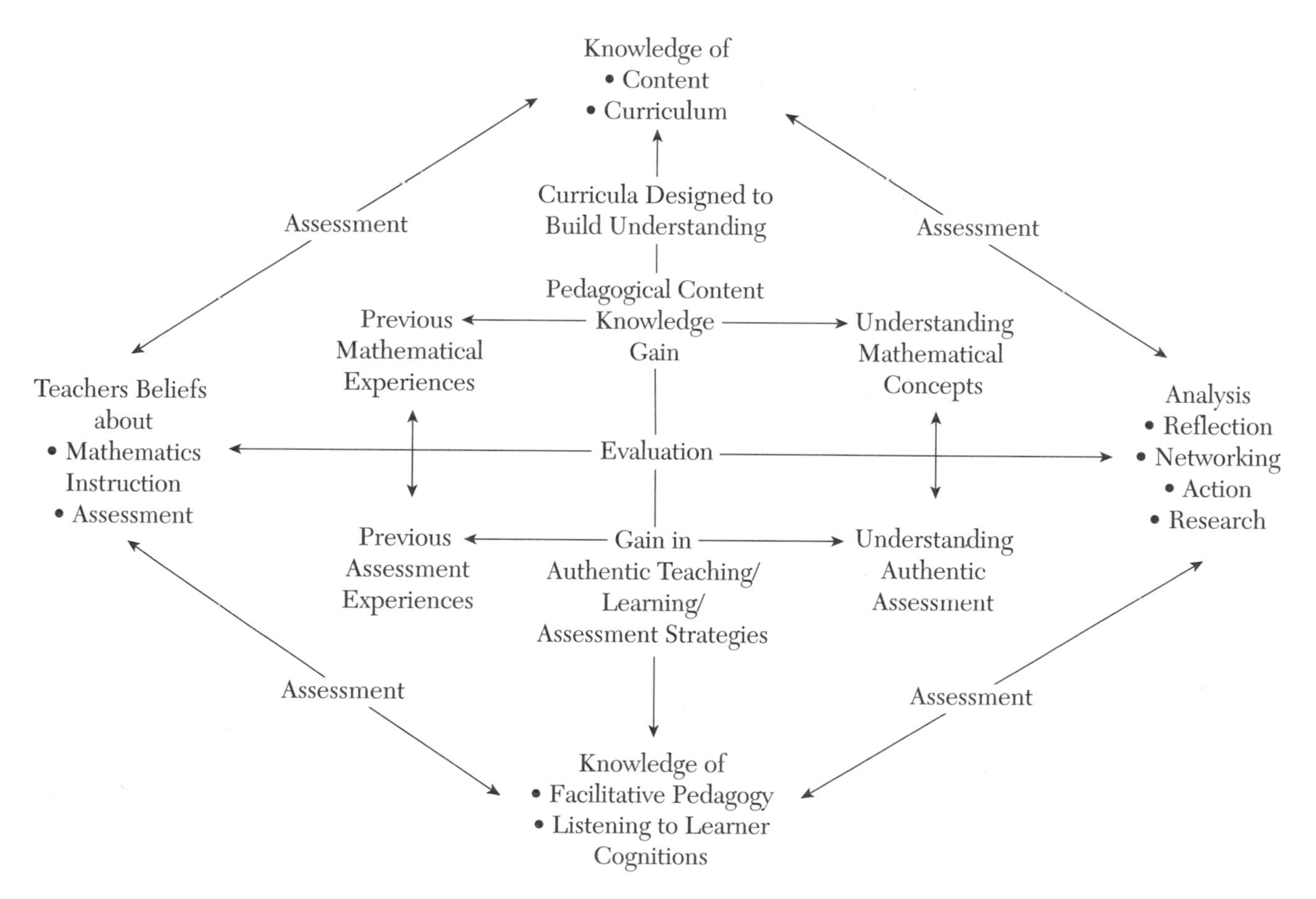

FIGURE 4.3 **Model of Professional Development Designed to Enhance Development of Teacher Knowledge of Mathematics**

From Grunow, J. E., *Using Concept Maps in a Professional Development Program to Assess and Enhance Teachers' Understanding of Rational Number,* p. 141. Copyright © 1999 by UMI Dissertation Information Service. Used with permission of author.

beliefs and moving teachers through experiences that addressed curricula, pedagogical content knowledge gain, facilitative pedagogy attuned to learner cognitions, and authentic teaching/learning/assessment strategies, she hoped for a result of reflective practitioners who would engage in action research, networking, analysis, and reflection. Finding positive results from the experience, she believes the model of development of teacher knowledge to be replicable and has effectively used it since in situations focused specifically on building teacher knowledge.

Teachers need to learn to model mathematical inquiry so that students can observe and participate in formulating questions, pursuing solutions, analyzing conclusions, and evaluating results.

Mathematics is changing. This places everyone in a learner's position—students, teachers, administrators, and parents. Teachers need to construct their own realizations of the newly available mathematics and technology so that they can be instrumental in helping their students acquire the similar new understandings. By the same token, teachers need to learn to model mathematical inquiry so that students can observe and participate in formulating questions, pursuing solutions, analyzing conclusions, and evaluating results.

Addressing knowledge acquisition with these parameters, teachers can become aware of the methodologies necessary to facilitate the development of understanding in their students. They can focus on offering environments that stimulate knowledge development, and they can give attention to becoming aware of how to facilitate discourse. To be able to listen to students to identify the understandings that exist and to make appropriate decisions for additional instruction based on those identified ideas can become a goal. Teachers need to work with representing problems and selecting contexts to help students experience meaningful associations to facilitate knowledge recall. They need to be able to select rich mathematical tasks that will help build conceptual understanding. They can give attention to modeling and to stimulating mathematical reasoning processes. Professional development experiences, to be instrumental in effective, real change, should address each of these components. (See Figure 4.4.)

What Assessment Techniques Can Be Used to Measure the Effects of Professional Development Experiences?

Assessment of Professional Development Experiences Has Been Inadequate

Alternative forms of assessment of professional development are needed.

One of the greatest deterrents to recognition of the importance of professional development and to subsequent funding is the lack of appropriate methods of determining the success of such efforts and the worth of investment in those efforts. It is clear that alternative forms of assessment of professional development are needed. Branham (1992) has noted that evaluation of staff development programs is often inadequate—there has been little evaluation, evaluation has not measured long-term changes, or evaluation occurs immediately following the program when "enthusiasm for use of the ideas is high" (24). Yet there is an increasingly growing interest in evaluation

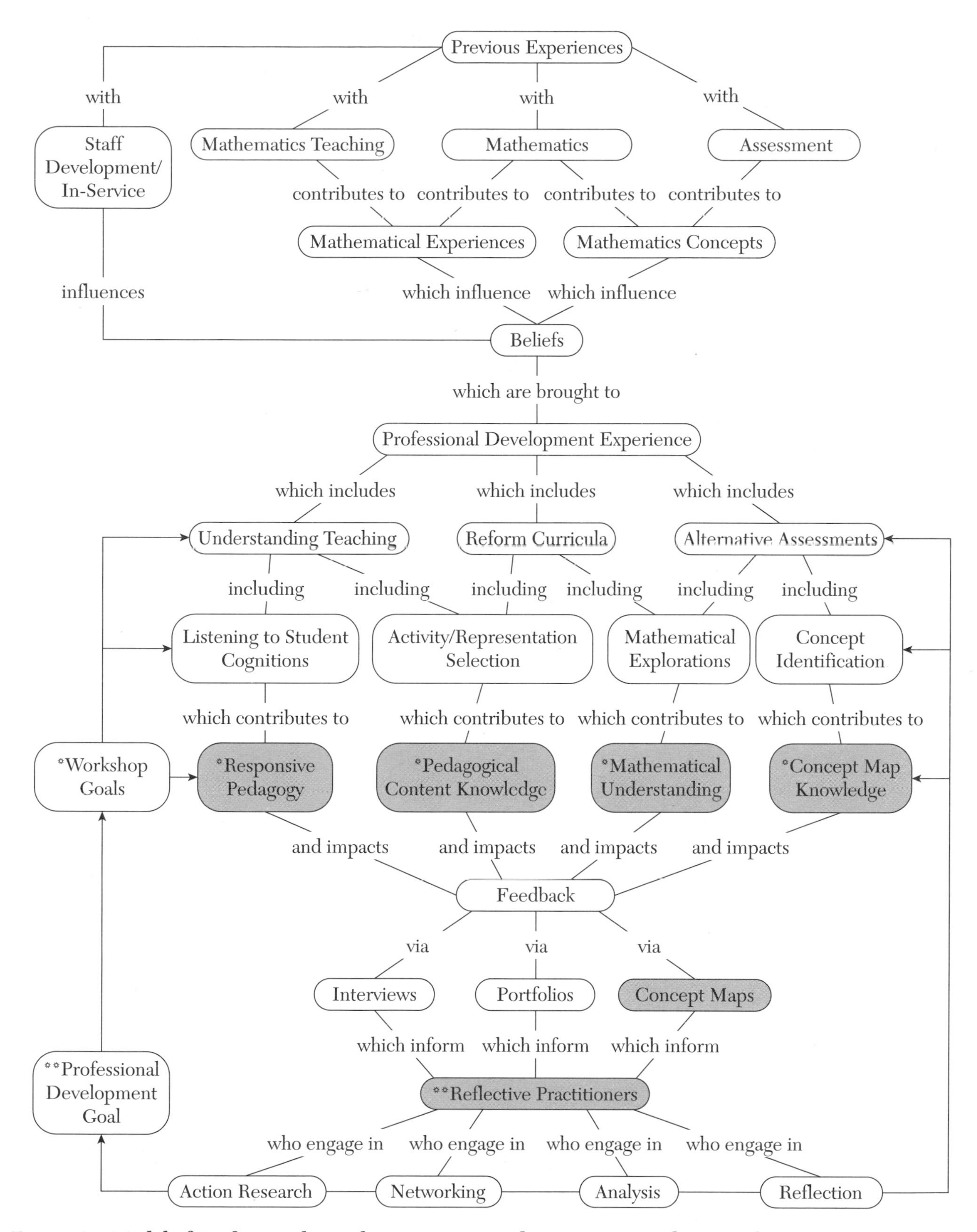

FIGURE 4.4 **Model of Professional Development Designed to Promote Mathematical Understanding.**
From Grunow, J. E., *Using Concept Maps in a Professional Development Program to Assess and Enhance Teachers' Understanding of Rational Number,"* p. 159. Copyright © 1999 by UMI Dissertation Information Service. Used with permission of author.

of professional development. Guskey (2000, 8) has found those reasons to be calls for a better understanding of the "dynamic nature" of professional development, recognition of professional development as "an intentional process," the need for better information to guide reform efforts, and increased pressure for "accountability."

There Are Assessment Measures Congruent With Teaching for Understanding

The *Assessment Standards for School Mathematics* (Assessment Standards Working Groups 1995) has called for the measurement of significant and correct mathematics, mathematical reasoning, connections among mathematical ideas, and communication about mathematics. To fully assess creativity and deep conceptual understanding, the process should allow for demonstration of understanding in new situations, situations that do more than call for recall. Grunow (1999) contends that "mathematics professional development should be assessed in the manner in which it is delivered." If the goal of professional development is "domain understanding as situated in a rich developmental context for building knowledge of teaching for understanding" (98), then the assessment needs to be congruent with the instruction. She suggests using concept mapping to achieve this end because the maps reflect participant understanding of the central concepts of the topic at hand, the relationships between and among those topics, and as well, the inconsistencies in understandings.

If the goal of professional development is domain understanding as situated in a rich developmental context for building knowledge of teaching for understanding, then the assessment needs to be congruent with the instruction.
—Grunow 1999, 98

Branham (1992) suggests assessment methods that measure the wider domain of systemic change. Alternative assessment methods suggested were (1) the case study model (Shulman 1992), an approach characterized by a thick description of ways the program is implemented and received (see *One District's Ongoing Story,* Appendix E, in this document); (2) the systems model (Marshall 1988), a method designed to determine if the manner in which staff development is structured and implemented brings about the greatest results; and (3) the goal-based model (Marshall 1988), a narrowly focused evaluation that aims to determine if the stated objectives for a specific staff development program were met, the criteria for evaluation having been built into the program objectives (25).

Focusing on evaluating professional development programs and activities, Guskey (2000, 272) offers several steps:

1. Clarify the intended goals.
2. Assess the value of the goals.
3. Analyze the context.
4. Estimate the program's potential to meet the goals.
5. Determine how the goals can be assessed.
6. Outline strategies for gathering evidence.
7. Gather and analyze evidence on participants' reactions.
8. Gather and analyze evidence on participants' learning.
9. Gather and analyze evidence on organization support and change.
10. Gather and analyze evidence on participants' use of new knowledge and skills.

Branham (1992) suggests assessment methods that measure the wider domain of systemic change.

11. Gather and analyze evidence on student learning outcomes.
12. Prepare and present evaluation reports.

Is Student Learning a Good Measure of the Effectiveness of Professional Development?

There is often an attempt to immediately correlate teacher learning and student learning. From a practical standpoint, there are far too many interceding variables for such conclusions to be valid. It is, therefore, important to measure results of the professional development on an individual basis and then to possibly draw connections, as self-described by the involved teachers, to classroom changes that occur. The results of those efforts can then be evaluated independently. Assessment efforts and reward systems that tie student results to teacher expertise must then take into account school support provided to the teacher, time allocations, teacher and student performances over extended periods of time, community/school/classroom environments, and so on. Such measures, to be reliable and valid, need to be carefully devised and all inclusive.

From a practical standpoint, there are far too many interceding variables for [immediate correlation of teacher learning and student learning] to be valid.

What Is the Ultimate Goal of Professional Development?

Reflective Practitioners Will Design and Contribute to Their Own Professional Development

A relatively consistent view of adult learning has emerged. Dewey (1938) has characterized it as "experiential learning." As dilemmas, problems, or difficulties arise, the individual locates and defines the dilemma and analyzes alternative solutions, choosing to act or not to act, as the case may be. Likewise, Argyris and Schon (1974, 1978) contend that learning in organizations takes place as nonroutine circumstances require heightened attention, experimentation, and problem analysis and solution formulation. Adults learn throughout their lives (Moll 1990). Adult learning occurs across settings and circumstances (Marsick and Watkins 1990), from formal, collective learning activities planned by others to informal, self-initiated, and self-directed activities. They learn from incidental and unintentional everyday experiences and from planned, well-devised activities.

Adult learning is problem oriented and occurs when problems relate in meaningful ways to adults' life situations.

Adults enter learning situations with accumulated knowledge, skills, and beliefs from past experiences (Knowles 1984). Prior knowledge and beliefs often affect current learning by serving as cognitive and normative schemata through which individuals perceive and interpret their situations, new information, and themselves as learners. These schemata may enhance or hinder learning (Moll 1990). Adult learning is problem oriented and occurs when problems relate in meaningful ways to adults' life situations. The problems may emanate from individuals' experiences, the social or organizational environments in which they work, or personal curiosity and self-initiated inquiry (Guskey and Huberman 1995). Adults can play an active role in their own learning and can be proactive and self-directed in searching for new learning opportunities and resources to apply to those opportunities (Knowles 1984).

Adults can play an active role in their own learning and can be proactive and self-directed in searching for new learning opportunities and resources to apply to those opportunities.

Action research as a major form of professional development is now seen as central to the restructuring of schools.

—Holly 1991, 133

Effective professional development capitalizes on facilitating active, self-directed learning and on helping adults realize their own orientations and beliefs and how they affect instructional approaches. Noting that research on the practice of teaching has shifted to understanding how teachers make sense of teaching and learning, Richardson (1994) has campaigned for teacher research that "gives voice to practitioners" (5). Teachers need to be aware of the formal research on teaching and learning so they can be aware of the big picture. Then they need opportunities to apply that research in their own classrooms. Teachers need to broaden their horizons by becoming familiar with what research is indicating, by trying out those ideas, by contemplating the effects of such trials, by adjusting and readjusting in accordance with results, and then by discussing results with colleagues as they, too, work through the same processes.

Holly (1991) characterizes such an approach as "action research" (95) and notes that "action research as a major form of professional development is now seen as central to the restructuring of schools" (133). The strength of the approach is that teachers either define the research questions or contribute to their definition in a meaningful way; "they have ownership over the process and are committed to promoting changes in practice that are indicated by their findings" (95). The research takes various forms: Teachers work together in collaborative teams of inquiry or with other researchers, often from universities or research centers; individual teachers pursue their own research studies and then discuss their findings with fellow teachers or researchers; or teachers examine relevant research and use it as a basis for collecting and analyzing data from their own classrooms.

Professional development experiences that allow teachers opportunities to continue to interact have yielded significant, positive results.

Key elements of action research are use of an action research cycle of planning, acting, observing, and reflecting; teachers linked with sources of knowledge and stimulation from outside their schools (Holly 1991); teachers working collaboratively (Oja and Smulyan 1989, 1–25); and documenting and sharing the learning gleaned from the action research (Loucks-Horsley, Harding, Arbuckle, Murray, Dubea, and Williams 1987). The value of collegiality is undisputed; teachers can learn together and from one another. Professional development experiences that allow teachers opportunities to continue to interact have yielded significant, positive results. (See exemplary professional development experiences in Appendix F in this document.)

Is There Anything Regarding Professional Development That Can Be Learned from the Experiences of Other Countries?

Teachers Are a Dedicated Lot the World Over

Care needs to observed when looking at mathematics experiences in other countries and then comparing them with those of teachers and students in the United States, but the global nature of our economy and the increasing interconnectedness facilitated by instantaneous communication make us aware of similarities and differences in cultures. Carefully constructed research efforts such as the *Third International Mathematics and Science Study* (TIMSS) (U.S. Department of Education National Center for Education Statistics 1996) can provide valuable information for comparison purposes. Some of the findings from the study that may have implications for professional development are:

Carefully constructed research efforts such as TIMSS can provide valuable information for comparison purposes.

1. Teachers in all countries routinely spend time outside the formal school day preparing and grading tests, reading and grading student work, planning lessons, meeting with students and parents, engaging in professional development or reading, keeping records, and completing administrative tasks.
2. The organization of the school day differs. Japanese teachers have the longest official workday (eight to nine hours; in the United States, seven to eight hours, and in Germany, five to five and a half hours) and the longest work year (240 days; in the United States and Germany, 180 to 184 days). Japanese teachers tend to have a broader range of responsibilities including supervising the playground and the lunch room and cleaning of a portion of the school each day (Stevenson and Nerison-Low 1997, 127–33).
3. U.S. teachers spend more time in direct instruction and less in settings that allow for professional development, planning, and collaboration (Continuing to Learn from TIMSS Committee, National Research Council 1999). Japanese teachers have more time to collaborate in the school day, more opportunity to observe each other's classes throughout the day (Kinney 1998, 223–27), and they have desks in a shared teachers' room with others who have common teaching assignments. Japanese teachers are involved in informal study groups and networks to strengthen professional ties. Teaching, study, research, travel, and hobbies are blended to augment professional growth for Japanese teachers. Germany requires teachers to spend only the morning at school; all additional activities are done at home.
4. Teacher preparation in the United States is relatively extended compared with international teacher education, though Germany puts teachers through an even longer period of formal preparation. Japan requires highly competitive examinations to qualify for a teaching position.
5. "The U.S. appears to offer a less formal or coherent system of professional development than do Germany and Japan" (Continuing to Learn from TIMSS Committee, National Research Council, 1999, 72). Germany and Japan have more formal systems of induction to teaching, including field-based student teaching experience and mentoring.
6. Japan has a formal, systematic professional development system over the career of a teacher, including rotation among schools every six years as well as changes in grade assignments. Mandatory professional development away from their schools is required in some areas in the sixth, tenth, and twentieth years (Kinney 1998, 207). Japan offers formal training at

Teaching, study, research, travel, and hobbies are blended to augment professional growth.

There are striking international similarities in how teachers view their work.

local resource centers, voluntary mentoring and peer observation, teacher-organized informal study and action-research groups—"a mode of professional development that teachers value highly" (73). National, regional, and local study grants send teachers traveling domestically and abroad. German teachers indicated that "an openness and willingness to learn" (Milotich 1996, 310) was important. They cited reading and journals as important professional development tools. There are state-sponsored institutions and academies offering continuing education courses for experienced teachers, and schools bring in experts to address issues identified as problems in schools.

7. Japanese teachers, compensated favorably compared with other professionals of the same educational level, worry about the erosion of respect for the profession. German teachers complained of being recognized as part-time workers, though their salaries were commensurate with professionals of the same educational level. Job security is a benefit; German teachers cannot be laid off. U.S. teachers earn less, on average, than professionals in other fields with comparable years of university and graduate education. Many teachers viewed the timeframe that coincided with that of their children to be a real benefit.
8. In European and Asian countries, teachers are the central investment; classroom teachers comprise 60 to 80 percent of education employees. In the United States, teachers comprise 43 percent of education employees.

There are also striking international similarities in how teachers view their work. Teachers are a dedicated lot. They are generally interested in improving their own personal knowledge and their collective knowledge of teaching. The uncertainties of an information society and the increased moves toward accountability tend to be "among the most troublesome demands currently made of teachers in all three countries" (Stevenson and Nerison-Low 1997, 140).

What Does the Research Show about Professional Development?

Education reform that makes a difference for students requires teachers and principals to respond in new ways to the need for change and to rebuild the very foundation of their thinking about teaching and learning.

—Hawley and Valli 1999, 127–50

In April 1999, the National Partnership for Excellence and Accountability in Teaching and the Learning First Alliance brought together representatives of schools and school districts where outstanding professional development was underway. The thrust-was analysis of the programs to determine how to best facilitate teacher learning that results in students learning and how to implement the necessary policies and practices. "Education reform that makes a difference for students requires teachers and principals to respond in new ways to the need for change and to rebuild the very foundation of their thinking about teaching and learning" (Hawley and Valli, 1999, 127–50). Expert teachers continually evaluate student learning and in so doing become aware of what they themselves need to learn. The new view of professional development, then, has teachers engaged in professional learning everyday, all day long. Additionally, teachers learn together and solve problems as teams; every teacher feels responsible for the success of every student. Professional development is learner centered.

There is impetus for change in professional development. The conference recognized professional development programs that are using learner-centered professional development. Professional and policy organizations are increasingly acknowledging the importance of professional development to school improvement. Research identifying the characteristics of effective professional development is growing. The National Partnership for Excellence and Accountability in Teaching has offered research-based principles for improving professional development. In-depth discussion of each principle can be found on the National Partnership for Excellence and Accountability in Teaching Web site: http://www.npeat.org/public-html/prefdev/pdprin.htm.

Every teacher feels responsible for the success of every student.

The content of professional development focuses on what students are to learn and how to address the different problems students may have in learning the material.

Professional development should be based on analyses of the differences between (1) actual student performance and (2) goals and standards for student learning.

Professional development should involve teachers in identifying what they need to learn and in developing the learning experiences in which they will be involved.

Professional development should be primarily school based and built into the day-to-day work of teaching.

Most professional development should be organized around collaborative problem solving.

Professional development should be continuous and ongoing, involving follow-up and support for further learning—including support from sources external to the school that can provide necessary resources and new perspectives.

Professional development should incorporate evaluation of multiple sources of information on (1) outcomes for students and (2) the instruction and other processes involved in implementing lessons learned through professional development.

Professional development should provide opportunities to understand the theory underlying the knowledge and skills being learned.

Professional development should be connected to a comprehensive change process focused on improving student learning.

There is nothing as exciting as learning.

We are all learners of mathematics. There is nothing as exciting as learning. Teachers chose their profession not only because they wanted to help students but also because they loved to learn and wished to continue learning for the rest of their lives. That love of learning is conveyed to their students. We owe teachers professional development experiences that will enhance understanding and that allow them to continue to learn.

We owe teachers professional development experiences that will enhance understanding and that allow them to continue to learn.

TO YOU

To sit and dream, to sit and read,
To sit and learn about the world
Outside our world of here and now—
Our problem world—
To dream of vast horizons of the soul
Through dreams made whole,
Unfettered free—help me!
All you who are dreamers, too,
Help me make our world anew.
I reach out my hands to you.

—*Langston Hughes (in Darling-Hammond 1997, 337)*

References

Argyris, C., and D. A. Schon. 1974. *Theory in Practice: Increasing Professional Effectiveness.* San Francisco: Jossey-Bass.

———. 1978. *Organizational Learning: A Theory of Action Perspective.* San Francisco: Jossey-Bass.

Assessment Standards Working Groups. 1995. *Assessment Standards for School Mathematics.* Reston, VA: National Council of Teachers of Mathematics.

Ball, D. L. 1991. Research on Teaching Mathematics: Making Subject Matter Knowledge Part of the Equation. In *Advances in Research on Teaching: Teachers' Subject Matter Knowledge and Classroom Instruction, 2,* edited by J. E. Brophy. Greenwich, CT: JAI Press.

Branham, L. A. 1992. An Update of Staff Development Evaluation. *Journal of Staff Development 13*(4): 24–28.

Clinchy, B. M. 1995. Goals 2000: The Student as Object. *Phi Delta Kappan 76*(5): 383, 389–92.

Cohen, D. K., and H. C. Hill. 1998. State Policy and Classroom Performance: Mathematics Reform in California. *Consortium for Policy Research in Education Policy Briefs, RB 23*(January), 1–16. Philadelphia, PA: Consortium for Policy Research in Education.

Continuing to Learn from TIMSS Committee, National Research Council. 1999. *Global Perspectives for Local Action: Using TIMSS to Improve U.S. Mathematics and Science Education.* Washington, DC: National Academy Press.

Darling-Hammond, L. 1997. *The Right to Learn: A Blueprint for Creating Schools That Work.* San Francisco: Jossey-Bass.

Darling-Hammond, L., and D. L. Ball. 1998. *Teaching for High Standards: What Policy-Makers Need to Know and Be Able to Do.* Philadelphia: CPRE Publications.

Darling-Hammond, L., and M. McLaughlin. 1995. Policies That Support Professional Development in an Era of Reform. *Phi Delta Kappan 76*(8): 597–604.

Dewey, J. 1938. *Experience and Education.* New York: Collier.

Fennema, E., and M. L. Franke. 1992. Teachers' Knowledge and Its Impact. In *Handbook of Research on Mathematics Teaching and Learning,* edited by D. A. Grouws. New York: Macmillan.

Ferguson, R. 1991. Paying for Public Education: New Evidence on How and Why Money Matters. *Harvard Journal of Legislation* 28: Summer.

Geeslin, W. E., and R. Shavelson. 1975. Comparison of Content Structure and Cognitive Structure in High School Students' Learning of Probability. *Journal for Research in Mathematics Education* 6:2 109–20.

Greeno, J. G. 1978. A Study of Problem Solving. In *Advances in Instructional Psychology,* edited by R. Glaser. Vol. 1. Hillsdale, NJ: Lawrence Erlbaum and Associates.

Grunow, J. E. 1999. *Using Concept Maps in a Professional Development Program to Assess and Enhance Teachers' Understanding of Rational Number.* UMI Microform 9910421. Ann Arbor, MI: UMI Dissertation Services.

Guskey, T. R. 2000. *Evaluating Professional Development.* Thousand Oaks, CA: Corwin Press.

Guskey, T. R., and M. Huberman, eds. 1995. *Professional Development in Education: New Paradigms and Practices.* New York: Teachers College Press.

Hawley, W. D., and L. Valli. 1999. The Essentials of Effective Professional Development: A New Consensus. In *The Heart of the Matter: Teaching in the Learning Profession,* edited by L. Darling-Hammond & G. Sykes. San Francisco: Jossey-Bass.

Hiebert, J., and T. P. Carpenter. 1992. Learning and Teaching with Understanding. In *Handbook of Research on Mathematics Teaching and Learning,* edited by D. A. Grouws. New York: Macmillan.

Hixson, J., and M. B. Tinzmann. 1990. *What Changes Are Generating New Needs for Professional Development?* Oak Brook, IL: North Central Regional Educational Laboratory.

Holly, P. 1991. Action Research: The Missing Link in the Creation of Schools as Center of Inquiry. In *Staff Development for Education in the '90s: New Demands, New Realities, New Perspectives,* edited by A. Lieberman and L. Miller. New York: Teachers College Press.

Jackson, B. T. 1993. Foreword to *In Search of Understanding: The Cases for Constructivist Classrooms*, edited by J. G. Brooks and M. G. Brooks. 5th ed. Alexandria, VA: Association for Supervision and Curriculum Development.

Kieren, T. E. 1993. Rational and Fractional Numbers: From Quotient Field to Recursive Understanding. In *Rational Numbers: An Integration of Research,* edited by T.P. Carpenter, E. Fennema, and T. A. Romberg. Mahwah, NJ: Lawrence Erlbaum.

Kinney, C. 1998. Teachers and the Teaching Profession in Japan. In *The Educational System in Japan: Case Study Findings*, edited by C. Kinney. Washington, DC: U.S. Department of Education.

Knowles, M. S. 1984. *The Adult Learner: A Neglected Species.* 3d ed. Houston: Gulf.

Leinhardt, G., and D. A. Smith. 1985. Expertise in Mathematics Instruction: Subject Matter Knowledge. *Journal of Educational Psychology* 77(3): 247–71.

Lieberman, A. 1995. Practices That Support Teacher Development. *Phi Delta Kappan* 76(8): 591–96.

Loucks-Horsley, S., C. K. Harding, M. A. Arbuckle, L. B. Murray, C. Dubea and M. K. Williams. 1987. *Continuing to Learn: A Guidebook for Teacher Development.* Andover, MA/Oxford, OH: The Regional Laboratory for Educational Improvement of the Northeast and Islands/National Staff Development Council.

Loucks-Horsley, S., P. W. Hewson, N. Love, and K. E. Stiles. 1998. *Designing Professional Development for Teachers of Science and Mathematics.* Thousand Oaks, CA: Corwin Press.

Marshall, J. 1988. A General Statement on Staff Development Evaluation. *Journal of Staff Development* 9(1): 2–8.

Marsick, V.J., and K. Watkins. 1990. *Informal and Incidental Learning in the Workplace.* New York: Routledge.

Milotich, U. 1999. Teachers and the Teaching Profession in Germany. In Draft Volume of *The Education System in Germany: Case Study Findings.* Ann Arbor: University of Michigan Center for Human Growth and Development.

Moll, L. C., ed. (1990). *Vygotsky and Education: Instructional Implications of Sociohistorical Psychology.* New York: Cambridge University Press.

National Commission on Mathematics and Science Teaching for the Twenty-First Century. 2000. *Before It's too Late: A Report to the Nation from the National Commission on Mathematics and Science Teaching for the* Twenty-First *Century.* Washington, DC: National Commission on Mathematics and Science Teaching for the Twenty-First Century.

National Commission on Teaching and America's Future. 1996. *What Matters Most: Teaching for America's Future.* New York: National Commission on Teaching and America's Future.

National Council of Teachers of Mathematics. 2000. *Principles and Standards for School Mathematics.* Reston, VA: National Council of Teachers of Mathematics.

Oja, S. N., and L. Smulyan. 1989. *Collaborative Action Research: A Developmental Approach.* Philadelphia: Falmer.

Osborn, J. M. 2001. Mathematics Knowledge Construction. In *Forward: A Guide for Program Development in Mathematics.* Madison: Wisconsin Department of Public Instruction.

Peterson, P. L., E. Fennema, T. P. Carpenter, and M. Loef. 1989. Teachers' Pedagogical Content Beliefs in Mathematics. *Cognition and Instruction* 6(1): 1–40.

Post, T. R., K. A. Cramer, M. Behr, R. Lesh, and G. Harel 1993. Curriculum Implications of Research on the Learning, Teaching, and Assessing of Rational Number Concepts. In *Rational Numbers: An Integration of Research,* edited by T. P. Carpenter, E. Fennema, and T. A. Romberg. Mahwah, NJ: Lawrence Erlbaum.

Richardson, V. 1994. Conducting Research on Practice. *Educational Researcher* 23(5): 5–10.

Romberg, T. A. 1991. *A Blueprint for Maths in Context: A Connected Curriculum for Grades 5–8.* Madison, WI: National Center for Research in Mathematics and Science Education.

Senge, P. 1990. *The Fifth Discipline: The Art and Discipline of the Learning Organization.* New York: Doubleday.

Shulman, J. H. 1992. *Case Methods in Teacher Education.* New York: Teachers College, Columbia University.

Shulman, L. S. 1986. Those Who Understand: Knowledge Growth in Teaching. *Educational Leadership* 15(2): 4–14.

———. 1987. Knowledge and Teaching: Foundations of the New Reform. *Harvard Educational Review* 57(1): 1–22.

Sparks, D. 1992. Merging Content Knowledge and Pedagogy: An Interview with Lee Shulman. *Journal for Staff Development* 13(1): 14–15.

———. 1998. Teacher Expertise Linked to Student Learning. In *Results*, edited by Joan Richardson. Oxford, OH: National Staff Development Council.

Sparks, D., and S. Hirsh. 1997. *A New Vision for Staff Development.* Oxford, OH: National Staff Development Council.

Sparks, D., and S. Hirsh. 2000. *A National Plan for Improving Professional Development.* http://www.nsdc.org/library/NSDCPlan.html (July 17).

Stevenson, H. W., and R. Nerison-Low. 1997. *To Sum It Up: Case Studies of Education in Germany, Japan, and the United States.* Washington, DC: U.S. Government Printing Office.

U.S. Department of Education, National Center for Education Statistics. 1996. *Pursuing Excellence*, NCES 97–198, by Lois Peak. Washington, DC: U.S. Government Printing Office.

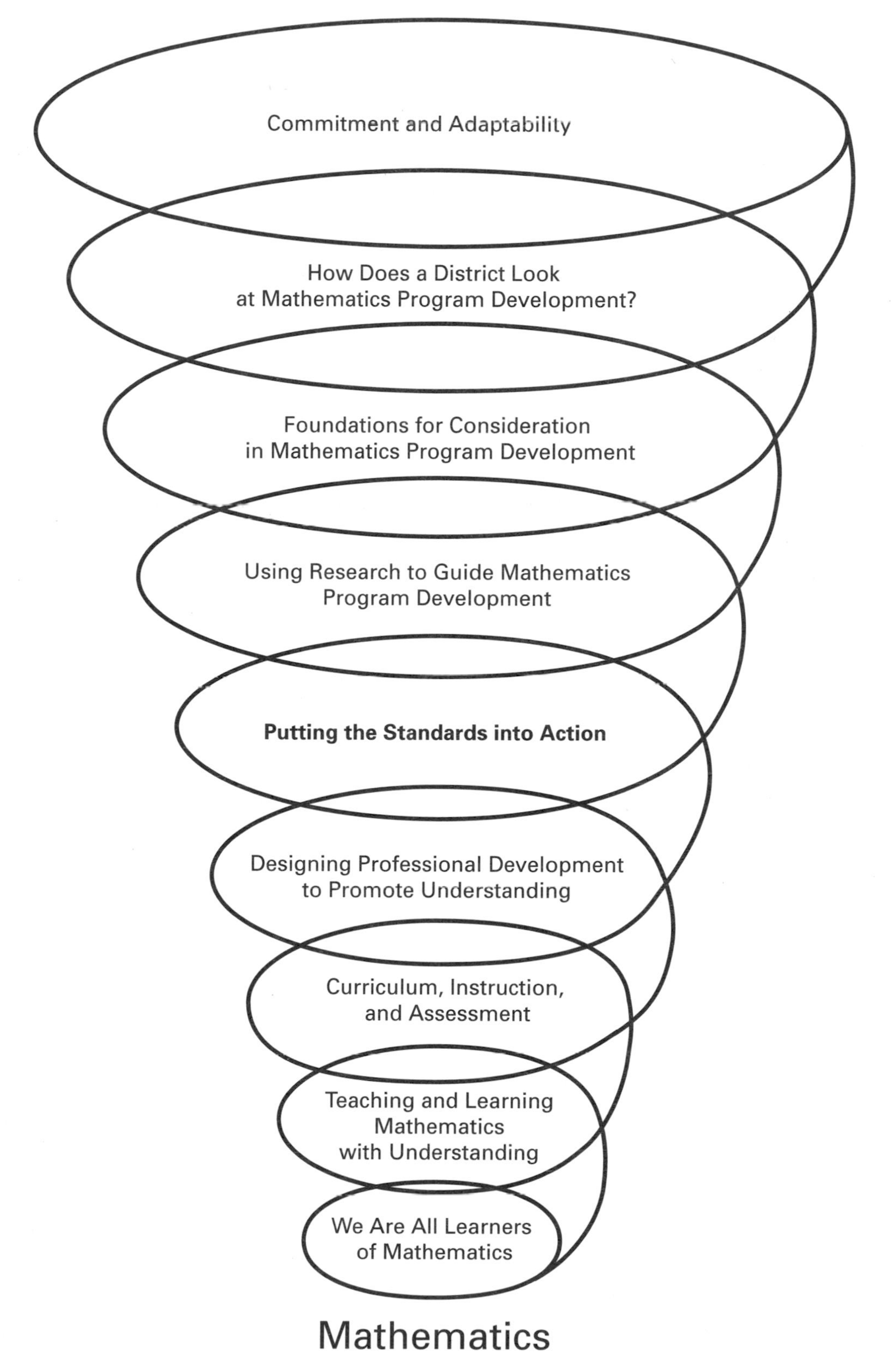

Spring (spring) *n.* 1. a source; origin; or beginning: *a spring into action.* 2. elasticity; resilience: *a spring forward with commitment.* (Morris, W., ed., 1971. *The American Heritage Dictionary of the English Language.* New York: American Heritage Publishing Company, Inc. and Houghton Mifflin Company, 1250, 2nd definition.)

Putting the Standards into Action

5

Putting the standards into action is the heart and soul of this document. In discussing district requests for help in implementing standards-led education, the task force felt that several components were essential: rationale for process and procedures, discussion of sequencing and timing of mathematics concepts reflective of *Wisconsin's Model Academic Standards for Mathematics* (WMAS) (Wisconsin Department of Public Instruction 1998) and *Principles and Standards for School Mathematics* (PSSM) (Standards 2000 Project 2000), and presentation of challenging mathematics investigations complete with hoped-for understandings, appropriate assessment suggestions, and extension questions. The introduction to the section gives an umbrella view of essentials of rich school mathematics programs as explicated through each of the WMAS strands. Grade-level sections follow and give overviews of ideas important to the levels; illustrative tasks for each strand; discussions of the tasks, student work, and standards addressed; connections to other standards and disciplines; extension questions; and other sample tasks. Believing this approach to be replicable in classrooms, it is carefully played out in each of the content strands at the various grade levels.

Mathematics today is seen as far more than a collection of skills and concepts; it is seen as a tool to solve problems, a science of pattern, an art, and a language.

Mathematics Content PK–12

What Is Mathematics?

The NCTM *Curriculum and Evaluation Standards for School Mathematics* (Commission on Standards for School Mathematics 1989) asked the profession to look long and hard at two fundamental questions: "What is mathematics?" and "What does it mean to know mathematics?" The continual reassessment of the answers to these questions drives the changes in curriculum, instruction, and assessment that are represented in WMAS (Wisconsin Department of Public Instruction 1998) and PSSM (National Council of Teachers of Mathematics 2000). Mathematics today is seen as far more than a collection of skills and concepts; it is seen as a tool to solve problems, a science of pattern, an art, and a language. With changes in perceptions and conceptions of mathematics, definition of mathematical concepts must be more clearly delineated. This document attempts to help in this effort by discussing each of the strands represented in WMAS (mathematical processes, number operations and relationships, geometry, measurement, statistics and probability, and algebraic relationships), by giving examples of those strands as exemplified in mathematical investigations, by considering strand distribution, and by suggesting possible grade-level benchmarks. As each of these pieces is discussed, it should be remembered that these subdivisions are done for clarification. They do not exist in isolation. The whole is greater than the sum of its parts. The strands are interconnected within mathematics and with other disciplines. A powerful way to build mathematics understanding is through connection of topics. Integration frequently creates new mathematical thrusts.

A powerful way to build mathematics understanding is through connection of topics.

How Is Mathematics Learned?

As the various content and process standards of mathematics are delineated in the following sections, it is important to realize that how mathematics is learned is inherently intertwined with what mathematics is learned. Each of the program objectives should be pursued in view of the following assumptions:

1. Expectations often dictate results. The curricula envisioned in the various standards offer rigorous goals for *all* students.
2. Instructional activities grow out of worthy problem situations. The contexts in which mathematical pursuits are situated often aid in the construction of knowledge and facilitate recall and transfer.
3. Technologies such as concrete materials, calculators, and computers help students build understanding and should be used as tools in appropriate situations, as means to an end and not ends in themselves. Students need to learn when and where to use which technological device.
4. Mathematics is a language; students need to speak the language to build understanding. Students need to communicate their understandings of mathematics in many forms—verbally, in written form, through pictures, charts, graphs, and so on.
5. Classroom tasks need to include open-ended investigations that allow students to build individual understandings, without intellectual constraints. Students need to be encouraged to use multiple strategies to solve the problems and need to be led to find multiple solutions.
6. Learning occurs in an integrated fashion. Connections among the strands of mathematics and applications of mathematics to other fields of study and real-world settings need to be emphasized to facilitate knowledge growth.
7. Mathematical tasks need to move the learner from the concrete to the abstract. Multiple representations facilitate visualization of mathematical ideas.
8. Instructional activities need to emphasize the development of both skills and understanding. Focus on isolated skills and procedures leads to fragmentation of knowledge and impedes transfer. Conceptual learning builds understanding and facilitates making ideas memorable.

How mathematics is learned is inherently intertwined with what mathematics is learned.

What Is the Process Strand?

WMAS (Wisconsin Department of Public Instruction 1998) places "Standard A: Mathematical Processes" at the beginning of the document. Believing skills such as "reasoning, oral and written communication, and the use of appropriate technology when solving mathematical, real-world, and nonroutine problems" (4) to be integral parts of every mathematical endeavor, these processes were grouped and placed as the first strand in WMAS.

The mathematical processes of reasoning, communication, and the use of appropriate technology need to be integral to the solving of every mathematical, real-world, or non-routine problem.

PSSM (Standards 2000 Project 2000, 402) separates the process standards and lists them as follows:

PROBLEM SOLVING

- Build new mathematical knowledge through problem solving.
- Solve problems that arise in mathematics and other contexts.
- Apply and adapt a variety of appropriate strategies to solve problems.
- Monitor and reflect on the process of mathematical problem solving.

REASONING AND PROOF

- Recognize reasoning and proof as fundamental aspects of mathematics.
- Make and investigate mathematical conjectures.
- Develop and evaluate mathematical arguments and proofs.
- Select and use various types of reasoning and methods of proof.

COMMUNICATION

- Organize and consolidate their mathematical thinking through communication.
- Communicate their mathematical thinking coherently and clearly to peers, teachers, and others.
- Analyze and evaluate the mathematical thinking and strategies of others.
- Use the language of mathematics to express mathematical ideas precisely.

CONNECTIONS

- Recognize and use connections among mathematical ideas.
- Understand how mathematical ideas interconnect and build on one another to produce a coherent whole.
- Recognize and apply mathematics in contexts outside of mathematics.

REPRESENTATION

- Create and use representations to organize, record, and communicate mathematical ideas.
- Select, apply, and translate among mathematical representations to solve problems.
- Use representations to model and interpret physical, social, and mathematical phenomena.

The content strands of number operations and relationships, geometry, measurement, statistics and probability, and algebraic relationships are presented in performance standards at the fourth, eighth, and twelfth grade in **Wisconsin's Model Academic Standards for Mathematics.**

What Are the Content Strands?

The content strands of number operations and relationships, geometry, measurement, statistics and probability, and algebraic relationships are presented in performance standards at the fourth, eighth and twelfth grade in WMAS. Each of the content strands includes emphasis at each level, and such listing implies that each strand needs to be taught at each grade, though with varying emphasis. State, national, and international assessments often identify weaknesses in students from the United States in measurement, geometry, statistics and probability, and number sense. Much of this lack is due to the heavy emphasis on the number and algebra strands and to mastery of skills and procedures, often to the exclusion of other areas. So that students can build the necessary understandings, both the national and state standards call for all five content strands at every grade level.

PSSM (Standards 2000 Project 2000, 392–401) identifies and explains the content strands as follows:

NUMBER AND OPERATIONS

Instructional programs from prekindergarten through grade 12 should enable all students to

- Understand numbers, ways of representing numbers, relationships among numbers, and number systems.
- Understand meanings of operations and how they relate to one another.
- Compute fluently and make reasonable estimates.

ALGEBRA

- Understand patterns, relations, and functions.
- Represent and analyze mathematical situations and structures using algebraic symbols.
- Use mathematical models to represent and understand quantitative relationships.

GEOMETRY

- Analyze characteristics and properties of two- and three-dimensional geometric shapes and develop mathematical arguments about geometric relationships.
- Specify locations and describe spatial relationships using coordinate geometry and other representational systems.
- Apply transformations and use symmetry to analyze mathematical situations.
- Use visualization, spatial reasoning, and geometric modeling to solve problems.

MEASUREMENT

- Understand measurable attributes of objects and the units, systems, and processes of measurement.

- Apply appropriate techniques, tools, and formulas to determine measurements.

Data Analysis and Probability

- Formulate questions that can be addressed with data and collect, organize, and display relevant data to answer them.
- Select and use appropriate statistical methods to analyze data.
- Develop and evaluate inferences and predictions that are based on data.
- Understand and apply basic concepts of probability.

PSSM offers a graph that shows suggested content strand distribution across grade levels and the relative emphasis that each strand might receive at the various levels (see Figure 5.00).

Principles and Standards for School Mathematics (Standards 2000 Project 2000) ***offers a graph that shows suggested content strand distribution across grade levels and the relative emphasis that each strand might receive at the various levels.***

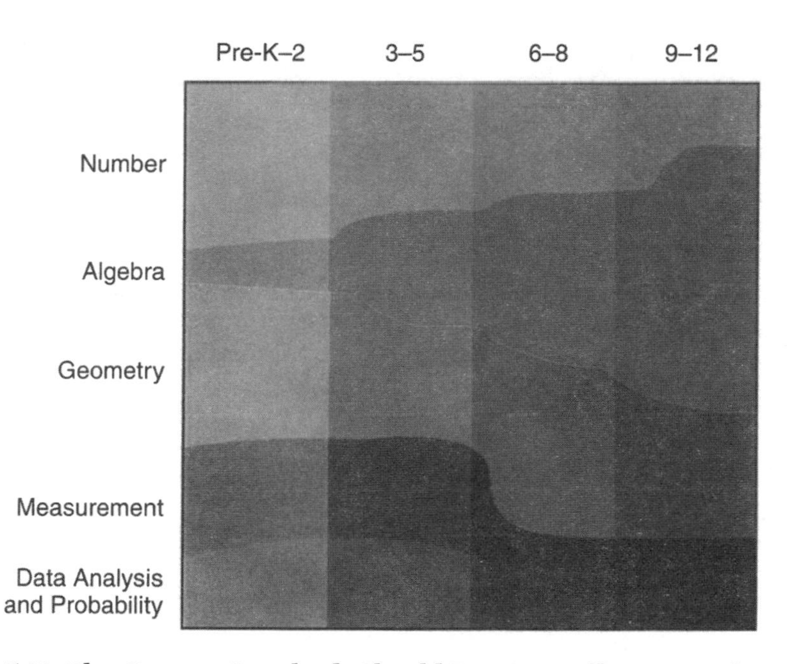

Figure 5.00 **The Content Standards Should Receive Different Emphases across the Grade Bands**

From Standards 2000 Project 2000, *Principles and Standards for School Mathematics* (Reston, VA: National Council of Teachers of Mathematics), 30. Copyright © 2000 by National Council of Teachers of Mathematics. Used with permission of author.

The section immediately following introduces each of the content standards as it exists across grade levels. The examples of classroom tasks given here offer "existence proof" that rich mathematical tasks can be addressed at various grade levels and that understanding is built from PK–12.

Standard A: Mathematical Processes

Students in Wisconsin will draw on a broad body of mathematical knowledge and apply a variety of mathematical skills and strategies, including reasoning, oral and written communication, and the use of appropriate technology, when solving mathematical, real-world, and nonroutine problems.

WMAS, Standard A, encompasses the process standards of problem solving, reasoning, communication, connections, and representations. Although these standards are addressed throughout the other strands, it is important to also pull them aside and look at them for the value they bring to a solid mathematics education. It is through the processes standard that we come to truly understand and to value mathematics learning. It is also through this standard that teachers can assess the value that students place on their learning and the depth of their knowledge.

It is through the processes standard that we come to truly understand and to value mathematics learning.

Problem solving is the primary reason to study mathematics. Problems should precede instruction to motivate the study of a topic, be an inherent component of instruction because students learn best within a context, and be the logical follow-up to instruction as students apply their knowledge. Through the selection of good problems and tasks for student work, the teacher sequences learning. Open-endedness in problems is important; problems should be selected that have many solutions and many ways to arrive at each one. Definitions, rules, formulas, and procedures should be results of activities, not starting points. Problems in the real world are frequently ill formed and missing information; work with problems similar to those that exist in the real world should be a part of the mathematics education of every student. Motivating problems are those that are relevant to students' lives.

The mathematics student needs to learn how to tackle a problem, how to employ multiple strategies, how to use algorithms that will make problem solving easier, and about how useful problem solving can be. The mathematical problem solver gains understanding and confidence in approaching problems and acquires skill in selection of approaches. Though the mathematics student should build a wide repertoire of experienced problems, the mathematics education program should be designed to prepare students to solve problems that have not been written yet, not just those that have been solved many times and have a "best" solution technique determined. Problem solving is what we do when we don't *know* what to do. Problem solving is a means and an end. It should be present in all strands of the curriculum.

Problems in the real world are frequently ill formed and missing information.

Problem solving is what we do when we don't know what to do.

Strategies for problem solving may be seen as worthy goals in their own right. Students need to see multiple strategies in action. Although different people may use other strategies to solve the same problem, the bigger the toolbox of strategies for each individual, the better problem solver he or she is. Strategies such as "draw a picture," "make a model," or "make a simpler problem," and so on are viable for the solution of problems posed at several grade levels, and are useful in problems of ever-increasing complexity. Flexibility comes with the availability of many strategies. Tasks need to be structured to help students acquire those strategies. When strategies are selected from a full repertoire, applied to an array of similar and different problems, and the solutions are evaluated for correctness and efficiency, students move from being answer finders to becoming problem solvers.

By the same token, students need to realize that realistic problems may not fit any particular situation and that new strategies may need to be developed to address those problems. One of the fallacies that students develop when strategies are too heavily emphasized is that all problems can be solved with a set group of approaches. Rather, it is important for students to realize that problem solving is just that—the application of creative and unique methods to address the situation. Problem solving is not a spectator sport but is an active and interactive human endeavor.

Problem solving is not a spectator sport but is an active and interactive human endeavor.

Problem: As five friends arrive at school each day, they like to greet one another with a handshake. Making sure that each friend shakes hands with each of the other friends in the group, how many handshakes will the group exchange as they begin their day?

The ever popular handshake problem can be handled at virtually all grade levels with increasing layers of sophistication. Primary grade school children might solve it by acting it out or by using modeling techniques. When asked to write a similar problem (see "Handshakes Revisited"), they might compare and contrast it to a problem such as "Five children want to exchange Valentines. How many Valentines will be exchanged?" Upper elementary and middle

---

Handshakes Revisited

Here are some problems that are similar to the Handshake Problem.

Find a rule to predict the total number of rectangles made by placing n squares in a row.

How many diagonals are there in a polygon with n sides?

If there are n points in a plane and three points lie on the same line, what is the maximum number of different lines that can be drawn between any two points?

In a large office, an intercom system is going to be installed so that each room can be reached by every other room. How many lines are needed?

What is the sum of the first 100 integers? n integers?

In a round robin tournament, every team plays every other team once. If ten soccer teams compete in such a tournament, how many games are played?

In an art class there are 10 different colors of paint. Additional colors can be made by mixing any two of the original colors together. How many new paints can be made?

If you have one of each of the following coins (penny, nickel, dime, quarter, and half dollar) what is the total of the different amounts of money you can obtain by forming all possible two-coin combinations?

How many segments are formed by n distinct points on a line?

Used with permission of E. Phillips, Mathematics Education Department, Michigan State University.

Reasoning needs to be a focus of the mathematics program at all levels.

school students can move the problem to larger numbers, such as 100 handshakes, and then generalize to a study of triangular numbers or to a consideration of the multiplication principles present in the Valentine problem. Middle school and high school students can use the handshake and similar problems to move into the study of sequences, series, difference equations, and permutations. Teachers need to develop repertoires of universal problems such as the handshake problem that can used to develop specific concepts.

Reasoning needs to be a focus of the mathematics program at all levels. Students' senses of logic, their abilities to make and defend valid conclusions, and their abilities to justify their answers need to be continually developed. Mathematics does deal with regularities and patterns. Showing that there is good reason for these regularities gives confidence that claims are true and that mathematics does not occur according to whim. From the earliest primary grades students should be asked, "Why do you think so?" and be told to convince the class. This process, over the course of time, should become gradually more sophisticated. Proofs may involve a demonstration, presentation of reasoning through a written paragraph, presentation of a counterexample, and so on. Early proof should be "that which convinces," but logical tools should be developed to facilitate progression to formalization of proofs. The repertoire of reasoning and proof strategies will become increasingly sophisticated as the student progresses in grade level.

Justification of thinking should be present in all instructional situations.

Justification of thinking should not simply be a unit within certain courses but should be present in all instructional situations. When there is an element of doubt, there is room for justification. But even when there is little or no doubt, the obvious may be used as an example of what justification or proof entails. For example, no one doubts that when looking at an isosceles triangle, the base angles are congruent. A valid argument could be made here on the basis of inspection. Surprising results, on the other hand, show need for justification. For example:

Problem: Using the following array of odd numbers, present a convincing argument that the middle number in a row is the square of the number of numbers in the row.

1

3 5

7 9 11

13 15 17 19

21 23 25 27 29

FIGURE 5.1

Many other similar questions can be asked about this array, such as: Find the sum of the numbers in the 50th row and explain how you reached that conclusion.

To increase the element of doubt, pattern problems that break down after a few steps should also be used.

To increase the element of doubt, pattern problems that break down after a few steps should be used.

Problem: Look at the following pictures. Note that each successive picture is generated by identifying one more point on the circle and drawing all possible chords between pairs of points and then counting the regions into which the circle is divided. Into how many regions would you expect the last figure to be divided when all the chords are drawn?

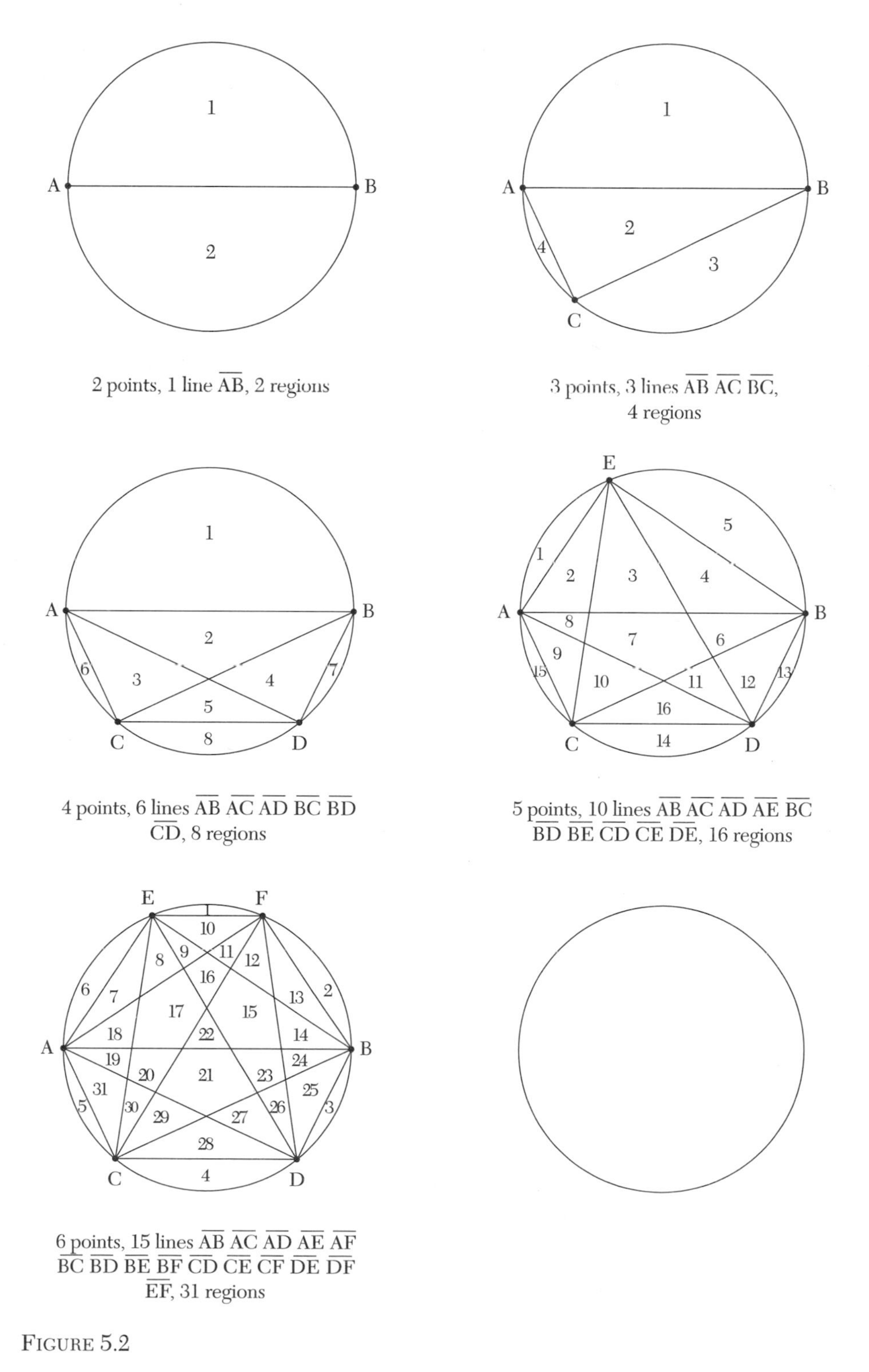

2 points, 1 line $\overline{AB}$, 2 regions

3 points, 3 lines $\overline{AB}$ $\overline{AC}$ $\overline{BC}$, 4 regions

4 points, 6 lines $\overline{AB}$ $\overline{AC}$ $\overline{AD}$ $\overline{BC}$ $\overline{BD}$ $\overline{CD}$, 8 regions

5 points, 10 lines $\overline{AB}$ $\overline{AC}$ $\overline{AD}$ $\overline{AE}$ $\overline{BC}$ $\overline{BD}$ $\overline{BE}$ $\overline{CD}$ $\overline{CE}$ $\overline{DE}$, 16 regions

6 points, 15 lines $\overline{AB}$ $\overline{AC}$ $\overline{AD}$ $\overline{AE}$ $\overline{AF}$ $\overline{BC}$ $\overline{BD}$ $\overline{BE}$ $\overline{BF}$ $\overline{CD}$ $\overline{CE}$ $\overline{CF}$ $\overline{DE}$ $\overline{DF}$ $\overline{EF}$, 31 regions

FIGURE 5.2

(Experimentation with various arrangements of the points around the circle will show that the maximum regions ever found will be 31, not the expected 32.)

Problems such as these help students see the need for justification beyond appearances.

The teacher's role is to facilitate the dialogue without ending it.

The communication standard is intertwined with the reasoning standard. As students are having mathematical dialogues, the subject matter will frequently center on convincing a second student of the correctness of the work. The teacher's role is usually to facilitate the dialogue without ending it. Steve Reinhart (2000), a Presidential Award–winning middle school mathematics teacher from Chippewa Falls, Wisconsin, says, "Never say anything a kid can say!" (See the article in full in Appendix E: Readings.)

Communiction often helps in clarification of thinking.

Communication of mathematical understandings is seen in many of the examples given here and in the grade-level sections. Written work, verbal explanations and answers, model creations, and development of visuals are all ways in which students can demonstrate understanding through communication. It is often through a natural course of events that teachers can use student communication, both verbal and written, to assess understandings as an integral part of the learning experience.

Communication provides linkage to the world of work. As students work in cooperative groups, divide up tasks, put together the results garnered by each team member, arrive at and agree upon a solution or solutions, and prepare a report stating and defending an answer and supporting the reasoning that led to it, they are modeling a workplace process model. In industry, problem solving is usually done in groups. Communication often helps in clarification of thinking. In explaining what has been done in solving a problem, the user gains insight him- or herself. Indeed, most teachers have probably said at some point in their lives, "I really learned that concept when I tried to teach it." In constructing an explanation for those who do not understand, a teacher or a student can create images and connections that allow for deeper understandings.

Connections within the field, connections with other disciplines, connections with the world of work, and connections with the important components of the world should also be consistently present and continually emphasized.

The connections standard must be pervasive through the grade levels and topics. Although this document does address number, geometry, measurement, probability, statistics, and algebra in different sections, it should always be obvious to the student that the real subject of study is mathematics and that the strands are interwoven. Connections within the field, connections with other disciplines, connections with the world of work, and connections with the important components of the world should also be consistently present and continually emphasized. When students can identify connectedness, transfer of knowledge is facilitated. Recent learning theory that conjectures about the connections between external and internal representations in the building of understanding is exciting for both students and teachers. The titles of many recent mathematics materials that incorporate the word *connections* give credence to the importance of its consideration. This guide is also predicated on the importance of connections.

When students can identify connectedness, transfer of knowledge is facilitated.

A final process standard, representation, has been added to problem solving, reasoning, communication, and connections in the NCTM *Principles and Standards for School Mathematics* (Standards 2000 Project 2000): "Representation refers to both process and to product—the act of capturing a mathematical concept or relationship in some form and the form itself" (67). Representation encompasses processes and products both external (observable) and internal (in the mind). Electronic technology has created an even greater need for attention to representation. PSSM suggests that instructional programs should enable all students to:

Electronic technology has created an even greater need for attention to representation.

- Create and use representations to organize, record, and communicate mathematical ideas.
- Select, apply, and translate among mathematical representations to solve problems.
- Use representations to model and interpret physical, social, and mathematical phenomena (67–71).

Hence, representation assists in problem solving, facilitates reasoning, provides means of communication, inherently connects mathematical components, and can provide excellent insights into internal composition of knowledge when translated into an external form. (See Cuoco and Curico, 2001.)

Standard B: Number Operations and Relationships

Students in Wisconsin will use numbers effectively for various purposes, such as counting, measuring, estimating, and problem solving.

> The true pedagogical order is to begin with numbers applied to tangible and present things; to proceed to numbers applied to familiar but absent things, and lastly to abstract numbers (Harris and Waldo 1911, 9).

Number work is the foundation for the mathematics curriculum PK–12. As a basis for patterns in algebra, for quantification in geometry and measurement, and for manipulations in data analysis, the essential nature of number skill and concept development in mathematics learning is pervasive. This does not imply, however, that number skill development is prerequisite to all other mathematical understanding. In fact, success in mathematics through visualization or problem solving can sometimes bring belief in self-worth that will initiate more success with number work. Numbers and the rest of mathematics are inherently intertwined, but the direction of entry for all students is not necessarily through number.

Numbers and the rest of mathematics are inherently intertwined, but the direction of entry for all students is not necessarily through number.

As Harris and Waldo so aptly put it in 1911, number work should begin with hands-on exercises, move to representations such as pictures and graphs, and finally come to algorithms and equations. This developmental sequence should be present in each investigation. As new number systems or operations are introduced, the learning sequence should be followed, whether we are working with four-year-olds on initial counting ideas or working with high school students on exponential growth. This sequential growth, in essence, flies in the face of the dichotomous argument regarding skills and procedures versus problem solving and conceptual development. Both need to be present in all instances of mathematics learning.

Sense making should be a central issue throughout the study of number.

Making sense of number and operations includes flexible and fluent use of number meanings and relationships, recognition of the size of numbers, understanding of the effects of operations on numbers (including estimation), and acquisition of referents for common units and measures. This sense making should be a central issue throughout the study of number. Regardless of the numbers being used, the commonalties of the definitions of the operations need emphasis. For example, many view the operations of addition and multiplication as very different when they are applied to the different number sets. For instance, when the addition operation takes place, the quantities are added and the "name" remains the same.

2 **cows** plus 3 **cows** equals 2 plus 3 (5) **cows**,
25 **eggs** plus 36 **eggs** equals (61) **eggs**

2/5 plus 1/5 equals (3) **fifths**;
3/5 plus 4/5 equals (7) **fifths** (or 1 whole plus 2/5)

0.6 plus 0.3 equals (9) **tenths**;
0.36 plus 0.95 equals (131) **hundredths** (or 1 whole plus 0.31)

$4x + 8x$ equals (12) x.

On the other hand, when multiplication takes place, the quantities are multiplied, but the names are changed to reflect the resulting entities:

3 **cm** times 4 **cm** equals (12) **cm**^{2}

2/5 times 1/5 equals (2) **twenty-fifths**;
3/5 times 4/5 equals (12) **twenty-fifths**

0.9 times 0.5 equals (45) **hundredths**;
0.36 times 0.95 equals (3420) **ten-thousandths**

$4x$ times $8x$ equals (32) x^2

Principles and Standards for School Mathematics (Standards 2000 Project 2000) *emphasizes that all students should know their basic number combinations and be fluent with computation.*

Emphasis on these commonalties across grade levels and topics eases the learning of each topic.

PSSM (Standards 2000 Project 2000) emphasizes that all students should know their basic number combinations and be fluent with computation. WMAS (Wisconsin Department of Public Instruction 1998) calls for students "in problem solving situations" to be able to "select and efficiently use appropriate computational procedures—addition, subtraction, multiplication, and division—using mental math, estimation, and selecting and applying algorithms" (6–7) with each of the number sets—whole numbers, integers, rational numbers, real numbers, and complex numbers. Students should be able to state basic (zero through nine added to zero through nine and the inverse facts) addition and subtraction facts with minimal hesitation and should understand the operations of addition and subtraction and the results of the operations by the end of second grade. Likewise, students should be able to state basic multiplication and division combinations and should understand the operations of multiplication and division by the end of fourth grade, certainly by the beginning of fifth grade. The middle school curriculum then consists of rational number, proportional reasoning, and algebraic and geometric concepts. Individual schools may wish to set intermediate subgoals such as addition by the end of first grade or multiplication by the end of third grade, but the overall goal should be considered the benchmark.

What it means to "know" number combinations and to state them with "minimal hesitation" needs clarification. Knowing a fact implies a thorough understanding of the definition(s) of the operation, recognition of the several embodiments of the fact, and grasp of the relationships that exist among facts. Minimal hesitation does *not* imply writing answers to 100 fact questions in two minutes but does imply that a student, when presented with a stimulus such as $8 + 7$, can respond readily with 15 (and would perhaps have time to mentally calculate something such as $10 + 5$ or $8 + 2 + 5$). A consistent message of understanding should be built with students through the elementary years. Reading students are consistently told to take time to understand what they read. The mathematical goal should be similar. Unfortunately, often the minimal hesitation goal is interpreted as meaning "Go faster. Get the answer first," a learning approach that is short sighted and fear inducing and that ill serves populations that value "think time."

As numbers become larger and problems become more complex, the response time implied by minimal hesitation should be proportionately longer. The mathematics program should be designed so that students can learn skills for generating facts they don't recall. If a student does not recall 7×8, he or she should not have to look at the back of the flash card to find the answer but should see the relationship to a known fact such as 7×7, should be able to visualize an array, or should know how to employ other useful strategies. Power and understanding, not speed, should be the mathematical goals. It is recognized that students will internalize their number combinations at different rates and that many diverse activities that capitalize on different approaches to learning combinations need to be experienced.

The mathematics program should be designed so that students can learn skills for generating facts they don't recall.

Fluency with computational procedures beyond facts also implies a close relationship to understanding the operation. The PSSM has stated: "Developing fluency requires a balance and connection between conceptual understanding and computational proficiency" (Standards 2000 Project 2000, 35). The PK–5 program should allow students to develop computational proficiency using a variety of tools, including mental, paper and pencil, and technology-related procedures. It is also important to recognize when, in a computational situation, an exact answer is needed and when an estimate is satisfactory. The instructional program should be designed with flexibility and choice of computational procedures in mind. The student cannot "select and efficiently use appropriate computational procedures" (Wisconsin Department of Public Instruction 1998, Standard B.4.5, 6) unless he or she has each of the tools read-

Fluency with computational procedures beyond facts also implies a close relationship to understanding the operation.

Instruction should center on the development of skills with mental, written, and calculator algorithms and estimation strategies.

ily available. Hence, instruction should center on the development of skills with mental, written, and calculator algorithms and estimation strategies. It is important to note the *s* on *algorithms.* Instruction that indicates to students that one single algorithm is always preferred limits creativity, may inhibit understanding, and makes mathematics a game in which symbols are moved around on a piece of paper. Flexibility and choice should rule.

The goal of producing computational answers efficiently is major. It is also important that the student be able to explain the procedure used and to couch this explanation in knowledge of the structure of the base 10 number system, definitions and properties of the operations, and other number relationships. By the end of fifth grade, students should compute fluently and make reasonable estimates in situations involving whole numbers.

Computational fluency with integers and rational numbers should be acquired by the end of Grade 8. Development of rational number understandings should center on conceptual understanding as prerequisite to computational fluency. Foci at the middle school level should be the relationships among the four operations, making sense of answers acquired, judging the reasonableness of results, and explaining methods of solutions. For example, if a student is asked to find the answer to $7/8-1/4$, he or she should also be able to respond to a statement such as "Show, using materials or pictures, why the answer to $7/8-1/4$ will undoubtedly be in eighths."

Standard C: Geometry

Students in Wisconsin will be able to use geometric concepts, relationships, and procedures to interpret, represent, and solve problems.

Emphasis on the deep, conceptual understanding of the properties of a wide variety of geometric figures and on the relationships between parts of figures and between various figures are the thrusts of the geometry strand. Students need cohesive school experiences with geometry, especially in the elementary and middle grades, that explore in depth the properties and relationships of the various components. Each investigation needs to explore extensively the addressed concept. Each subsequent activity needs to build on the understandings developed in the previous activities so that there is continual progress of knowledge. There needs to be conscious effort to make sure that geometry is included in each year's curriculum. Geometry is one of the areas that shows up as a weakness in many mathematics assessments. Often the topic appears at the end of the book and is not reached. Overreliance on broad but general ideas is typical of many texts.

"I do and I understand" should characterize an ideal school program in geometry.

"I do and I understand" should characterize an ideal school program in geometry. Students need to explore two- and three-dimensional figures and their properties and relationships by handling, drawing, constructing, measuring, folding, cutting, sliding, flipping, turning, and superimposing them. In addition, students need to talk about what they have discovered and should make conjectures about possible relationships and speculate on generalizations regarding relationships.

To achieve the desired depth of geometric understanding, geometry in middle school and high school should revisit many of the topics and figures studied at the more intuitive level in the earlier grades, this second time at a more analytic level. Another avenue of investigation that needs to be pursued is intensive use of geometric software to repeat earlier experiments, this time gathering more data, making additional conjectures, and investigating more broadly.

In the examples that follow, the big ideas are continuously revisited in slightly different contexts and at different suggested grade levels.

Geometry in middle school and high school should revisit many of the topics and figures studied at the more intuitive level in the earlier grades, this second time at a more analytic level.

Primary

- Work with a partner to see how many different figures you can make by putting the two triangle shapes together. Use the dot paper to make a record of what you did.

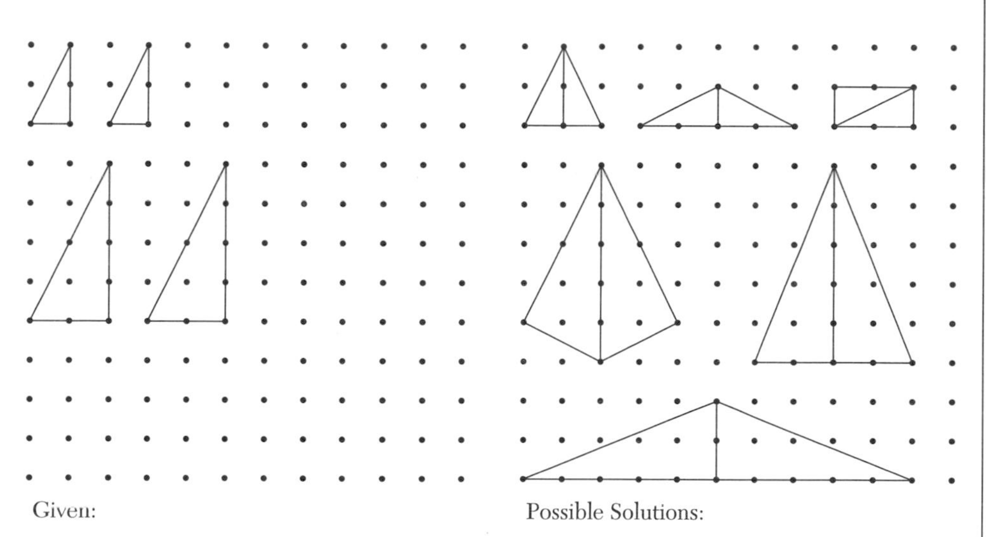

FIGURE 5.3A

- Try it again, this time using four triangle shapes.

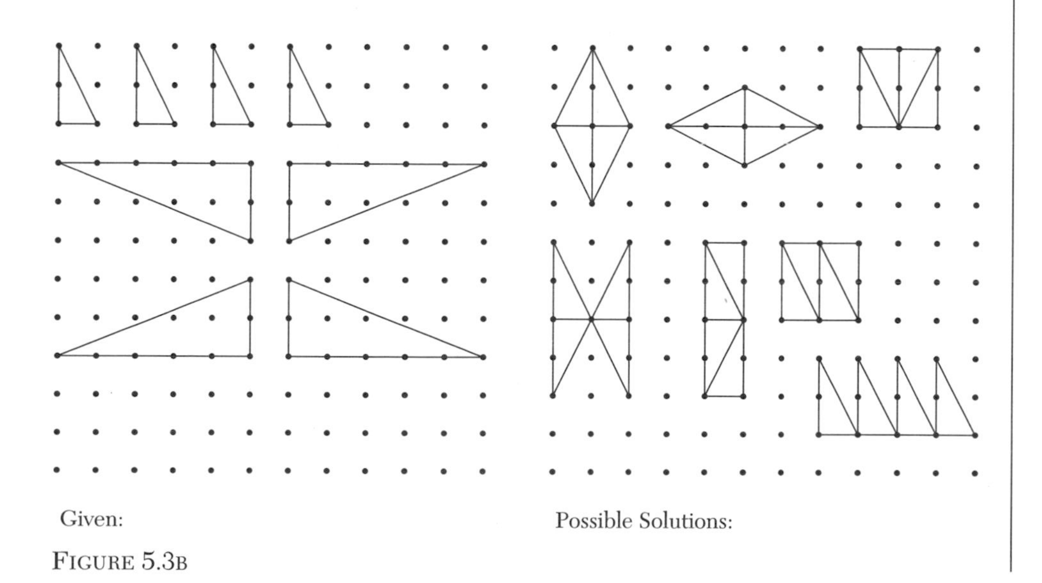

FIGURE 5.3B

Lower Intermediate

This is a project for small groups:

- Use cardboard or stiff paper to make two copies of a triangle. Put those two triangles together in as many different ways as possible to make a new figure. Use tracing paper to make a record of what you did.

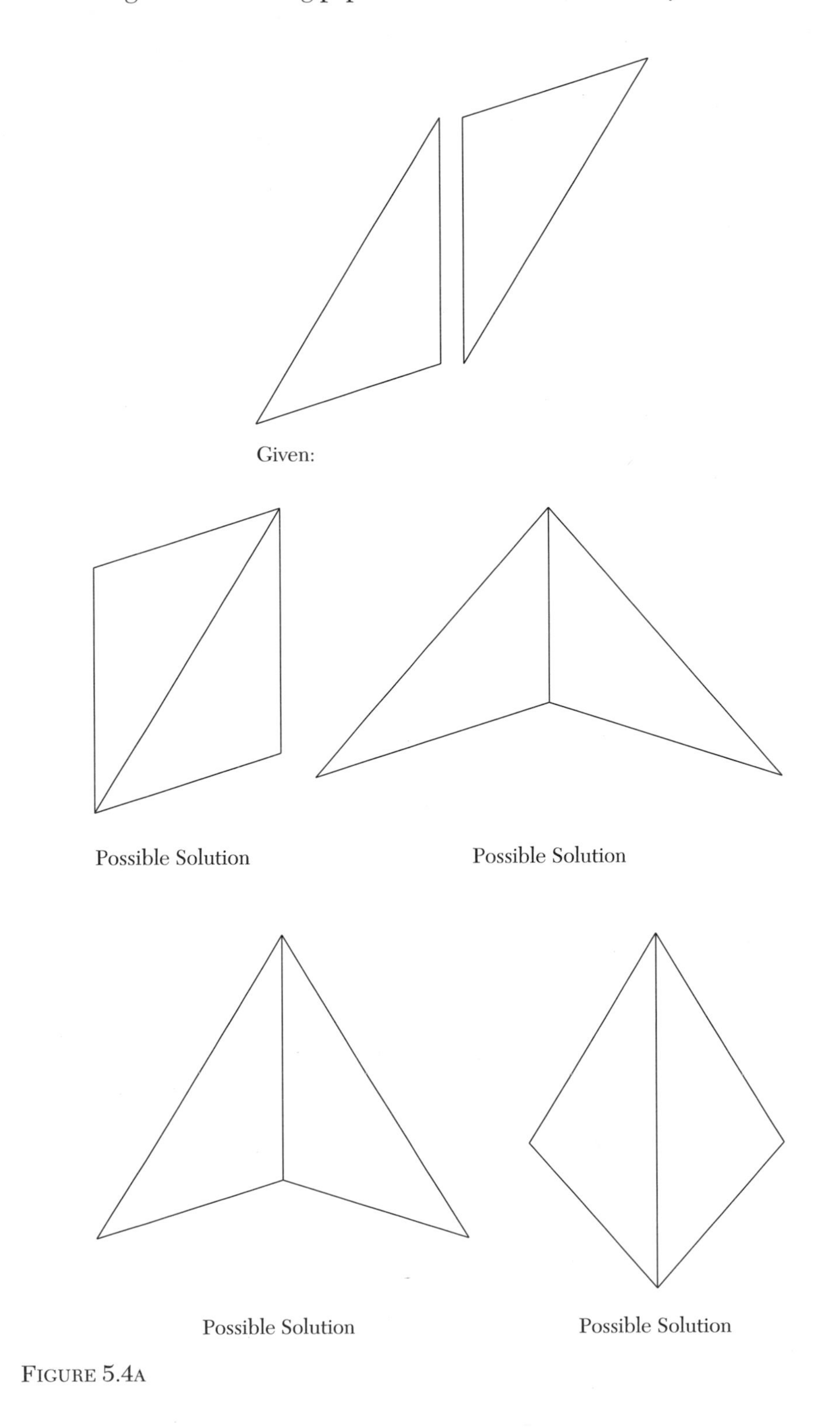

FIGURE 5.4A

- Repeat these directions with a new pair of triangles. In your group, do this at least 20 times. Be sure to begin with different types of triangles—acute, right, obtuse, scalene, isosceles, and equilateral.

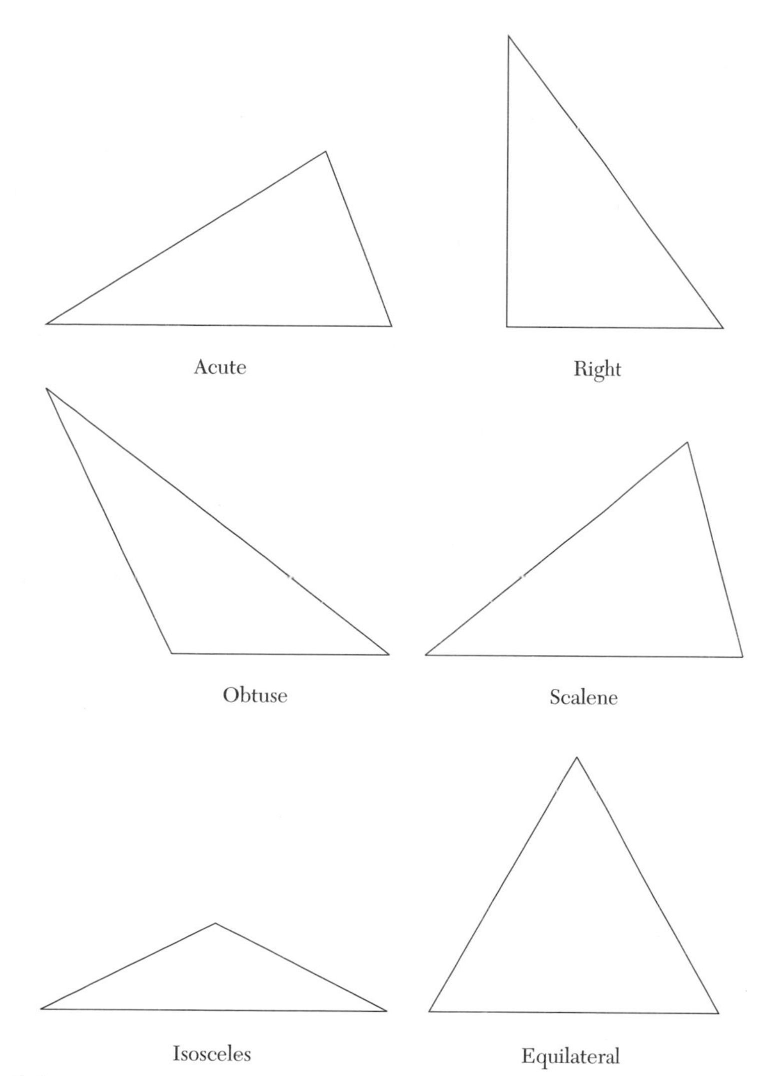

FIGURE 5.4B

- Repeat the above two steps, this time with four copies of your original triangle.

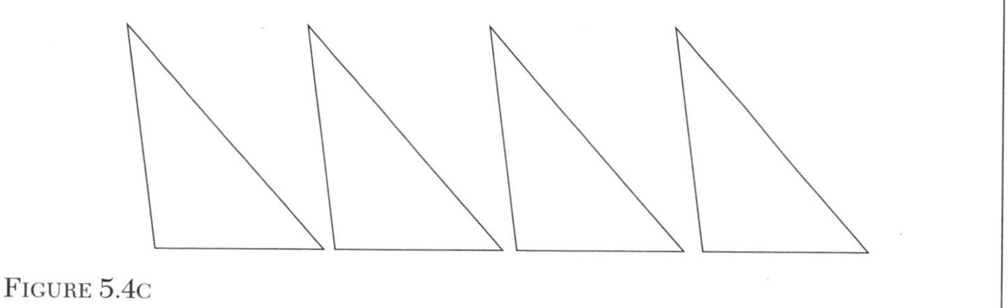

FIGURE 5.4C

Intermediate

- With the help of a ruler and protractor, draw a pair of intersecting segments that meet at right angles. Then connect the endpoints of those segments in order to form a four-sided figure (quadrilateral).

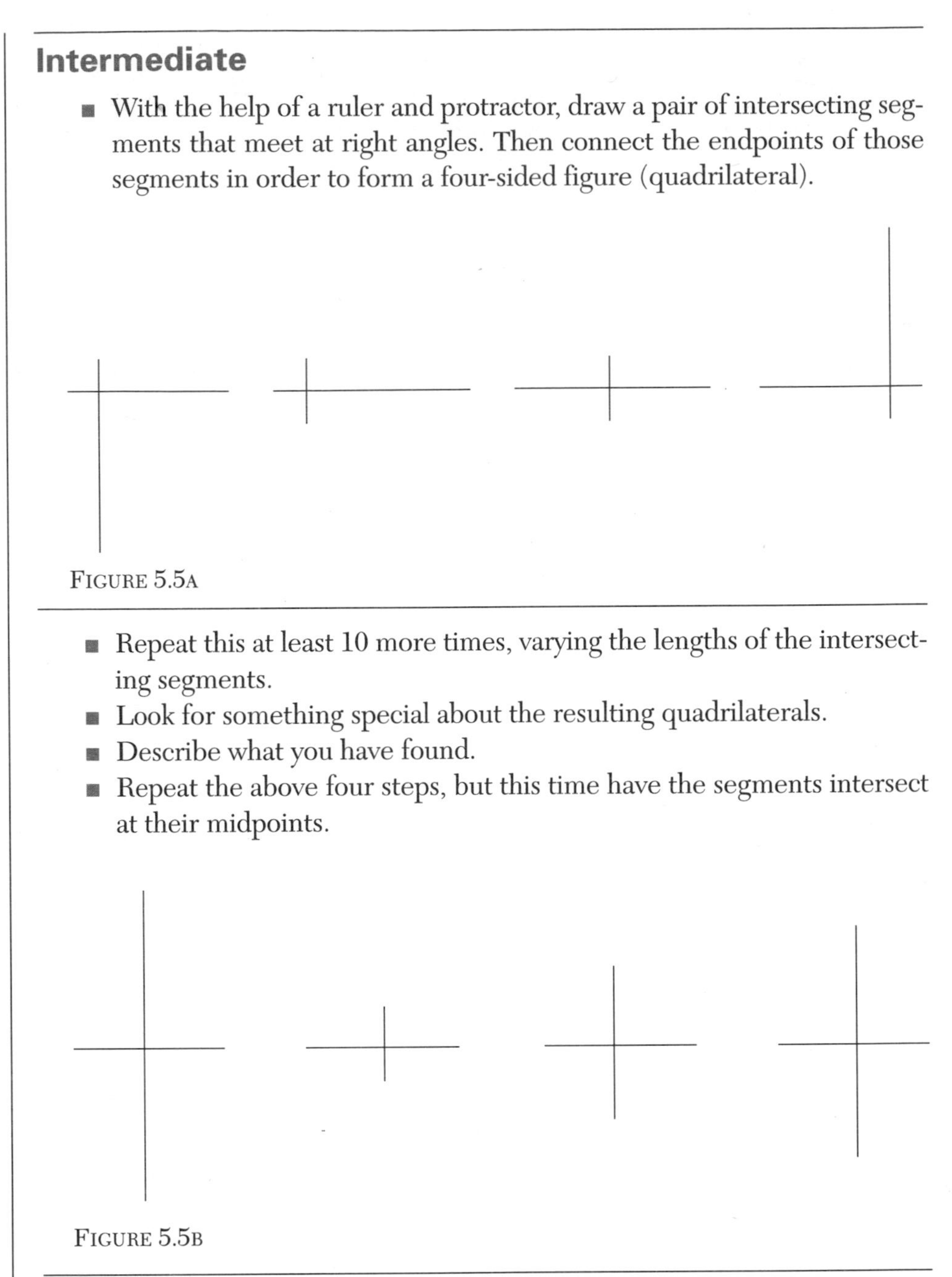

FIGURE 5.5A

- Repeat this at least 10 more times, varying the lengths of the intersecting segments.
- Look for something special about the resulting quadrilaterals.
- Describe what you have found.
- Repeat the above four steps, but this time have the segments intersect at their midpoints.

FIGURE 5.5B

Early Middle

This is a kite-flying project for small groups.

- Use a traditional kite as a model. Use suitable lightweight materials to construct a kite. Test your construction by attempting to fly it out-of-doors.

Middle

- Use a compass and straightedge to construct the perpendicular bisector of a given line segment.

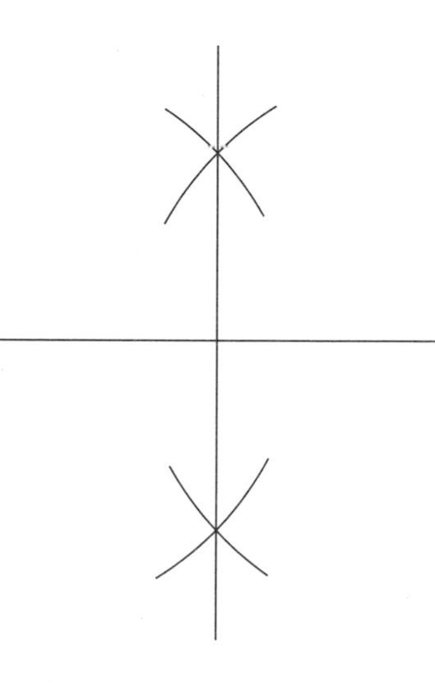

FIGURE 5.6

This is a tessellating project to try

- Prepare some sort of demonstration to convince a classmate that any triangle-shaped tile can be used to cover the entire plane.
- Prepare some sort of demonstration to convince a classmate that any quadrilateral-shaped tile can be used to cover the entire plane.

Late Middle or Early Secondary

Using technology to gather data:

- Use the *Geometer's Sketchpad* (Key Curriculum Press 2001, www.kcpress.com) to investigate if any three line segments can be used to make a triangle. Make a conjecture regarding what you think will be your conclusion. Test your hypotheses. Then explain what happened and why.

Secondary

- Prove: All points on the perpendicular-bisector of a segment are equidistant from the endpoints of that segment.
- Present a convincing argument to prove the following: If the diagonals of a quadrilateral bisect each other and are perpendicular to each other, that quadrilateral is a square.
- The family of four-sided figures has many members—quadrilateral, square, rectangle, rhombus, trapezoid, and parallelogram. Prepare some sort of diagram or family tree to show how they are related to each other.

Standard D: Measurement

Students in Wisconsin will select and use appropriate tools (including technology) and techniques to measure things to a specified degree of accuracy. They will use measurements in problem-solving situations.

Learning about measurement requires attention throughout the entire school mathematics program.

The importance of measurement skills and understanding cannot be overemphasized. When one intends to use mathematics to solve a real-world problem, it is often prudent to begin by creating a quantitative mathematical model. Two things are required to do this. The first is to represent the entities of the problem in the question numerically; the essence of that representational process is measurement. The second aspect of the model is to represent the relevant properties and relationships inherent in the problem; the essence of that is the use of appropriate mathematical symbols, essentially algebra.

The breadth of potential real-world problems is vast—scientific, economic, social, artistic, civic, athletic, and medical, to name just some of the possible categories. Learning about measurement appropriate to these fields requires attention throughout the entire school mathematics program. Unfortunately, many school programs equate the study of measurement with the study of geometry. Doing so does a grave injustice to both! There must be more than measurement of length, perimeter, area, and volume of simple geometric figures in much the same way that careful study of geometric properties and relationships must involve much more than just the easily identified measurable properties of those figures.

One of the key ideas that should permeate the entire measurement strand is the development of measurement sense.

One of the key ideas that should permeate the entire measurement strand is the development of measurement sense. As an illustration, consider this question:

How far is it from here to __X__?

- "1,238,416 inches" is not a sensible response. Inches are not used as units to measure large distances.
- "37.41286 miles" is not a very sensible response either. This pseudo-precision of using five decimal places implies that one hundred thousandths of a mile was used as the unit of measure. Hardly likely!
- "About 48 miles" or "between 45 and 50 miles" are both sensible responses. Almost all measurements are approximations, and because X may be the name of the city and not an exact location within that city, these responses are entirely sensible.
- "Three hours" can be a very sensible response even though an hour is technically a unit used to measure the attribute of duration of an event, not of length (distance). This reply might be more sensible than say "14.5 miles" if the distance to be covered by the questioner is in an urban area at morning rush hour.
- The distance does not become greater if 81 kilometers is reported rather than 50 miles. One needs to have a sense about the relative sizes of units being used.
- "About 50 miles" is not sensible if the responder knows that location X is closer than location Y and also knows that location Y is only about 35

miles distant. Having a sense about known measurement benchmarks is important when making estimates.

Throughout the school years, students must increase their skills in determining appropriate measurements. Work begins with crude instruments and sometimes arbitrary units to get at easily identified, directly measurable attributes and moves in later years to use of formulas, conversion techniques, and algebraic methods to determine more complicated, indirect measures.

Students must increase their skills in determining appropriate measurements.

Problem: How much water is wasted in a home with a leaky faucet?

Solution: Define *leaky.*

Decide on what unit to use in reporting results. Quarts per hour? Gallons per day (week, month)?

What data need to be gathered? How to gather this data? (Suggestion: Simulate varying degrees of leakiness by pricking holes of varying sizes in different water-filled, half-pint milk containers.)

Students must learn measurement skills and measurement sense in natural ways by investigating things of interest to them, and to the extent possible, by measuring actual objects within their classroom. As a final example, let's consider the question of learning about metric units. Ask anyone who has traveled or lived extensively in another country where the metric system (or a foreign language) is used exclusively: "You learn quickly in order to survive."

Suggestions for "living" with metric units:

Temperature: Gather daily highs and lows in Celsius degrees for your city over the period of several weeks or months. Compute daily differences. Plot the data. (See how quickly and easily children learn about negative numbers!)

Assign each student a city from around the world and do as just described, gathering data via the Internet.

Length: Students periodically measure their own height using centimeters as the unit of measure. Study small animal or small plant growth, gathering and plotting periodic data using centimeters or millimeters.

Mass (weight): Duplicate or combine (for bivariate data) with the previous length suggestions, using grams or kilograms as appropriate.

Standard E: Statistics and Probability

Students in Wisconsin will use data collection and analysis, statistics, and probability in problem-solving situations, employing technology where appropriate.

Cursory examination of daily newspapers and weekly magazines leaves little doubt about the importance of data analysis and the study of chance. Consider *USA Today,* in essence our only national newspaper. A graph is run daily in the lower left-hand corner of the front page, and other graphs and tables are seen in other locations throughout each issue with great regularity. The mis-

Cursory examination of daily newspapers and weekly magazines leaves little doubt about the importance of data analysis and the study of chance.

sion of *USA Today* is to convey a lot of news in a little space, and the publication recognizes the power of using graphs in place of many words.

As the pace of our societal life increases, undoubtedly visual displays will be used with increasing frequency to quickly convey large amounts of information. It has become crucial for students to gather data in ways that make it analyzable, to summarize data numerically and visually, to draw and support conclusions from data, and to critically analyze statistical reports of others. PK–12 instructional programs should strive for these goals with a constant eye on sense making. The most elaborate of statistical techniques or the fanciest graph cannot make poorly collected data meaningful. As students acquire increasingly sophisticated computational techniques, they must continually ask if the numbers they are addressing are worth crunching. The data units in the curriculum can be some of the best places for students to use, practice, and review computational procedures learned under the number strand. Letting students collect data that is used for computational practice puts relevance and personalization into what is frequently seen as tedious. Friel and Corwin (1990, 73) state that "learning statistics can be a vehicle to give meaning to computational skills that are too often isolated from relevant contexts."

Learning statistics can be a vehicle to give meaning to computational skills that are too often isolated from relevant contexts.

Likewise, the study of chance and the judging of likelihood done in probability units leads students to interpret the many probabilistic statements they hear daily. From "The chance of rain is 40 percent" to using cards or spinners to determine groups in class to "The odds are 5 to 2 for ______ in the Kentucky Derby" to statements of how unlikely it is that one could win a huge lottery prize, students regularly hear probabilities stated or implied. Consequently, they must learn developmentally, over time, to evaluate probability claims and to accurately make such statements themselves. Misconceptions about probability should be a part of the curriculum, and students should confront and overcome such misconceptions as "Three hundred people died in that plane crash last week. The chances of being killed in a plane are too high. I'm going to drive to Orlando."

Through the study of statistics and probability, students learn to ask and answer questions, to make and evaluate decisions, and to make predictions. These skills are very important in the overall school goal of developing a literate, informed citizenry who can carry out our democratic ideals. Mathematics is seen here as greatly impacting the future lives of our students. Although calculators and computers may compute many answers, decision making under uncertainty requires skills, concepts, practice, and the ability to evaluate options and choose the best for the circumstances.

Through the study of statistics and probability, students learn to ask and answer questions, to make and evaluate decisions, and to make predictions.

As with each of the strands, the probability and statistics strand should be present throughout the PK–12 curriculum. It is important that attention be given not only to incorporation of the strand in the K–5 and 6–8 sequence but in the high school curriculum as well. Although statistics is one of the mentioned curricular inclusions under state law, along with algebra and geometry, it is often not present in the traditional Algebra I, Geometry, Algebra II, and Precalculus programs that prevail. New, integrated high school curricula do ensure this offering, and many school districts do offer a statistics course. Curriculum developers need to be aware of this and work toward an appropriate balance of strands in the curriculum. It should be noted at the high school level that Advanced Placement Statistics offers a very viable route of pursuit for students to gain college credit for high school work, because it seems that at least as many college-entering students are required to take a statistics course as those who are asked to take a calculus course.

At least as many college-entering students are required to take a statistics course as those who are asked to take a calculus course.

Data-gathering activities should begin with the earliest activities that students do in primary grades. The sorting and classification tasks that are regularly done at these grades can fit in this domain very well as the students describe their sorts and draw conclusions about category placement. A daily or bidaily activity many teachers are finding to be successful is known as "Kid-Pins." A small headshot picture of each child in the class is attached to a clothespin. Then a topic that the class will research that day is selected. For example, an early task might simply be "boys and girls." A cardboard picture of a boy and one of a girl is placed in front of the room. Each child clips his or her pin to the appropriate picture. Then a whole-class discussion of what conclusions can be drawn from the results is held. Eventually, statements such as "In Mrs. Olson's first-grade class at Evergreen Elementary School, there are more girls than boys" are made, discussed, and defended. As the categories increase with topics like "How did you get to school today? Walk? Ride the bus? Ride in a car? Ride your bike?" several subconclusions are drawn, debated, and analyzed. These kinds of activities lay the groundwork for increasingly more sophisticated data gathering as students progress through the school sequence. The basic ideas of identification of the population of interest; of developing mutually exclusive but exhaustive categories; of drawing clear, substantiated conclusions; and of defending statements continue to persist in future investigations.

Construction of visual displays of data and interpretation of data displays should also be done across the grade levels. An activity that can be done with grade-level appropriate graphs follows. Students are paired up, and each child in the pair is given a different graph. (These graphs can be cut from *USA Today*, magazines, or other sources.) The graphs are not shown to the partners. Each child writes a paragraph about his or her graph, using as much information as possible: What kind of graph is it? What information does it convey? How is it arranged? Once these descriptions are completed, the children exchange the paragraphs, but still do not show the graphs. The partner draws what he or she believes the "mystery" graph looks like, based on the given paragraph. The partners then display their graphs and discuss what clues gave them the most information, what statements threw them off, what other information was needed, and so on. This activity has students work with visual information, conversion of that information to written descriptions, translation of written descriptions to visual displays, and oral discussion of the entire process.

Construction of visual displays of data and interpretation of data displays should also be done across the grade levels.

Standard F: Algebraic Relationships

Students in Wisconsin will discover, describe, and generalize simple and complex patterns and relationships. In the context of real-world problem situations, the student will use algebraic techniques to define

and describe the problem to determine and justify appropriate solutions.

Algebraic reasoning is a skill that needs to be carefully nurtured and developed through well-chosen, sequential activities that carry across grade levels and progress from intuitive to formal and abstract.

The emphasis of the algebra strand is on the thinking that is done as one recognizes patterns, generalizes, deals with inverse relationships, works with properties of mathematics, reasons proportionally, deals with notions of balance, reasons inductively or deductively, or uses function relationships. As students at the prekindergarten and kindergarten level recognize and label AB or ABB patterns, as a primary grade student recognizes that not only is $3 + 5 = 5 + 3$, but $a + b = b + a$ for all a and b, as an early grades student recognizes that a missing addend problem may be solved by subtraction, the essential character of algebraic thinking is developing.

The standards focus is on instructional tasks and assessments that are based on deep conceptual understanding of mathematics. The use of and communication of algebraic reasoning gives evidence of this understanding and is a skill that needs to be carefully nurtured and developed through well-chosen, sequential activities that carry across grade levels and progress from intuitive to formal and abstract. From the use of boxes and blanks to the use of letters, from the recognition of patterns to the explanations of why the patterns exist, algebraic reasoning is developed over time and across strands of the curriculum.

Note in the following examples how the requirement for what constitutes a convincing argument changes and is strengthened as the knowledge of mathematics increases, the ability to abstract grows, and connections are made.

Primary

1. Find, color in, and extend patterns on 100's charts, addition tables, and so on.
2. Describe orally and in writing why extension of the pattern is correct.
3. Present an argument using manipulatives for showing that the sum of four numbers in a two-by-two square extracted from an addition table is four times the number in the upper-right corner of the square.

   7, 8
   8, 9

4. Relate patterns on the addition table to the meaning of the addition operation.

Intermediate

1. Color in and describe visually and numerically patterns that exist in the body of the multiplication table for multiples of 2, 3, 4, 5, 6, 7, 8, and 9. Relate these patterns to the multiplication operation.
2. By constructing and displaying rectangular arrays, describe the inverse relationship of multiplication and division, justify the commutative property of multiplication, and demonstrate other general properties of multiplication and division.

3. Using rectangular arrays and organized lists, present a convincing argument orally or in writing on which counting numbers have an odd number of divisors, two divisors, four divisors, and so on.

Middle

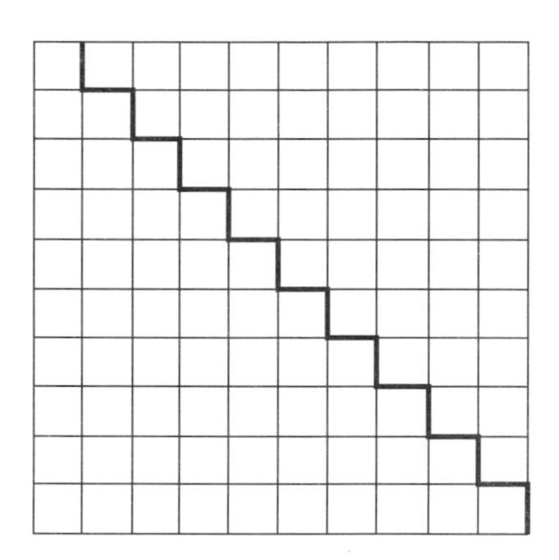

1. Using pattern searches and organized listings, develop some problems such as the following:

   Find the sum of the first n counting numbers ($1 + 2 + 3 + \ldots n$). (How does this relate to the handshake problem?)

   Find the twentieth term of (use graph of stair steps.)
2. Use variables to model and justify patterns on the addition table.
3. Use symbolic notation to describe counting numbers that have any given number of factors. For example: Numbers that have exactly four factors are either of the form $p \times q$ where p and q are both primes or the form p^3 where p is prime.

   2×3 3×5 7^3 11^3

High School

1. Present a numerical and visual argument that $1/2 + 1/4 + 1/8 + 1/16 + \ldots$ (continuing forever) must approach 1.
2. Identify and justify patterns on Pascal's triangle and use these generalizations in probability work and binomial expansions.
3. If f_n is the nth term of the Fibonacci sequence, find and justify a one-term expression for $f_1 + f_2 + f_3 + \ldots + f_{2n-1}$.

A problem such as the following shows the growth of algebra and algebraic thinking across the years:

Art: 8 × 8 grid

Problem: If we color just the border squares on this grid, how many squares will be colored?

At the elementary level, the discussion would center around multiple ways of counting this and why particular students counted it as they did. Possible solutions might include:

$4 \times 8 - 4$
$4 \times 6 + 4$
$8 + 7 + 7 + 6$
4×7
$64 - 36$
$8 + 8 + 6 + 6$

Students would then be asked to try a 10×10 grid using one of the various six methods.

During the middle school years, the problem would be generalized to an $n \times n$ grid, and the conversation would center around the parts of the expressions that do not change as the number of squares on the sides change.

High school students could be assigned the task of showing that $n^2 - (n - 2)^2$ is equivalent to $4n - 4$ or other expressions. Algebraic simplification within the context of a meaningful problem is then practiced.

The approaches presented here of valuing thinking that may diverge, of developing notation that models the thinking, and of justifying the statements mathematically is an approach that should be followed to allow students to see the need for algebraic manipulations that are designed to mirror their own thinking.

Early Beginnings in Mathematics

Young children are curious about their world. They are intrigued by quantitative events and spatial arrangements in their physical environment and social interactions. This early mathematics learning grows naturally from children's experiences as they develop a fairly complex set of intuitive and informal mathematical ideas. Children's informal knowledge is relatively powerful and serves as the foundation for further learning of mathematics (Baroody 2000; Ginsburg and Baron 1993). Early childhood centers and preschool programs need to provide all children with rich experiences and high-quality programs that include significant mathematics presented in a manner that respects both the mathematics and the nature of young learners (Standards 2000 Project 2000).

Early mathematics learning grows naturally from children's experiences as they develop a fairly complex set of intuitive and informal mathematical ideas.

Young Children's Learning

The learning of children from birth to age four is a time of profound developmental change. At no other time in a child's life is cognitive growth so remarkable. The brain literally grows new connections with stimulation from the child's environment as it attempts to make meaning of its world. The greater the number of connections and associations that one's brain creates, the more firmly the information is woven in neurologically (Jensen 1998). Children contribute to their own learning and development as they struggle to make sense of and give meaning to their daily experiences.

Children contribute to their own learning and development as they struggle to make sense of and give meaning to their daily experiences.

Enriched environments contribute to children's learning and development. The National Association for the Education of Young Children (1990) has identified seven basic assumptions about learning and teaching that inform their view of developmentally appropriate practice for young children.

- Children learn best when their physical needs are met and they feel psychologically safe and secure.
- Children construct knowledge.
- Children learn through social interaction with adults and other children.
- Children's learning reflects a recurring cycle that begins in awareness and moves to exploration, to inquiry, and finally to utilization.
- Children learn through play.
- Children's interests and need to know motivate learning.
- Human development and learning are characterized by individual variation.

The cycle of learning begins with an awareness of events, objects, people, or concepts that comes from children's experiences. Through play, children explore, experiment, and manipulate their environment as they bring their own personal meaning to their experiences. Opportunities for exploration allow children to pursue their interests and motivate learning. Play also promotes inquiry as children construct understanding of commonalties across

Children explore, experiment, and manipulate their environment as they bring their own personal meaning to their experiences.

events, objects, people, or concepts and develop generalizations. The social interaction among young children allows them to test their developing ideas and hypotheses against the thinking of other people. Finally, children apply or make use of their understanding.

Mathematical Programs for Young Learners

Mathematical programs for young learners need to maximize children's learning with appropriate practices (Bredekamp and Copple 1997; National Association for the Education of Young Children 1990; Richardson and Salkeld 1995). They need to build on and provoke children's natural curiosity, place an emphasis on informal and conceptual understanding of mathematical ideas, and nurture children's innate interest and ability in solving problems. The goal of these programs for young learners should be to develop mathematical power (Baroody 2000; Richardson and Salkeld 1995). Mathematical power includes a positive disposition to learn and use mathematics, understanding of mathematics, and the ability to engage in the processes of mathematical inquiry.

Programs that promote the development of mathematical power are purposeful, meaningful, and inquiry based. Purposeful instruction includes the use of everyday activities to provide opportunities to learn mathematics, uses children's questions as teachable moments, uses games as interesting and structured ways to explore mathematics, and uses literature as a source of rich problems. Meaningful instruction fosters and builds on children's informal mathematical knowledge and focuses on helping children see patterns and relationships. Inquiry-based instruction poses thought-provoking tasks for children to examine and solve, prompts children to reflect on their processes and solutions, and encourages peer-to-peer dialogue (Baroody 2000).

The teacher's role is to find ways to engage children in thinking about mathematical ideas by purposefully selecting tasks and materials that prompt and support children in the development of particular ideas.

The teacher's role is to find ways to engage children in thinking about mathematical ideas by purposefully selecting tasks and materials that prompt and support children in the development of particular ideas. Thus, teachers need to know what mathematical concepts children should become aware of, interact with, and think about as they explore their environment. In these environments, children need to be actively engaged with mathematics, and teachers must assess what they are really thinking and understanding rather than simply making sure that the children complete assigned tasks (Richardson and Salkeld 1995).

Mathematics programs for young learners are often limited in their visions of what children can explore (Greenes 1999). Early childhood and preschool programs need to expand the mathematics content and more clearly define a cohesive and comprehensive mathematics program for their young learners. When WMAS was published in 1998, the Early Learning Center (ELC) in Sheboygan, Wisconsin, reviewed its curriculum and instruction in relationship to the standards. The ELC was one of the first districts in Wisconsin to recognize the unique needs of early childhood and to subsequently provide early childhood education to large numbers of children in a public school setting. Using identified characteristics of early childhood programming (Perry and Duru 2000), the ELC revised its curriculum to provide its students with a solid foundation of early mathematics skills.

Important Ideas in Early Mathematics

Young children are more complex mathematical thinkers than previously thought (Greenes 1999; Ginsburg and Baron 1993). The WMAS, as well as national mathematics standards (Standards 2000 Project, 2000), broadens our view of the important mathematical ideas that young children are capable of exploring and considering. Rather than comprising a collection of unrelated activities, early childhood and preschool mathematics programs need to plan for the cohesive and coherent development of the important ideas of number, geometry, measurement, data, and algebraic relationships (Greenes 1999; Standards 2000 Project 2000). The development of these ideas must be integrated with the processes of learning mathematics. The following summarizes the important ideas for mathematical processes and content that provide children with a solid foundation of early mathematics. These ideas are adapted from Miller and Weber (2000).

Young children are more complex mathematical thinkers than previously thought.

Mathematical Processes

To make connections between mathematics and the real world, young children need experiences using the processes and language of mathematics in real-life contexts. Explorations should grow from children's daily experiences and current interests and stimulate them to make conjectures about observations, to make decisions about what to do and how to do it, to create ways to represent their ideas and findings, and to reflect on and talk about their methods and conclusions (Greenes 1999). These types of explorations engage children in the processes of mathematics—problem solving, reasoning, communication, connections, and representations.

Explorations engage children in the processes of mathematics—problem solving, reasoning, communication, connections, and representations.

Classroom activities that provide inquiry-based experiences and real-life problem solving are essential. Children's mathematical play with manipulative materials such as blocks and the use of various measuring tools develop an experience base necessary for mathematics learning. Children's play also contributes to the development of representational thought (National Association for the Education of Young Children 1990).

Providing opportunities to develop mathematical ideas and learn skills such as counting in a wide variety of contexts helps children understand the breadth and applicability of those ideas and skills. The information gathered through direct experiences can then begin to be internalized and expanded upon by having children talk about what they are doing and why. Discussing numbers, shapes, location, size, symbols, charts, and graphs are ways children begin to see how mathematics can give them a way to describe and ultimately solve real-life problems. For example, children use the processes of mathematics when they collect data and then discuss whether more children like green apples or red apples and why they think these results occurred.

Children's familiarity with quantities and counting begins as they start to use numbers to group, measure, and estimate.

Number Operations and Relationships

Understanding number and operations requires much more than verbal counting. It also includes the ability to determine the total number of objects in a set and reason about that numerosity using number relationships (Payne and Huinker 1993). Children's familiarity with quantities and counting begins as they start to use numbers to group, measure, and estimate.

Children at a very young age can learn to see groups of twos and threes and begin reasoning with these quantities. This ability, called *subitizing,* is the basis for enabling children to think in groups and is the foundation for learning number relationships. Children need encouragement and numerous opportunities, such as use of dot patterns, to subitize quantities (Clements 1999). This leads to a part whole understanding of a number in which children can think of a number as both a whole amount and as being comprised of smaller groups or parts. The construction of a part whole concept, the understanding of how a whole is related to its parts, is an enormously important achievement (Baroody 2000). This allows children to understand that six can be comprised of five and one more or be a combination of three and three, thus forming the foundation for learning basic number combinations and for understanding the operations.

Other distinctions children at this young age are beginning to grasp include understanding that the number of objects is not related to the physical characteristics of the objects or that the order in which the objects are counted does not affect the quantity. Numerous opportunities in working with everyday experiences and objects having varying characteristics such as shape, size, and color characteristics are needed. For example, bags with the same number of objects in each but with different mixes of objects can be provided for the children to group and quantify. Children at a young age should also be encouraged to tell quantitative stories about their everyday experiences or about the objects in the bags. These stories can be acted out with objects and discussed allowing children to further develop understanding of operation relationships.

Geometry

Continual exploration of the objects in their environment is characteristic of young children. The natural curiosity of three- and four-year-olds automatically exposes them to a variety of two- and three-dimensional shapes as they look more closely at the physical world in which they live. Exposure and direct experience in working with shapes and examining their properties can occur through play and more structured activities. Puzzles, card games, block printing, shape walks, matching games, and shape gardens are just a few of the many activities that engage children in exploring shapes.

Children should be asked to compare, sort, and classify two- and three-dimensional shapes and then discuss their observations. These types of experiences prompt children to focus on the properties of shapes. Children should also discuss uses of shapes in their lives. For example, why do they think cans are in the shape of cylinders and not cubes? As young children talk about their observations, they should be encouraged to use their own informal language to describe the shapes and their properties. Teachers can help children connect their informal language to more conventional terminology. However, terminology itself should not be the focus of the program. The goal for early experiences with geometry is to begin developing spatial reasoning and understanding of the two- and three-dimensional world in which we live.

The goal for early experiences with geometry is to begin developing spatial reasoning and understanding of the two- and three-dimensional world in which we live.

Measurement

Children are fascinated with filling various-sized cups with water or determining how many steps it takes to walk across the room. They are interested in how big, how long, and how much. Measurement experiences arise through children's spontaneous interaction with their environment, support the development of other mathematical topics, and are easily connected to other subject areas and real-life experiences.

Children are fascinated with how big, how long, and how much.

Young children begin developing measurement ideas by comparing objects using various attributes such as length, volume, weight, and area. They develop an understanding of attributes by looking at, touching, or comparing, directly or indirectly, two or more objects. In these situations, children use the language of comparative words such as longer, shorter, heavier, and lighter as a basis for describing their measurements. For example, a child might stand two teddy bears next to each other to see which is taller. Numerous informal measurement activities ranging from using balance scales to compare items in the classroom grocery store to comparing the length of apple peels when preparing applesauce need to be provided for children.

Later children examine items to determine how long, how high, how heavy, or how big. They should be given opportunities to figure out their own ways of measuring using a variety of nonstandard units. This leads children to use the process of measurement in which they choose attributes to measure, select appropriate units of measure, cover or balance objects with the units, and finally count the number of units. For example, a child might use paper clips or interlocking cubes to measure the height of the table or of their favorite stuffed animal, or they might use pieces of paper to measure the area of the meeting area.

Statistics and Probability

Collecting, organizing, and interpreting data are ongoing activities in our everyday lives. Children can collect data from a variety of sources in an early childhood environment that involve them in answering real questions. They may want to find out how many children like pepperoni and how many do not like pepperoni on their pizza. They may vote to see whether the class will take their bingo game home today or play it one more time tomorrow in school.

Organizing the data involves sorting and grouping it into similar categories.

Organizing the data involves sorting and grouping it into similar categories. Teachers can facilitate this process by providing the children with the opportunity to represent data in different ways. Tally sheets using real-life objects such as shoes or blocks may be one of the first experiences young children have in learning to gain information from charts, graphs, or tables. Using tally marks can come next. Eventually numbers can be used to record data the children have collected. Once the data is organized, children can use it to make decisions, examine it for trends, or predict other events. For example, after determining that more children in their class like red apples, they can predict whether more children in another class will prefer red or green apples.

Algebraic Relationships

The beginnings of algebraic reasoning emerge when young children sort, classify, and order objects.

In this area of mathematics, the early childhood teacher serves as a facilitator in helping children explore their world and see it in the context of mathematics. The beginnings of algebraic reasoning emerge when young children sort, classify, and order objects. The ordering of objects and the noticing of regularity in their world lead to an understanding of patterns. The recognition, comparison, and analysis of patterns requires children to reason inductively, thus preparing them for later work with functions (Greenes 1999; Standards 2000 Project 2000).

Children's work with patterns should involve replication, completion, prediction, extension, description, and generalization. Beyond the more common patterns involving color, shape, number, and texture, children's explorations should also include movement, rhythm, song, literature, and pitch patterns. Exploring patterns in different contexts with different stimuli helps children build robust understanding of the concepts related to patterns. Teachers should help children form generalizations about patterns by asking them to describe and compare observed patterns.

Conclusions

Teachers need to draw on their knowledge of mathematics content and processes and be responsive to the context of children's experiences, interests, and ideas.

As teachers plan and implement a coherent early childhood mathematics program, they need to draw on their knowledge of mathematics content and processes and be responsive to the context of children's experiences, interests, and ideas (Bredekamp and Copple 1997). Nurturing a young child's appreciation and acquisition of mathematics is not the exclusive job of the school. Family support and encouragement are also critical in the process. Having discussions among family members about ways mathematics is used at home, helping children communicate mathematically, and most importantly, conveying a positive attitude toward mathematics are three key ways families can support mathematics development in their young children. The following are some specific suggestions that can be shared with parents on how they can be involved and supportive of their child's mathematical development (adapted from Mokros 1996).

- Do mathematics with your child every day.
- Let your child hear and see you doing mathematics.
- Point out to your child when you are using mathematics in your everyday life at home and at work.
- Explore mathematics in the books you read with your children.
- Let your children lead the process of exploring mathematical ideas.
- Ask your child to explain their reasoning and ideas to you.

Learning to use mathematics to solve problems should be a lifelong process in today's world.

Learning to use mathematics to solve problems should be a lifelong process in today's world. Providing a solid foundation of support at school and at home offers children in their early years a good start on this journey.

Learning Mathematics in the Elementary Grades

Mathematics programs in the elementary grades must build upon the intuitive mathematical knowledge that children bring with them to school. Children's informal knowledge is then expanded and formalized with increasing sophistication. Throughout the elementary grades, the focus of teaching and learning must be on helping children make sense of mathematical ideas. The development of a positive disposition toward mathematics and confidence in themselves as learners of mathematics is established in these grades through a foundation of understanding and mathematical inquiry.

The focus of teaching and learning must be on helping children make sense of mathematical ideas.

Children develop beliefs about the nature of mathematics and mathematics learning from their experiences in school. What is mathematics? What does it mean to know and do mathematics? All children need to see themselves as young mathematicians capable of exploring and making sense of their mathematical world. Mathematicians (1) look for mathematics in the world around them; (2) ask questions, pose problems, and make conjectures; (3) think about number and shape; (4) want to know why; (5) work together to investigate mathematics; and (6) use technology as a tool to explore mathematics (Banchoff 2000). Teachers can stimulate and nurture these characteristics of mathematicians in all their students.

Children are active, resourceful individuals who construct, modify, and integrate ideas through interactions in their physical and social world. They need to be challenged with intellectually stimulating and worthwhile tasks and need to explore these tasks in collaboration with peers. Mathematical programs should foster a sense of inquiry and exploration of mathematical ideas and problems. In order to foster such a learning environment, teachers should ask questions such as the following (National Council of Teachers of Mathematics 1991; Sheffield 2000):

Mathematical programs should foster a sense of inquiry and exploration of mathematical ideas and problems.

- What do others think about what she said? Do you agree? Disagree?
- Can you convince the rest of us that your solution makes sense?
- Why do you think that is true? How did you reach that conclusion?
- Does that always work? Can you think of a counterexample?
- What would happen if . . . ? What are some possibilities here?
- Can you predict the next one? What about the last one?
- How did you think about the problem?
- What decision do you think he should make? Why?
- What is alike and what is different about your method of solution and John's strategy?
- What ideas that we have learned before were useful in solving this problem?

In the elementary grades, a coherent and comprehensive mathematics program includes more than just numbers and operations; it is critical to ensure emphasis is also given to the development of mathematical processes and the content of geometry, measurement, data analysis and probability, and algebraic reasoning. Children need a deeper and broader understanding of

A coherent and comprehensive mathematics program includes more than just numbers and operations.

mathematics from the very beginning of their education. The goal of elementary mathematics programs must be to establish a solid foundation of mathematical understanding as understanding is the basis for all learning.

Learning Mathematical Processes in the Elementary Grades

Helping children make connections between the real world and the mathematics learned in school makes mathematics a set of related ideas.

In striving to make sense of their world, young children use reasoning, make connections, and communicate their ideas with others. In developing reasoning abilities, children engage in activities that allow them to develop clear thinking by making and investigating conjectures. Young children use a variety of forms to communicate their understanding of mathematics and strategies for solving problems. The use of words, numbers, models, pictures, and diagrams allows students to clarify and deepen their understanding of mathematical ideas and allows their ideas to be more easily discussed with others. Helping children make connections between the real world and the mathematics learned in school makes mathematics a set of related ideas. These mathematical processes are the tools that are the basis for all learning of mathematics.

National Council of Teachers of Mathematics Process Standards, *Principles & Standards for School Mathematics* (Standards 2000 Project 2000)

The NCTM Process Standards, PK–12, are offered as five strands: problem solving, reasoning and proof, communication, connections, and representation. (See introduction in this section, p. 89.) WMAS (Wisconsin Department of Public Instruction 1998, 4–5) enfolds the processes into a single standard that encompasses using reasoning abilities, communicating mathematical ideas, connecting mathematical learning with other experiences, using mathematical language with understanding, and explaining and supporting mathematical conclusions.

Wisconsin's Model Academic Standards for Mathematics, Standard A: Mathematical Processes

Students in Wisconsin will draw on a broad body of mathematical knowledge and apply a variety of mathematical skills and strategies, including reasoning, oral and written communication, and the use of appropriate technology, when solving mathematical, real-world and nonroutine problems. (Wisconsin Department of Public Instruction 1998, 4–5).

TABLE 5.1 **Wisconsin's Model Academic Standards for Mathematical Processes**

Wisconsin performance standards By the end of grade 4 students will:	Elaboration
A.4.1 Use reasoning abilities to: ■ perceive patterns, ■ identify relationships, ■ formulate questions for further exploration, ■ justify strategies, and ■ test reasonableness of results.	Elementary mathematics experiences need to capitalize on the informal learning about mathematics that young students bring with them to school. Subsequent investigations can be developed so students work with augmenting that knowledge through patterning, identifying attributes, and relating present experiences to previously investigated ideas. As students build understanding from subsequent mathematical investigations, they should be encouraged to continue to test their knowledge by checking reasonableness, telling why, and formulating extension questions.
A.4.2 Communicate mathematical ideas in a variety of ways, including words, numbers, symbols, pictures, charts, graphs, tables, diagrams, and models.	Students come to recognize multiple representations as "mathematics" and as ways to show and explain mathematics as well as ways to clarify thinking about mathematics.
A.4.3 Connect mathematical learning with other subjects, personal experiences, current events, and personal interests: ■ see relationships between various kinds of problems and actual events, and ■ use mathematics as a way to understand other areas of the curriculum (e.g., measurement in science, map skills in social studies).	Conceptual understanding is facilitated through connections with meaningful contexts and situations. Transfer is strengthened when connections can be made between and among concepts. When children view mathematics as a tool for use in many contexts, learning is augmented.
A.4.4 Use appropriate mathematical vocabulary, symbols, and notation with understanding based on prior conceptual work.	Mathematics is a language. Children need to speak and listen to the language and to read and write it.
A.4.5 Explain solutions to problems clearly and logically in oral and written work and support solutions with evidence.	Explaining solutions helps children clarify their personal thinking. Communicating solutions and justifying them helps other students gain access to additional ways of thinking about mathematics.

Additional Standards Information

Additional support information can be found in the NCTM *Curriculum and Evaluation Standards*, pp. 23–35, and in the NCTM PSSM, 116–141 and 182–209.

Illustrative Task

Three-Cube Towers

Build as many different three-cube towers as you can when you have two different colors of cubes to work with.

Then write a letter to Jaime, who is absent today. The letter should describe what you did in class today with the cubes. Be sure to describe all the different three-cube towers that you built when you had two colors of cubes to work with. Then explain why you are sure that you made every possible tower and had not left any out.

Source: Mathematical Sciences Education Board, *Measuring Up: Prototypes for Mathematics Assessment* (Washington, DC: National Academy Press, 1993).

Discussion of the Task

This task was investigated by a class of third grade students. Each pair of students was given two colors of interlocking cubes. The teacher began by having the students work with their partners to build two-cube towers with two colors. They were to try to build as many different towers as possible. As the teacher interacted with small groups of students, it was apparent that a dilemma had emerged for the students. Some pairs said there were three different two-cube towers; others found four different towers, and some could not decide.

Using orange and green cubes for the whole-class discussion, the class agreed on two of the towers, either both orange or both green. The students were not sure whether the tower with green on the bottom and orange on the top should be considered the same as the tower with orange on the bottom and green on top. Some students argued that because it could just be flipped around, it should count as the same tower. Others argued that because they were to build towers that have a top and a bottom, they should be considered different. The teacher asked the students to discuss the two views further in their pairs and to formulate an explanation of why they supported one view or the other.

Most, but not all, pairs agreed that they should be considered different arrangements after further discussion. Several pairs presented their reasoning. The class voted, and it was nearly unanimous that the green top–orange bottom and orange top–green bottom should be counted as two different two-color towers. Those who did not vote for this position did agree that they would consider them different for now. However, the teacher did encourage these students and the others to view the problem from both perspectives supporting their intuitive notions of permutations and combinations. "If the order or orientation matters and we don't allow them to be flipped, then we think the cubes can be arranged to make four different two-color towers. If we allow flips, then we think we are only able to make three different towers."

Next the teacher asked, "Let's say we don't allow flips. Are you really sure you found all thc ways? How do you know?" In pairs the students further examined their arrangements and tried to make additional towers. They also tried to explain how they knew they had all the towers. Several of the pairs then shared their reasoning with the whole class.

The teacher perceived that the class was ready to move on to a more challenging task. She presented the three-cube tower problem. Each pair was to build as many different three-cube towers as possible using two colors of cubes. Then each pair wrote a letter which included a description of what they did, of all the different towers they found, and why they thought they knew they had found all the towers. Two of the student letters are shown in figures 5.7 and 5.8. (See p. 124.)

Standards Addressed

A.4.1 Use reasoning abilities.

A.4.2 Communicate mathematical ideas in a variety of ways.

A.4.4 Use appropriate mathematical vocabulary, symbols, and notation with understanding based on prior conceptual work.

A.4.5 Explain solutions to problems clearly and logically in oral and written work and support solutions with evidence.

Connections to Other Standards

This problem addresses the following standards: number operations and relationships (B.4.2), geometry (C.4.2), and statistics and probability (E.4.4).

Extension Questions

Reading *A Three Hat Day* by Laura Geringer (Harper Collins 1985) would provide a literature connection. Rather than building towers, the students could find all the different ways that R.R. Pottle could wear his three hats. Then the students could be asked to consider the following question: Let's pretend that R.R. Pottle bought the sombrero that he was trying on in the hat shop. If he always wore the sombrero as the top hat, on top of the other three hats, how many different ways would he now be able to arrange his hats?

Older students could also be asked to find all the ways to make four-cube towers or larger towers. The students could also look for patterns in the number of arrangements of various sized-towers. For example, how many ways are there to build a one-cube tower? A two-cube tower? A three-cube tower? A four-cube tower? And so on.

Today in math class, we built towers of 3 blocks using 2 different colors. We found 8 different ways to combine them. Here are the ways:

■ = Blue
□ = Brown

We are sure we found them all because using 2 different colored blocks the tower can be made up of one color, or a combination of colors. There are only 2 different one-colored towers, all blue and all brown. The two-colored towers consist of 2 blocks of one color and a single block of the other color. The single block can only be on the bottom row, the middle row, or the top row, and there are 2 towers of each of these combinations, toatalling 6 two-colored towers. This equals a total of 8 different towers.

FIGURE 5.7

Today in math class I made towers of three counters. We had two colors making diffrent patterns trying not to repeat. I found eight diffrent ways and this is how I found them

First I started with three blue ones, then two with one green one and then one with two green ones after that all greens, then two greens with one blue, after that one green two blues, then one blue, green and blue then green blue green

P.S. I'm sure that I found them all because thats all I can find. Can you find more?

FIGURE 5.8

Other Sample Tasks

Sour Ball Guesses

Your teacher has a glass container filled with sour ball candies. The student who makes an estimate that is closest to the actual number of pieces of candy wins the container and its contents. You may ask questions to get information to help make an estimate. Here are the rules.

- No more than 5 questions.
- You cannot ask how many candies there are altogether.
- At most 2 questions that have a specific number in them (Example: "Is the number between 40 and 60?").

List your questions. Then explain how you will use the answers to your questions to make your estimate. Give examples.

Source: Wisconsin Department of Public Instruction 1995.

Arranging 15 Objects

Show all the ways that 15 objects can be put into four piles so that each pile has a different number of objects in it. How many different ways did you find? Do you think you found them all? Why or why not? Describe the strategy you used.

Source: C. Storey. "How do you do it?" *Wisconsin Teacher of Mathematics*, 45 (3): 18–19.

How Many Wheels?

The children in Johnsonville visited the local park and noticed that there were 15 wheels at the bike rack. There are bicycles (with 2 wheels) and tricycles (with 3 wheels). How many bicycles and tricycles could there be?

This is an example of a problem with multiple solutions. Children can be given manipulatives or pictures of bicycles and tricycles in order to solve this problem. Variations of the problem might involve increasing the number of wheels or adding other things with wheels, such as a wagon or unicycle.

Extension: Create a data table to record the data. Look for patterns and relationships between the total number of wheels and the "vehicle." Discuss the effect of the number of wheels on a tricycle on the total number of wheels. Students should notice that an odd number of tricycles is needed when the number of wheels is odd.

Mystery Number

Using number cards 0–9, students try to determine the mystery number after carefully listening to three clues.

1. The number is between 0 and 9 (eliminate 0 and 9).
2. The number times 3 is not an even number (eliminate 2, 4, 6, and 8).
3. Thirty-five is not a multiple of this number (eliminate 1, 5, and 7).
   (Answer: 3)

This activity can be used to assess understanding of concepts such as even and odd, multiples and factors, and so on. Variations of the activity might involve using a hundreds chart and creating clues for any number from 1 to 100. Students can also be encouraged to write their own "Mystery Number" clues.

Learning Number Operations and Relationships in the Elementary Grades

Number sense includes knowing when to find an exact answer and when an estimate is appropriate as well as selecting and using appropriate computational procdures.

As children begin school, they bring their everyday experiences with counting and numbers. During the elementary grades, this informal knowledge is built upon and extended as students develop understanding of whole numbers, fractions, and decimals. The development of number sense or children's ability to use numbers flexibly and meaningfully and with confidence grows throughout the elementary grades. Children need many opportunities to use physical materials and pictorial representations to give meaning to symbols, to examine number relationships, and to explore operations with whole numbers, fractions, and decimals. Number sense includes knowing when to find an exact answer and when an estimate is appropriate as well as selecting and using appropriate computational procedures. Children need to develop a variety of strategies for use in problem-solving situations, including mental math, algorithms, and using a calculator. When developing algorithms, the goal is for children to obtain computational fluency by having efficient and accurate methods for computing that they understand, can explain to others, and can use in problem-solving situations.

National Council of Teachers of Mathematics Number and Operations Standard, *Principles and Standards for School Mathematics* (Standards 2000 Project 2000)

Instructional programs from prekindergarten through Grade 12 should enable all students to:

- Understand numbers, ways of representing numbers, relationships among numbers, and number systems.
- Understand meanings of operations and how they relate to one another.
- Compute fluently and make reasonable estimates.

Wisconsin's Model Academic Standards for Mathematics, Standard B: Number Operations

Students in Wisconsin will use numbers effectively for various purposes, such as counting, measuring, estimating, and problem solving. (Wisconsin Department of Public Instruction 1998, 6–7.)

TABLE 5.2 **Wisconsin's Model Academic Standards for Number Operations and Relationships**

Wisconsin performance standards **By the end of grade 4 students will:**	**Elaboration**
B.4.1 Represent and explain whole numbers, decimals, and fractions with: ■ physical materials, ■ number lines and other pictorial models, ■ verbal descriptions, ■ place-value concepts and notation, and ■ symbolic renaming (e.g., 43=40+3=30+13).	This standard gets at the beginning of "number sense." Students should be comfortable with dealing with numbers and simple fractions in a variety of ways. At first, the work should be highly manipulative, including pictures of sets, both grouped and ungrouped. The presentation process goes both ways: (1) given a physical or pictorial representation, tell the number, and (2) give a symbolic number (written or spoken), and create a physical or pictorial representation. In the later elementary years, students should be facile with a variety of symbolic representations.
B.4.2 Determine the number of things in a set by: ■ grouping and counting (e.g., by threes, fives, and hundreds), ■ combining and arranging (e.g., all possible coin combinations amounting to 30 cents), and ■ estimation, including rounding.	This standard describes meaningful counting as a way of telling how many objects are in a set. Counting must progress well beyond a sing-song one-by-one recitation of the string of counting numbers. It is also important to note that in many meaningful problem situations, an estimate or an approximation by rounding of the number of objects is perfectly acceptable.
B.4.3 Read, write, and order whole numbers, simple fractions (e.g., halves, fourths, tenths, unit fractions) and commonly used decimals (monetary units).	The ability to read and write numbers is obviously something students must have to succeed in arithmetic; teaching these skills is also something careful teachers have been doing for a long time. Learning to order whole numbers comes through familiarity with beginning counting and, later on, competence with and understanding of place value. Fractions and decimals, on the other hand, take work. Although nothing is specifically mentioned in the text of this standard about written expressions of order of numbers, students should certainly begin to use the standard symbols associated with these relationships.
B.4.4 Identify and represent equivalent fractions for halves, fourths, eighths, tenths, and sixteenths.	Proficiency with this standard is expected only in the realm of physical and pictorial representations. By no means should teachers attempt to teach the symbolic, algorithmic procedures for generating equivalent fractions.
B.4.5 In problem-solving situations involving whole numbers, select and efficiently use appropriate computational procedures such as: ■ recalling the basic facts of addition, subtraction, multiplication, and division, ■ using mental math (e.g., 37 + 25, 40 × 7), ■ estimation, ■ selecting and applying algorithms for addition, subtraction, multiplication, and division, and ■ using a calculator.	This standard deals directly with whole number computation. Its meaning and intent should not be misconstrued. The expectation is that the study of *all* of whole number arithmetic, including algorithms for all four operations with any size number, should be completed by the end of fourth grade and that students should be able to use all of the different methods listed. When computational algorithms are contemplated, less efficient ones, or ones of a student's own invention, are perfectly acceptable.

TABLE 5.2 **(Continued)**

Wisconsin performance standards

By the end of grade 4 students will:		**Elaboration**
B.4.6	Add and subtract fractions with like denominators.	These two standards can be discussed together. Neither contains the magnitude or import of standard B.4.5. The intent is that students should begin to work with computation with fractions and decimals. Both suggest only simple situations. Both do not necessarily imply use of abstract algorithms. At this point, use of manipulatives or pictures is perfectly acceptable, although the relative simplicity of content would suggest that many students could perform well without such aids.
B.4.7	In problem-solving situations involving money, add and subtract decimals.	

Additional Standards Information

Additional support information can be found in the NCTM *Curriculum and Evaluation Standards,* 36–47 and 57–59, and in the NCTM PSSM, 78–88 and 148–156.

Illustrative Task

Baskets of Cookies

The fourth graders at Franklin Elementary made baskets of cookies for the kindergarten, first, second, and third grade classes. They made 12 baskets with 24 cookies in each basket. How many cookies would the fourth graders need to make? Show two different ways to solve this problem.

Discussion of the Task

This task was given to fourth grade students early in the school year. It was used as an assessment tool to examine student strategies for solving multidigit multiplication problems prior to formal instruction of multiplication algorithms. The students had previous instruction focusing on conceptual understanding of multiplication. The teacher wanted to assess what students knew about multiplication and what they would do with a situation that involved larger numbers than those with which they had been working in class. The teacher used the information from this task to guide instruction. It allowed her to find out where her students were and where she needed to go next in linking students' current understandings and strategies to more formal work. Specifically, she wondered:

- Are students able to extend their conceptual understanding of multiplication to situations with larger numbers?
- Can students build upon their informal strategies and prior knowledge to solve a contextual multiplication problem?
- Can students solve the problem in two different ways?

Four student work samples are shown in figures 5.9 to 5.12. All four students used repeated addition in different ways to solve the problem. Tamara wrote a column of twelve 24s and added them up. John added a column of 24s, but then as a second method he grouped pairs of 24s together to get six

48s and then added these together. Alicia successively added 12s to the cumulative total and then as a second method successively added 24s to the cumulative total. Erin showed several ways to think about and solve the problem. She used repeated addition to solve the problem, as well as more fluent methods, such as using a known fact of 12×12 and then doubling this amount. Some students in the class also used pictures to represent 12 baskets and then drew 24 dots in each basket.

From the responses, the teacher was able to see that almost all students could extend their conceptual understanding and use informal strategies to solve the contextual problem. Many, but not all, students could solve the problem in two ways. Most of the strategies involved repeated addition. From this assessment and from observations of the students, the teacher felt the students were ready to begin developing more fluent strategies for multidigit multiplication. The assessment also helped the teacher identify students who needed more support and those who needed more challenge. Furthermore, the assessment task became a starting point for the class to share strategies and discuss the need for more fluent methods when working with larger numbers.

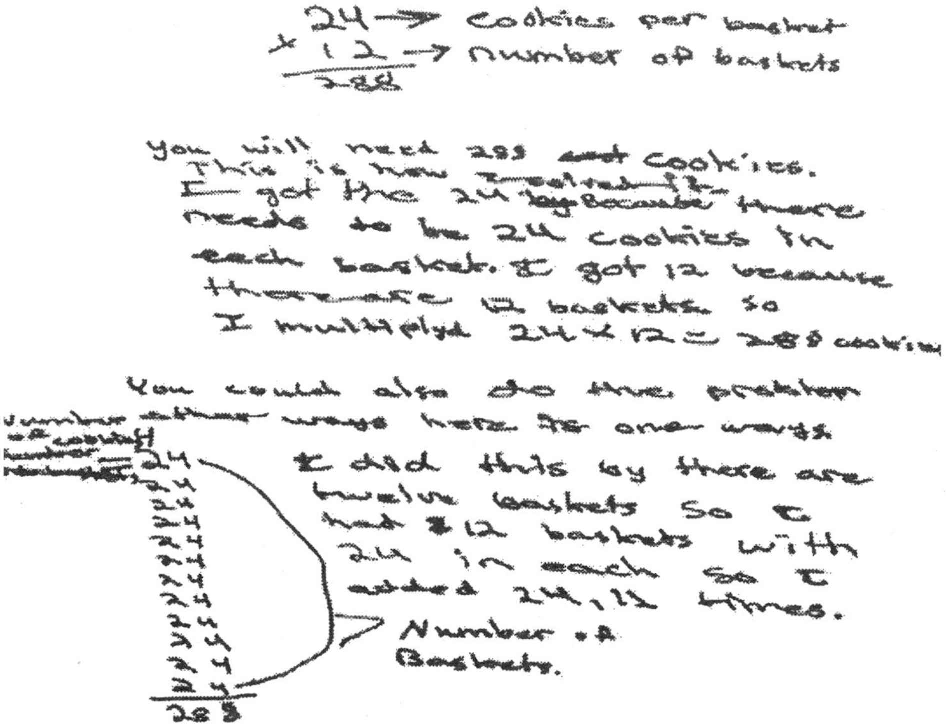

FIGURE 5.9 **Tamara's Work**

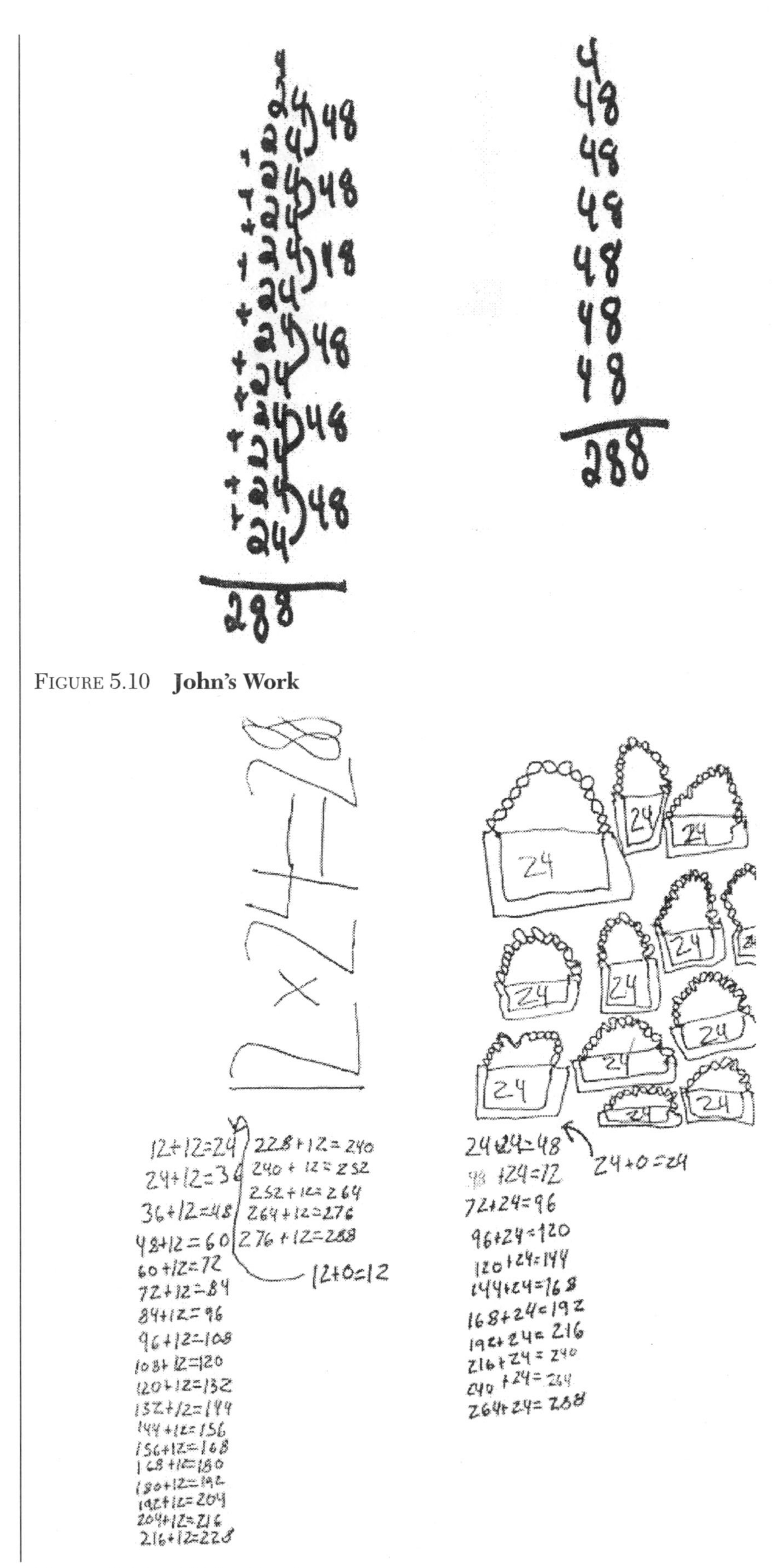

FIGURE 5.10 **John's Work**

FIGURE 5.11 **Alicia's Work**

12 × 12 = 144 I cut 24 in half so I could do 12×12. That equal
× 2 144. The I dubble that. Finally I got 288.
288

Add 24 12's.
12+12+12+12+12+12+12+12+12+12+12+12+12+12+12+12+12+12+12+12+12+12+12+12

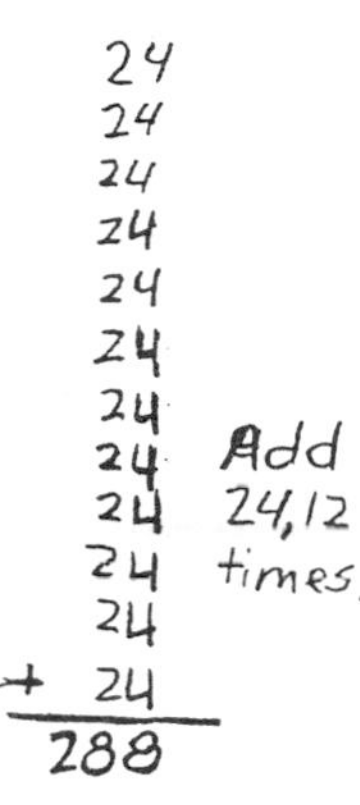

FIGURE 5.12 **Erin's Work**

Standards Addressed

B.4.2 Determine the number of things in a set.

B.4.5 In problem-solving situations involving whole numbers, select and efficiently use appropriate computational procedures.

Connections to Other Standards

This problem addresses the following standards: mathematical processes (A.4.2, A.4.3, A.4.5) and algebraic relationships (F.4.6).

Extension Questions

The students could also be asked to estimate the number of cookies needed to fill the baskets. A similar task can be used later in the school year to assess students' growth in using more fluent strategies. The task could be altered by changing the context or changing the numbers.

In the earlier grades, smaller numbers can be used or the situation can be altered to examine students' understanding of and strategies for addition or subtraction. For older students, the situation could be altered to examine students' understanding of and strategies for division.

Other Sample Tasks

What Do I Do All Day?

Have students keep track of their activities for one weekday. They should determine what part of the usual weekday they spend on various activities such as school, eating, sleeping, watching TV, playing outside, playing on the computer, and homework. As a class you might decide on some of the categories and then let students select the others. They should make a chart that lists the activity, the number of hours, and the fraction of the day for that particular activity. Students could also keep track of their activities for one day on a weekend and make comparisons.

Source: M. Burns. *The Good Time Math Event Book* (Sunnyvale, CA: Creative Publications, 1977).

No Change

A shop has a notice that reads: No change given here. What amounts can be paid using these four coins: a quarter, a dime, a nickel, and a penny?

Source: K. Wood. "Who's Thinking in Your Classroom?" *Wisconsin Teacher of Mathematics* 45 (3): 16–17.

Becca's Dad's Problem

Becca was doing her math homework one night when she called her dad into the room. "Dad, do you think 36×8 is the same as 48×6?" Before her dad even had time to think, Becca said, "Well, it is." And she showed her dad her work. She asked her dad, "Why does this work? Are there other numbers that work this way?" Help Becca's dad answer these questions for his daughter. Explain why the first pair of factors have the same product. Then find two other pairs of numbers that work this way.

Source: D. Huinker. "Performance Assessment Tasks," *Wisconsin Teacher of Mathematics* 46(1): 32–33.

Calendar

Activities with the calendar can nurture and develop number operations and relationships at all elementary levels with varying degrees of complexity. Students can do simple counting and skip counting or look for patterns and relationships. They can be asked to predict what the date will be one week from today or two weeks from yesterday. Increasing the difficulty, students can determine how many days remain before a vacation by counting up from the present date or counting back from the vacation date, showing the inverse relationships of addition and subtraction. Students enjoy making equations for the current date. The possibilities are endless.

Learning Geometry in the Elementary Grades

Children's interest in the world around them is the start of their learning of geometry. From observations of their environment, children can identify two- and three-dimensional shapes and begin examining their attributes and properties. Throughout the elementary grades, children's mathematical vocabulary is expanding as they hear and use these terms in the context of their investigations and discoveries. They learn to recognize and describe figures by their characteristics and use relationships among figures to solve geometric problems. Spatial reasoning ability develops as children construct shapes with physical materials, draw pictures and diagrams of shapes, and visualize shapes, relationships, and geometric motions or transformations. Children can also begin to formulate reasons and mathematical arguments about geometric relationships. These early experiences and explorations with geometry lay the foundation for more formal work with geometry in the later grades.

Children's interest in the world around them is the start of their learning of geometry.

National Council of Teachers of Mathematics Geometry Standard, *Principles and Standards for School Mathematics* (Standards 2000 Project 2000)

Instructional programs from prekindergarten through Grade 12 should enable all students to:

- Analyze characteristics and properties of two- and three-dimensional geometric shapes and develop mathematical arguments about geometric relationships.
- Specify locations and describe spatial relationships using coordinate geometry and other representational systems.
- Apply transformations and use symmetry to analyze mathematical situations.
- Use visualization, spatial reasoning, and geometric modeling to solve problems.

Wisconsin's Model Academic Standards for Mathematics, Standard C: Geometry

Students in Wisconsin will be able to use geometric concepts, relationships and procedures to interpret, represent, and solve problems. (Wisconsin Department of Public Instruction 1998, 8–9).

TABLE 5.3 **Wisconsin's Model Academic Standard for Geometry**

Wisconsin performance standards	
By the end of grade 4 students will:	**Elaboration**
C.4.1 Describe two- and three-dimensional figures (e.g., circles, polygons, trapezoids, prisms, and spheres) by: ■ naming them, ■ comparing, sorting, and classifying them, ■ drawing and constructing physical models to specifications, ■ identifying their properties (e.g., number of sides or faces, two- or three-dimensionality, equal sides, and number of right angles), ■ predicting the results of combining or subdividing two-dimensional figures, and ■ explaining how these figures are related to objects in the environment.	This standard essentially aims at getting students familiar with a wide range of geometric figures. The various bulleted substatements characterize the many different ways in which students can describe their nascent understanding of these figures and their properties.
C.4.2 Use physical materials and motion geometry (such as slides, flips, and turns) to identify properties and relationships, including but not limited to: ■ symmetry, ■ congruence, and ■ similarity.	The intent of this standard is to have students physically manipulate models of two-dimensional figures to develop and show their understanding of some basic and important geometric properties and relationships. Formal terminology associated with motion (or transformational) geometry need not be stressed here; what is important is the hands-on introduction to these very important concepts.
C.4.3 identify and use relationships among figures, including but not limited to: ■ location (e.g., between, adjacent to, interior of), ■ position (e.g., parallel, perpendicular), and ■ intersection (of two-dimensional figures).	This standard emphasizes some simple and basic ideas that can easily be ignored by some instructional programs; namely, that where a figure is placed in relation to one or more other figures is an important way of characterizing that figure.
C.4.4 Use simple two-dimensional coordinate systems to find locations on maps and to represent points and simple figures.	Describing the location of an object using a formalized coordinate system is an extremely important skill, not just in mathematics but in many real-life applications (e.g., finding one's seat in a theater or stadium). Development of this skill must begin in the early grades.

Additional Standards Information

Additional support information can be found in the NCTM *Curriculum and Evaluation Standards,* 48–50, and in the NCTM PSSM, 96–101 and 164–169.

Illustrative Task

String Polygons

The purpose of this activity is for students to investigate forming polygons. Each group of students is given a large loop of string and they use their hands to serve as vertices of polygons. The students explore the effect of changing the size of an angle and increasing or decreasing the number of sides while the perimeter is unchanged. Each group makes a list of their observations using words and diagrams.

Discussion of the Task

This class of fourth grade students is experimenting with making polygons and exploring the effects of changing characteristics of shapes. This activity promotes students' understanding of the properties and relationships of polygons, provides an opportunity to further develop mathematical vocabulary in context of their explorations, and develops spatial sense. Through teacher questioning, the investigation pushes students beyond just identifying shapes to an analysis of the characteristics of polygons and to formulating informal proofs or justifications.

T: Today we're going to be exploring polygons. I want you to turn to your neighbor and tell each other how you know when a shape is a polygon.

As the students discussed, the teacher moved about the room listening and getting an idea of students' understanding and use of geometric terms for describing polygons.

T: So, what is a polygon?
S: It's a shape with straight sides.
S: The shape has to be closed.
T: What does it mean to be closed?
S: There can't be any openings.
T: How many sides does a polygon have?
S: Four.
S: No, well, it can have four, but it can have more than four, too.
S: It can have any number of sides.
T: Any number? Can it have one thousand?
S: Yes, but it would have to be really big.
S: Well, it could be small, but it would be hard to really see the tiny sides.
T: So, let's see. A polygon is a closed shape with straight sides, and it can have any number of sides.
S: It can't have one side.
T: That's interesting. What do you mean it can't have one side?
S: It would just be a straight line, and that wouldn't be a polygon. It can't have two sides either.
T: It can't? What do others think about this? Can a polygon have two sides? Talk to your neighbor and then draw a polygon with two sides on your paper.

The students engaged in a lively discussion. They attempted to draw a polygon with two sides. Some commented that it was impossible, and the teacher prompted them to try to explain why they thought it was impossible. Others were still not sure.

T: I want you to think more about how many sides a polygon can have throughout today's math class, and we'll talk more about it later.

The teacher wrote the following questions on the board: How many sides can a polygon have? Can a polygon have two sides? Why or why not?

T: Today we're going to explore polygons. I'm going to give each group a large loop of string to use to make polygons. I need three people to help me show you what we're going to do today. (Three students came to the front of the room. The teacher had each student grab onto the string with one hand and then step back until the string was tight and formed a polygon.) What shape did they make, and how do you know?
S: It's a triangle because it has three sides.
T: What else do you notice about the triangle?
S: It has one short side and two longer sides.
T: So you noticed something about the length of the sides. What other things do you notice?
S: That one angle looks like it might be a right angle.
T: So you noticed something about the angles. I wonder what kind of triangle this is? We're not going to answer this right now, but in your small groups you will be exploring triangles and other polygons. Use your hands to make the vertices and then pull the string tight to make the sides just as we did here. Form four different polygons, record the name of the shape, make a sketch of it, and then describe how you made the shape. On the table are some meter sticks if you would like to use them. If there are other tools you need as you work, you may get them from the materials area.

As the students worked in small groups of three or four students, the teacher observed and questioned them. For example, some of the questions asked of students included: Are you sure it's a rectangle? What has to be true for it to be a rectangle? What do you call a big angle like that? How could you decide if that is a right angle? What else can you observe about this shape? How could you change that quadrilateral so that it would be a parallelogram? How could you change that triangle to make a different type of triangle?

T: Let's share some of your observations. What shape did some groups make, and how did you make it?
S: We made a square. Each of us grabbed onto the string with one hand. There were four of us. So that gave us the four corners. Then we moved our hands until all the sides were the same size.
T: I'd like everyone to think about what needs to be true if a polygon is a square. Then talk with your partner about what questions we could ask this group to determine whether or not they really made a square.

As the students talked this over with their neighbor, the teacher asked the group of four students to get up and make a square with their string.

T: What has to be true if this shape is really a square? What questions do you have for this group?
S: How do you know for sure that all the sides are the same length?
S: Well, they look like they are the same length.
T: What suggestions do you have for how we could check to see if the sides are the same length?
S: Use a meter stick.
S: Or they could just match up the string; that's what our group did.
T: We know that a square needs to have four sides that are equal in length, and we have two ideas of how we could check to make sure. What else needs to be true if the polygon is a square?
S: The corners need to be right angles.
T: What's a right angle, and how could you check the angles?
S: A right angle is like a square corner and it's 90 degrees. Our group used our library books and put one at each corner and made sure that the string lined up with the sides of the book.
T: Is there anything else that needs to be true?
S: The sides across from each other have to be like railroad tracks.
T: What do you mean by that?
S: Well, the sides, if they kept on going, would never touch.
T: Does someone remember what we call it when two lines or sides are like this and how could you check?
S: They're parallel, and we could use the meter stick.
T: I'd like each group to use their string to make a square. Think about the ideas we just shared and figure out how to make sure that your polygon is a square.

All the groups worked to make squares. Most of them used meter sticks to measure the sides, but some just matched the sides and made adjustments as needed. All of the groups used books or folders or paper to make right angles. The teacher visited the groups, asking, "How are you sure you have a square?" For the remainder of the class, the students took turns demonstrating a polygon for the class, discussing the characteristics of it and how to check for these characteristics, and then returning to their small groups to make the polygons. Some of the polygons they explored included a right triangle, a parallelogram, and a pentagon. For homework, the students wrote about the questions that had been written on the board earlier regarding the number of sides that a polygon can have.

Standards Addressed

C.4.1 Describe two-dimensional figures.
C.4.2 Use physical materials and motion geometry to identify properties and relationships.

Connections to Other Standards

This problem addresses the following standards: mathematical processes (A.4.1, A.4.2, A.4.4, A.4.5); measurement (D.4.1, D.4.3, D.4.4).

Extension Questions

Put several polygon tiles or pattern blocks in a bag for each group of students. Each group begins by making any type of quadrilateral. Without letting go of the string, someone pulls a shape out of the bag. They must now change or transform their current shape into the new shape trying to preserve as many characteristics or attributes of the previous shape and make only those changes that are necessary. Then they pull out another shape and continue in a similar manner. Another option is to simply call out the shape that all groups should make or the attribute they should change.

In the earlier grades, students might explore making different triangles with a focus on spatial visualization and development of intuitive and informal ideas about shape characteristics. Then on another day they could explore four-sided figures.

Other Sample Tasks

Pattern Block Symmetry

Direct students to create a symmetrical pattern using only red, blue, green, and yellow pattern blocks. The design may have rotational symmetry, mirror symmetry, or both. The design must also fit on a 9-inch-by-12-inch sheet of construction paper. When students have completed making their design, have them transfer it to triangle paper and color it. Finally, students write a paragraph that describes their design. The descriptions must contain discussion of the kind of symmetry used and justification or proof that it is that type. It should also describe the design in enough detail that a reader would be able to match the paragraph to the design.

Source: L. Williams. "Descriptions by Design," *Wisconsin Teacher of Mathematics* 49(3): 16–17.

The Four-Triangle Problem

Give each group of students two different colors of 3-inch squares. They will need about 20 squares of each color. The students fold and cut each square in half along a diagonal. Using these right triangles, each group of students tries to find all the possible arrangements when using four triangles to make each shape. They should tape together each arrangement they find. In each arrangement, each triangle must touch the side of at least one other triangle and follow the rule that sides that touch must be the same length and match exactly. Students should be prepared to discuss why they think they have found all the arrangements.

Source: C. Rectanus. *Math by All Means: Geometry Grades 3–5* (Sausalito, CA: Math Solutions Publications, 1994).

Geoboard Simon Says

Without showing the students, describe how you are putting a geoband on a geoboard. Students should follow and do the same on their geoboards. For younger children, start with simple paths such as designs, letters, numerals, or combinations of shapes. For older children, emphasize geometric properties, such as "I am forming a shape in which the opposite sides are parallel." Students can then make and describe their own designs to each other.

Source: Bruni, J., & Seidenstein, R. (1990). Geometric concepts and spatial sense. In J. N. Payne (Ed.), *Mathematics for the young child* (pp. 202–227). Reston, VA: National Council of Teachers of Mathematics.

Quick Images

Draw some simple two-dimensional designs on transparencies such as a stack of squares or connecting triangles. Flash one of the images for three seconds on an overhead projector. Students are to draw what they saw. The goal is for students to draw the design from their mental image of it, thus enhancing their spatial reasoning. They quickly learn to study the figure carefully and note certain features of it. Then flash the image again for another three seconds and allow students to revise their drawings. Next reveal the image so that all students can complete or revise their drawings. Finally, have students discuss the various ways they saw the images and how they used this to help them make their drawings.

Source: C. Tierney and S. J. Russell. *Ten-Minute Math* (Parsippany, NJ: Dale Seymour Publications, 2001).

Learning Measurement in the Elementary Grades

Measurement is often overlooked as part of the mathematics curriculum, and yet it is a topic that permeates so many other areas of mathematics and of real-world experiences. Young children should be involved with a wide variety of measurement activities with arbitrary and standard units. They should experience informal measurement by looking, guessing, moving, and comparing objects. These types of experiences assist children in understanding and recognizing measurable attributes. Later, children will determine an appropriate unit and use measuring instruments. Students should come to realize the need for standard units and become familiar with metric and U.S. customary units to evaluate different objects, amounts, and situations.

Young children should be involved with a wide variety of measurement activities with arbitrary and standard units and they should experience informal measurement by looking, guessing, moving, and comparing objects.

National Council of Teachers of Mathematics Measurement Standard, *Principles and Standards for School Mathematics* (Standards 2000 Project 2000)

Instructional programs from prekindergarten through Grade 12 should enable all students to:

- Understand measurable attributes of objects and the units, systems, and processes of measurement.
- Apply appropriate techniques, tools, and formulas to determine measurements.

Wisconsin's Model Academic Standards of Mathematics, Standard D: Measurement

Students in Wisconsin will select and use appropriate tools (including technology) and techniques to measure things to a specified degree of accuracy. They will use measurements in problem-solving situations. (Wisconsin Department of Public Instruction 1998, 10–11).

TABLE 5.4 **Wisconsin's Model Academic Standards for Measurement**

Wisconsin performance standards	
By the end of grade 4 students will:	**Elaboration**
D.4.1 Recognize and describe measurable attributes, such as length, liquid capacity, time, weight (mass), temperature, volume, monetary value, and angle size and identify the appropriate units to measure them	This standard gets at the beginning of measurement sense. Students need to recognize that things can be measured in many different ways. For example, a classmate has height, weight, age, number of siblings, and distance traveled to school, to name just a few attributes or characteristics. To measure that classmate's height, one does not use pounds or years; rather, a unit that itself has length is appropriate. Yet it is inappropriate to use miles or even meters to measure the height. Inches, or a combination of feet and inches, makes more sense.
D.4.2 Demonstrate understanding of basic facts, principles, and techniques of measurement, including: ■ appropriate use of arbitrary and standard units (metric and U.S. Customary), ■ appropriate use and conversion of units within a system (such as yards, feet, and inches; kilograms and grams; gallons, quarts, pints, and cups), and ■ judging the reasonableness of an obtained measurement as it relates to prior experience and familiar benchmarks	This is closely related to D.4.1 in that it, too, gets at the development of measurement sense. Of particular importance is the ability to judge reasonableness of measurements. For example, if a student hears "My brother is five and a half inches tall," he or she should recognize the mistake, knowing either that an inch is only as long as part of my thumb or that the speaker probably means five and a half feet because he or she knows that mother is just a little bit taller than five feet.
D.4.3 Read and interpret measuring instruments (e.g., rulers, clocks, thermometers)	A student will learn to use measuring instruments effectively only if given ample opportunity to use them in realistic, hands-on situations. Only looking at pictures in workbooks and text books will not get the job done!

D.4.4	Determine measurements directly by using standard tools to these suggested degrees of accuracy: ■ length to the nearest half-inch or nearest centimeter, ■ weight (mass) to the nearest ounce or nearest five grams, ■ temperature to the nearest five degrees, ■ time to the nearest minute, ■ monetary value to dollars and cents, ■ liquid capacity to the nearest fluid ounce	The degrees of accuracy stated here are, as stated, only suggestions. Teachers should feel free to use their own good judgement in this regard, varying expectations with students of differing levels of manual and visual skills.
D.4.5	Determine measurements by using basic relationships (such as perimeter and area) and approximate measurements by using estimation techniques	This standard marks the beginning of learning the skills and concepts associated with indirect measurement. Many attributes and units of measurement do not lend themselves to the reading of some instrument. Area is a good example. Although, in theory, one could cover a surface with cutouts of square inches (feet, meters, etc.) and then count the number of units used, it simply is not practical to do so. It is also important to recognize that in many practical, real-life situations, an approximate measurement is good enough to solve the problem that gave rise to the need for having a measurement in the first place.

Additional Standards Information

Additional support information can be found in the NCTM *Curriculum and Evaluation Standards,* 51–53, and in the NCTM PSSM, 102–106 and 170–175.

Illustrative Task

Tornado Safety Plan

After the new addition to Eastside School is completed and more students are enrolled, the Tornado Safety Plan will need to be updated. Mr. Brown, the principal, has asked the students in Mrs. K's class to determine how many fourth graders will be able to duck and cover in the back hallway. Your group is responsible for coming up with an answer and final recommendation regarding the Tornado Safety Plan. Be sure to include data, drawings, calculations, and conclusions. Also include a thorough explanation of how you arrived at your recommendation.

Discussion of the Task

The authentic situation for this task arose from the renovation of Eastside School. Mrs. K discussed the reasons for the addition of more rooms to their school and some of the implications for school policies. Then she posed the task to her fourth grade students. The students worked on the task in small groups for four days.

On the first day, each group developed a plan for investigating the situation. In small groups, students discussed these two questions: "What do you need to know?" and "What do you need to do?" Then through whole-class discourse, it was agreed that they needed to measure the length of the hallway and that they would measure using meters. The students also discussed that they needed to take into consideration how much space each student would require to duck and cover. Because the students would be lined up single file next to the wall, they realized they only needed to measure the length and not the area needed by each student to sit on the floor. The students now returned to their small groups to further develop their plans and to begin working.

Some students began by measuring the hallway while others began measuring students. The students realized that different-sized students would require different amounts of space. Someone like Jenny would take up less space than others like Robert. Two approaches emerged. Some groups worked on determining how much space a typical student would require. They faced the dilemma, "How to find the typical student length?" This provided an opportunity to reinforce the use of median as a measure of a typical student. The students first measured their own group members and then measured students from other groups so that they would have a more representative sample. One group decided that a student would require 35 centimeters. This group had measured the hallway in meters. Now they had to resolve their use of different units by converting units within the metric system.

The other groups approached the task by figuring out how many students could fit in a meter. These groups laid a meter stick on the floor and then had students sit next to it. They tried a mixture of larger and smaller students and also checked with other groups to see if they came up with the same number of students. As the students worked, the teacher assessed each group to see if they understood the task, if they had a plan, and if they were ready to carry out their plan.

During the second day, the students reexamined the task and their plans. The teacher realized that many of her students still did not completely understand the problem and all the facets that it entailed. They often think they know what to do, but to fully understand the task, they had to mess around a bit and do some initial measuring. Now the teacher assisted the students in further clarifying aspects of their plan by questioning them about what they were going to do and why. The groups continued their measuring from the previous day. Then the students used their measurements to determine how many students could duck and cover in the hallway and prepared reports of their recommendations.

On the third day, the students presented their recommendations to each other. Several issues emerged. Can a student sit in front of a doorway? Will the fourth graders be able to sit on both sides of the hallway or only on one side? Why did groups have different measurements for the length of the hallway? Can two or three students fit in a meter? How much space should be designated for each fourth grader? How many fourth graders will there be next year? Two of the student written reports are shown in figures 5.13A and 5.13B. The teacher evaluated each report for an understanding of the problem, ability to communicate a plan and a recommendation, and the reasonableness of the recommendation.

On the fourth day, the class examined the issues that had emerged during the presentations. Through further discussion and investigation, the students reached consensus on the issues and prepared a class recommendation to the principal.

Tornado Safety Plan

about 3 kids can fit in one meter

Are goal is

Are goal is to find out how many kids can fit in the hall way. And if the forth grade can fit in the hall way.

There are about 65 kids in the forth grade. Next year there will be one more class. So next year there will be about 21 more kids so 65 + 21 = 86

We just found out that about 80 kids can fit in the hall way becaus there are about 40 meters with 2 rows of kids. Here is our work:

卌 卌 卌 卌 卌 卌 卌 卌

5 10 15 20 25 30 35 = 40

About to meters can fit in the hall way plus we have to − about 3 meters because there is a door at one end of the hall way. So −3 for the other row to so −6 in all. $\begin{array}{r}40\\ -6\\ \hline 34\end{array}$ meters

So now we have 34 meters.

So if we can fit 80 kids when we had 40 meters we have to −6 kids from 80 so $\begin{array}{r}80\\ -6\\ \hline 74\end{array}$ So now we can fit 74 kids.

FIGURE 5.13A

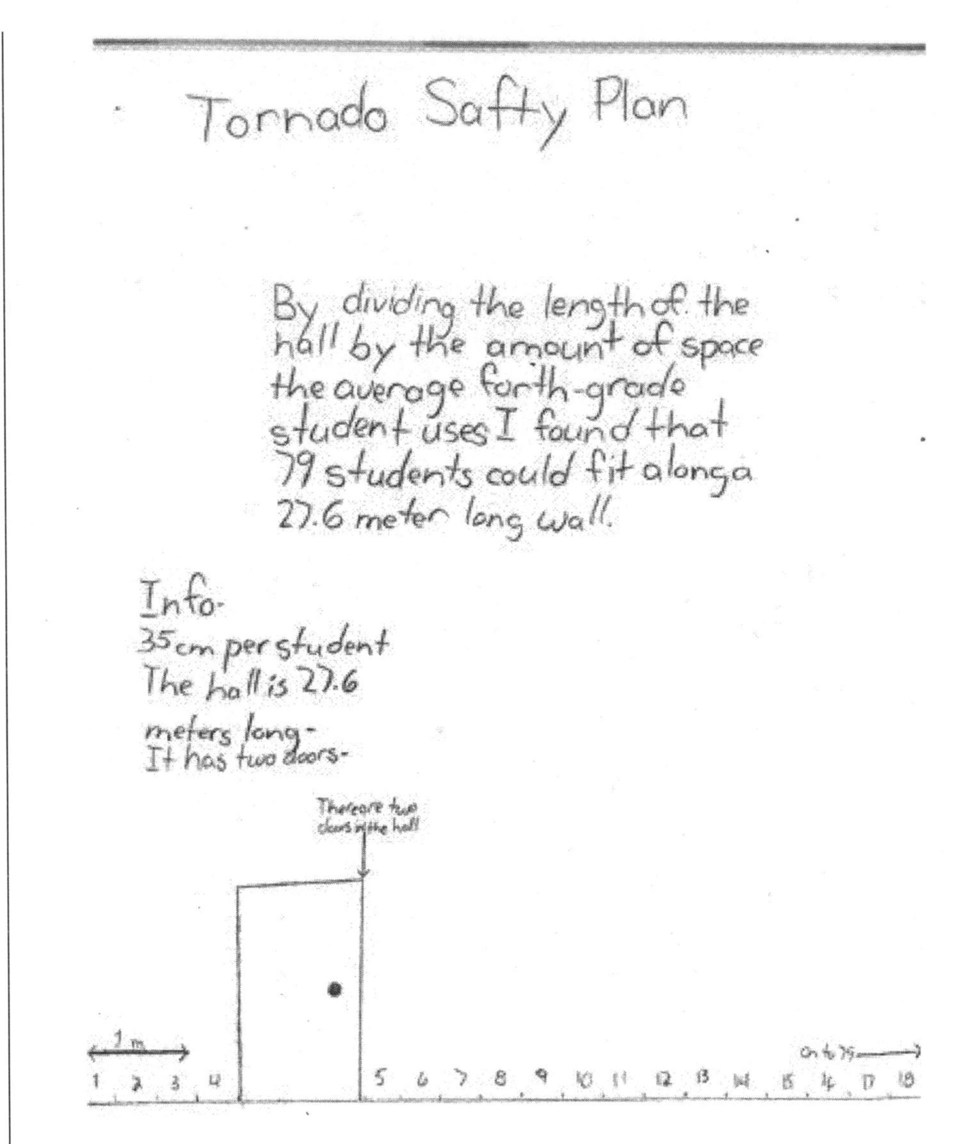

FIGURE 5.13B

Standards Addressed

D.4.1 Recognize and describe measurable attributes.

D.4.2 Demonstrate understanding of basic facts, principles, and techniques of measurement

D.4.3 Read and interpret measuring instruments.

D.4.4 Determine measurements directly by using standard tools.

Connections to Other Standards

This problem addresses the following standards: mathematical processes (A.4.2, A.4.3, A.4.5), number operations and relationships (B.4.5), and statistics and probability (E.4.1, E.4.2).

Extension Questions

A related data investigation would involve measuring the space needed by each student in the class to duck and cover. This data could then be graphed, and the class would determine the typical space needed by a student by finding the median.

Students could also investigate other school and nonschool situations that involve setting capacity limits. For example, what is the capacity of a classroom, the lunchroom, the gym, the auditorium, an elevator, or a meeting room? The students could also research fire codes and how capacity limits are established.

Other Sample Tasks

Shopping with Ads

Using the advertisements from the local newspaper, students can compare prices among different vendors. The goal is to compare quantities and prices to shop for the best deals. The quantities should include liquid measurements, linear measurements, and price offerings such as "39 cents each or 3 for $1.00. What's the better deal?" This activity will take several class periods to complete, but the resulting discussions among students represent wisely spent, high-quality learning time. Students should be encouraged to use calculators to aid in the comparisons and to communicate their thinking and their results in both written and verbal forms. A valuable source of assessment information is having students write out their shopping lists with the "best deals" and the rationale behind choosing one product over another.

Area of My Hand and Foot

Materials: centimeter grid paper and graph paper.

1. Students will trace one hand and one foot on a piece of centimeter grid paper. Then students should count the total area of each, taking into consideration partial squares. The total area should be to the nearest square centimeter.
2. Graph the results on an Area of Hands versus Area of Feet graph.
3. Compare results. If data is collected from a wide range of ages and sizes, a best-fit line can be drawn, and students can analyze the data and make predictions. The question can be posed: "Suppose a new student came into our class, what would you predict that his/her hand and foot area would be?"

A similar activity can be done with height and arm span.

Design a Tile Floor

Students use pattern blocks to design a floor of a six-foot-by-eight-foot room and determine how many blocks of each kind are needed to complete the floor. Use a scale (a one-inch block equals one foot) or design a part of the floor at the actual size. Tiles may have to be cut to fit the edge of the room.

Source: M. M. Lindquist. "The Measurement Standards" *Arithmetic Teacher* 37(2): 22–26.

How Much Taller is a Fourth Grader Than a First Grader?

Students measure their own heights and then the heights of students in a class of the other grade level. Students need to determine an effective way to represent the data from the two classes. This may include data tables, bar graphs, stem-and-leaf plots, and other representations. In this activity, students will consider ideas such as "typical" or "average" height of students and various ways to represent and analyze two sets of data.

Source: S. J. Russell, R. B. Corwin, A. Rubin, and J. Akers. *The Shape of the Data*. (Menlo Park, CA: Dale Seymour Publications, 1998).

Learning Statistics and Probability in the Elementary Grades

Children should formulate questions, collect and represent data, and then determine whether they can reach a reasonable conclusion or need to collect more or different data.

Early experiences are the basis for children's understanding of statistics and probability. They develop concepts for decision making through comparing, sorting, counting, organizing, and creating representations of data. Children come to realize that data can provide them with interesting and important information as they collect and organize data about themselves, their school, their families, and their community. Children should formulate questions, collect and represent data, and then determine whether they can reach a reasonable conclusion or need to collect more or different data. Data investigations in the upper elementary grades should also involve students in the comparison of related data sets. The concept of probability is developed from children's informal experiences in describing everyday events as likely or not likely and from their exploration with experiments. In the upper elementary grades, children learn to quantify and predict the likelihood of future events.

National Council of Teachers of Mathematics Data Analysis and Probability Standards, *Principles and Standards for School Mathematics* (Standards 2000 Project 2000)

Instructional programs from prekindergarten through Grade 12 should enable all students to:

- Formulate questions that can be addressed with data and collect, organize, and display relevant data to answer them.
- Select and use appropriate statistical methods to analyze data.
- Develop and evaluate inferences and predictions that are based on data.
- Understand and apply basic concepts of probability.

Wisconsin's Model Academic Standards of Mathematics, Standard E: Statistics and Probability

Students in Wisconsin will use data collection and analysis, statistics and probability in problem-solving situations, employing technology where appropriate. (Wisconsin Department of Public Instruction 1998, 12–13).

TABLE 5.5 **Wisconsin Model Academic Standards for Statistics and Probability**

Wisconsin performance standards **By the end of grade 4 students will:**	**Elaboration**
E.4.1 Work with data in the context of real-world situations by: ■ formulating questions that lead to data collection and analysis, ■ determining what data to collect and when and how to collect them, ■ collecting, organizing, and displaying data, and ■ drawing reasonable conclusions based on data.	This standard addresses the notion of exploratory data analysis. It cannot be emphasized too strongly that the intent is to work with situations that are real and meaningful to students, in which they, the students, for the most part collect and analyze actual data. Use of "canned" data sets where everything turns out "just right" should be minimized.
E.4.2 Describe a set of data using: ■ high and low values and range, ■ most frequent value (mode), and ■ middle value of a set of ordered data (median).	The expectation is that students learn to use these concepts while using the appropriate terminology.
E.4.3 In problem-solving situations, read, extract, and use information presented in graphs, tables, or charts.	One of the important skills of data analysis is to read and interpret data generated by someone else. Because those data are often summarized in tables, charts, or graphs, students must be comfortable with all of those modes of presentation.
E.4.4 Determine if future events are more, less, or equally likely, impossible, or certain to occur.	Probability is the measurement of a future event where the attribute in question is expected frequency of its occurrence. This standard begins the work of acquiring this fundamental concept.
E.4.5 Predict outcomes of future events and test predictions using data from a variety of sources.	The future events envisioned in this standard are simple ones. Some examples might be tosses of coins or dice, results of a class or school survey, outcomes of a sporting event, or weather conditions. Ideally, the choice of events to study would promote actual data collection by the students, although externally generated data are acceptable. In fact, choice of what data to use in testing of a prediction is an integral part of this standard.

Additional Standards Information

Additional support information can be found in the NCTM *Curriculum and Evaluation Standards,* 54–56, and in the NCTM PSSM, 108–114 and 176–181.

Illustrative Task

Kids' Feet and Adults' Feet

Part 1. After reading, "How Big Is a Foot?" by Rolf Myller, students measure the length of their feet. The foot lengths are listed on the chalkboard. Working with a partner, students organize the numbers on a line plot so that they can answer this question: "What is the typical foot length in our class?"

Part 2. Students measure the foot length of at least one adult outside of school. For each measurement, the person's name, age, and foot length are recorded. As a class, organize the data on a line plot. Discuss what is typical about the foot length of adults.

Part 3. Compare the two line plots. Discuss the similarities and differences between the length of student feet and the length of adult feet.

Source: K. Economopolous, J. Mokros, and S. J. Russell, *From Paces to Feet* (Menlo Park, CA: Dale Seymour Publications, 1998).

T: Yesterday we took off our shoes and measured our feet. Here are all of your foot lengths. In pairs, organize this data in a way that makes sense to you. The question that I want you to be able to answer after you organize your data is "What is typical about our foot lengths?"
S: What does *typical* mean?
T: Does anyone have an idea what this means?
S: I think it means to describe what is interesting about our foot lengths.
T: It means more than just interesting, what else?
S: It's what you notice about a lot of kids in our class.
T: Once your data is organized, describe what you notice is typical or similar about a lot of kids in our class.

Some of the students were unsure of the meaning of typical. *This is a key idea and something the teacher assessed as she interacted with small groups of students while they organized the data.*

T: Does someone have some ideas about how to organize your data?
S: Make a line plot.
T: Would that help you organize your data?
S: Yes!
T: Remember that a line plot is a quick sketch; it does not have to be detailed and beautiful.

The students worked in pairs to make line plots. Some of the measurements included fractional amounts. Thus, students had to decide how to plot the fractions. Once students decided how to mark the intervals on their line and how to plot the fractions, one student read off the data as the other student plotted it. Here is the line plot of the foot lengths of the third grade students.

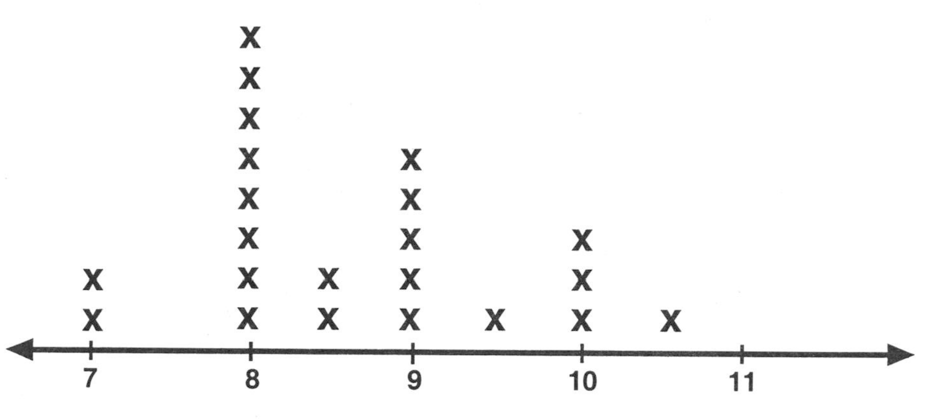

FIGURE 5.14

T: Let's look at our data. Can we answer our question now? What is typical about the foot lengths in our class, or what do you notice about a lot of students in our room?

S: A lot of kids have the same-sized foot. They have 8 or 9.

T: So you noticed that a lot of us have foot lengths of 8 or 9 inches. So are you saying it is typical for kids in our room to have 8- or 9-inch feet? Does anyone want to add more to what she said or say something different?

S: Not everyone has the same length feet.

T: Is there anything unusual about the data? Do you notice any outliers?

S: Seven.

T: Are there people with 7-inch feet?

S: Yes.

T: How many?

S: Two.

T: So you are saying that the two people in the classroom with 7-inch feet don't have typical feet. Why?

S: They are kind of far away from the others.

T: Where is most of the data clumped or clustered? *(This question helped to refocus the students back on thinking about what is typical in the data.)*

S: Around 8 and 9.

T: Can you come up and show us with your hands where most of the data is clumped? *(The student comes up to the board and shows that part of the line plot.)*

T: I'm seeing this bump right here at 8; what can you learn from this bump?

S: Lots of people have 8-inch feet.

The teacher then asked the students to get out their homework. They were each to measure the foot length of at least one adult outside of school. First the teacher led a discussion on how to mark the interval on the line. It was decided to begin at 8 and end at 13. Next they created a class line plot of all the data. As each student reported his or her data, the teacher marked it on the line plot.

T: What do you notice about this data?

S: There are two modes.

T: What do you mean by *modes*?
S: There are two highest numbers, 10 and 11.
T: So those are called *modes* because they have the highest number of people or votes. What do you notice about a lot of the data? What do you notice about the foot lengths of adults?
S: Most of the adults have 10-, 11-, or 12-inch feet.

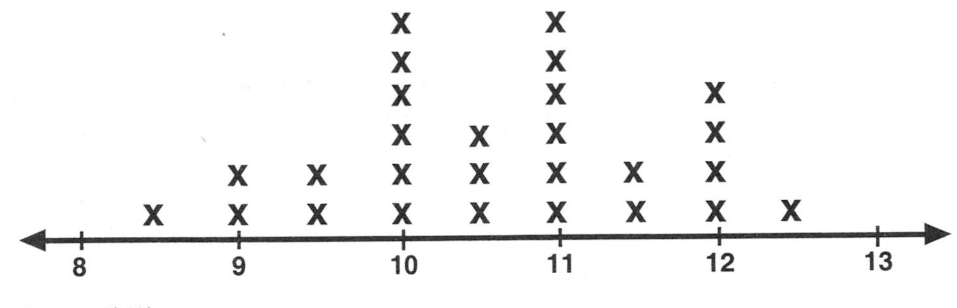

FIGURE 5.15

T: Let's show both line plots, yours and the adults. Make a comparison. Can you find anything different about third grade foot length and adult foot length?
S: In our class 8 has more than the adults have at 10 or 11.
T: What else do you notice about this data?
S: Some of the kids have bigger feet than the grownups.
T: Come up and show us.
S: If you're a 10-inch-foot kid, then you're bigger than these adults (pointing to the data points at 8.5, 9, and 9.5 inches).
T: You noticed that some of the third graders in this class have bigger feet than some of the adults. What else do you notice? Can you tell us about the shape of the data?
S: The mode on the adult is higher than on our graph. The mode for us is a lower number.
T: What else do you notice?
S: When you grow up, your foot lengths are more spread out, and when you are a kid, your foot lengths are all about the same.
T: We've discussed several things about the foot lengths of third graders and of adults. In your math journal, respond to these three questions: (1) What is typical about the foot lengths of third graders in our class? (2) What is typical about the foot lengths of adults? (3) Make three comparisons of the length of student feet with the length of adult feet.

Discussion of the Task

This third grade class is investigating the foot lengths of third graders and comparing this with the foot lengths of adults. The students collect, organize, and display data. Then they discuss the shape and important features of the data and compare related data sets.

In this lesson, the students were analyzing and summarizing data. The teacher challenged the students to capture the essence of the data by asking them to describe what was "typical" about each data set and to compare the data sets. The written task at the end of the lesson allowed the teacher to further assess each individual student's understanding of summarizing and comparing the data sets.

Standards Addressed

E.4.1 Work with data in the context of real-world situations.
E.4.2 Describe a set of data.

Connections to Other Standards

This problem addresses the following standards: mathematical processes (A.4.2, A.4.3), number operations and relationships (B.4.1), and measurement (D4.3, D4.4).

Extension Questions

At the earlier grades, students might trace their footprints on construction paper, cut them out, and then measure the lengths. A pictograph of their foot lengths could then be created with the cutout prints. Another option would be to have students create a real or human graph based on the lengths of their feet.

In the later grades, students could discuss what appears to be typical foot lengths of adults and children and then find the median and mean foot lengths. Then they could use these measures of center to compare the data sets.

Other Sample Tasks

How Long is a Word?

Have students work in pairs to explore the typical lengths of words. Each pair needs a book, such as a library book. They should choose one page in the book. On this page they need to determine how many words have only one letter, how many have two letters, three letters, and so on. They should make a table or graph to keep track of their data. Each pair of students should then analyze their data to determine the typical length of a word. Depending upon the grade level of the students, they might just inspect the data to determine what is typical, or they could find the median. All the data can then be combined to create a class chart or graph, and further analysis can be conducted. The students should also discuss if their findings are generalizable. Is their data biased in anyway? Would this be the typical length of word in any book?

Rolling Totals

Students will explore the concept of chance and make predictions based on probabilities.

Materials: Two 10-sided dice with numbers 0–9 per pair of students (spinners can be substituted), graph paper, and blank paper.

1. What are all of the possible sums from the two dice?
2. What sums do you think will occur most often? Why do you think these sums are most likely?
3. Roll two dice 100 times. Graph the sum of the two numbers as you roll the dice or spin the spinner.
4. Make a table to list all of the possible combinations of numbers for each sum.
5. How does your data relate to the T-chart?
6. Compile all of the data from the class. What similarities and differences do you see?

It's in the Bag

Show one red, one green, and two blue counters to the class. Ask these questions:

- How many counters do you see?
- How many different colors do you see?

Place the counters in a paper bag. Ask the following questions:

- What color do you think I will probably pull out of the bag? Why?
- Does each color have an equally likely chance of being pulled from the bag? Why or why not?
- How could we make this activity one with equally likely outcomes?
- Can you describe a certain event, using these counters? How about an impossible event, using the counters?
- What fraction of the counters is green? Red? Blue?

Give each group a paper bag with one red, one green, and two blue counters. The students first predict which color they think will be selected most often in 20 draws from the bag. Each time a counter is drawn, the color is recorded, and then the counter is returned to the bag, which is shaken before another counter is selected. Then each group should discuss the following questions:

- Which color was selected most? Least? Did you expect this result? Why or why not?
- About what fraction of the counters drawn were blue? Were red? Were green? What do you think about this result?
- If you pulled a counter from the bag 100 times, how many times would you expect to select a blue counter? A green counter?

Source: F. Fennell, "Probability," *Arithmetic Teacher* 38 (4): 18–22.

Spinning Sums

In this activity, students begin to explore probability based on possible occurrences. Sara, Jimmy, and Pat are playing a board game using spinners. Each spinner is numbered with sections 1 to 10. Sara went first and spun a 3 and a 5, so she got to move 8 spaces. Jimmy spun a 4 and a 7, so he got to move 11 spaces. Pat got to move 7 spaces.

- What numbers could Pat have spun to move 7 spaces?
- What is the largest sum a player can spin?
- Which is more likely to happen: spinning a sum of 7 or spinning a sum of 17? Explain your thinking.

Even or Odd

You and a partner are going to play a game with two dice, each numbered 1–6. One person will get a point each time the product of the two numbers rolled is even. The other person will get a point each time the product of the two numbers rolled is odd. If the goal is to be the person to score the most points, is it better to be the even or the odd person, or is there no difference? Explain why you think this game is fair or not. How would you change the game?

Learning Algebraic Relationships in the Elementary Grades

As children enter the primary grades, they have already had a wide variety of experiences with patterns and relationships. Rhyming songs, building blocks, toys, and other experiences in their environments add to the child's abilities to recognize the patterns and make sense of the world around them. In the early primary grades, students explore patterns with their bodies, actions, and words. As they continue to explore patterns with objects and pictures, they reinforce their ability to recognize and generalize patterns in other contexts. Working with patterns nurtures the kind of mathematical thinking that empowers children to solve problems confidently and relate new situations to previous experiences.

Working with patterns nurtures the kind of mathematical thinking that empowers children to solve problems confidently and relate new situations to previous experiences.

Algebraic thinking in the primary grades lays critical groundwork for more advanced thinking in mathematics and is best developed from continuous experiences with patterns and relationships. Fostering the habit of looking for patterns and relationships between variables begins to give students the necessary background for algebraic thinking. Even in the primary grades where algebra is generally an intuitive process, algebraic thinking involves opportunities for students to recognize, visualize, verbalize, symbolize, and analyze patterns. Children need multiple opportunities to use manipulatives to look for patterns, make conjectures, provide reasons for their conjectures, and represent their patterns and reasoning in pictorial and symbolic ways. At the same time, building an understanding that the equal sign symbolizes equivalence and balance is best achieved through interactive experiences. Children begin to experience and develop an understanding of = as a repre-

Algebraic thinking in the primary grades lays critical groundwork for more advanced thinking in mathematics and is best developed from continuous experiences with patterns and relationships.

sentation for "the same as." These early informal experiences with patterns, functions, and algebra provides a solid base in the development of understanding of important concepts. Eventually, algebra emerges as a way to generalize and represent ideas.

National Council of Teachers of Mathematics Algebra Standards, *Principles and Standards for School Mathematics* (Standards 2000, Project 2000)

Instructional programs from prekindergarten through Grade 12 should enable all students to:

- Understand patterns, relations, and functions.
- Represent and analyze mathematical situations and structures using algebraic symbols.
- Use mathematical models to represent and understand quantitative relationships.
- Analyze change in various concepts.

Wisconsin's Model Academic Standards of Mathematics, Standard F: Algebraic Relationships

Students in Wisconsin will discover, describe, and generalize simple and complex patterns and relationships. In the context of real-world problem situations, the student will use algebraic techniques to define and describe the problem to determine and justify appropriate solutions. (Wisconsin Department of Public Instruction 1998, 14–15).

TABLE 5.6 **Wisconsin's Model Academic Standards for Algebraic Relationships**

Wisconsin performance standards

	By the end of grade 4 students will:	Elaboration
F.4.1	Use letters, boxes, or other symbols to stand for any number, measured quantity, or object in simple situations (e.g. $N + 0 = N$ is true for any number).	This standard is intended to suggest that the important skill of algebraic representation should be introduced in an intuitive way at the early levels of instruction. Two key ideas should be noted: (1) The representation may denote a generalization that is true for a wide variety of situations, such as the ABAB... pattern, and (2) the representation may stand for a single specific value as in the open number sentence $\_ + 5 = 11$. It is not expected that each of the three types of representation (letters, boxes, or other symbols) must be used with each of the three entities described (number, measured quantity, or object).
F.4.2	Use the vocabulary, symbols, and notation of algebra accurately (e.g., correct use of the symbol =, effective use of the associative property of multiplication).	Although work at this level should be at the informal, intuitive level, students should still learn correct and standard vocabulary and symbols.

TABLE 5.6 **(Continued)**

Wisconsin performance standards

By the end of grade 4 students will:	**Elaboration**
F.4.3 Work with simple linear patterns and relationships in a variety of ways, including ■ recognizing and extending number patterns, ■ describing them verbally, ■ representing them with pictures, tables, charts, and graphs, ■ recognizing that different models can represent the same pattern or relationship, and ■ using them to describe real-world phenomena.	Linear patterns and relationships are essentially number patterns in which the numerical difference between each pair of successive elements of the pattern is the same. Examples are {3,6,9,12,15,...} or {2,7,12,17,22,...}.
F.4.4 Recognize variability in simple functional relationships by describing how a change in one quantity can produce a change in another (e.g., the number of bicycles and the total number of wheels).	The simple functional relationships mentioned here are essentially the linear relationships described in F.4.3. As an example, a student should be able to recognize that if candy costs 5¢ a piece, then 3 pieces will cost 15¢, whereas 7 pieces will cost 35¢.
F.4.5 Use simple equations and inequalities in a variety of ways, including ■ using them to represent problem situations, ■ solving them by different methods (e.g., use of manipulatives, guess-and-check strategies, recall of number facts), and ■ recording and describing solution strategies.	This standard marks the beginning of a long developmental sequence in which the student becomes more and more proficient with the powerful tools of algebraic equations and inequalities. Although the precise definition of *simple* is properly left to the wisdom of individual teachers and school districts, a suggested upper limit might be basic two-step equations with positive whole number solutions (e.g., $2N + 5 = 13$) or one-step inequalities (e.g., $3N < 18$ or $N - 4 \neq 10$). It must be carefully noted that there is no suggestion here that solution methods be formal ones; rather, the emphasis is on informal, intuitive methods reinforced with manipulative materials.
F.4.6 Recognize and use generalized properties and relationships of arithmetic (e.g., commutativity of addition, inverse relationship of multiplication and division).	As students become proficient with computational arithmetic (mental math, basic facts, standard or nonstandard algorithms), their work is made easier through the recognition of the properties and relationships of number operations (commutativity, associativity, identities [addition of 0, multiplication by 1]). As in F.4.2, it is important to use correct and standard vocabulary and symbolism.

Additional Standards Information

Additional support information can be found in the NCTM *Curriculum and Evaluation Standards,* 60–62, and in the NCTM PSSM, 90–95 and 158–163.

Illustrative Task

What's My Rule

The goal of this activity is for students to figure out a rule created by the teacher, such as "All numbers are multiples of 4." The teacher lists a set of numbers on the board that fit the rule, for example, 4, 16, 40. Students then guess a number or several numbers they think fit the rule. The teacher records the guesses on the board and lets the class know whether or not they fit the rule. As a final test of their conjectures, the teacher lists several numbers that may or may not fit the rule, and students apply their conjectures to determine which ones satisfy the rule. Finally, students are asked to describe the rule.

Source: T. G. Coburn, *Patterns. Addenda Series, Grades K–6* (Reston, VA: National Council of Teachers of Mathematics, 1993).

Discussion of the Task

Whether it is noticing patterns and relationships among numbers or solving simple equations, algebra is a vital part of the elementary mathematics curriculum. "What's My Rule" encourages students to make conjectures about relationships among numbers by recognizing number patterns and then allows them to test their conjectures.

In a fourth grade classroom, the teacher listed the following set of numbers on the board: 2, 4, and 8. The students then tried to determine the rule that applied to this set of numbers. They were to give a set of three numbers that might fit the rule based on a conjecture they had. The teacher recorded the guesses on the chalkboard and indicated either yes, it fit the rule, or no, it did not fit the rule. Here is how the investigation began.

TABLE 5.7

Guesses			Response
2	4	8	Yes
4	8	12	No
5	10	15	No
5	10	20	Yes

It works best not to disclose the rule too quickly. By having students name numbers that they predict will fit the rule rather than naming the rule itself, more students are kept engaged in the thinking process for a longer period of time. The teacher was able to observe and informally assess students' understanding throughout the activity. When the teacher sensed that many students had figured out the rule, she asked the students to write their conjectures on paper.

TABLE 5.8

Guesses			Response
8	16	24	
13	26	52	

To then test their conjectures, the teacher wrote some more sets of numbers on the chalkboard, as shown in Table 5.8. The students were to apply their conjectures to determine which sets of numbers satisfied the rule.

By having the students record these sets of numbers and their responses on paper, the teacher was able to walk around the classroom and quickly assess the students' understanding. The written answers, conjectures, and rules were also handed in as a more formal assessment.

Finally, the students discussed their conjectures as a class. Have you figured it out yet? In this case, the teacher was using the rule "Each number is twice the value of the previous number."

Standards Addressed

F.4.1 Use letters, boxes, or other symbols to stand for any number, measured quantity, or object in simple situations (e.g., $N + 0 = N$ is true for any number).

F.4.3 Work with simple linear patterns and relationships in a variety of ways, including:

- Recognizing and extending number patterns
- Describing them verbally
- Representing them with pictures, tables, charts, and graphs
- Recognizing that different models can represent the same pattern or relationship
- Using them to describe real-world phenomena

F.4.4 Recognize and use generalized properties and relationships of arithmetic (e.g., commutativity of addition, inverse relationship of multiplication and division).

Connections to Other Standards

Also addresses the following standards: mathematical processes (A.4.1, A.4.2, A.4.5), number operations and relationships (B.4.1), and geometry (C.4.2, C.4.3).

Extension Questions

Here are some other rules that you might want to try:

- Each number is three more than the previous number.
- All numbers are multiples of eight.
- All numbers are odd.
- The third number is the sum of the first two.

In the earlier grades, students may be more prone to pattern recognition involving shape, size, and rhythm rather than number value. This is also valuable to build the algebraic foundation. For example, a teacher may play a patterning game of "What Comes Next?" using sounds such as "clap, clap, snap, clap, clap, snap, _____, _____" or may have the students stand in a boy-girl-boy-girl line.

Later, as students become more familiar with numbers, the teacher may begin by introducing a simpler version of "What's My Rule?" with single numbers or pairs of numbers in an "input-output machine" such as this:

TABLE 5.9

In	Out
1	4
4	7
5	8

Students build their reasoning skills and can handle increasingly more complex rules as they progress into the upper elementary grades and beyond. They will be thrilled to be the leader who establishes the rule, too, as they become more familiar with these games.

Other Sample Tasks

How Many Buttons?

How many buttons are in this arrangement?

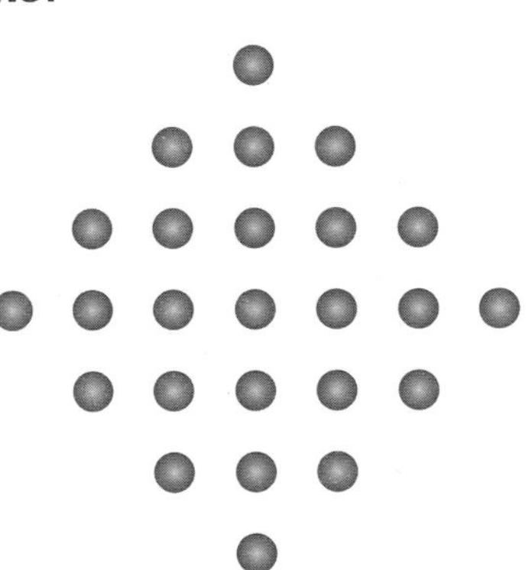

Try to find your answer in as many different ways as possible.

Make a record of each of these ways.

This activity is valuable on a number of different levels. In the early primary grades, children are encouraged to solve this problem in as many ways as possible. Early primary children may simply count each button, whereas others may begin looking for patterns. Counting strategies, divergent strategies, and spatial visualization are inherent in this problem. Children should also represent their solutions using number sentences. A discussion of order of operations will naturally evolve as students "simplify" their number sentences (see examples below).

Solutions include

$1 + 3 + 5 + 7 + 5 + 3 + 1 = 25$

$(2 \times 1) + (2 \times 3) + (2 \times 5) + 7 = 25$

$1 + 8 + 7 + 8 + 1 = 25$

$(2 \times 1) + (2 \times 8) + 7 = 25$

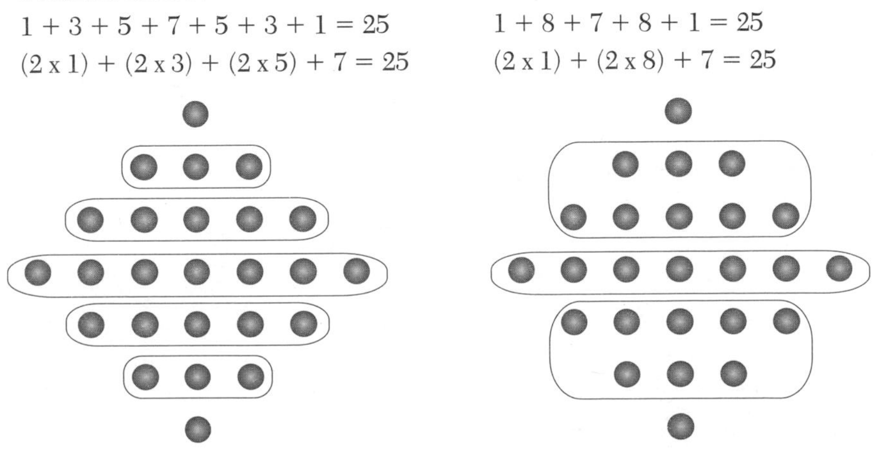

The Swimming Pool

Tat Ming is designing square swimming pools. Each pool has a square center that is the area of the water. Tat Ming uses blue tiles to represent the water. Around each pool there is a border of white tiles. Here are pictures of the three smallest square pools that he can design with blue tiles for the interior and white tiles for the border.

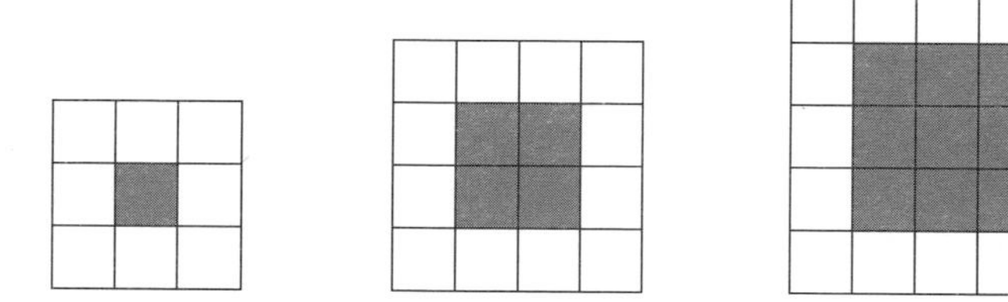

Questions for Grades K–2

- For each square pool, sort the tiles into blue tiles for the water and white tiles for the border. Count how many tiles are in each pile. Are there more white tiles or blue tiles?
- Build each of the three blue squares. How many blue tiles are in each square? Build the next-biggest square that you can make out of the blue tiles. Then build the next. Count the squares in each. What patterns do you see? What is a square?
- Build the three pools using blue and white tiles to show the water and the border tiles. Record the information for each pool in a table. How many tiles will be in the next largest pool? Check your answer by building the square. Describe your methods for counting the different tiles. What patterns do you see?

Questions for Grades 3–4

- Build the first three pools and record the number of blue and white tiles for each pool in a table. Continue the table for the next two squares. How do you know your answers are correct?
- If there are 32 white tiles in the border, how many blue tiles are there? Explain how you got your answer.
- Can you make a square with 49 blue tiles? Explain why or why not.
- Can you make a square with 12 blue tiles? Explain why or why not.
- In each of the first three square pools, decide what fraction of the square's area is blue for the water and what fraction is white for the border. What patterns do you see?

Source: J. Ferrini-Mundy, G. Lappan, and E. Phillips, "Experiences with Patterning," *Teaching Children Mathematics* 3 (1997): 262–68.

True and False Number Sentences

Present number sentences with the equal sign in various locations, similar to those listed here, to your students and ask them to discuss whether the number sentences are true or false. They should be encouraged to use various approaches to prove their assertions. These discussions help children develop an understanding of the equals sign as a notion of equality rather than as a directive to compute.

$4 + 5 = 9$	$12 - 5 = 9$	$7 = 3 + 4$
$8 + 2 = 10 + 4$	$7 + 4 = 15 - 4$	$8 = 8$

Source: K. L. Falkner, and T. P. Carpenter, "Children's Understanding of Equality: A Foundation for Algebra," *Teaching Children Mathematics* 6 (1999): 232–36.

Performance Assessment Task

Assessment needs to mirror instruction. Assessment of conceptual understanding requires a different look at student knowledge. Performance assessment asks students to construct responses: "Students produce an answer from their own knowledge and in their own words" (Wisconsin Department of Public Instruction 1995, 2). The Wisconsin Center for Education Research (WCER) at the University of Wisconsin–Madison, in cooperation with the Wisconsin Department of Public Instruction (DPI), developed several performance assessment items from 1994 through 1998. The "Hot Dog" task, complete with scoring rubric and student work, follows as an example of an item that allows students to take in information, to analyze it, and to explain their reasoning. It is the kind of item that could be used to prepare students for the constructed response component of the Wisconsin Knowledge and Concepts Examination (WKCE). The item is used with permission from the Office for Educational Accountability, Wisconsin Department of Public Instruction.

Hot Dog Problem

	Mr. Douglas	Ms. Sabroff
Number of students	27	22
Number of adult aides	3	2
Number of teachers	1	1

Ms. Sabroff and Mr. Douglas are having a picnic with their classes. The chart gives information about who is invited.

Your job is to decide the number of packages of hot dogs and the number of packages of buns to buy.

Here is what you need to know:

More than half of the people will eat at least two hot dogs.
There are 10 hot dogs in a package.
Buns come 8 to a package or 12 to a package.

Use the working space below to write down ideas, numbers and possible plans.

1. How many packages of hot dogs will you buy? ____________
2. How many, and what size packages of buns will you buy? ____________

____

3. Explain how you decided on the number of packages of hot dogs and the number of packages of buns to buy. ______________________________

____

____

____

____

____

____

____

____

Elements of a Proficient Response

- Total number of attendees = 56. One-half of 56 is 28. 56 + 28 = 84.
- If all persons eat at least one hot dog, a minimum of 84 hot dogs will be eaten.
- If all persons eat 2 hot dogs, a maximum of 112 hot dogs will be eaten unless some have more than 2.

Question 1

- Answers may vary. But if above assumptions hold, a minimum of 9 packages need to be bought.

Question 2

- Answers may vary. But if one assumes that every hot dog will be eaten with a bun, a minimum of 84 buns are needed. See Question 3 for some possible combinations of packages.

Question 3

- Explanations will vary, but the explanation should be consistent with answers provided for Questions 1 and 2.
- Possible combinations of buns to get a minimum of 84 servings:

12-Bun Pack	8-Bun Pack	Extra Buns
7	0	0
6	2	4
5	3	0
4	5	4
3	6	0
2	8	4
1	9	0
0	11	4

Major Flaws

- Deals only with 56 people and ignores fact that at least 28 of them will eat two hot dogs.

- Justification for decision has little or no relation to answers given for Questions 1 and 2.

Minor Flaws

- Minor calculation error.

Potential Elements of an Advanced Response

- Justifies more than 84 by explicitly stating divergent assumptions, such as, there may be leftovers, more than just one-half will eat 2, some will eat 3, adults eat more, etc.
- Correctly justifies solution using accurate and reasonable estimation techniques.
- Correctly justifies solution for less than 84 by assuming some might eat none.

Student Responses

See examples of proficient and advanced student work.

Example of an Advanced Response

	Mr. Douglas	**Ms. Sabroff**
Number of students	27	22
Number of adult aides	3	2
Number of teachers	1	1

Ms. Sabroff and Mr. Douglas are having a picnic with their classes. The chart gives information about who is invited.

Your job is to decide the number of packages of hot dogs and the number of packages of buns to buy.

Here is what you need to know:

More than half of the people will eat at least two hot dogs.
There are 10 hot dogs in a package.
Buns come 8 to a package or 12 to a package.

Use the working space below to write down ideas, numbers and possible plans.

27
+ 22
49
+ 3
2
54
+ 1
1
56
x 2
112

10)112 = 11 R.2
10
12
10
2

8)112 = 14
8
32
32
0

12)112 = 9 R.3
108
3

1. How many packages of hot dogs will you buy? 12.

2. How many, and what size packages of buns will you buy?

~~8~~ 14 of the eights.

3. Explain how you decided on the number of packages of hot dogs and the number of packages of buns to buy.

First I added to see how ~~many~~ ~~pe~~ many people there would be, and then I doubled the number, since more than half would have seconds. Then I divided that by ten. Next I tried dividing the numbers of buns into my answer, and came up with: 12 buns gives you a remainder, but 8 doesn't. If I knew the cost of everything, I'd probobly change my answer, because I would want to keep my picnic low on cost, so I'd buy the 12 to a package buns, if that made it cheaper. It would change my asnswer though, if somebody was absent.

Example of a Proficient Response

	Mr. Douglas	Ms. Sabroff
Number of students	27	22
Number of adult aides	3	2
Number of teachers	1	1

Ms. Sabroff and Mr. Douglas are having a picnic with their classes. The chart gives information about who is invited.

Your job is to decide the number of packages of hot dogs and the number of packages of buns to buy.

Here is what you need to know:

More than half of the people will eat at least two hot dogs.
There are 10 hot dogs in a package.
Buns come 8 to a package or 12 to a package.

Use the working space below to write down ideas, numbers and possible plans.

1. 27 + 22 + 5 + 2 = 56

2. 56 ÷ 2 = 28 + 4 = 32 students and teachers eating 2 hot dogs.

3. 32 x 2 = 64 hot dogs

4. 56 − 32 = 24 kids that eat 1 hot dog.

5. 64 + 24 = 88 hot dogs.

88 ÷ 8 = 11 packs of buns

88 ÷ 10 = 8.8 hot dogs

7

7. 83 ÷ 12 = 6 groups of 12 and 11 singles.

I would buy 7 packs of 12 hot dog buns because it wouldn't hurt to have 1 extra

1 hot dog extra

1. How many packages of hot dogs will you buy?

2. How many, and what size packages of buns will you buy?

11 packs of 8.

3. Explain how you decided on the number of packages of hot dogs and the number of packages of buns to buy.

First I figured out how many students there was and teachers. Then I divided the class in ½ and then added 4 people to 1 ½. Then same ½ I doubled because they are eating 2 hot dogs. Then I figured out how many students and teachers are left. Then I added the the 2 answers together and came up with 88 hot dogs. Then I went to figureing out buns and packs of hot dogs. For the buns I went 88 ÷ 8 = 11 packs of hot dog buns. Then for hot dogs I did 88 ÷ 10 = 8.8 hot dogs. So I rounded 8.8 hot dogs and bought 9 packs.

The Young Adolescent Learner of Mathematics

Students entering middle school think very concretely, whereas those leaving middle school reason more abstractly; this cognitive growth is dramatic for many youth. In 1994, Armstrong quoted Gardner as saying, "It is of utmost importance that we recognize and nurture all of the varied human intelligences, and all of the combinations of intelligences. If we recognize this, I think we will have at least a better chance of dealing appropriately with the many problems that we face in the world" (Armstrong 1994, 1). Advances in brain research and knowledge of how children learn support the notion that students must be actively engaged in learning. If knowledge is going to be retained and understood, then students must use it in a demonstration of complex performance (Caine and Caine 1991; Perkins 1992). Maximizing brain growth requires that students are challenged through problem solving, critical thinking, relevant projects, and complex activity and provided with learner feedback that is specific, multimodal, timely, and learner directed.

Students entering middle school think very concretely, whereas those leaving middle school reason more abstractly; this cognitive growth is dramatic for many youth.

On the basis of the work completed at the Center for Early Adolescence the following list of physical, social, emotional, and cognitive characteristics exists for young adolescents. They include diversity, self-exploration and self-definition, meaningful participation in school communities, social interaction with peers and adults, physical activity, a sense of competence and achievement, and structure with clear limits (Clewell, Anderson, and Thorpe 1992, 18–19). Recommendations for improving performance of minorities and females in middle school mathematics include (1) instruction that is activity based, proceeding from concrete to abstract; (2) students seeking solutions to problems through hands-on experiences where teachers relate problems to students' interests and experiences; (3) students interacting with one another as a means to enhance achievement; (4) efforts to combat the perception that mathematics is a white male domain and stressing the usefulness of mathematics and technology; (5) teachers having high expectations for every student and confidence in each student's ability to learn; and (6) teachers adapting teaching methods to accommodate varying stages of cognitive development (Clewell, Anderson, and Thorpe 1992, 27–28). Integrating these concepts into all instruction will ensure that all students will learn mathematics.

In their *Curriculum and Evaluation Standards*, the NCTM made it clear that all students need to learn more and often different mathematics than they had been taught previously, and that instruction of mathematics must be significantly revised. They also made it clear that a broad range of teaching strategies are necessary to support students' learning; students learn to reason and communicate mathematically only if they have the opportunity to do so. NCTM and WMAS explicitly redefine, expand, and deepen the mathematics curriculum content for grades 5 through 8. A basic premise is that what a student learns depends to a great degree on how he or she has learned it.

The NCTM made it clear that all students need to learn more and often different mathematics than they had been taught previously, and that instruction of mathematics must be significantly revised.

In *Turning Points, Preparing American Youth for the 21st Century*, the Carnegie Council on Adolescent Development recommends that schools should "staff middle grade schools with teachers who are expert at teaching young adolescents and who have been specifically prepared for the assign-

Schools should staff middle grade schools with teachers who are expert at teaching young adolescents and who have been specifically prepared for the assignment to the middle grades.

To learn mathematics as we were taught is hard enough; to learn to teach in the ways envisioned in the new math standards is harder still.

—Ball 1996

ment to the middle grades" (1989, 1). There is a growing recognition that teachers, like their students, bring with them experiences and prior understandings that profoundly shape their learning. To learn mathematics as we were taught is hard enough; to learn to teach in the ways envisioned in the new math standards is harder still (Ball 1996). Learning of this kind involves more than knowledge and skills. It includes personality qualities such as patience, curiosity, confidence, trust, and imagination (Lampert and Eshelman 1995). Most teachers currently in the classroom have not experienced for themselves instruction that addresses the vision of the NCTM standards or WMAS. This clearly indicates that appropriate professional development activities must be provided for these teachers. These experiences should immerse the teacher in activities that integrate theory, demonstration, and real experiences. Feedback from these activities will provide valuable information, which will help teachers know what changes to make in their teaching and how to go about making those changes.

The world is constantly transformed by science and technology in ways that have profound effects on economic well-being. Students will require much technical competence and a great deal of flexibility: not just one set of skills acquired early and essential for life but adaptability to an evolving body of knowledge. Successful participation in a technically based and interdependent world economy will require that students be more skillful and adaptable than ever before. Only by providing rich, challenging, and varied learning experiences that are designed to meet the unique needs of the young adolescent learner will teachers develop students who are indeed more skillful and adaptable than ever before.

We can no longer allow middle grades mathematics programs to contain very little new content as documented in the Flanders study in 1987 (see Figure 5.16). Today's standards require that middle school students go far beyond pre-

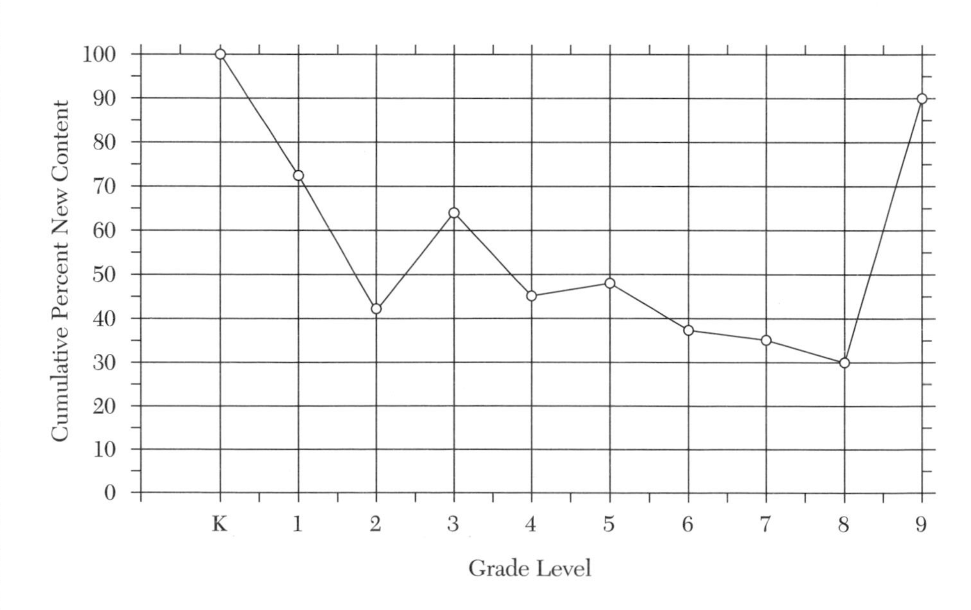

FIGURE 5.16 **Percentage of new content in three mathematics text series.** From "How Much of the Content in Mathematics textbooks is New?" by J. R. Flanders, 1987, in *Arithmetic Teacher,* pp. 18–23. Copyright by National Council of Teachers of Mathematics. Used with permission of NCTM.

vious expectations. To adequately achieve the goals and standards as established for middle-level mathematics and have them taught in meaningful ways to students, appropriate time allocations must be provided to teachers and students. No longer can a 35- or 40-minute class period per day adequately allow teachers and students to achieve the challenging curriculum as outlined in current standards. Longer class periods will be required. The following is a position statement from NCTM: "The National Council of Teachers of Mathematics believes that schools must restructure the use of teacher time to enable teachers to implement a variety of activities aimed at enhancing students' learning. These activities should provide opportunities for teachers to reflect on their practice, support teachers' ongoing professional development, and support a culture of professionalism" (http://www.nctm.org/position-statements/teachertime.htm).

Appropriate time allocations must be provided to teachers and students.

Longer class periods will be required.

The NCTM, in its book *Mathematics Assessment—Myths, Models, Good Questions, and Practical Suggestions,* states that "as the forms of mathematics teaching become more diverse—including open-ended investigations, cooperative group activity, and emphasis on thinking and communication—so too must the form of assessment change" (Stenmark 1991, 3). The 1989 NCTM *Curriculum and Evaluation Standards* state that "evaluation is fundamental to the process of making the *Standards* a reality. Just as the Curriculum Standards propose changes in K–12 content and instruction, the Evaluation Standards propose changes in the processes and methods by which information is collected." (Commission on Standards for School Mathematics, 1989, 190). "Assessment increases and improves the gathering of relevant, useful information. Assessment and program evaluation practices must change along with the curriculum" (Stenmark 1991, 189).

Using Mathematical Processes in the Middle Grades

WMAS states: "In order to participate fully as a citizen and a worker in our contemporary world, a person should be mathematically powerful. Mathematical power is the ability to explore, to conjecture, to reason logically, and to apply a wide repertoire of methods to solve problems. Because no one lives and works in isolation, it is also important to have the ability to communicate mathematical ideas clearly and effectively" (Department of Public Instruction 1998, 33). The mathematical processes as described by the Wisconsin Performance Standards and the PSSM outline the essential tools of mathematical processes that allow students to learn and comprehend all other aspects of mathematics. These tools are essential to the learning of all mathematics content. All mathematics instruction, regardless of the specific content being addressed, should incorporate the use of appropriate mathematical processes. Students should frequently be asked to explain their strategies, justify their reasoning and solutions, and defend the reasonableness of an answer. Students should frequently be asked to write about their mathematics learning.

All mathematics instruction should incorporate the use of appropriate mathematical processes.

Problem solving is the process by which students experience the power and usefulness of mathematics in the world around them. Content knowledge, and the ability to use that knowledge in a variety of situations, is re-

Reasoning is an integral part of doing mathematics.

Communication is an essential feature in classrooms where students are challenged to think and reason about mathematics.

Connections are inherent in learning mathematics; connections can also be made to the rich problem contexts that arise in other disciplines, as well as the different real-world and daily life experiences that interest middle school students.

Use of representations helps students make sense of mathematical concepts and relationships.

quired for students to solve challenging problems. Reasoning is an integral part of doing mathematics. It connects to all other processes of mathematics. Communication is an essential feature in classrooms where students are challenged to think and reason about mathematics and express the results of their thinking to others orally and in writing. In the middle grades, students should develop a keener sense of audience in communicating their reasoning formally and informally. Connections are inherent in learning mathematics. Looking for connections is part of what it means to think mathematically, and making connections builds mathematical knowledge. The middle grades offer new opportunities to use and build connections. Connections can also be made to the rich problem contexts that arise in other disciplines, as well as the different real-world and daily life experiences that interest middle school students. Use of representations helps students make sense of mathematical concepts and relationships. In the middle school grades, where fractions, decimals, percents, the relationships among them, and proportional reasoning receive major attention, students learn to recognize, compare, contrast, and work with these ideas by exploring a variety of representations for them.

National Council of Teachers of Mathematics Process Standards, *Principles and Standards for School Mathematics* (Standards 2000 Project 2000)

The Process Standards suggested by NCTM are: problem solving, reasoning and proof, communication, connections, and representation. (See introduction in this section, p. 89.)

Wisconsin's Model Academic Standards for Mathematics, Standard A: Mathematical Processes

Students in Wisconsin will draw on a broad body of mathematical knowledge and apply a variety of mathematical skills and strategies, including reasoning, oral and written communication, and the use of appropriate technology, when solving mathematical, real-world and nonroutine problems (Wisconsin Department of Public Instruction 1998, 4–5).

Wisconsin Performance Standards

By the end of grade 8 students will be able to do the following:

A.8.1 Use reasoning abilities to:

- Evaluate information.
- Perceive patterns.
- Identify relationships.
- Formulate questions for further exploration.
- Evaluate strategies.
- Justify statements.
- Test reasonableness of results.
- Defend work.

A.8.2 Communicate logical arguments clearly to show why a result makes sense.

A.8.3 Analyze nonroutine problems by modeling, illustrating, guessing, simplifying, generalizing, shifting to another point of view, etc.

A.8.4 Develop effective oral and written presentations that include:

- appropriate use of technology.
- the conventions of mathematical discourse (e.g., symbols, definitions, labeled drawings).
- mathematical language.
- clear organization of ideas and procedures.
- understanding of purpose and audience.

A.8.5 Explain mathematical concepts, procedures, and ideas to others who may not be familiar with them.

A.8.6 Read and understand mathematical texts and other instructional materials and recognize mathematical ideas as they appear in other contexts.

Additional Standards Information

Additional support information can be found in the NCTM *Curriculum and Evaluation Standards,* 75–86, and in the NCTM PSSM, 256–285.

Sample Problem with Commentary

Students are given a carefully drawn picture of a roller-coaster track with points along the track identified in Figure 5.17.

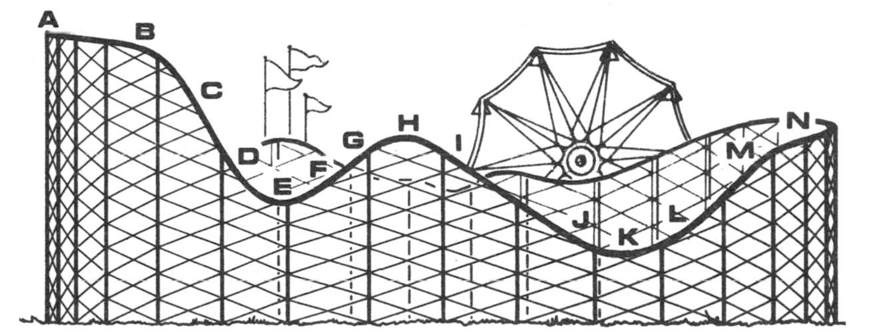

FIGURE 5.17 **Roller coaster problem.**
From Commission on Standards for School Mathematics. (March 1989). *Curriculum and evaluation standards for school mathematics,* p. 83. Copyright by National Council of Teachers of Mathematics. Used with permission of NCTM.

1. Students are then directed to write a description of the speed of the roller coaster car as it travels along the track.
2. Students are challenged to sketch a graph to represent the speed of the roller coaster car as it travels along the track.

Comments about the Problem

The class should discuss what riding a roller coaster is like for those who have not actually experienced such a ride. This problem allows students to demonstrate and communicate their reasoning through writing. It further challenges them to represent their thinking in the form of a second graph.

Extensions to the Problem

Students may be given a graph and asked to write a story line that matches what is depicted in the graph. They may be further challenged to write their own story line about a situation of their choice and then graph that situation.

Suggested Assessment Items

Design your own roller coaster and draw a picture of it. Construct a graph showing how the speed of the roller coaster changes as the car makes one complete run around the track. Write a paragraph describing in detail how the speed of a car traveling your roller coaster changes during the ride.

Write a paper describing in detail how you would explain this problem to a younger student. Be sure to explain how the speed changes over time using the diagram and the graph.

Make a journal entry explaining which part of this problem was the most difficult to understand.

Significant Mathematics/Standards Addressed

Within this problem there is ample opportunity for students to write about (A.8.2) and explain their reasoning (A.8.1). They will also be demonstrating that they can recognize mathematics as it appears in a different context (A.8.6). They will be describing a real-world phenomenon that is represented by a graph (F.8.2).

Additional Resources

Additional rich problems such as these can be found in sources such as:

1. The NCTM Standards and Addenda Series books.
2. Professional journals:
   - *Wisconsin Teacher of Mathematics,* published by the Wisconsin Mathematics Council.
   - *Mathematics—Teaching in the Middle School,* published by NCTM.
3. Problems found in the curricula developed under the auspices of the National Science Foundation (see appendix A of this volume).

Learning Number Operations and Relationships in the Middle Grades

Concrete materials should be used as tools to help students develop understandings of rational numbers.

In grades before Grade 6, students will understand and work with whole numbers by learning basic facts and applying computational strategies. They should have an understanding of number classes and their properties, such as odd, even, prime, and composite. Students need to understand many interpretations and models of fractions (e.g., parts of a whole, measures or quantities, indicated division, decimals, percents, operations, rates, ratios, and parts of proportions). Concrete materials should be used as tools to help students develop these understandings. Over the course of the middle school years, teachers need to provide students with problem situations that help students use their understandings of these interpretations and models.

Students should also represent fractions and decimals as numbers, with emphasis on how they are related to whole numbers and one another. Understanding operations with rational numbers is emphasized in the middle grades as students expand their work with fractions, decimals, and percents. They should develop flexibility in moving from one representation to another.

The development of proportional reasoning should be an area of major emphasis in the middle grades. Proportionality forms the basis for many algebraic concepts, such as linear functions and equations. Middle school students bring to the classroom an intuitive understanding of proportionality based upon their real-world experiences with proportional situations. Teachers should help students develop and build upon these experiences as they work to help students become proportional thinkers.

The development of proportional reasoning should be an area of major emphasis in the middle grades.

Fraction and decimal computation is a focus in the middle grades. Unlike operations on whole numbers, adding, subtracting, multiplying, and dividing rational numbers and integers requires students to operate on abstract relationships rather than on concrete objects. This is because students cannot understand the meaning of an integer like −2 unless they realize that −2 represents an abstract quantity that cannot be touched or felt.

Operations with rational numbers and integers require students to operate on abstract relationships rather than on concrete objects.

Instruction in rational numbers and integers should begin with concrete activities. They should be designed to help students build the abstractions needed to understand rational numbers and integers. As these abstract concepts gradually become established over the course of the middle school years, students become more and better equipped to interpret and solve ever more complex problem situations involving integers, fractions, decimals, and percents.

National Council of Teachers of Mathematics Number and Operations Standard, *Principles and Standards for School Mathematics* (Standards 2000 Project 2000)

Instructional programs should enable all students to:

- Understand numbers, ways of representing numbers, relationships among numbers, and number systems.
- Understand meanings of operations and how they relate to one another.
- Compute fluently and make reasonable estimates.

Wisconsin's Model Academic Standards for Mathematics, Standard B: Number Operations and Relationships

Students in Wisconsin will use numbers effectively for various purposes, such as counting, measuring, estimating, and problem solving (Wisconsin Department of Public Instruction 1998, 6–7).

TABLE 5.10 **Wisconsin's Model Academic Standards for Number Relationships**

Wisconsin performance standards

	By the end of grade 8 students will:	Elaboration
B.8.1	Read, represent, and interpret students' various rational numbers (whole numbers, integers, decimals, fractions, and percents) with verbal descriptions, geometric models, and mathematical notation (e.g., expanded, scientific, exponential).	Here the domain of the number sense is expanded to include all of the rational numbers. But, it is more than just the "fractions-decimals-percents" content of traditional middle school programs of our recent past. Note the integers and use of exponents, content that had often been reserved for beginning algebra courses. Though not explicitly stated, the ability to represent and explain various kinds of numbers supposes facility with various modes of representation as well as "translating" from one mode to another.
B.8.2	Perform and explain operations on rational numbers (add, subtract, multiply, divide, raise to a power, extract a root, take opposites and reciprocals, determine absolute value).	Having number sense involves more than knowing numbers and how to represent them. As with standard B.8.1, this goes beyond just computation with whole numbers and fractions-decimals-percents; it also includes work with numbers and operations commonly associated with the study of algebra.
B.8.3	Generate and explain equivalencies among fractions, decimals, and percents.	Here the expectation is that students can work with equivalencies at the symbolic, algorithmic level. Though not explicitly mentioned, mixed numbers are included.
B.8.4	Express order relationships among rational numbers using appropriate symbols ($<$, $>$, $\leqslant$, $\geqslant$, $\neq$).	This statement should have included the word "determine" in order to more clearly express its meaning. The intent is certainly there. For example, in order to express the order relationship between, say, 3/5 and 4/7, a student must first figure out which of the two is larger than the other, assuming that they are not equal.
B.8.5	Apply proportional thinking in a variety of problem situations that include, but are not limited to: ■ ratios and proportions (e.g., rates, scale drawings, similarity), and ■ percents, including those greater than 100 and less than one (e.g., discounts, rate of increase or decrease, sales tax).	The key to this standard is "Apply ... in a variety of problem situations." There is a great deal of conceptual and skill development associated with the ability to effectively use ratios, proportions, and percents. But it must not end there. Most importantly, students need to appreciate the power of these concepts and skills to help solve an incredibly wide variety of problems.
B.8.6	Model and solve problems involving number-theory concepts such as: ■ prime and composite numbers, ■ divisibility and remainders, ■ greatest common factors, and ■ least common multiples.	It is hoped that this standard is perfectly clear. It deals with content familiar to students and teachers at these grade levels.
B.8.7	In problem-solving situations, select and use appropriate computational procedures with rational numbers such as: ■ calculating mentally, ■ estimating, and ■ creating, using, and explaining algorithms* using technology (e.g. scientific calculators, spreadsheets).	Computing with rational numbers involves more than mechanistic application of symbolic algorithms. For example, one should be able to reason that the sum of 11/12 and 13/14 is pretty close to 2, but less than 2, since both fractions are close to but less than 1. That is both estimating and "calculating" mentally. The other notable part of this standard is "creating ... and explaining algorithms...." More than just applying and using algorithms, students need to know and describe why they work the way they do.

Additional Standards Information

Additional support information can be found in the NCTM *Curriculum and Evaluation Standards,* 87–97, and in the NCTM PSSM, 214–221. Both NCTM standards documents contain the elaboration on the standard and identify focus areas for the middle grades.

Sample Problem with Commentary

Each year the sixth grade students at Wisconsin Middle School go on an outdoor education camping trip. During the trip students study nature and participate in recreational activities. Everyone pitches in and helps with cooking and cleanup.

David and Allison are in charge of making orange juice for the campers. They make the juice by mixing water and orange juice concentrate. To find the mix that tastes best, David and Allison decided to test some recipes on a few of their friends. They have four recipe mixes:

TABLE 5.11

Mix A	Mix B
2 cups concentrate 3 cups cold water	1 cup concentrate 4 cups cold water
Mix C	**Mix D**
4 cups concentrate 8 cups cold water	3 cups concentrate 5 cups cold water

1. Which recipe will make the juice that is the most "orangey"? Explain your answer.
2. Which recipe will make the juice that is the least "orangey"? Explain your answer.
3. Assume that each camper will get 1/2 cup of juice. For each recipe, how much concentrate and how much water are needed to make juice for 240 campers? Explain your answers.

Problem adapted from Lappan, G., Fey, J. T., Fitzgerald, W. M., Friel, S. N., and Phillips, E. D. 1998. Comparing and scaling, 26–29. *Connected Mathematics Project.* Menlo Park, CA: Dale Seymour.

Comments about the Problem

As students approach these questions, they will use a variety of approaches. Some will examine the ratios of concentrate to water, whereas others will examine the ratio of concentrate to the whole mixture. In either of these cases, students must decide which recipe has the greatest ratio of concentrate for the amount of water. Some students may examine this by creating equivalent fractions, some will convert the ratio to decimal form, and others will find the percent of the mixtures that are concentrate.

Some students will look at the relationship of how much water is put with one cup of concentrate or how much concentrate is used for each one cup of water. Some students will use drawings, pictures, and a table as a tool.

All of these approaches are valid and should be thoroughly discussed as a whole class once the initial activity and answer is determined. The real development of true understanding in students comes through the discussion of the various strategies students used. Giving validity to all approaches that work is important and will help develop student confidence and mathematical power.

Suggested Assessment

The completion of the activity where students explain their reasoning and response is in itself a form of assessment. Students might also be asked to write about what they learned as a result of the sharing of strategies used by other students.

Extensions

Follow-up questions might include: (1) How did you use ratios in solving the problem? and (2) For each recipe, how much concentrate and how much water is needed to make one cup of juice?

Significant Mathematics/Standards Addressed

Completion of this problem requires students to read, represent and interpret various rational numbers in a variety of representations (B.8.1, B.8.3, and B.8.4). Students will also apply proportional thinking (B.8.5) and perform calculations or estimations (B.8.7). In addition, students will use a variety of mathematical processes including reasoning abilities (A.8.1), communicating why an answer makes sense (A.8.2), and writing about their reasoning as they explain their answers (A.8.4). The class discussion involves explaining mathematical concepts and procedures (A.8.5).

Additional Resources

Additional rich problems such as these can be found in sources such as:

1. The NCTM Addenda Series books *Understanding Rational Numbers and Proportions* and *Developing Number Sense in the Middle Grades.*
2. Professional journals:
   - *Wisconsin Teacher of Mathematics,* published by the Wisconsin Mathematics Council.
   - *Mathematics—Teaching in the Middle School,* published by NCTM.
3. Examples given in NCTM Standards books.
4. Problems found in the curricula developed under the auspices of the National Science Foundation (see appendix A of this volume).

Learning Geometry in the Middle Grades

"Students should come to the study of geometry in the middle grades with informal knowledge about points, lines, planes, and a variety of two- and three-dimensional shapes; with experience in visualizing and drawing lines, angles, triangles, and other polygons; and with intuitive notion about shapes, built from years of interacting with objects in their daily lives."

"Geometry provides a rich context for the development of mathematical reasoning, including inductive and deductive reasoning, making and validating conjectures, and classifying and defining geometric objects."

—Standards 2000 Project 2000, 233

"In middle-grades geometry programs based on these recommendations, students investigate relationships by drawing, measuring, visualizing, comparing, transforming, and classifying geometric objects. Geometry provides a rich context for the development of mathematical reasoning, including inductive and deductive reasoning, making and validating conjectures, and classifying and defining geometric objects" (Standards 2000 Project 2000, 233).

In the middle grades, the study of geometry also provides representational tools for understanding other topics in mathematics. Geometry has great interrelatedness with other branches of mathematics, including measurement (Wisconsin Department of Public Instruction, 1986). The rectangular coordinate system should be used as a tool for solving a wide variety of problems. Making and validating conjectures further develops students' inductive and deductive reasoning skills, providing valuable background for more formal validation and proof that will come in later studies.

National Council of Teachers of Mathematics Geometry Standard, *Principles and Standards for School Mathematics* (Standards 2000 Project 2000)

Instructional programs should enable all students to:

- Analyze characteristics and properties of two- and three-dimensional shapes and develop mathematical arguments about geometric relationships.
- Specify locations and describe spatial relationships using coordinate geometry and other representational systems.
- Apply transformations and use symmetry to analyze mathematical situations.
- Use visualization, spatial reasoning, and geometric modeling to solve problems.

Wisconsin's Model Academic Standards for Mathematics, Standard C: Geometry

Students in Wisconsin will be able to use geometric concepts, relationships and procedures to interpret, represent, and solve problems (Wisconsin Department of Public Instruction 1998, 8–9).

TABLE 5.12 **Wisconsin's Model Academic Standards for Geometry**

Wisconsin's performance standards	Elaboration
By the end of grade 8 students will:	
C.8.1 Describe special and complex two- and three dimensional figures (e.g., rhombus, polyhedron, cylinder) and their component parts (e.g., base, altitude, and slant height) by: ■ naming, defining and giving examples, ■ comparing, sorting, and classifying them, ■ identifying and contrasting their properties (e.g., symmetrical, isosceles, regular), ■ drawing and constructing physical models to specifications, and ■ explaining how these figures are related to objects in the environment.	This standard continues the developmental work begun in Standard C.4.1. Here in the middle grades, the skill and conceptual level is elevated. More complex figures are considered and more sophisticated ways of describing them are expected.
C.8.2 Identify and use relationships among the component parts of special and complex two- and three-dimensional figures (e.g., parallel sides, congruent faces).	In much the same way that Standard C.8.1 extends the work begun in Standard C.4.1, this standard extends Standard C.4.2. Of important note is that this standard looks not only at properties and relationships of figures, but also at component parts of those figures.
C.8.3 Identify three-dimensional shapes from two dimensional perspectives and draw two-dimensional sketches of three dimensional objects preserving their significant features.	As important as it is to have students manipulate actual two or three dimensional figures, of equal importance is their ability to understand the representations of these figures drawn on a two-dimensional page or computer drawing.
C.8.4 Perform transformations on two-dimensional figures and describe and analyze the effects of the transformations on the figures.	The intent here is a much more formal study of geometric transformations, including correct terminology. To do this properly, sufficient instructional time must be given over a period of several school years.
C.8.5 Locate objects using the rectangular coordinate system.	This skill of using the two-dimensional coordinate system goes both ways. Given an object located on a four-quadrant (i.e., using positive and negative numbers) coordinate grid, the student must correctly give its two coordinates as an ordered pair. And conversely, given an ordered pair, the student must correctly place an object on the grid.

Additional Standards Information

Additional support information can be found in the NCTM *Curriculum and Evaluation Standards,* 112–115, and in the NCTM PSSM, 232–239. Both NCTM standards documents contain the elaboration on the standard and identify focus areas for the middle grades.

Sample Problem with Commentary

In "How Many Quadrilaterals?" students are given a sheet of modified dot paper with dots forming a three-by-three array (figure 5.18). They are challenged to draw as many quadrilaterals as they can using these five rules:

- Each figure must be made up of four line segments.
- Each segment must begin and end at a point.
- Stay inside the dots.
- Each figure must be closed.
- Figures are the same if you can rotate, flip, or slide one to look like another.

A few questions naturally arise concerning the definition of quadrilateral and what it means to be the "same." These discussions help students build the definition and also encourage them to think of congruence and similarity in terms of transformations. Students generally begin working independently to find as many quadrilaterals as possible and later begin to compare with partners or team members. It is rare for an individual to come up with all 16 different quadrilaterals, but some good group discussions about whether figures are the same are generated as students begin to compare papers. After students have had a sufficient amount of time to work and discuss the problem, they can be invited to draw a quadrilateral (on an overhead transparency) that has not yet been shown or discussed.

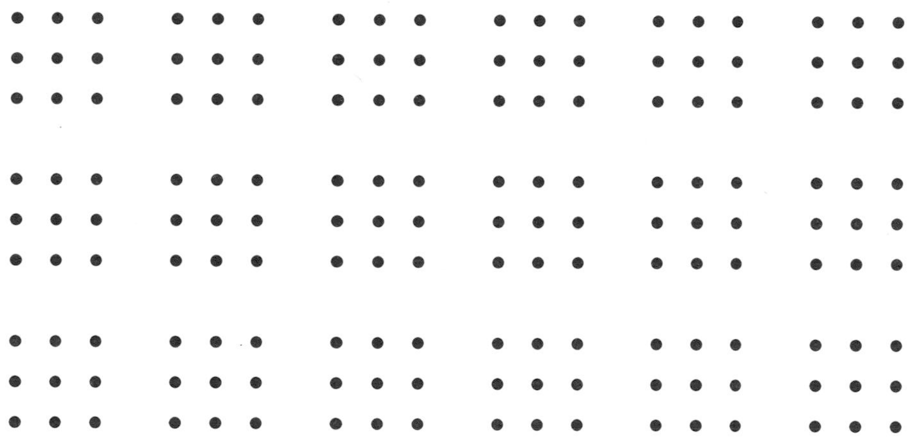

FIGURE 5.18

Solutions

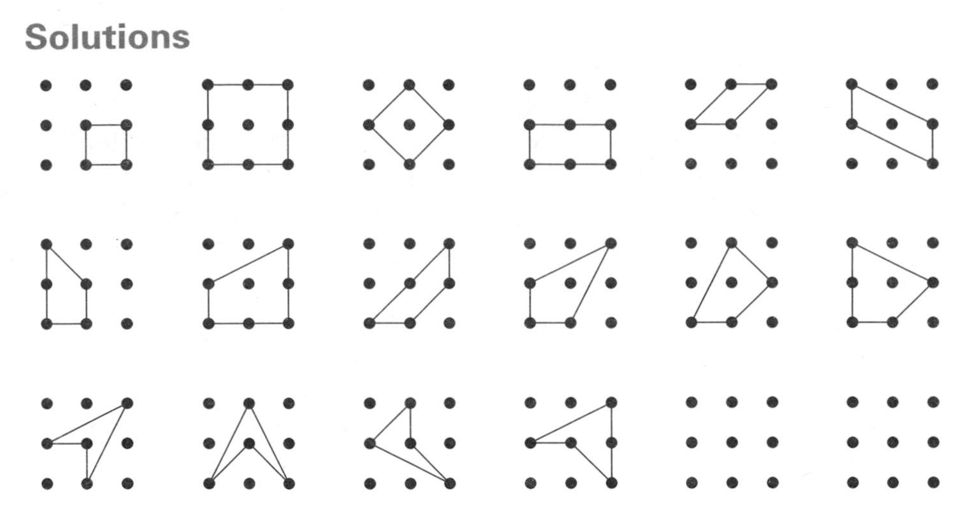

FIGURE 5.19

Discussion Questions

Once all 16 are displayed (see Figure 5.19), the following questions are asked to help students define and classify quadrilaterals:

- What is a square?
- How many of the figures are squares? (Three)
- How many rectangles? A frequent response is one, which leads to the next series of questions.
- Is a square a rectangle?
- Does a square meet all of the requirements in our definition of *rectangle?*
- Are all squares rectangles?
- Are all rectangles squares?

Similar discussions will result from the following questions:

- What is a parallelogram?
- How many figures are parallelograms? (Six)
- Are squares parallelograms?
- Are rectangles parallelograms?
- Are parallelograms squares? rectangles?
- What is a trapezoid?

Note: This definition varies from text to text. One defines it as a quadrilateral with at least one pair of parallel sides, whereas another requires exactly one pair of parallel sides. Once this has been decided, students may be able to answer the remaining questions.

- How many figures are trapezoids?
- Are squares, rectangles, or parallelograms trapezoids?
- What is a quadrilateral?
- How many quadrilaterals are there?

This problem was contributed by task force member Steve Reinhart.

Suggested Assessment Items

The completion of the activity where students explain their reasoning and response is in itself a form of assessment. Ask students to write about what they learned as a result of the sharing of strategies used by other students.

Write six questions that could be included on a quiz that would follow the "quadrilaterals" problem. Be sure to provide answers.

Extensions

Have students find the area of each figure drawn. This activity can also lead to developing Pic's Theorem, an algebraic relationship between the number of points on the border (B) of a figure, the number of points in the interior (I), and the area (A) of the figure. By drawing several figures; collecting data on B, I, and A; and building a table, students can discover patterns that lead to the formula: $A = (B/2) - 1 + I$.

Significant Mathematics/Standards Addressed

This problem is an engaging way to lead students to the discussion of classifying quadrilaterals (C.8.1). As students draw these figures themselves, they are involved in a problem-solving process (A.8.1 and A.8.3). They feel ownership, and the resulting discussions can be quite rich in nature. Students develop spatial and visual skills (C.8.2), both inductive and deductive reasoning (A.8.1), and use and further develop their knowledge of similar figures and transformations (C.8.3). Using the problem extension leads to other significant mathematics. These include working with patterns (F.8.2) and describing a relationship and generating a rule (F.8.3).

Additional Resources

Additional rich problems such as these can be found in sources such as:

1. The NCTM Addenda Series book *Geometry in the Middle Grades.*
2. Professional journals:
   - *Wisconsin Teacher of Mathematics,* published by the Wisconsin Mathematics Council.
   - *Mathematics—Teaching in the Middle School,* published by NCTM.
3. Examples given in NCTM Standards books.
4. Problems found in the curricula developed under the auspices of the National Science Foundation (see appendix A of this volume).

Learning Measurement in the Middle Grades

Measurement activities in context provide us with an abundance of wonderful examples that help teachers show the natural connections among the strands of mathematics as well to other disciplines. Students best develop measurement skills and an understanding of concepts through experiences with applied problems. For these reasons, measurement should be emphasized throughout the middle level years rather than taught as a separate unit of study (NCTM 2000, 232).

Students best develop measurement skills and an understanding of concepts through experiences with applied problems.

Formulas should be the result of instruction, not the focus.

Students enter the middle years having had numerous experiences with measurement, both in and out of the classroom. They should continue to gain experiences with both metric and customary systems of measure. Problems presented should require students to measure both directly and indirectly using attributes including length, perimeter, circumference, area, surface area, volume, mass, time, temperature, ratios and proportions, and angles. Benchmarks should be developed for angle measure (e.g., right angles; straight angles; 30, 45, and 60 degree), length, area, and volume for both U.S. and metric standard units. Students should be able to use these benchmarks to help them make reasonable estimates.

Formulas are important tools in determining measurements and deserve special attention in the middle grades. It is important to note that PSSM emphasizes the development of formulas. **Formulas should be the result of instruction, not the focus.** Students should not be given formulas to memorize and use without a complete understanding of the measure. Rather, they should be provided with many informal experiences that lead to understanding attributes and to the discovery and development of methods, procedures, and formulas.

National Council of Teachers of Mathematics Measurement Standard, *Principles and Standards for School Mathematics* (Standards 2000 Project 2000)

Instructional programs should enable all students to:

- Understand measurable attributes of objects and the units, systems, and processes of measurement.
- Apply appropriate techniques, tools, and formulas to determine measurements.

Wisconsin's Model Academic Standards of Mathematics, Standard D: Measurement

Students in Wisconsin will select and use appropriate tools (including technology) and techniques to measure things to a specific degree of accuracy. They will use measurements in problem-solving situations (Wisconsin Department of Public Instruction 1998, 10–11).

TABLE 5.13 **Wisconsin's Model Academic Standards for Measurement**

Wisconsin's performance standards	Elaboration
By the end of grade 8 students will:	
D.8.1 Identify and describe attributes in situations where they are not directly or easily measurable (e.g., distance, area of an irregular figure, likelihood of occurrence).	The key in using measurement skills and concepts in a problem-solving situation is to first know exactly what it is one wants to measure. This standard is a natural extension of ideas covered in D.4.1 and D.4.5; here we consider more difficult attributes or those that are difficult attributes or those that are not immediately discernible to the visual or tactile senses.
D.8.2 Demonstrate understanding of basic measurement facts, principles, and techniques including the following: ■ approximate comparisons between metric and U.S. units (e.g., a liter and a quart are about the same, a kilometer is about six tenths of a mile.), ■ knowledge that direct measurement produces approximate, not exact, measures, and ■ the use of smallest units to produce more precise measures.	This standard emphasizes that there is much more to measurement than simple physical, manipulative skills. Acquisition of measurement sense requires understanding of the concepts listed here.
D.8.3 Demonstrate measurement directly using standard units (metric and U.S.) with these suggested degrees of accuracy: ■ lengths, to the nearest millimeter or 1/16 of an inch, ■ weight (mass), to the nearest 0.1 gram or 0.5 ounce, ■ liquid capacity, to the nearest milliliter ■ angles, to the nearest degree, and ■ temperature, to the nearest C° or F°.	As with standard D.4.4, the degrees of accuracy listed here are only suggestions. Teachers should feel free to vary these according to the abilities of their students.
D.8.4 Determine measurements indirectly using estimation: ■ conversion of units within a system (e.g., quarts to cups, millimeters to centimeters), ■ ratio and proportion (e.g., similarity, scale drawings), and ■ geometric formulas to derive lengths, areas, volumes of common figures (e.g., perimeter, circumference, surface area).	As this standard suggests, indirect measurements can be obtained in a wide variety of ways. Clearly, the geometric methods described here cannot be accomplished unless a school district is committed to a comprehensive program of instruction in geometry in the elementary and middle school grades.

Additional Standards Information

Additional support information can be found in the NCTM *Curriculum and Evaluation Standards*, 116–119, and in the NCTM PSSM, 240–247.

Sample Problem with Commentary

Conducting experiments and collecting data is a powerful way for students to learn about mathematical relationships in real-world situations. For example, measuring the circumference and diameter of several cylinders or circular items, making a table of the class data, and graphing circumference versus diameter gives students the opportunity to explore and make sense of an important relationship. Representing this relationship with concrete models, tables, graphs, language, and finally with symbols gives all students the necessary time to develop deeper understandings.

In a similar fashion, students can investigate the relationship between the area *(A)* of a circle and the radius *(R)*. Students are given a sheet of centimeter graph paper on which each draws three circles, one small, one medium, and one large. Students estimate the area of each by counting, record the area versus radius on a table, and construct a graph. (The class data may be displayed on the chalk board and graphed on large graph paper to provide additional visuals of the relationship.) Typical data is shown in Figure 5.20 below.

The following questions arise and are discussed before proceeding. What relationships or patterns do you observe? Is the graph a straight line or a curve? Does the graph pass through the point (0,0)? How do you know? What is the area of a circle with radius 2? 3? Explain how you know. What is the radius of a circle with an area of 25? 70? Explain how you know.

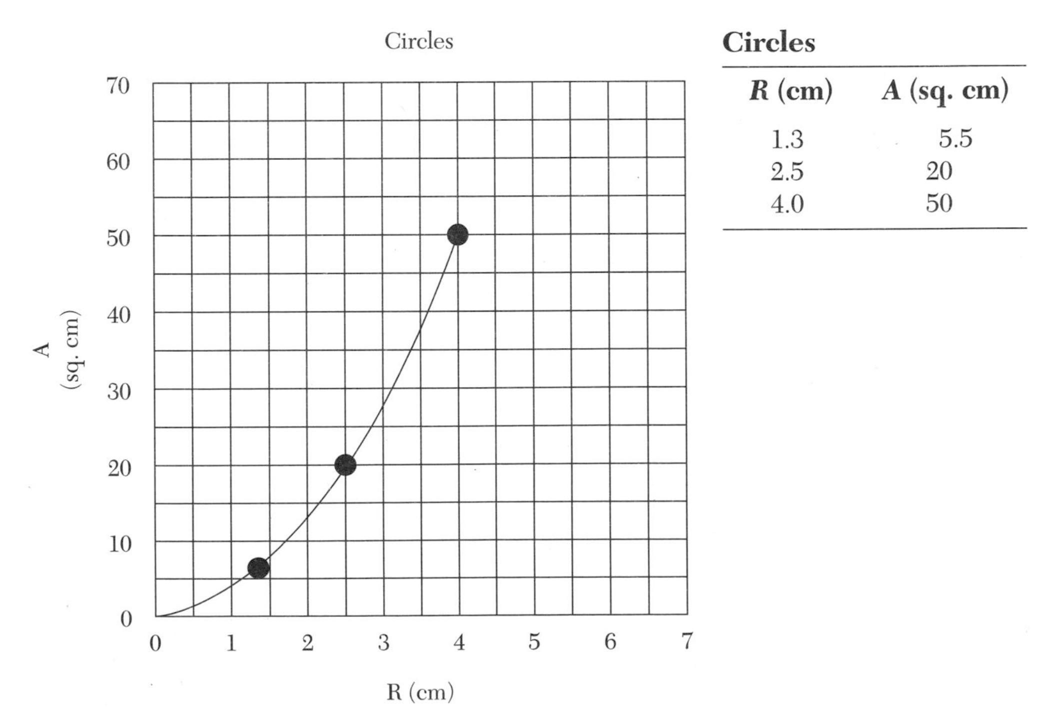

Circles

R (cm)	*A* (sq. cm)
1.3	5.5
2.5	20
4.0	50

FIGURE 5.20

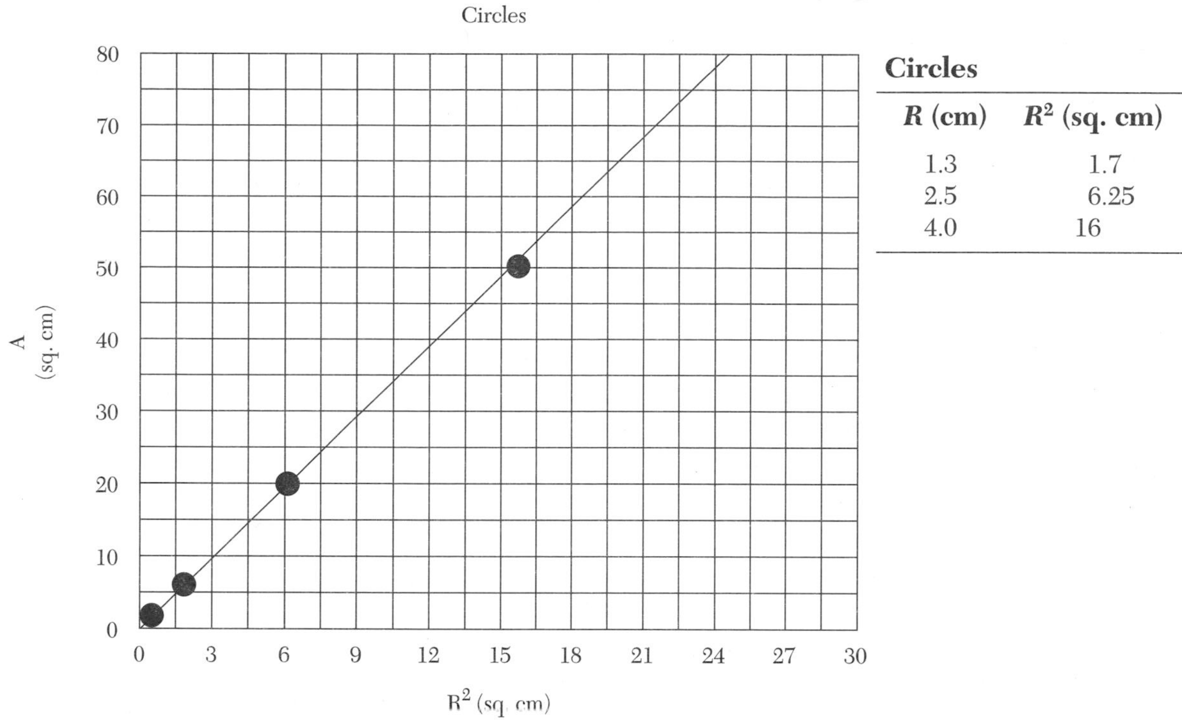

Circles

R (cm)	R^2 (sq. cm)	A (sq. cm)
1.3	1.7	5.5
2.5	6.25	20
4.0	16	50

FIGURE 5.21

Students are then asked to build a table and plot A versus R^2 and many are surprised to see the resulting straight line. Typical data is shown in Figure 5.21 above.

From the new table and graph, students will note that the area of a circle is about 3 times R^2 or that the A/R^2 is approximately 3. Use the second graph to find the area of a circle with a radius of 5, 7, and 9. From conducting this experiment, many students will be able to verbalize the formulas for the circumference and area of a circle. The step from words to symbolic representation, $C = \pi D$ and $A = \pi R^2$, is simpler and more meaningful to students after a thorough investigation of the relationships.

Extensions

A discussion of square roots follows naturally when students are asked to find the radius of a circle with a given area. The slope of the line in the second graph is about 3 π and takes on meaning in this contextual situation. The first graph also prompts a discussion of quadratic relationships and graphs.

Suggested Assessment Items

Select one of the following writing prompts:

1. Write everything you know about the relationship between the area and the radius of a circle.
2. Write everything you know about the relationship between the circumference and the diameter of a circle.

Significant Mathematics/Standards Addressed

While the focus of this problem is on measurement (D.8.3) and developing an understanding of the relationship between area and radius (D.8.4), it is also an algebra problem (F). Students are asked to estimate area, collect and display data in tables and graphs (F.8.2), and use square roots in linear and quadratic relationships (F.8.3).

Additional Resources

Additional rich problems such as these can be found in sources such as:

1. The NCTM Addenda Series book *Measurement in the Middle Grades.*
2. Professional journals:
   - *Wisconsin Teacher of Mathematics,* published by the Wisconsin Mathematics Council.
   - *Mathematics—Teaching in the Middle School,* published by NCTM
3. Examples given in NCTM Standards books.
4. Problems found in the curricula developed under the auspices of the National Science Foundation (see appendix A of this volume).

Learning Statistics and Probability in the Middle Grades

Students should engage in and employ the full process of data investigation: posing questions; collecting, organizing, analyzing, and interpreting data; and answering questions.

The study of data analysis and statistics in the middle grades provides opportunities for students both to use what they have learned elsewhere in their study of mathematics and also to learn new ideas. In the middle grades, students should engage in and employ the full process of data investigation: posing questions; collecting, organizing, analyzing, and interpreting data; and answering questions. These experiences are most meaningful when students generate data about themselves or develop their own questions to explore. Data from sources such as newspapers, magazines, and Web sites are readily available and can be easily brought into the classroom for investigation and discussion. Data analysis in the middle grades provides a rich context for students to solve problems. It also allows them to work with large numbers, measurement, statistical displays, measures of central tendency, and variance. As students work with data analysis, they apply their knowledge of ratios, fractions, decimals, graphs, and measurement. The results of statistical analysis can be used to make predictions.

In grades 6 through 8, students' experiences with data should be intertwined with their study of probability. In their study of theoretical and experimental probability in the middle grades, students learn to use the idea of relative frequency to determine probabilities. Students need considerable time and many experiences to build a solid conceptual foundation of probability. For example, students need to understand concepts of equal likelihood, fairness, randomness, and relationships among events, such as independent events and dependent events, complementary events, and mutually exclusive events. Their experiences prepare students for the further study of statistics and probability in high school.

National Council of Teachers of Mathematics Data Analysis and Probability Standard, *Principles and Standards for School Mathematics* (Standards 2000 Project 2000)

Instructional programs should enable all students to:

- Formulate questions that can be addressed with data and collect, organize, and display data to answer them.
- Select and use appropriate statistical methods to analyze data.
- Develop and evaluate inferences and predictions that are based on data.
- Understand and apply basic concepts of probability.

Wisconsin's Model Academic Standards of Mathematics, Standard E: Statistics and Probability

Students in Wisconsin will use data collection and analysis, statistics, and probability in problem-solving situations, employing technology where appropriate (Wisconsin Department of Public Instruction 1998, 12–13).

TABLE 5.14 **Wisconsin's Model Academic Standards for Statistics and Predictability**

Wisconsin's performance standards	Elaboration
By the end of grade 8 students will:	
E.8.1 Work with data in the context of real-world situations by: ■ formulating questions that lead to data collection, ■ designing and conducting a statistical investigation, and ■ using technology to generate displays, summary statistics, and presentations.	This standard is part of a continuum beginning with Standard E.4.1 and ending with Standard E.12.1. It goes beyond the ground work laid in the elementary school by requiring students to design and conduct experiments and to use technology as part of data analysis and presentation.
E.8.2 Organize and display data from statistical investigations using: ■ appropriate tables, graphs, and/or charts (e.g., circle, bar, or for multiple sets of data), and ■ appropriate plots (e.g. line, stem-and-leaf, box, scatter).	This is a skill-oriented standard that delineates the types of visual presentations students are expected to be able to generate.
E.8.3 Extract, interpret, and analyze information from organized and displayed data by using: ■ frequency and distribution, including mode and range, ■ central tendencies of data (mean and median), and ■ indicators of dispersion (e.g., outliers).	As with Standard E.8.1, this standard is part of the continuum defined in Standards E.4.3, E.8.3, and E12.3. The words "extract" and "using" imply that students can generate statistics such as range, mode, mean, or median as well as apply them in their analyses.
E.8.4 Use the results of data analysis to: ■ make predictions, ■ develop convincing arguments, and ■ draw conclusions.	Students must go beyond data collection, presentation, and summarization. This standard gets at the heart of why people carry out data analysis.

TABLE 5.14 **(Continued)**

Wisconsin's performance standards	Elaboration
By the end of grade 8 students will:	
E.8.5 Compare several sets of data to generate, test, and, as the data dictate, confirm or deny hypotheses.	Up to this point, the standard statements have dealt with various aspects of data collection and analysis with respect to a single set of data. Here we go beyond that to begin comparisons of more than one set of data.
E.8.6 Evaluate presentations and statistical analyses from a variety of sources for: ■ credibility of source, ■ techniques of collection, organization and presentation of data, ■ missing or incorrect data, ■ inferences, and ■ possible sources of bias.	Very often, data are presented to a reader or consumer for the purpose of convincing that reader to adopt a point of view or to purchase a particular product. Here, the student is required to make a qualitative analysis of data analyses, essentially to question whether conclusions drawn by the original data presenter are valid or not valid.
E.8.7 Determine the likelihood of occurrences of simple events: ■ using a variety of strategies to identify possible outcomes (e.g., lists, tables, tree diagrams), ■ conducting an experiment, ■ designing and conducting simulations, and ■ applying theoretical notions of probability (e.g., that four equally likely events have a 25 percent chance of happening).	In Standard E.4.4, we characterized probability as a measure of a future event. Here, the students learn how to generate such a numerical measure. It is important for them to understand that such measures can only be approximations, particularly when probabilities are empirically determined (as in experiments and simulations). Of course, it is true that almost all measurements are, in fact, approximations.

Additional Standards Information

Additional support information can be found in the NCTM *Curriculum and Evaluation Standards,* 105–111, and in the NCTM PSSM, 248–255.

Sample Problem with Commentary

Most students are very interested in knowing more about their classmates. They can move from their individual experiences, collect data, tally those data, and represent and combine the information. There are many basic questions that students could address. The question may be one that is developed by the students themselves. The following is given as one example:

1. What is an average member of our class like?

To address this broad question, the class must make a number of decisions about obtaining rather specific information about the members of the class. They will need to address such questions and perform such tasks as the following:

- What kind of information do we want to know about the students in the class, and how can we obtain the information we need?
- How do we develop appropriate survey questions? (Perhaps the number of questions should be limited to five or six depending on the age and nature of the class.)

- How do we determine whom to survey (or conduct a census of all students)?
- How do we conduct a survey?
- How do we organize and represent the data?
- Write a written report of the results, including measures of central tendency and the existence of outliers.
- Use the information from one class to make predictions about another class in the same grade or in the same school.

Comments about the Problem

Different groups of students will develop different questions to ask based on their ages and experiences. The teacher must facilitate the discussion to channel the students so that their task is one that is manageable. The value of the activity comes from the fact that students are addressing questions in which they are interested. The process of gathering data, displaying, and analyzing it is paramount.

Suggested Assessment Items

Working in cooperative groups, students can prepare an oral presentation to the class complete with displays.

Write a survey question (or several questions) on a topic of your choice. Test them on several people, including friends and relatives. Be sure to ask adults as well as students. Rewrite your questions to make them more clear and usable. Write a paper describing the responses you received and how they helped you write your final draft questions. Describe the qualities of a good survey question in your paper. Include any interesting information you may have learned about your topic if you wish.

Significant Mathematics/Standards Addressed

By posing their own questions, students work with a real-world situation in which they are interested. They formulate questions and design and conduct a statistical investigation for data collection and analysis (E.8.1). When students display the results of their investigation, they must organize and display the data in a meaningful way using appropriate tables, graphs, or plots (E.8.2). In developing their own survey questions, students must evaluate the validity of their information by examining their techniques of collection, organization, and presentation of the data (E.8.6). By composing a written report, students will demonstrate the ability to develop convincing arguments and draw conclusions (E.8.4). Using their data to make predictions about another class leads into the concept of likelihood of occurrences (E.8.4 and E.8.7). Students also have the option of using a computer to generate displays of the data via technology.

Other Sample Problems

1. Any initial question can be generated to create a project such as the one described previously. One example might be "Should a traffic light be installed on the street by our school?

2. Having students use source books or Web sites to gather data can have benefits. They become familiar with other sources and also broaden their basic knowledge base. While looking for information on one topic, students often become involved in information displayed in a variety of formats (tables, graphs, plots, and so on). Besides learning about statistical principles, students have a real opportunity to integrate mathematics with other subjects.

Additional Resources

Additional rich problems such as these can be found in sources such as:

1. The NCTM Addenda Series book *Dealing with Data and Chance.*
2. Professional journals:
   - *Wisconsin Teacher of Mathematics,* published by the Wisconsin Mathematics Council.
   - *Mathematics—Teaching in the Middle School,* published by NCTM.
3. Examples given in NCTM Standards books.
4. Problems found in the curricula developed under the auspices of the National Science Foundation (see appendix A of this volume).

Learning Algebra in the Middle Grades

Narrow curricular expectations of memorizing isolated facts and procedures must give way to developing algebraic thinking as a connected whole with an emphasis on conceptual understanding, multiple representations, mathematical modeling, and problem solving.

The selection of rich mathematics problems that address several aspects of the standards within the same problem is extremely valuable.

Students in the middle grades should learn algebra both as a set of concepts and competencies tied to the representation of quantitative relationships and as a style of mathematical thinking for formalized patterns, functions, and generalizations (NCTM 2000, 223). Narrow curricular expectations of memorizing isolated facts and procedures must give way to developing algebraic thinking as a connected whole with an emphasis on conceptual understanding, multiple representations, mathematical modeling, and problem solving. Students in the middle grades develop a sound conceptual foundation by moving from intuitive to formal thinking through concrete models and real-world problems. Early in the middle school experience, emphasis should be on generalizing patterns and functions by recognizing, describing, and representing them using multiple representations, manipulative materials, pictures, words, tables, and graphs.

As students move through the middle grades, they progress from number sentences to proportional reasoning and to linear and nonlinear relationships. The selection of rich mathematics problems that address several aspects of the standards within the same problem is extremely valuable. Many of these rich problems offer good contexts from which students may work, provide connections to other mathematics, and provide connections to other content areas.

"The middle school mathematics curriculum is, in many ways, a bridge between the concrete elementary school curriculum and the more formal mathematics curriculum of the high school. One critical transition is between arithmetic and algebra. It is essential that students in grades 5 through 8 explore algebraic concepts in an informal way," and build to an in depth understanding of algebra per se. "Informal explorations should emphasize physical models, data, graphs, and other mathematical representations." Formal algebraic understanding should be based on solving rich mathematical problems. "Students should be taught to generalize number patterns to model, represent, or describe physical patterns, regularities, and problems. These explorations of algebraic concepts should help students gain confidence in their abilities to abstract relationships from contextual information and use a variety of representations to describe those relations" (NCTM Standards 1989, 102). The transition from arithmetic to algebraic thinking and algebra should be a continuous process starting in early grades and integrated into all instruction.

National Council of Teachers of Mathematics Algebra Standards, *Principles and Standard for School Mathematics* (Standards 2000 Project 2000)

Instructional programs should enable all students to:

- Understand patterns, relationships, and functions.
- Represent and analyze mathematical situations and structures using algebraic symbols.
- Use mathematical models to represent and understand quantitative relationships.
- Analyze change in various contexts.

Wisconsin's Model Academic Standards of Mathematics, Standard F: Algebraic Relationships

Students in Wisconsin will discover, describe and generalize simple and complex patterns and relationships. In the context of real-world problem situations, the student will use algebraic techniques to define and describe the problem to determine and justify appropriate solutions (Wisconsin Department of Public Instruction 1998, 14–15).

TABLE 5.15 **Wisconsin's Model Academic Standards for Algebraic Relationship**

Wisconsin's performance standards	Elaboration
By the end of grade 8 students will:	
F.8.1 Work with algebraic expressions in a variety of ways, including: ■ using appropriate symbolism, including exponents, and variables, ■ evaluating expressions through numerical substitution, ■ generating equivalent expressions, and ■ adding and subtracting expressions.	Much of the content of this standard involves work with expressions that are found in a traditional first-year algebra course. A key item to note is the mention of exponents; students must be able to use them effectively and correctly.

TABLE 5.15 **(Continued)**

Wisconsin's performance standards	Elaboration
By the end of grade 8 students will:	
F.8.2 Work with linear and nonlinear patterns and relationships in a variety of ways, including: ■ representing them with tables, with graphs, and with algebraic expressions, equations, and inequalities, ■ describing and interpreting their graphical representations (e.g., slope, rate of change, intercepts*), ■ using them as models of real-world phenomena, and ■ describing real-world phenomenon that a given graph might represent.	See F.4.3 for the characterization of linear patterns and relationships. Students should also be comfortable with exponential relationships (e.g., {1, 2, 4, 8, 16, 32, ... [i.e., 2^x]}, {2, 6, 18, 54, 162, ... [i.e., 2×3^x] and quadratic relationships (e.g., {5, 10, 20, 40, 80, 160, ... [i.e., 5×2^x]}. The importance of connecting these kinds of relationships, both linear and non-linear, with real-world phenomena cannot be over-emphasized.
F.8.3 Recognize, describe, and analyze functional relationships by generating a rule that characterizes the pattern of change among variables. These functional relationships include exponential growth and decay (e.g. cell division, depreciation).	This standard is connected with the one just above. Students cannot truly understand and work with relationships unless they deal with the notion of variability. But the key is to appreciate that however random a change in one of two related variables may be, the very nature of a relationship tells us that the change in the other related variable is predictable and easily determined.
F.8.4 Use linear equations and inequalities in a variety of ways, including: ■ writing them to represent problem situations and to express generalizations, ■ solving them by different methods (e.g., informally, graphically, with formal properties, and with technology), ■ writing and evaluating formulas (including solving for a specified variable), and ■ using them to record and describe solution strategies.	The intention of this standard should be made clear. All of the work with using and solving linear equations and inequalities that in the past were typically contained in a single beginning algebra typically taught to most students at the ninth grade level is now expected by the end of eighth grade. While the manner of implementation is left to individual teachers and school districts, the suggestion is offered to continue the work begun with standard F.4.5 with a year-by-year program in the middle grades that gradually, but carefully, increases the complexity of types of equations and inequalities as well as the sophistication and formality of solution methods.
F.8.5 Recognize and use generalized properties and relations, including: ■ additive and multiplicative properties of equations and inequalities, ■ commutativity and associativity of addition and multiplication, ■ distributive property, ■ inverses and identities for addition and multiplication, and ■ transitive property.	This standard speaks to the desire for a balance in the school algebra program between skill development and conceptual understanding. In the earlier middle grades, work with these properties and relations can be informal and intuitively based. But, by the end of eighth grade, students are expected to use correct and standard vocabulary and symbolism.

Additional Standards Information

Additional support information can be found in the NCTM *Curriculum and Evaluation Standards,* 98–104, and in the NCTM PSSM, 222–231.

Sample Problem with Commentary

A supporting beam used in construction is frequently made of triangles formed by using steel rods that are the same length. The triangles are in an alternating pattern as shown in the diagram. Contractors have a need to know the amount of material (rods) needed to construct a beam.

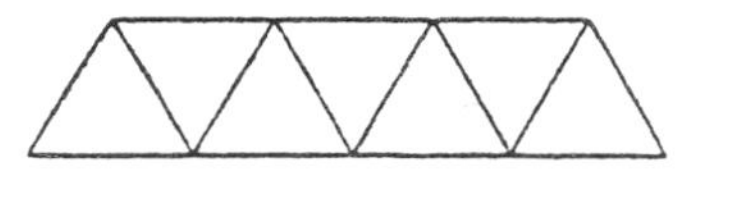

FIGURE 5.22

1. How many rods will it take to construct a beam of length 4? Can you find more than one method to find that number of rods?
2. Write an expression using numbers to illustrate how you solved part 1.
3. Write an equation relating the bottom length of the beam to the total number of rods needed for any beam.

Comments about the Problem

Students may approach the problem from a variety of perspectives, all of which allow them to think about the relationships between the two variables: length of the beam and the total number of rods. The teacher should allow students to enter the problem from their own perspective before doing any additional activities related to the problem. Initially students begin by just counting and then looking for ways to increase the efficiency of their counting. They use number expressions and make verbal descriptions before they write the symbolic representation.

Some students will count the three beams across the top, the four beams across the bottom, and the eight bars through the middle. The resulting number sentence is $3 + 4 + 8 = 15$.

Other students will multiply the number of triangles pointing up by three and add the top bars. This results in the numerical statement $4(3) + 3 = 15$.

Still other students will multiply the number of triangles pointing down by three and add the bottom bars and the ends. The corresponding number sentence is $3(3) + 4 + 2 = 15$.

Some students will count the total number of triangles by three and subtract the shared bars, resulting with $7(3) - 6 = 15$.

Some students may see the problem as partial triangles and count two sides of each triangle. They would then multiply the number of triangles times two and add the end piece. The resulting number sentence is $7(2) + 1 = 15$.

Once students have counted the bars needed for a length of four, they should be challenged to reexamine the problem and look at the number of rods needed for a beam of any length. One approach might be to take their procedure for part 1 and generate a table for beams of lengths 1, 2, 3, ... The ability to verbalize the patterns shown is important. Formal algebraic representation may be developed later.

For example the table for approach 1 will look as follows:

TABLE 5.16

Beam Length	Top Rods	Bottom Rods	Middle Rods	Total
1	0	1	2	3
2	1	2	4	7
3	2	3	6	11
4	3	4	8	15

Students should be able to verbally state a generalization for each of the columns in the table. Then students need to write these generalizations in algebraic representations. In the example above the length column would define the variable (r), the number of top rods would be represented with $(r - 1)$, the number of bottom rods as r, the number of middle rods is $2r$, and the total number of rods is $(r - 1) + r + 2r$. Each of the approaches that students might use as they initially approach the problem will produce different generalization patterns, allowing students to recognize equivalent algebraic expressions. Graphing calculators can be used to find that equivalent expressions have the same tables and graphs.

Problem adapted from Romberg, T. A., et al. (1996). Sampler, 46–49. *Mathematics in Context (MiC)*. Chicago, IL: Encyclopedia Britannica Eduational Corporation.)

Extension to the Problem

Students can translate the data from their table to a graph and determine if the relationship is linear, exponential, or quadratic. As the relationship is linear, the students have the opportunity to investigate the concept of slope and y-intercept and the slope-intercept form of linear equations.

Suggested Assessment Items

Write a letter to a classmate who was gone today explaining how you arrived at the total number of pieces in the beam and how that method can be used to find the number of pieces in a beam of any length.

Select any two algebraic expressions developed today and show how they are equivalent.

In your journal, complete the following prompt: Today I learned ...

Significant Mathematics

Within this problem the students were given the opportunity to use appropriate symbolism while generating equivalent expressions (F.8.1), work with a linear pattern using a real-world model by representing it in a variety of forms (F.8.2), describe and generalize a functional relationship by generalizing a pattern (F.8.3), write an equation to represent a problem situation (F.8.4), and identify properties and relations through the extension (F.8.5). There are natural connections between this problem and other content and performance standards. These include mathematical processes (A), number operations and relationships (B), geometry (C), and measurement (D).

Other Sample Problems

Examples of other rich problems such as this can be found from a number of sources including the NCTM standards, the NCTM Addenda Series for the Middle Grades, and problems from curricula produced through the projects funded by the National Science Foundation. For instance, an example of exponential growth and decay can be found in the NCTM Addenda Series book *Patterns and Functions* (Investigations 1–5 on pages 6 through 20). This source will provide problems which students can do and written discussion of the problems including appropriate tables and graphs.

Additional Resources

Additional rich problems such as these can be found in sources such as:

1. The NCTM Addenda Series book *Patterns and Functions* and *Algebra in a Technological World.*
2. Professional journals:
   - *Wisconsin Teacher of Mathematics,* published by the Wisconsin Mathematics Council.
   - *Mathematics—Teaching in the Middle School,* published by NCTM.
3. Examples given in NCTM Standards books.
4. Problems found in the curricula developed under the auspices of the National Science Foundation (see appendix A in this volume).

Performance Assessment Task

Assessment of conceptual understanding calls for opportunity for students to take in information, to weigh it, and to respond with explanation. Performance assessment tasks allow students to "produce an answer from their own knowledge and in their own words" (DPI 1995, 2). The Wisconsin Center for Education Research (WCER) at the University of Wisconsin-Madison, in cooperation with the Department of Public Instruction, developed several performance assessment items from 1994–1998. The "Picky Patterns" task, complete with scoring rubric and student work, follows as an example of such assessment. It is the kind of item that could be used to prepare students for the constructed response component of the *Wisconsin Knowledge and Concepts Examination* (WKCE). The item is used with permission from the Office for Educational Accountability, Wisconsin Department of Public Instruction.

Performance assessment tasks allow students to "produce an answer from their own knowledge and in their own words."

—Wisconsin Department of Public Instruction 1995, 2

Picky Patterns

1. On dot paper draw different figures using these guidelines:
   - Use only straight lines to connect dots.
   - All sides must begin and end at a dot.
   - All figures must be simple and closed.
   - No dots can be in the interior of a figure.

Examples

1.

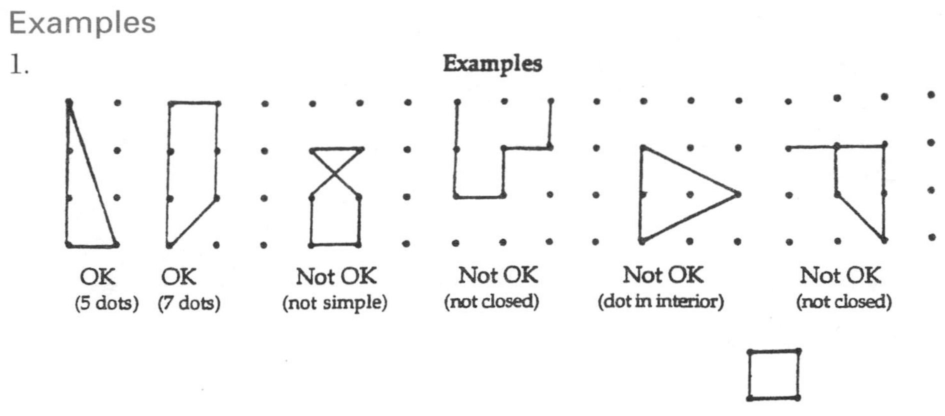

FIGURE 5.23

2. Using data from your drawings, create a table relating the number of dots on a figure to the area of that figure.
3. What would be the area of a figure with 100 dots? _______
4. Explain in detail how you found the area of the 100-dot figure (see Figure 5.25).

Scoring

Major Flaws

- Inaccurate drawings. (Points in the center, not closed, not simple)
- Fails to organize data in table form.

Minor Flaws

- Incomplete organization of data.
- Gives area as 50 or 48 in Question 3.
- Inaccurate algebraic expression where one is attempted.
- Insufficient number of drawings. At least five drawings should be included.

Potential Elements of an Advanced Response

- An algebraic expression for the area may be expressed as (n/2)-1 or (n-2)/2. Other variables are acceptable (such as D for number of dots.)
- Includes the algebraic expression and clearly explains how it was derived and what each part of the expression represents.
- Gives a very clear and insightful explanation of the relationship between the number of dots and the area of a figure.

Student Responses

See the following example of an advanced student performance.

Examples

1.

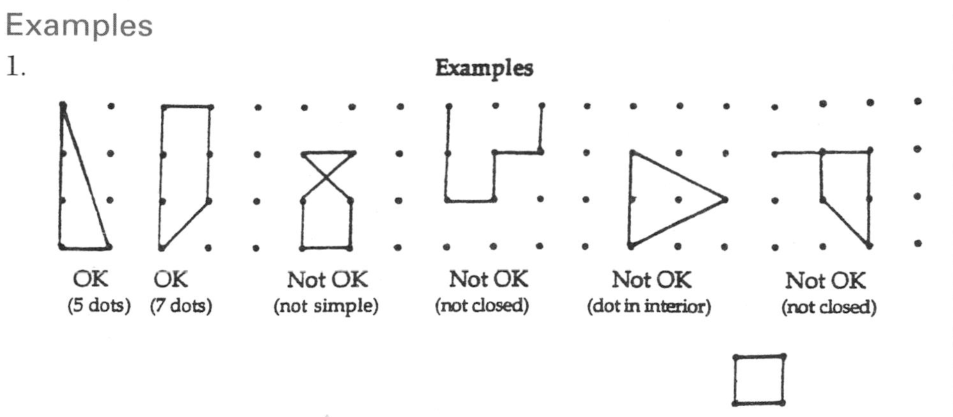

FIGURE 5.24

2. Using data from your drawings, create a table relating the number of dots on a figure to the area of that figure.

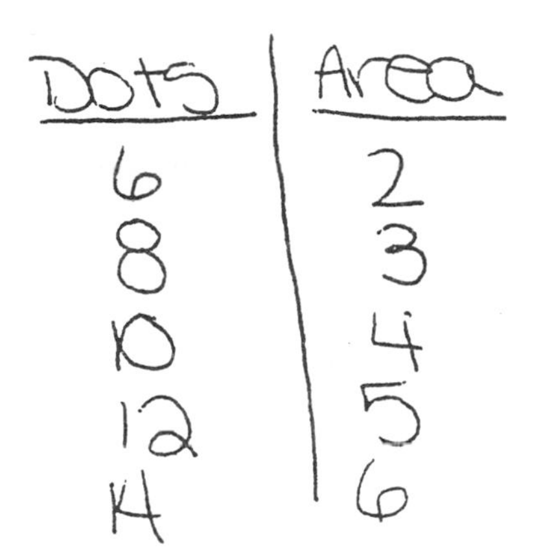

Dots	Area
6	2
8	3
10	4
12	5
14	6

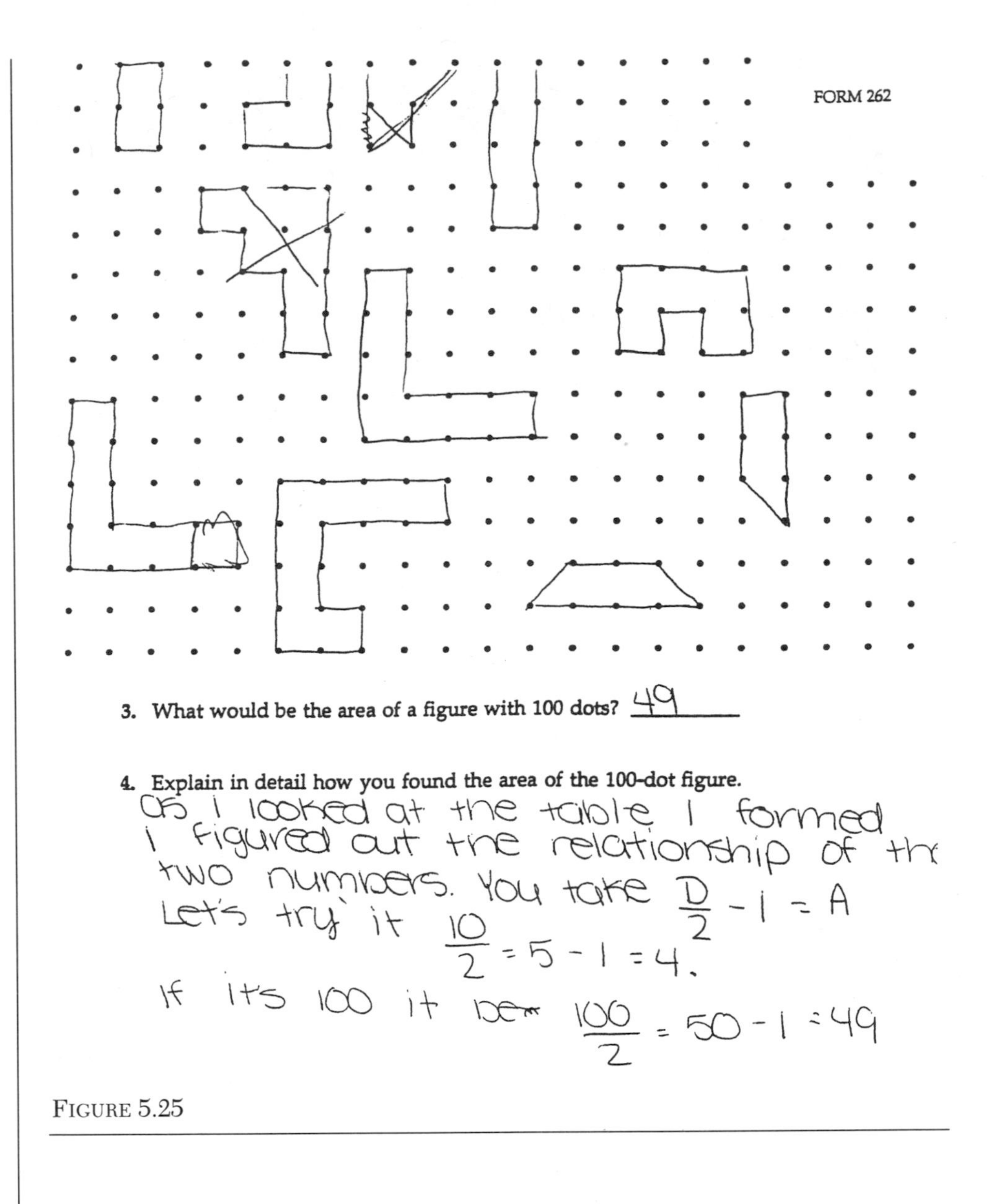

FORM 262

3. What would be the area of a figure with 100 dots? 49

4. Explain in detail how you found the area of the 100-dot figure.

as I looked at the table I formed I figured out the relationship of the two numbers. You take $\frac{D}{2} - 1 = A$ Let's try it $\frac{10}{2} = 5 - 1 = 4$.

If it's 100 it be $\frac{100}{2} = 50 - 1 = 49$

FIGURE 5.25

The Secondary Learner of Mathematics

Students entering high school today are much different than students in the past, because they enter with much stronger mathematical backgrounds due to the fact that they have had greater experiences with problem solving, reasoning, communication, algebraic thinking, and geometric concepts. As a result, schools at the high school level need to continue to challenge these students, as well as, help prepare them for the future. All mathematics programs at the high school level must contain challenging mathematics, including, not only algebra and geometry, but probability, statistics, indirect measurement, discrete mathematics, and the basic elements of calculus.

All mathematics programs at the high school level must contain challenging mathematics, including, not only algebra and geometry, but probability, statistics, indirect measurement, discrete mathematics, and the basic elements of calculus.

Students should be encouraged to take four years of mathematics in high school whether they plan to pursue further study of mathematics, to enter the workplace, or to look at other post-secondary options. Additional time will allow students to learn not only more concepts but those concepts in more depth. Teachers will have time to better facilitate mathematical explorations, problem solving, and communication. Districts will need to find creative solutions in designing their mathematics programs so that four meaningful years of mathematics are possible for all students. This may mean creating new courses such as integrated mathematics, so that students who choose not to pursue the traditional junior and senior year mathematics sequence (i.e., Algebra II, Pre-Calculus/Trigonometry, and Calculus) still have the opportunity to continue learning challenging mathematics.

High school students of today have grown up in a technological world. Technology should be available to students as needed. Graphing calculators and computer software need to serve as tools in helping students build understanding of mathematical concepts and for multiple representations that can be used to solve problems.

Assessment needs to inform instruction and to facilitate student achievement. Alternative types of assessment such as portfolios, performance assessments, journals, and observations, as well as traditional types of assessment such as homework, quizzes, and tests, need to be used. Assessments should be substantive and parallel instruction. Even when using technology, students should explain what their reasoning was in solving the problem, including how and why the technology was used. More detailed information about assessment is available in the assessment section of this document.

Using Mathematical Processes in Secondary Schools

WMAS (1998) stated, "In order to participate fully as a citizen and a worker in our contemporary world, a person should be mathematically powerful. Mathematical power is the ability to explore, to conjecture, to reason logically, and to apply a wide repertoire of methods to solve problems. Because no one lives and works in isolation, it is also important to have the ability to communicate mathematical ideas clearly and effectively." The mathematical processes as

Mathematical power is the ability to explore, to conjecture, to reason logically, and to apply a wide repertoire of methods to solve problems.

described by the Wisconsin Performance Standards and the *Principles and Standards for School Mathematics* (Standards 2000 Project 2000) outlined the essential components of mathematical processes that allow students to learn and comprehend all other aspects of mathematics. These tools are essential to the learning of all mathematics content.

The process standard acts as an umbrella (see Figure 5.26) over all the content standards:

Mathematical Processes
Number Operations and Relationships
Geometry
Measurement
Statistics and Probability
Algebraic Relationships

FIGURE 5.26

Every effort needs to be made to incorporate all the parts of the process standard into each mathematical endeavor! Mathematical processes permeate all mathematics instruction, regardless of the specific content being addressed. In addition to reading and understanding mathematical statements, students need to be able to demonstrate those understandings by speaking about them, writing about them, and using them when problem solving. Students should frequently be asked to explain their strategies, justify their reasoning and solutions, and defend the reasonableness of an answer. The essence of mathematics can be thought of as a way of thinking, a language. The 9–12 experience transitions students from the communication skills learned in earlier grades to the communications skills that are necessary in today's workplace.

Every effort needs to be made to incorporate all the parts of the process standard into each mathematical endeavor.

National Council of Teachers of Mathematics Process Standard, *Principles and Standards for School Mathematics* (Standards 2000 Project 2000

National Council of Teachers of Mathematics Process Standards include five strands of processes: problem solving, reasoning and proof, communication, connections, and representation (See introduction in this section, p. 89). The *Wisconsin Model Academic Standards for Mathematics* (Wisconsin Department of Public Instruction, 1998, 4–5) offers a single process standard which, because it impacts all other strands, is offered as Standard A: Mathematical Processes and encompasses the use of reasoning and logic, logical argument, analysis of nonroutine problems, effective oral and written presentations, succinct organization of mathematical procedures, and the ability to read and understand written mathematics.

Wisconsin's Model Academic Standards for Mathematics, Standard A: Mathematical Processes

Students in Wisconsin will draw on a broad body of mathematical knowledge and apply a variety of mathematical skills and strategies, including reasoning, oral and written communication, and the use of appropriate technology, when solving mathematical, real-world, and non-routine problems (Wisconsin Department of Public Instruction 1998, 4–5).

Wisconsin Performance Standards

By the end of grade 12 students will:

A.12.1 Use reason and logic to:
- evaluate information,
- perceive patterns,
- identify relationships,
- formulate questions, pose problems, and make and test conjectures, and
- pursue ideas that lead to further understanding and deeper insight.

A.12.2 Communicate logical arguments and clearly show:
- why a result does or does not make sense,
- why the reasoning is or is not valid, and
- an understanding of the difference between examples that support a conjecture and a proof of the conjecture.

A.12.3 Analyze nonroutine problems and arrive at solutions by various means, including models and simulations, often starting with provisional conjectures and progressing, directly or indirectly, to a solution, justification, or counter-example.

A.12.4 Develop effective oral and written presentations employing correct mathematical terminology, notation, symbols, and conventions for mathematical arguments and display data.

A.12.5 Organize work and present mathematical procedures and results clearly, systematically, succinctly, and correctly.

A.12.6 Read and understand:
- mathematical texts and other instructional materials,
- writing about mathematics (e.g., articles in journals), and
- mathematical ideas as they are used in other contexts.

Additional Standards Information

Additional support information can be found in the NCTM *Curriculum and Evaluation Standards,* 137–149, and in the NCTM PSSM, 334–364.

Sample Problem with Commentary

Students who enter high school with a firm foundation in mathematical understanding will better be able to augment their knowledge. Understanding that mathematical ideas need verification, students will continue to seek proof for advanced mathematical concepts. Knowing that mathematics has an experimental aspect that helps predict truth, students will also know that experimentation is not a formal mathematical proof. Technology can often be used to develop and enhance the intuitive understanding of mathematical concepts. Formal proofs for experiments done in elementary and middle school will be part of their secondary school experience.

Reasoning and proof and representations are two of the process standards from the NCTM *Principles and Standards for School Mathematics.* Students in grades 9–12 develop an understanding that mathematical argument depends on certain assumptions. Encouraging the understanding of mathematics through patterns and conjectures is a sound preliminary educational approach, but formal axiomatic development helps avoid false conclusions. For example, to show that the sum of the angles of a triangle is 180 degrees, axioms or properties of the plane, such as any two distinct points have a unique line through them, can be used. However, when working with a sphere (see Figure 5.27), two points on the equator, together with the point at the north pole, form a triangle that will have an angle sum of more than 180 degrees because the shape is not a plane figure. The proof that the angle sum in a triangle in a plane is 180 degrees uses the fact that two distinct points have a single line through them; this is not true for a sphere. The use of one set of axioms and assumptions will determine specific outcomes, whereas use of a different set of axioms and assumptions will result in a different conclusion—one which may be equally valid.

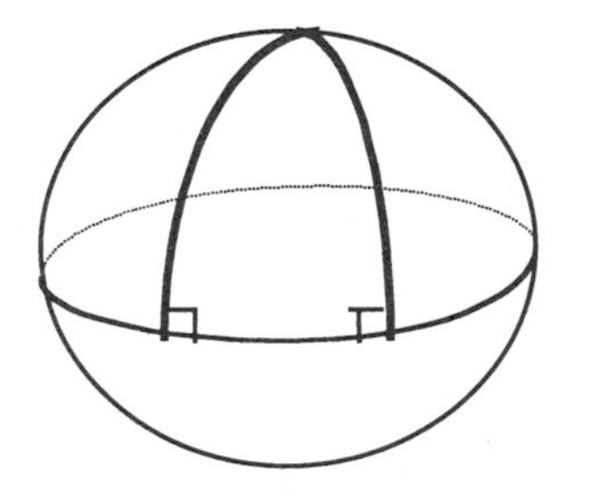

Figure 5.27

Additional Resources

Additional rich problems can be found in sources such as:

1. The 1989 NCTM Standards and Addenda Series books.
2. Professional Journals:
   - *Wisconsin Teacher of Mathematics* published by the Wisconsin Mathematics Council.
   - *Mathematics Teacher* published by National Council of Teachers of Mathematics.
3. Problems found in the curricula developed under the auspices of the National Science Foundation.

Learning Number Operations and Relationships in Secondary Schools

Secondary students who feel comfortable with numbers and number operations will have better skills and more self-confidence in mathematics. As a result, they are at an advantage to learn other mathematical concepts. In grades 9–12, students will be dealing with the number properties at a formal level. For example, the distributive property will be emphasized and fully developed since it plays a critical role in the understanding of mathematics at the high school level. The use of this property, along with many other properties, will be expanded and applied to the set of real numbers.

Students at the 9–12 level need to use and understand a variety of number systems, acceptable limits of accuracy, various number scales, reasonableness of results, exact versus rounded values, numerical arguments, and more challenging patterns of numbers. Secondary mathematics can lead to students seeing the beauty of numbers, as well as, the usefulness of mathematical applications. Use of technology should be encouraged for experimentation and enhancing inductive reasoning. Communication is an extremely important part of the program and allows students to formulate conjectures, give logical arguments why generalizations are true, and to supply counter examples to support or to question conjectures.

Students at the 9-12 level need to use and understand a variety of number systems, acceptable limits of accuracy, various number scales, reasonableness of results, exact versus rounded values, numerical arguments, and more challenging patterns of numbers.

National Council of Teachers of Mathematics Number and Operations Standard, *Principles and Standards for School Mathematics* (Standards 2000 Project 2000)

Mathematics instructional programs should foster the development of number and operation sense so that all students:

- Understand numbers, ways of representing numbers, relationships among numbers, and number systems.
- Understand the meaning of operations and how they relate to each other.
- Compute fluently and make reasonable estimates.

Wisconsin's Model Academic Standards for Mathematics, Standard B: Number Operations and Relationships

Students in Wisconsin will use numbers effectively for various purposes, such as counting, measuring, estimating, and problem solving (Wisconsin Department of Public Instruction 1998, 6–7).

TABLE 5.17 **Wisconsin's Model Academic Standards for Number Questions and Relationships**

Wisconsin's performance standards	Elaboration
By the end of grade 12 students will:	
B.12.1 Use complex counting procedures such as union and intersection of sets and arrangements (permutations and combinations) to solve problems.	The intent is to use these procedures in an informal and intuitive way. While the concepts behind them are important, students, and teachers need not become involved with the "heavy" and formal symbolism that is often associated with these ideas. Factorial notation is probably needed, but that is not such a formidable idea.
B.12.2 Compare real numbers using order relations and transitivity, ordinal scales including the logarithmic scale, arithmetic differences, and ratios, proportions, percents, rates of change.	Here the number domain of interest is expanded to include irrational numbers. One should not read too much into use of logarithmic scales. Yet, students should be made aware of their applied use. Remember that students are expected to be proficient in use of exponential and scientific notation as ways to represent numbers. Students should be aware of the different uses of numbers such cardinality, ordinality, or identification.
B.12.3 Perform and explain operations on real numbers (add, subtract, multiply, divide, raise to a power, extract a root, take opposites and reciprocals, determine absolute value).	This standard directly parallels standard B.8.2. The only difference is the expansion of the number domain to include irrational numbers. The expectation is that student will be able to deal with numbers expressed in radical notation.
B.12.4 In problem-solving situations involving the application of different number systems select and use appropriate computational procedures, properties, and modes of representation.	This statement gives a wide-ranging summarization of what it means for a student to have number sense. It goes beyond just the ability to compute, important though that skill is. Recognizing properties of number operations as well as using different ways to represent and think about numbers is part of the repertoire of a numerate person.
B.12.5 Create and critically evaluate numerical arguments presented in a variety of classroom and real-world situations.	This is simply another aspect of numeracy. Here the number-sensible student is able to apply what he or she knows about a number (relative size, properties, computation) to numerical statements made by others in a wide variety of contexts.
B.12.6 Routinely assess the acceptable limits of error when evaluating strategies, test the reasonableness of results, and using technology to carry out computations.	And, finally, this standard addresses yet another mark of a person with number sense. Far from being the exact science that some people believe it is, mathematics, especially the applied portion of it deals with real problems, is a way of handling the variability and imprecision that naturally arises from computing with numerical data derived from measurements or from approximations and estimates. Here, the student needs to be able to determine reasonable confidence intervals that are consistent with the requirements of the original problem situation.

Additional Standards Information

Additional support information can be found in the NCTM *Curriculum and Evaluation Standards,* 184–186, and in the NCTM PSSM, 290–294.

Sample Problem 1 with Commentary

Building on activities in elementary school where students work with fraction concepts at an intuitive level and on investigations in middle school where students apply rational numbers and proportional reasoning, including rational number summations, the following problem then extends and formalizes those understandings. What is the sum of the series:

$$(1) \quad \frac{1}{3}+\frac{1}{3^2}+\frac{1}{3^3}+\frac{1}{3^4}+\ldots?$$

Discussion

Using a calculator, enough terms can be added to make a prediction or approximation of the value of the infinite sum. Defining the partial sum S(n) to be the sum of the first n terms, and plotting those terms, the graph approaches the value of (1). Next, working algebraically, one gets

$$S(1)=\frac{1}{3},\ S(2)=\frac{4}{9},\ S(3)=\frac{13}{27},\ S(4)=\frac{40}{81},\ \ldots$$

The Question

What is the general formula? The denominator of $S(n)$ is 3^n and the numerator is the denominator minus 1 then divided by 2. Thus,

$$S(n)=\lfloor(3^n-1)/2\rfloor/3^n=\frac{1}{2}-\frac{1}{(2\cdot 3^n)}.$$

Introducing the concept of limit, it is clear that the limit of S(n) as n goes to infinity is $1/2$. Another way to express the problem: let

$$S=\frac{1}{3}+\frac{1}{3^2}+\frac{1}{3^3}+\frac{1}{3^4}+\ \ldots\ .$$

Calculate

$$\frac{1}{3}S=\frac{1}{3^2}+\frac{1}{3^3}+\frac{1}{3^4}+\ldots\ .$$

Comparing these expressions it can be observed that

$$\frac{1}{3}S+\frac{1}{3}=S.$$

The solution to this equation

$$S=\frac{1}{2}.$$

More generally, for the sum of an infinite geometric series where the absolute value of the common ratio (r) is less than one, it is possible to find the formula for the arbitrary geometric series. $S = a + ar + ar^2 + ar^3 + \ldots$. By calculating rS, noting $rS + a = S$, and solving for S, one arrives at the formula

$$S = \frac{a}{1-r}.$$

Sample Problem 1 was an original problem contributed by J. Marshall Osborn. The extension can be found in "Using three representations to present a new concept," by J. F. Marty. *Wisconsin Teacher of Mathematics*, vol. 47, no. 3 (1996): 28–30. The application came from the New Standards Project.

Standards Addressed

A.12.1 Use reason and logic to perceive patterns and pursue ideas that lead to further understanding and deeper insights.

A.12.5 Organize work and present mathematical procedures and results clearly, systematically, succinctly, and correctly.

B.12.2 Compare real numbers using order relations and transitivity, ordinal scales including the logarithmic scale, arithmetic differences, and ratios, proportions, percents, rates of change.

B.12.3 Perform and explain operations on real numbers (add, subtract, divide, raise to a power, extract a root, take opposites and reciprocals, determine absolute value)

B.12.5 Application of different number systems, rational numbers in this case, and that students use number properties and different modes of representation.

Connection

A discussion of a similar problem with the common ratio equal to $1/2$ can be found in the article titled "Using Three Representations to Present a New Concept," published in the *Wisconsin Mathematics Teacher,* Fall 1996, Vol. 47, No. 3. In the article, the author describes how the problem

$$S = \frac{1}{2} + \frac{1}{4} + \frac{1}{8} + \frac{1}{16} + \ldots$$

was examined from a numerical, geometric, and symbolic perspective in a classroom setting.

Numerical Perspective

See below for a program for a graphing calculator. The program merely required inputting the initial term and the common ratio. Using a PAUSE statement allows students to see the effect of adding each term to the sum (see Figure 5.28). Please note that the sum not only converges to 1, but actually assumes the value of 1. The fact that this is an infinite loop implies that there is an infinite series. Not only do students see the series numerically converge to a sum of 1, but the program also leads to an interesting discussion of rounding, errors resulting from rounding, and the implication of such errors.

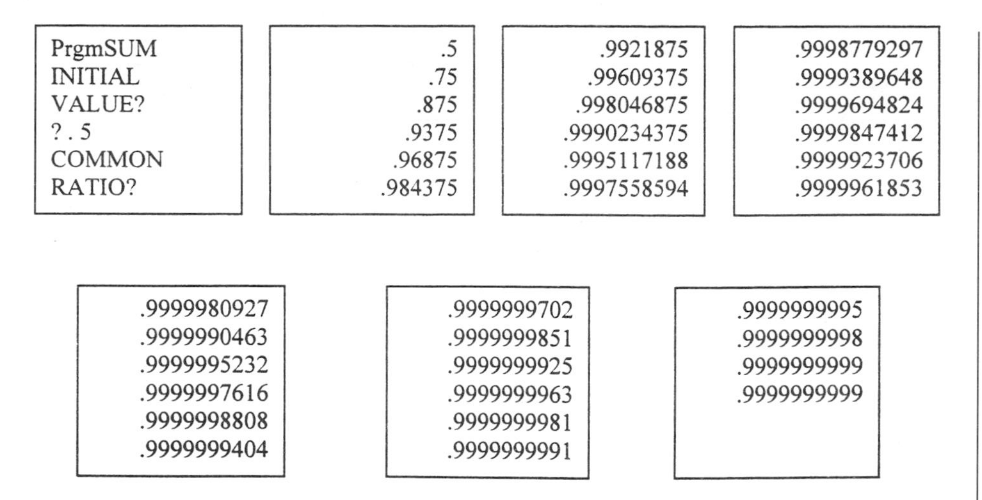

FIGURE 5.28

```
0 → SUM
Disp "INITITAL VALUE?"
Input
Disp "COMMON RATION?"
Input R
I → SUM
Disp SUM
Lbl A
I * R → I
SUM + I → SUM
Pause
Go to A
```

Geometric Perspective

Take a unit square and divide it into two congruent rectangles, each with an area of one-half. Divide one of the rectangles again into two congruent rectangles (See Figure 5.29). Continue to divide each successive rectangle in half several more times. Each rectangle is then the next term in a series in a sample series creating a spiral pattern. The area left after each division approaches zero, so each tiny addition of area forces the area to get closer to the area of the original square.

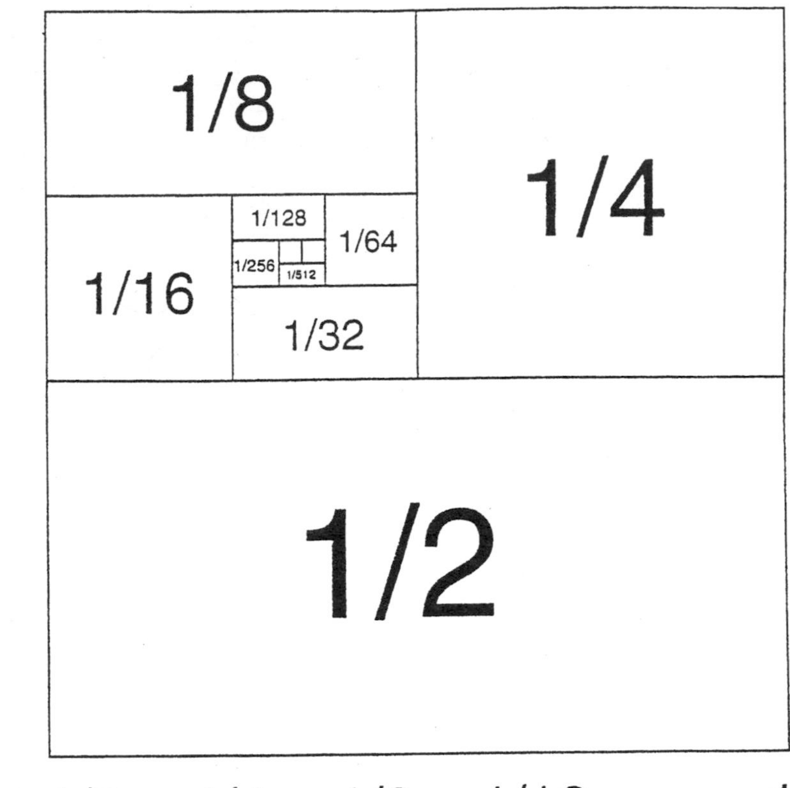

FIGURE 5.29

Symbolic Perspective

As n gets larger, the value of

$$(1/2)^n$$

gets smaller. More specifically, as n gets infinitely large,

$$(1/2)^n$$

approaches zero. Limit notation can be used

$$\lim_{n \to \infty} (1/2)^n = 0$$

The formula can be reexamined for the nth partial sum of a geometric series.

$$S_n = a_1 \frac{1-r^n}{1-r}$$

As n becomes very large, r^n approaches zero. The formula for the sum of the infinite series (S) approaches

$$\frac{a_1}{1-r}$$

where a_1 is the initial term of the series and r is the common ratio. Some limit notation can again be used.

$$\lim_{n \to \infty} S_n = a_1 \frac{1}{1-r}$$

Thus, algebra can be used to derive a formula for the sum of an infinite geometric series, namely

$$S = \frac{a_1}{1-r}$$

Application/Extension

A bank vault contains a large amount of gold and you are told that you may keep all that you carry out, under the following conditions:

a. On the first trip you may take out only one pound.
b. On each successive trip you may take out half the amount you carried out on the previous trip.
c. Each complete trip takes one minute.

Explain how much gold you can carry out, and how long it will take to do it. Also, determine your hourly rate of earnings if you "work" for 15 minutes. Assume that the value of the gold is $350 per ounce. What would be your hourly rate if you "worked" for 20 minutes? What if you "worked" for an hour?

This contextual example of a converging infinite series was published by the *New Standards Project* (1997, 58–60).

Sample Problem 2 with Commentary

Although the previous examples involve geometric series, students should also be familiar with finding a formula for the nth term and the nth partial sum of an arithmetic series (linear expressions) and a series that can be described by quadratic models. An example of the later is figurate numbers (numbers that form a geometric figure), for example, a pentagon. The number (see Table 5.18) represents the number of dots in the picture. The pentagonal numbers are whole numbers that are represented by pentagonal shapes. The first four pentagonal numbers are shown in Figure 5.30.

TABLE 5.18

Number	Number of Dots (Pentagon Numbers)
1	1
2	5
3	
4	
5	

Pentagonal Numbers

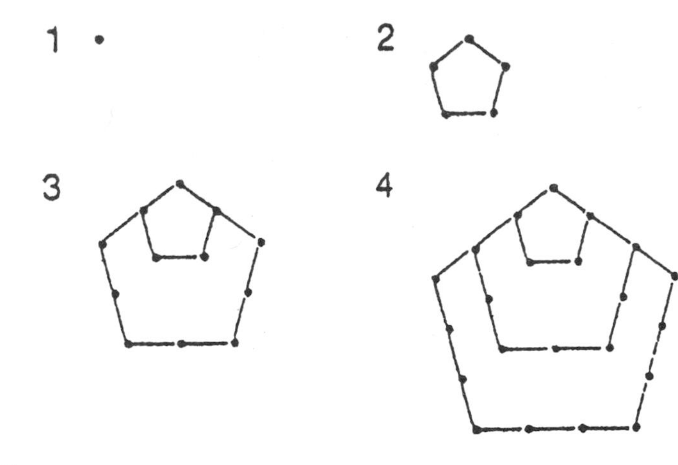

FIGURE 5.30

Pentagonal Numbers

Answer the following questions.

a. Complete the table and look for a pattern.
b. Make a sketch to represent the fifth pentagonal number.
c. How many dots are in the diagram representing the ninth pentagonal number?
d. Is there a pentagonal number that has 200 dots in its shape? If so, which one?
e. Write a formula for the *n*th pentagonal number.

Discussion

By taking the difference of successive pentagonal numbers, one gets the sequence 4,7,10,13, ... Because this is not a constant sequence, pentagonal numbers do not form a linear sequence. By taking the difference of the numbers in this new sequence, we get 3,3,3, ... which is a constant sequence. Thus, the pentagonal sequence must be a quadratic sequence, i.e., if y_n is the *n*th pentagonal number, then for some constants *a, b,* and *c,* $y_n = an^2 + bn + c$. The values of *a, b,* and *c* can be found by studying three special cases of this equation:

n = 1: $1 = a + b + c$
n = 2: $5 = 4a + 2b + c$
n = 3: $12 = 9a + 3b + c$

Solving these equations simultaneously yields

$$a = \frac{3}{2},\ b = -\frac{1}{2},\ \text{and } c = 0.$$

Thus, the formula is

$$y_n = \frac{3}{2}n^2 - \frac{1}{2}n.$$

Examples of other sets of numbers that can be used to develop sequences and series of this nature include the square numbers, the cubic numbers, the triangular numbers, the rectangular numbers, and the hexagonal numbers. Other rich sources for working with numbers include Pascal's Triangle (Seymour, 1986), Fibonacci Numbers, and Lucas Numbers (Bezuska & Kenney, 1982).

Sample Problem 2 was contributed by James F. Marty.

Standards Addressed

B.12.5 Create and critically evaluate numerical arguments presented in a variety of classroom and real-world situations (e.g., political, economic, scientific, social).

The various perspectives represented in this section address the representation process standard from NCTM *Principles and Standards School Mathematics:* Select, apply and translate among mathematical representations to solve problems.

Sample Problem 3 with Commentary

Assuming every registered voter is keeping up on current events, a poll of 100 registered voters designed to find out how voters kept up with current events revealed the following facts:

- Sixty-five watched the news on television.
- Thirty-nine read the newspaper.
- Thirty-nine listened to radio news.
- Twenty watched TV news and read the newspaper.
- Twenty-seven watched TV news and listened to radio news.
- Nine read the newspaper and listened to radio news.
- Six watched TV news, read the newspaper, and listened to radio news.

a. How many of the 100 people surveyed kept up with current events by some means other than the three sources listed?

b. How many of the 100 people surveyed read the paper but did **not** watch the TV news?

c. How many of the 100 people surveyed used only one of the three sources listed to keep up with current events?

Discussion

While there are various methods to solve this problem, one option is to a use a table or a Venn Diagram.

Here is a possible solution to this problem using a Venn Diagram.

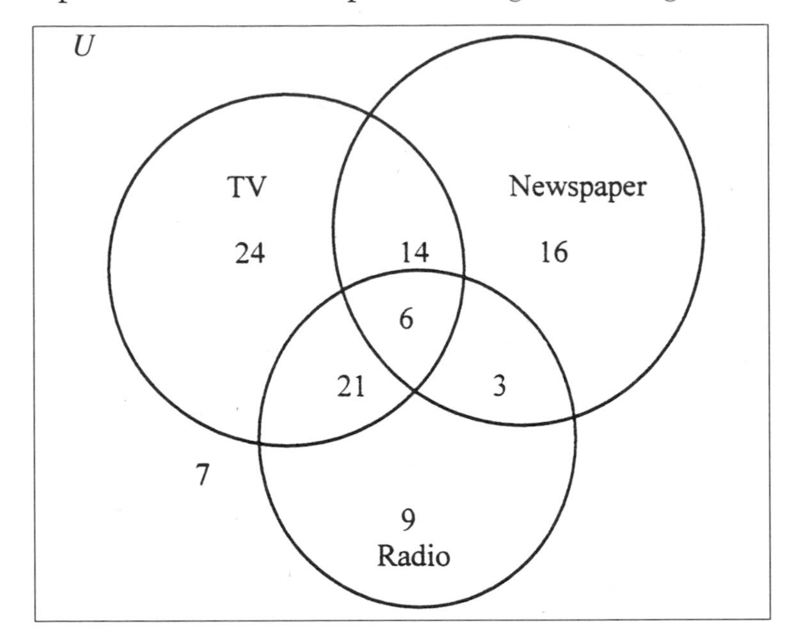

FIGURE 5.31

Sample Problem 3 was contributed by James F. Marty.

Standards Addressed

B.12.3 Perform and explain operations on real numbers (add, subtract, divide, raise to a power, extract a root, take opposites and reciprocals, determine absolute value).

Sample Problem 4 with Commentary

Most people read about Richter scale readings when an earthquake occurs, understand that a measurement of the pH tells how acidic a liquid or solid is, or know how many decibels a loud music sound is, but few people understand that these numbers come from formulas involving logarithms. The common logarithm of a number is the power of ten to which the number is equal. For example, log 1,000 = 3 because $10^3 = 1{,}000$, whereas log 0.01 = –2 because $10^{-2} = 0.01$. If something is equal to the logarithm of the intensity, then the log changing by 1 means the intensity changed by a factor of 10.

A logarithmic scale describes the intensity of sound. The quietest sound that a human can hear has an intensity of about 10^{-12} watts per square meter (w/m^2). The human ear can also hear sounds with intensities as large as 10^2 w/m^2. Because the range from 10^{-12} to 10^2 is so large, it is convenient to use another unit, the decibel (dB), to measure sound intensity. The decibel is $^1/_{10}$ of a bel, a unit named after Alexander Graham Bell, the inventor of the telephone.

Table 5.19 gives the decibel and the corresponding w/m^2 values for some common sounds.

Table 5.19

Watts/Square Meter	Type of Sound	Decibels
10^{2}	jet plane (30 m away)	140
10^{1}	pain level	130
10^{0}	amplified rock music (2 m)	120
10^{-1}		110
10^{-2}	noisy kitchen	100
10^{-3}	heavy traffic	90
10^{-4}		80
10^{-5}		70
10^{-6}	normal conversions	60
10^{-7}	average home	50
10^{-8}		40
10^{-9}	soft whisper	30
10^{-10}		20
10^{-11}		10
10^{-12}	barely audible	0

As the decibel values in the right column increase by 10, the corresponding intensities in the left column multiply by 10. Thus, if the number of decibels is increased by 20, the sound intensity is multiplied by 100. If you increase the sound intensity by 40 dB, you multiply the watts per square meter by 10,000. In general, an increase of n dB multiplies the intensity by $10^{n/10}$. The 120 dB intensity of loud rock music is $10^{60/10} = 10^{6} = 1{,}000{,}000$ times the 60 dB intensity of normal conversation.

Logarithmic scales are different from linear scales. On a linear scale, the units are spaced so that the difference between successive units is the same. On the logarithmic scale, the units are spaced so the *ratio* between successive units is the same. Logarithmic scales are often used to model data with a very wide range of values.

Use Table 5.19 to answer the following questions:

1. What is the intensity of a sound barely audible to human beings?
2. Give an example of a sound that is 100 times more intense than a noisy kitchen.
3. How many times more intense is normal conversation than a soft whisper?
4. The intensity of a jet plane at 600 m is 20 dB more than that of a pneumatic drill at 15 m. How many times more intense is the sound of the jet?

Answers to Practice Problem 4:

1. $10^{-12} w/m^2$
2. amplified rock music
3. 1000 times
4. 100

Sample Problem 4 can be found in the *University of Chicago School of Mathematics Project: Advanced Algebra*, by Z. Usiskin (Glenview, IL: Scott, Foresman and Company, 1990) pp. 500–502. Both extensions were contributed by James F. Marty.

Standards Addressed

B.12.2 Compare real numbers using ordinal scales including logarithmic (e.g,. Richter, pH rating).

In order to fully understand what these readings in problem 4 mean, a little bit must be known about logarithms. It is not the intent of this standard to have everyone learn all there is to know about logarithms, but merely to know that a change of 1 means something different in a different context, namely when something is defined in terms of a logarithm.

Extension 1

The loudness of sound, as experienced by the human ear, is based upon intensity levels. A formula used for finding the decibel level α which corresponds to a sound intensity I is $\alpha = 10\log(I/I_0)$ decibels where I_0 is a special value of I agreed to be the weakest sound that can be detected by the ear under certain conditions. (Note: log refers to the common logarithm, base 10.)

Find α if:

1. I is 10 times as great as I_0.
2. I is 1,000 times as great as I_0.
3. I is 10,000 times as great as I_0. (This is the intensity level of the average voice.)

Answers to Extension 1:

1. because $I/I_0 = 10$ and $\log 10 = 1$, $\alpha = 10 \cdot 1 = 10$ decibels.
2. because $I/I_0 = 1{,}000$ and $\log 1{,}000 = 3$, $\alpha = 10 \cdot 3 = 30$ decibels.
3. because $I/I_0 = 10{,}000$ and $\log 10{,}000 = 4$, $\alpha = 10 \cdot 4 = 40$ decibels.

Standards Addressed

This topic goes beyond the Number Operations and Relationships Standard. The last example presented in the Statistics and Probability Standard section of this document illustrates that when given a power function, the graph of $\log y$ versus $\log x$ will produce a linear graph. Given an exponential function, the graph of $\log y$ versus x, will also produce a linear graph. See Extension 2 of Sample Problem 3 in the Statistics and Probability Standard for further elaboration on these ideas.

Extension 2

Ideas from the Algebraic Relationships strand of the WMAS.

Assume we have an exponential function	$y = a \cdot 10^{mx}$
Take the logarithms of both sides	$\log y = \log(a \cdot 10^{mx})$
Expand using properties of logarithms	$\log y = \log a + mx \cdot \log 10$
	$\log 10 = 1$ since $10^1 = 10$
	$\log y = mx + \log a$
$\log a$ is a constant, call it b, and let $y' = \log y$	$y' = mx + b$
The graph of $\log y$ versus x is linear	

Assume we have a power function	$y = a \cdot x^m$
Take the logarithms of both sides	$\log y = \log(a \cdot x^m)$
Expand using properties of logarithms	$\log y = \log a + m \log x$
log a is a constant, call it b, let $y' = \log y$, and $x' = \log x$	$y' = m x' + b$
The graph of log y versus log x is linear.	

Three Important Points Related to Standards

1. WMAS must be viewed as a complete document, all the standards are interrelated as is mathematics as a whole.
2. WMAS is to be viewed as foundations on which mathematical understanding can be built. Certainly teachers and students are expected to extend beyond just the topics illustrated in the document.
3. Such topics as linear functions may seem quite elementary in an algebraic sense because so many phenomena in our world require nonlinear mathematical models; however, their true value is reflected in their close ties to the more powerful aspects (mathematical modeling) that appear later in the development of mathematics.

Additional Resources

Additional rich problems such as these can be found in sources such as:

1. The 1989 NCTM Standards and Addenda Series books.
2. Professional Journals:
   - *Wisconsin Teacher of Mathematics* published by the Wisconsin Mathematics Council.
   - *Mathematics Teacher* published by NCTM.
3. Problems found in the curricula developed under the auspices of the National Science Foundation.

Learning Geometry in Secondary School

Geometric applications are extremely practical in many real-life situations; the understanding of geometric and visualization skills is essential. These skills are also prerequisite for further study of mathematics (for example, in understanding graphs). Many students encounter mathematics that does not primarily deal with numbers through geometry. On an intuitive level, students need to experience explorations of geometric objects, measurement in geometry, geometric transformations, congruence, and similarity before they begin to construct formal logical arguments. Formal argument in geometry serves two purposes for the student. First, it enhances and expands on the student's reasoning ability, helping the student to find the solution to problems of a type that have not been encountered before. Secondly, it prepares the student for the more formal approach to the learning of mathematics through the deductive reasoning process. Quality high school mathematics programs incorporate informal geometry, synthetic geometry, coordinate geometry, and transformational geometry.

The understanding of geometric and visualization skills is essential for further study of mathematics.

Geometry is an integral part of trigonometry and calculus courses, where the student needs to know the geometric properties of figures and be able to visualize such things as solids of revolution and minimizing distances. Additionally, students with a good background in geometry are most successful in their study of physics, especially when working with such topics as vectors and derived attributes such as velocity and acceleration. Beyond academia, the understanding of geometry is needed in many occupations such as carpentry, landscaping, and decorating.

National Council of Teachers of Mathematics Geometry Standard, *Principles and Standards for School Mathematics* (Standards 2000 Project 2000)

Mathematics instructional programs should include attention to geometry and spatial sense so that all students:

- analyze characteristics and properties of two- and three-dimensional geometric shapes and develop mathematical arguments about geometric relationships,
- specify locations and describe spatial relationships using coordinate geometry and other representational systems,
- apply transformations and use symmetry to analyze mathematical situations, and
- use visualization, spatial reasoning, and geometric modeling to solve problems.

Wisconsin's Model Academic Standards for Mathematics, Standard C: Geometry

Students in Wisconsin will be able to use geometric concepts, relationships, and procedures to interpret, represent, and solve problems. (Wisconsin Department of Public Instruction 1998, 8–9).

Table 5.20 **Wisconsin's Model Academic Standards for Geometry**

Wisconsin's performance standards	**Elaboration**
By the end of grade 12 students will:	
C.12.1 Identify, describe, and analyze properties of figures, and relationships among their parts.	This standard builds upon the groundwork laid in Standards C.4.1 and C.8.1. The important difference here is the inclusion of word "analyze" which opens up the use of logic and reasoning to examine the why's and why-not's of the existence of familiar properties and relationships. Also important is the use of computer software in the analysis process that strongly suggests the use of inductive reasoning as well as the more traditional deductive logic.
C.12.2 Use geometric models to solve mathematical and real-world problems.	This process-oriented standard is included to emphasize to all readers that the study of geometry at this level should not deal just with abstract concepts and skills. Rather, an appeal to the practical uses of geometry is made.
C.12.3 Present convincing arguments by means of demonstration, informal proof, counter-examples, or any other logical means to show the truth of statements and generalizations.	The importance of logical thinking cannot be over-emphasized. But, it should not be limited to the strictly formal type of logic and the "two-column" proofs which has been the heart of traditional, college-preparatory geometry courses at the secondary level. This standard argues for use of many different modes of logical analysis.
C.12.4 Use the two-dimensional rectangular coordinate system and algebraic procedures to describe and characterize geometric properties and relationships such as slope, intercepts, parallelism, and perpendicularity.	Here the skills of coordinate systems learned in elementary and middle school are applied to the study of geometric properties and relationships. The beauty of the Cartesian coordinate system is the integration of algebraic and geometric methods. This should be made accessible to all students.
C.12.5 Identify and demonstrate an understanding of the three ratios used in right-triangle trigonometry (sine, cosine, tangent).	The intent of this standard is to introduce students to the practical use of the three trigonometric ratios. It is definitely not intended to require knowledge of the more formal aspects of trigonometry found in an advanced course for college-intending students.

Additional Standards Information

Additional support information can be found in the NCTM *Curriculum and Evaluation Standards,* 157–162, and in the NCTM PSSM, 308–318.

Sample Problem 1 with Commentary

A horse is secured by a 50-foot-long rope that is tied to a stake. The stake is placed 10 feet from the corner of a 20-foot-by-40-foot barn. A line from the stake to the corner makes a 135-degree angle with the sides of the barn. Under these conditions, how much area does the horse have to graze in?

The following is one student's solution to this problem (see Figures 5.32 and 5.33).

The lettering below refers to lettering on the following student work.

A The student organized the task by clearly identifying five regions (labeled A to E) whose area needed to be found. From this point on, the response represents a continual process of setting up a relationship and using it to

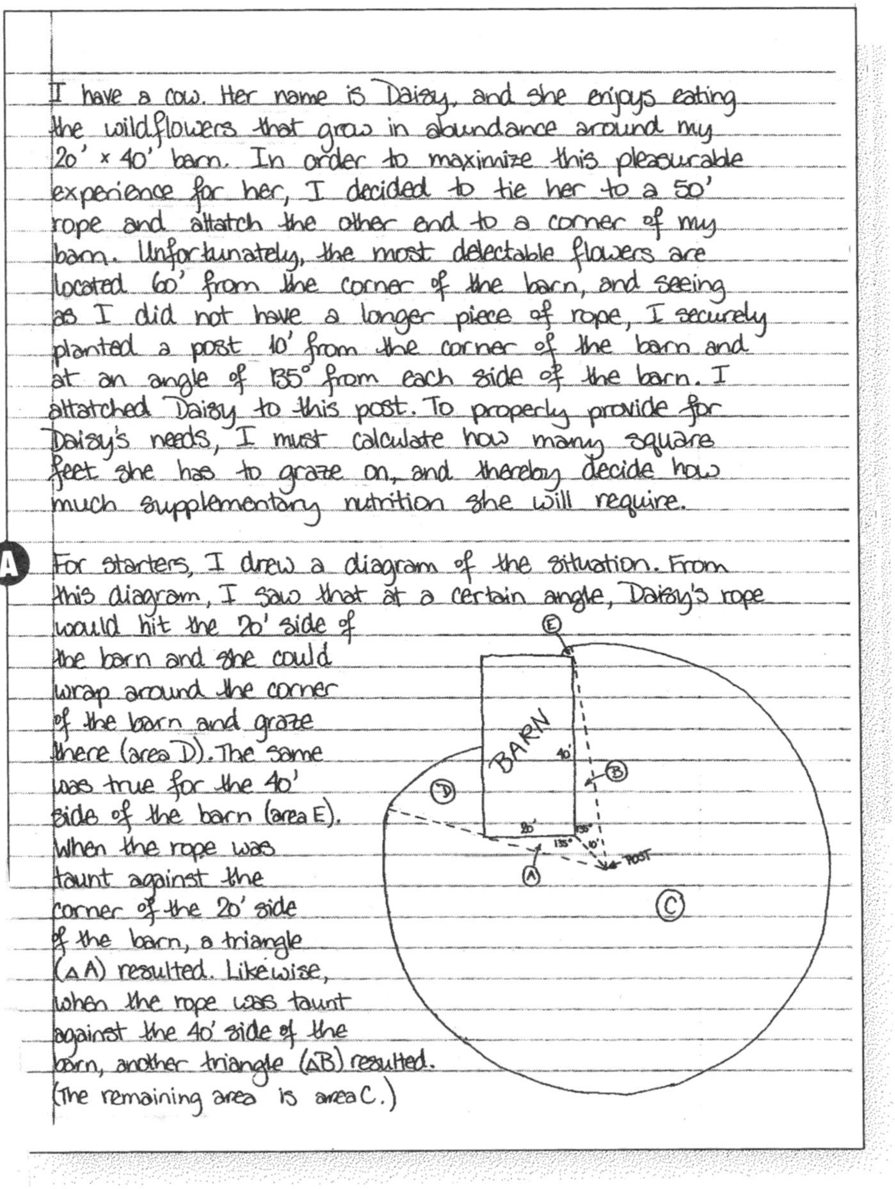

I have a cow. Her name is Daisy, and she enjoys eating the wildflowers that grow in abundance around my 20' × 40' barn. In order to maximize this pleasurable experience for her, I decided to tie her to a 50' rope and attatch the other end to a corner of my barn. Unfortunately, the most delectable flowers are located 60' from the corner of the barn, and seeing as I did not have a longer piece of rope, I securely planted a post 10' from the corner of the barn and at an angle of 135° from each side of the barn. I attatched Daisy to this post. To properly provide for Daisy's needs, I must calculate how many square feet she has to graze on, and thereby decide how much supplementary nutrition she will require.

A For starters, I drew a diagram of the situation. From this diagram, I saw that at a certain angle, Daisy's rope would hit the 20' side of the barn and she could wrap around the corner of the barn and graze there (area D). The same was true for the 40' side of the barn (area E). When the rope was taunt against the corner of the 20' side of the barn, a triangle (△A) resulted. Likewise, when the rope was taunt against the 40' side of the barn, another triangle (△B) resulted. (The remaining area is area C.)

BARN
40'
20'
135°
10'
POST
A
B
C
D
E

FIGURE 5.32A

find an unknown, then setting up a new relationship using what was just found to find another unknown, and so on for many steps.

A The response was built on a complex and effective geometric model of the problem situation. The model consisted of a division of the region into five separate regions, each of which consists of a triangle or a circle sector, and the introduction of techniques to find the area of each.

Throughout the work, the student used knowledge of two-dimensional figures and their properties. For example:

B Knowing two angles of a triangle, the student found the third.

C Knowing the angle of a sector of a circle, the student found the angle of the complementary sector.

In order to procede in my calculations, I deemed it necessary to find the measures of the angles (in both ΔA & ΔB) opposite the 10' side. I also needed to know the length of rope that it took to make up the third sides of the triangles. In order to do this, I made the unknown "rope-sides" of the triangles into hypotenuses of right triangles. The 45-45-90 triangle that appeared allowed me to calculate the altitude of my original triangle (shaded), along with the hypotenuse of my new triangle (or the rope-side of my original triangle). In 45-45-90 triangles, the sides opposite the ∠45° = x and the side opposite the

G ∠90° = x√2. Thus the altitudes of each of the triangles (ΔA & ΔB) were 5√2, and their lengths were 20+5√2 (ΔA) & 40+5√2 (ΔB). The hypotenuse length for each triangle was calculated using the trig properties. I then calculated

E the areas of triangles A & B using the formula for area: area of Δ = ½(base)(height).

I also calculated the measurement of the angles of the triangles that are opposite the 20' & 40' barn walls. These were calculated with

B the equation stating that the sum of the 3 angles of a triangle add up to 180° I added these

ΔA

$\tan\theta = \frac{opp}{adj} = \frac{5\sqrt{2}}{20+5\sqrt{2}}$

$\theta = \tan^{-1} .2612$

$\theta = 14.64°$

$\sin\theta = \frac{opp}{hyp}$

$\sin 14.64 = \frac{5\sqrt{2}}{hyp}$

$hyp \doteq 28$ ft.

ΔB

H $\tan\theta = \frac{opp}{adj} = \frac{5\sqrt{2}}{40+5\sqrt{2}}$

$\theta = \tan^{-1} .15022$

$\theta = 8.54°$

$\sin\theta = \frac{opp}{hyp}$

I $\sin 8.54 = \frac{5\sqrt{2}}{hyp}$

$hyp \doteq 47.6$ ft.

area $\Delta A = \frac{1}{2}bh = \frac{1}{2}(20)(5\sqrt{2}) = 70.71$ ft^2

area $\Delta B = \frac{1}{2}bh = \frac{1}{2}(40)(5\sqrt{2}) = 141.42$ ft^2

ΔA: $\angle x = 180 - (135° + 14.64°) = 30.36°$

ΔB: $\angle x_2 = 180 - (135° + 8.54°) = 36.46°$

FIGURE 5.32B

D The angle θ in triangle A is actually supplementary to the *sum* of 90 degrees and the angle of arc D. (Similarly for triangle B and the angle of arc E.)

Throughout the work, the student used knowledge of area formulas. For example:

E Knowing the base and height of a triangle, the student found its area.
F Knowing the angle and radius of a sector of a circle, the student found its area. This is a key part of the response, and the student managed it nicely. The result being used is that the area of a sector of a circle with angle θ (in degrees) and radius r is $\theta/360\ \pi\ r^2$.
G Knowing the lengths of two sides of a right triangle (or knowing the length of the hypotenuse of an isosceles right triangle), the student used the Pythagorean Theorem to find the length of the third side. The re-

C two angles together and subtracted their sum from 360° to get the measurement of the angle made by circle C. This angle was divided by **F** 360° to find the fraction of a circle that C made up. I then multiplied this fraction by the area of the circle with radius 50' (area = πr^2). Area C ended right before the rope wrapped around the barn on either side and created smaller arcs with shorter radii (areas D & E). For sections D & E, the rope left over (creating a new radius) was the 50' rope minus the hypotenuses of the two triangles A & B. I also found the angles θ that I had found earlier helpful in calculating the measures of the angles which were parts of the arcs of areas D & E. The angle θ (in both D & E) was **D** supplementary to the 90° angle the barn made and the angle of arc D or E. Thus I calculated the percentages of a circle that each area (D or E) was and multiplied this fraction by the area of a complete circle of that radius.
To find the total grazing area, I simply added up all of my individual segments, triangles A & B, and circle-pieces C, D & E

C

BARN

C's angle

$C\text{'s } \angle = 360 - (x_1 + x_2)$
$= 360 - (30.36° + 36.46°)$
$= 293.18°$

$\frac{293.18}{360}$ = % of circle grazed
= .8144 or 81.44%

area of $C = \pi r^2 \times .8144 = \pi \cdot 50^2 \cdot .8144$
$= 6396.2\ ft^2$

D

BARN

$\angle y_1 = 180 - (90° + 14.64°) = 75.36°$
% of circle $= \frac{75.36}{360} = 20.93\%$
$r = 50 - hyp = 50 - 28 = 22'$

area of $D = \pi r^2 \cdot .2093 = \pi \cdot 22^2 \cdot .2093$
$= 318.3\ ft^2$

BARN

$\angle y_2 = 180 - (90° + 8.54°) = 81.46°$
% of circle $= \frac{81.46}{360} = 22.63\%$
$r = 50 - hyp = 50 - 47.6 = 2.4$ ft.

area of $E = \pi r^2 \cdot .2263 = \pi \cdot 2.4^2 \cdot .2263$
$= 4.09\ ft^2$

A =	70.71
B =	141.42
C =	6396.20
D =	318.30
+ E =	4.09

TOTAL AREA: $6930.72\ ft^2$

FIGURE 5.33A

sponse cites and uses a specific rule about 45°–45°–90° triangles: the hypotenuse of an isosceles right triangle is $\sqrt{2}$ times the leg. This rule can be derived using the Pythagorean Theorem.

H Knowing the length of two sides of a right triangle, the student used the inverse of the tangent function to find the acute angle. This is the one place in the solution where use of trigonometry was necessary. The student found an acute angle of a right triangle by using a calculator and the inverse tangent function to solve for θ in the defining formula for the tangent: tan θ = opp./adj. This is possible because the opposite (opp.) and adjacent (adj.) sides are both known. (The calculation is shown in the response for both triangles A and B, though the step is not mentioned in the prose explanation.)

I Here the student used trigonometry again, this time to find the hypotenuse, knowing the angle and the opposite side. The hypotenuse could have also have been found without trigonometry by the

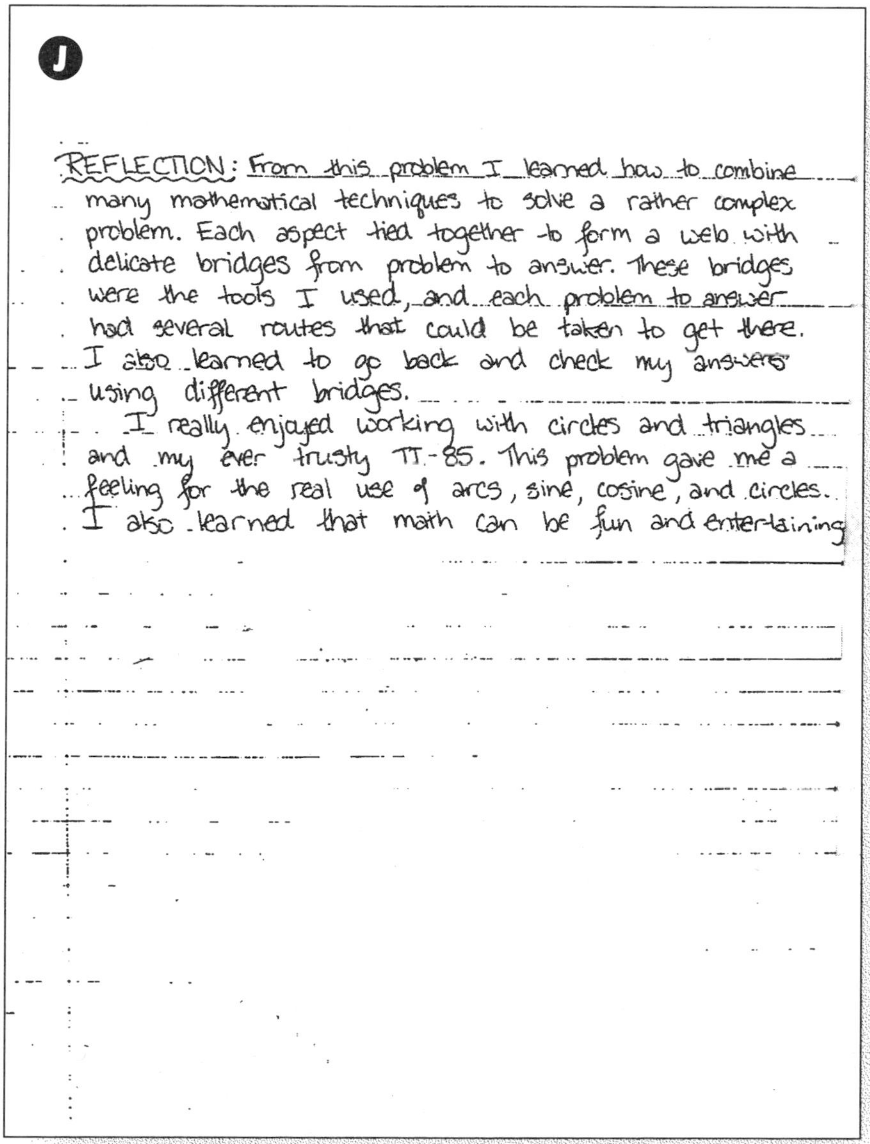

J

REFLECTION: From this problem I learned how to combine many mathematical techniques to solve a rather complex problem. Each aspect tied together to form a web with delicate bridges from problem to answer. These bridges were the tools I used, and each problem to answer had several routes that could be taken to get there. I also learned to go back and check my answers using different bridges.

I really enjoyed working with circles and triangles and my ever trusty TI-85. This problem gave me a feeling for the real use of arcs, sine, cosine, and circles. I also learned that math can be fun and entertaining

FIGURE 5.33B

Pythagorean Theorem, because both the opposite and the adjacent sides are known. (The hypotenuse lengths of triangles A and B are not used until later when triangles D and E are treated.)

The response used the symbol ≐ to mean, "is approximately equal to," in cases where decimals are rounded off. Actually, this symbol should have been used in more of the cases here, because all the decimals have been rounded off.

J The student indicated that the experience of working on this extended task was a rich and rewarding one and echoed the language of the standard.

Although the requirements of the problem are quite clear, solving the problem required some visualization on the student's part. Knowledge of plane figures and their properties is required. It also entails the concept of indirect measurement, which may be exhibited through the use of such things as the Pythagorean Theorem, the trigonometric ratios, and the area formulae. The problem also emphasized problem solving and communication.

Sample Problem 1 can be found in the *New Standards Project* 1995, pp. 49–51.

Standards Addressed

A.12.2 Communicate logical arguments.

C.12.1 Students had to identify, describe, and analyze properties of figures, relationships among figures, and relationships among parts of figures.

C.12.2 Students were required to use geometric models.

C.12.5 Students identified and demonstrated an understanding of the three ratios used in right-triangle trigonometry.

D.12.3 Students determined measurements indirectly by using techniques of algebra, geometry, and right-triangle trigonometry.

F.12.4 Students modeled and solved a variety of mathematical and real-world problems by using algebraic expressions, equations, and inequalities.

The problem was a rich problem in that it encompasses many of the standards within a single task.

Sample Problem 2 with Commentary

The following problem provides practice in two- and three-dimensional geometry and with the Pythagorean Theorem. A little house is to be constructed from cardboard that has the form of a cube 4 inches on a side topped by a pyramid with base edges of 4 inches and height 1 inch is shown in Figure 5.34a.

1. What is the distance from a corner of the base of the roof to the tip of the roof?
2. What is the area of the roof of the house?
3. What is the volume of the inside of the house, including the part up under the roof?

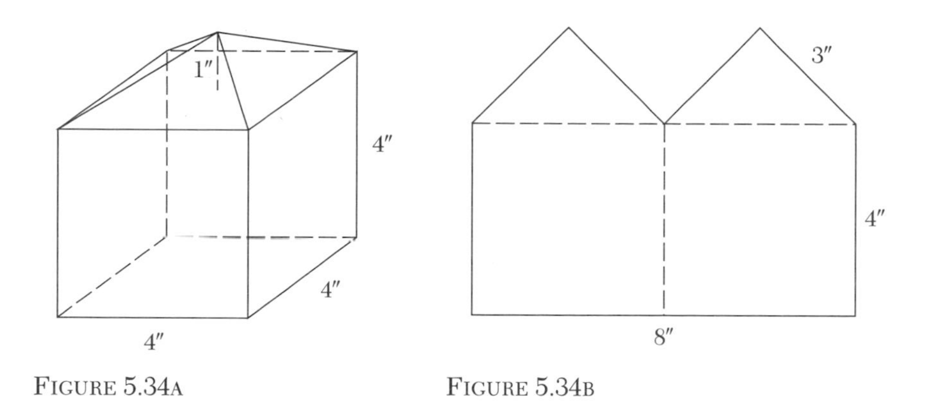

FIGURE 5.34A FIGURE 5.34B

4. The house is to be constructed out of two pieces of cardboard, each of which has the form of a rectangle with an 8-inch base and 4-inch height, which has two triangles on top of it (see Figure 5.34B). Bending each piece on the dotted lines, we can put the two pieces together to form the house (the edges will be taped, so there will be no overlap). Can these two pieces of cardboard be cut out of a single rectangular piece of cardboard that is 9 inches by $11\,^1/_2$ inches?
5. What are the dimensions of the cardboard rectangle of smallest area out of which these two pieces for the house can be cut?

The answer to part 1 can be found to be 3 by two uses of the Pythagorean Theorem. For part 2 we note that each piece of the roof is a triangle with base of 4 inches and two other sides of 3 inches. The height of this triangle can be found to be $\sqrt{5}$ inches, so the area of the triangle is $2\sqrt{5}$ square inches. Hence the area of the four roof sections is $8\sqrt{5}$ square inches. To answer part 3, we observe that the volume under the roof is a pyramid with base of 16 square inches and height 1, so the volume is

$$\frac{1}{3}\times 1\times 16=\frac{16}{3}=5\frac{1}{3}.$$

Thus, the whole volume of the house is

$$64+5\frac{1}{3}=69\frac{1}{3}$$

cubic inches.

Part 4 can be solved in a very concrete fashion by cutting out two pieces of cardboard as in figure 5.34B, cutting out a rectangle of cardboard, 9 inches by $11\,^1/_2$ inches, and then placing the first two pieces on the rectangular piece so that the two do not overlap. This can also be done by placing the two cardboard pieces shaped like figure 5.34B with their slanting edges touching, as in figure 5.34C, but with the line determined by segment A 1 inch to the left of segment B. Then one measures that the height is less than $11\,^1/_2$. For students with enough background, these calculations can be done by drawing pictures rather than using pieces of cardboard.

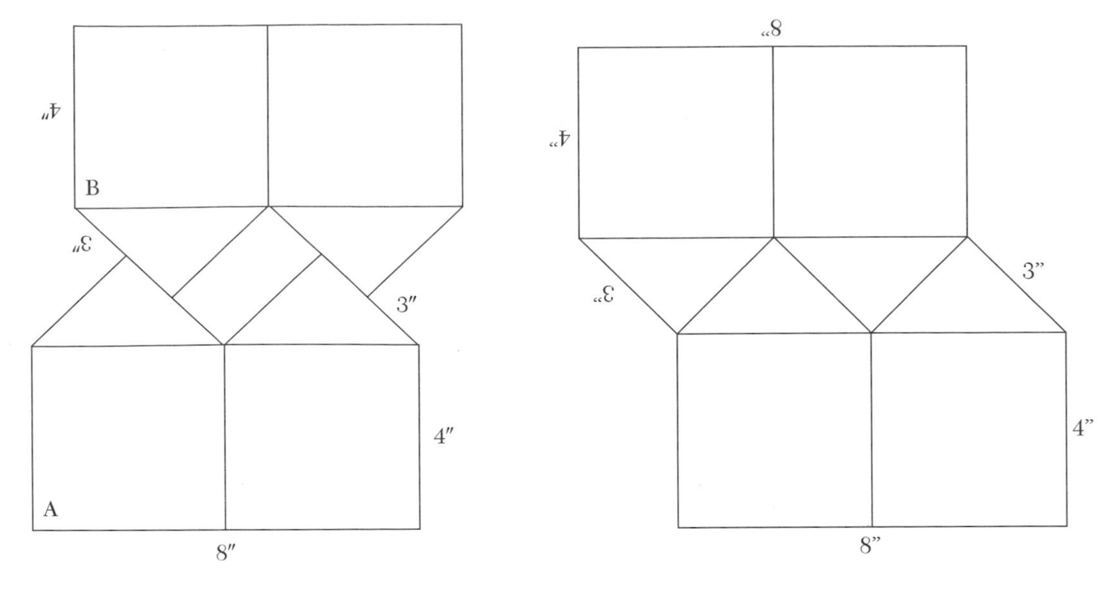

FIGURE 5.34C

FIGURE 5.34D

For part 5, the student can put the two pieces of cardboard shaped as in figure 5.34B together as in figure 5.34C. By trying several cases where the slanted sides of the two pieces touch more or less, the student will discover that the rectangle needed to cut out the two figures will have minimum area when the two pieces are pushed completely together, as in figure 5.34D. In this case the width of the rectangular piece of cardboard that they cut out is 10 inches and the length is $8 + \sqrt{5}$ inches. For the student who is doing these calculations with pencil and paper, it may be helpful to note that the distance from the line through segment A to the segment B is exactly one-third of the length of the overlap.

Sample Problem 2 was contributed by J. Marshall Osborn, University of Wisconsin–Madison.

Sample Problem 3 with Commentary

Begin by drawing any triangle using the segment tool (F2) (see Figure 5.35). It might be wise to start with an acute triangle, but it is not absolutely necessary. Your observations and conjectures should apply to any triangle.

A median in a triangle is a segment that joins the midpoint of a side to the opposite vertex. Use the midpoint (F4) and segment (F2) tools to construct the three medians of your triangle (see Figure 5.36). Notice where the medians intersect (or are concurrent); this point is known as the centroid. Use the label tool (F7) to label the centroid M. In Physics, the centroid is also known as the center of mass or center of gravity. If your triangle was laminated with uniform thickness and density, then this is the point at which you can balance the triangle on your finger, a pencil point, or something similar. You could use the distance tool (F6) to observe that the centroid is located 2/3 of the distance from the vertex to the midpoint. Use the hide tool (F7) to hide your medians and midpoints (see Figure 5.37).

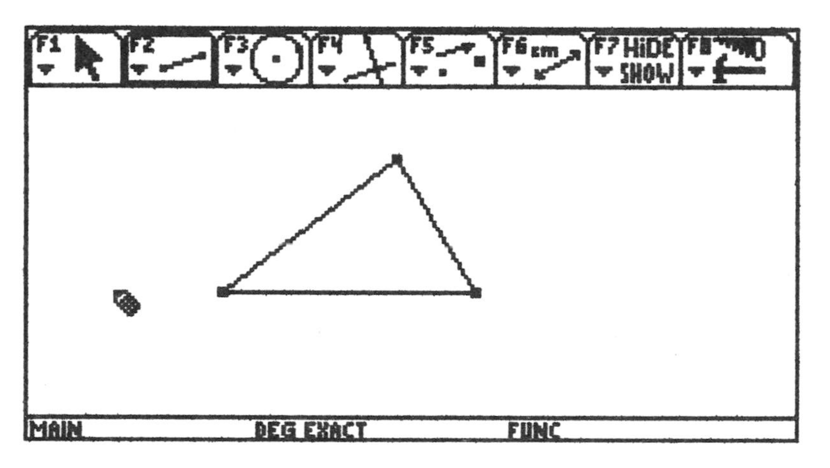

FIGURE 5.35 **Any triangle**

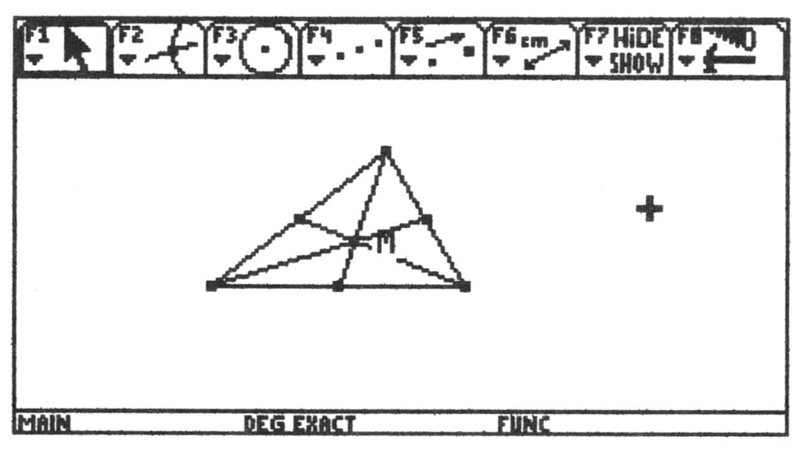

FIGURE 5.36 **Triangle containing the medians**

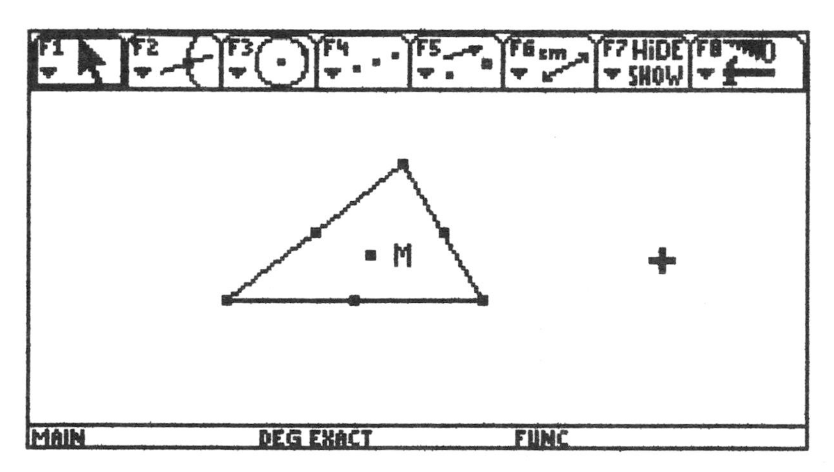

FIGURE 5.37 **Triangle containing centroid**

Use the construction tool (F4) to draw the three perpendicular bisectors for the sides of your triangle (see Figure 5.38). Are the perpendicular bisectors concurrent? This point is called the circumcenter because if you draw a circle with its radius running from the circumcenter to any vertex of the tri-

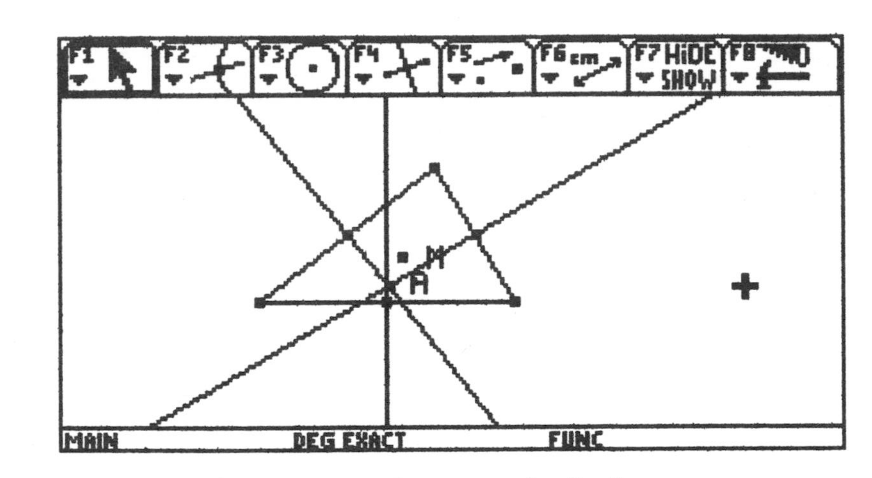

FIGURE 5.38 **Triangle containing the perpendicular bisectors**

angle, you should notice something about the triangle's vertices. You may want to use the circle tool (F3) to help you generate some ideas. Your observation allows us to say the circle is circumscribed about the triangle. Check out your observations using different types of triangles, namely acute, right, or obtuse. What do you observe?

Use the label tool (F7) to label the circumcenter A. Then use the hide tool (F7) to hide the three perpendicular bisectors (see Figure 5.39).

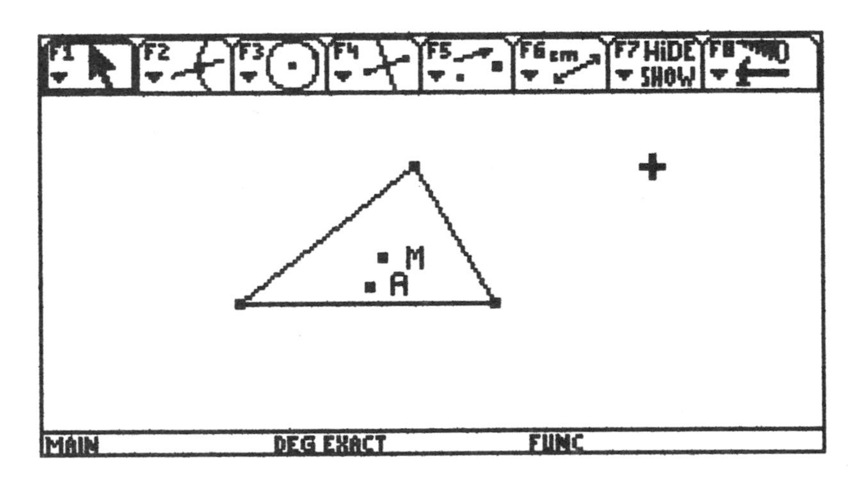

FIGURE 5.39 **Triangle with circumcenter and centroid**

An altitude goes from a vertex and is perpendicular to the line that contains the opposite side. The altitudes intersect at a point called the orthocenter. Use your construction tools (F7) to draw the three altitudes (see Figure 5.40). (In an obtuse triangle, you will need to put lines (F2) over the sides of your triangle since the altitude may not hit the opposite side.) Depending upon whether the triangle is acute, right, or obtuse, where is the orthocenter located relative to the triangle? Label (F7) the orthocenter T and hide (F7) the altitudes (see Figure 5.41).

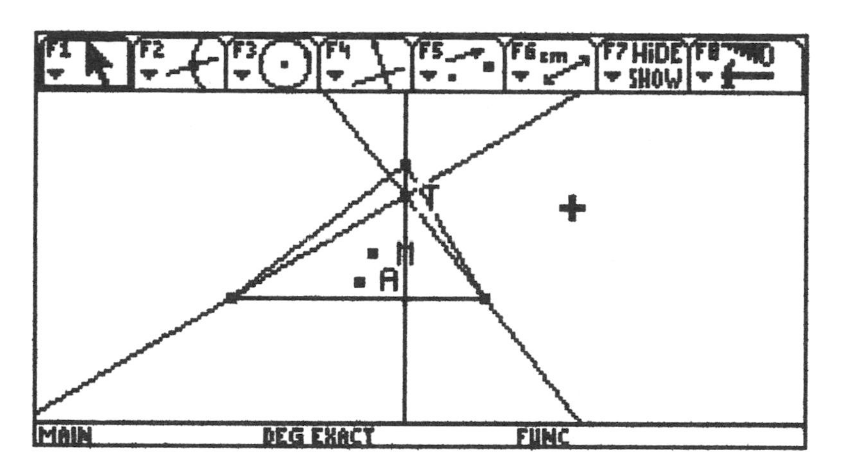

FIGURE 5.40 **Triangle with altitudes**

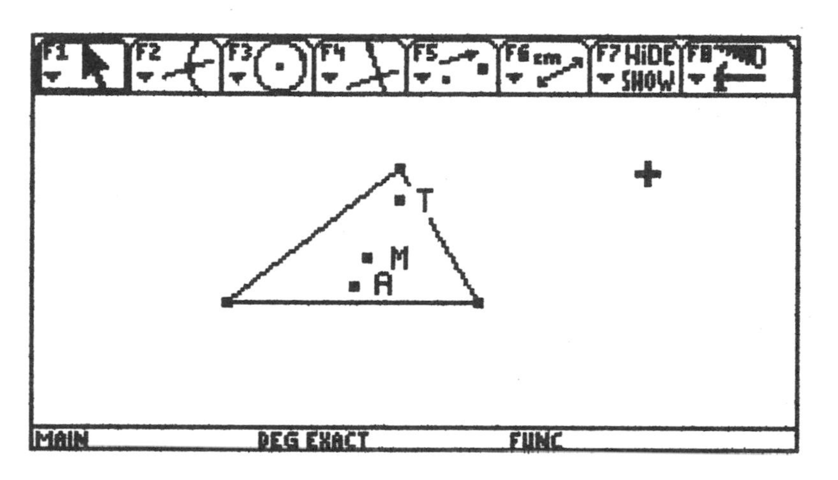

FIGURE 5.41 **Triangle with orthocenter, circumcenter, and centroid**

Use the pointer (F1) and grab (the little hand) any vertex and slide it around the plane. What happens to the three centers? Two examples of sliding a vertex are shown in Figures 5.42 and 5.43. Note: If a point's label does

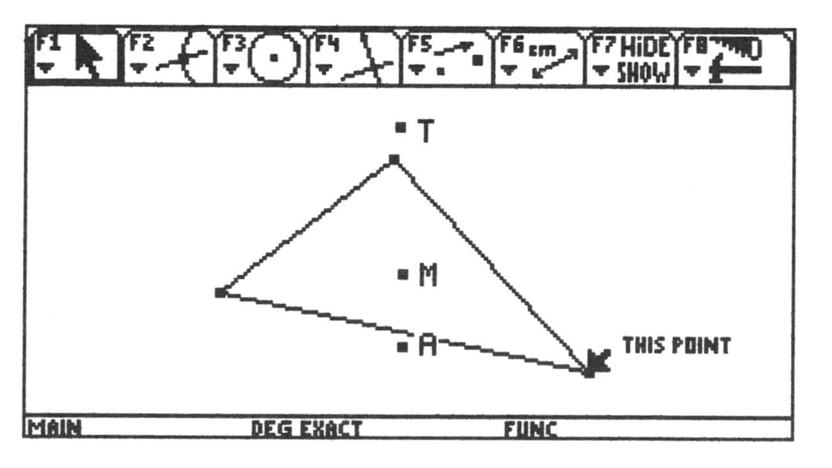

FIGURE 5.42 **Figure 5.41 with a vertex moved**

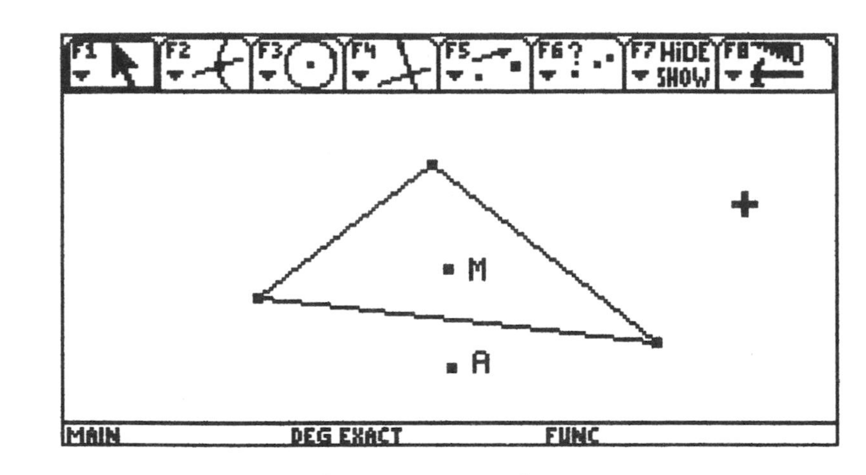

FIGURE 5.43 **Figure 5.42 with a vertex moved**

not appear, it is because it is behind the label for another point or off the screen. Use the pointer (F1) and grab (the little hand) the label so you can slide one of the labels to another location.

If points M, A, and T are the centroid, circumcenter, and orthocenter, respectively, you should observe something about these centers. You may want to run some tests using the Check Property tool (F6) (see Figure 5.44). We call this Euler's Line.

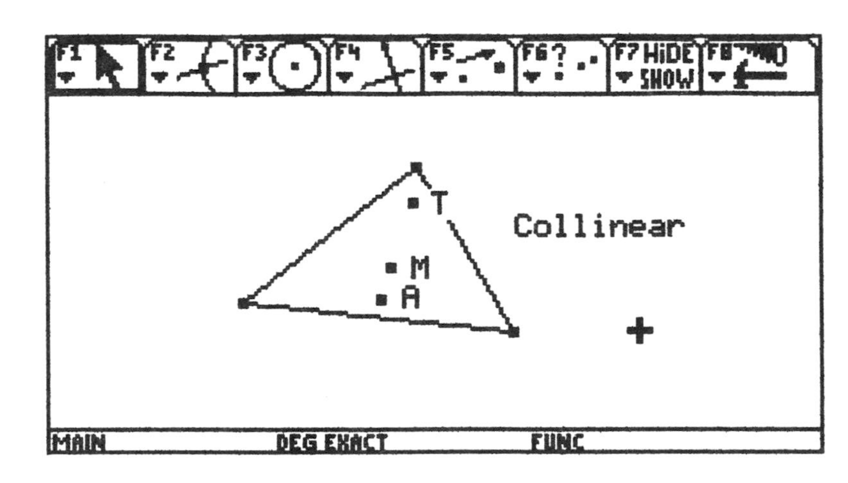

FIGURE 5.44 **A check for collinearity**

Use your construction tool (F4) to construct the angle bisectors of each of the three angles (see Figure 5.45). The angle bisectors should be concurrent (or intersect) at a point called the incenter. Why did you think this is called the incenter? Can you use the circle tool (F3) to illustrate this? Then hide (F7) the angle bisectors (see Figure 5.46). Is the incenter on the Euler's Line like the centroid, circumcenter, and orthocenter were? (See Figure 5.47.)

Final questions to investigate are these: Are these centers always distinct points? Does it make a difference if the triangle is acute, right, or obtuse? Is the answer different for scalene, isosceles, or equilateral triangles? Another perti-

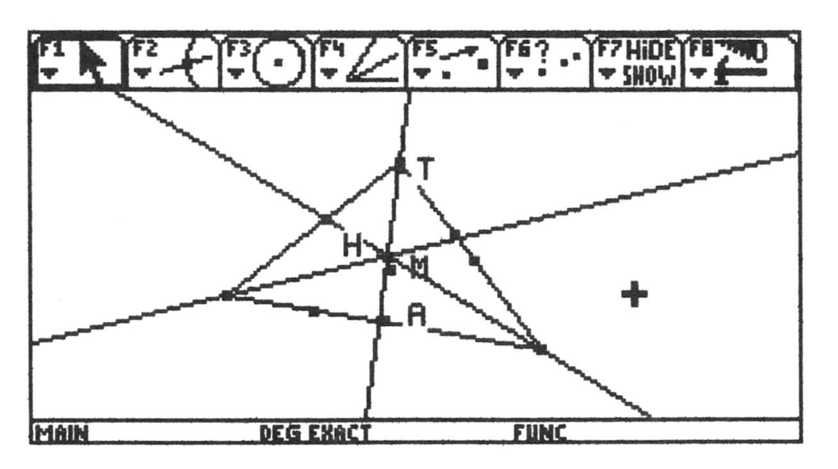

FIGURE 5.45 **Triangle with angle bisectors**

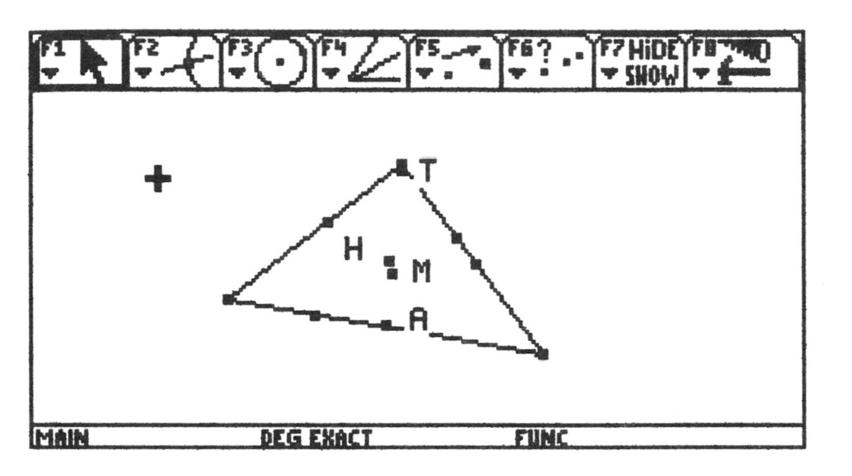

FIGURE 5.46 **Triangle with incenter, orthocenter, circumcenter, and centroid**

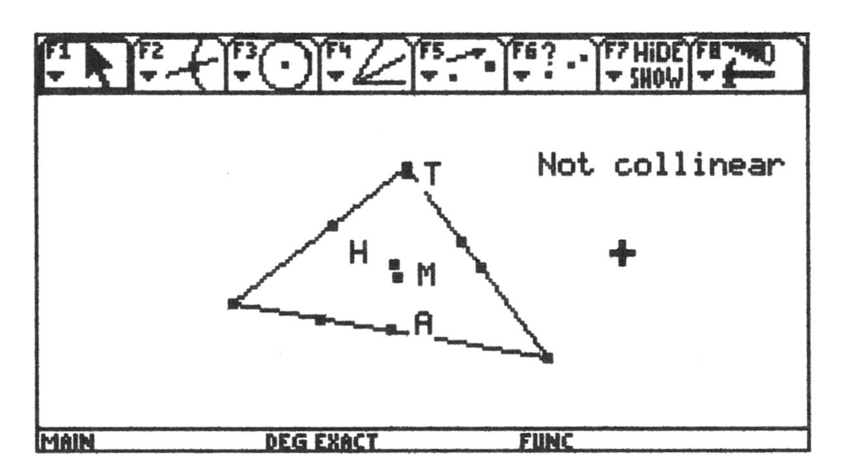

FIGURE 5.47 **Check for collinearity**

nent question: Are any of these centers serving another role in any of these triangles? For example, where is the circumcenter located in a right triangle?

Sample Problem 3 can be found in "Centers of Triangles Exploration," by J. F. Marty *Wisconsin Teacher of Mathematics,* vol. 48, no. 2 (1997): 30–33. This example appeared as an article published in the *Wisconsin Teacher of Mathematics,* Spring 1997, Vol. 48, No. 2. It was entitled "Centers of Triangles Exploration on the TI-92." Although the article deals with centroids, orthocenters, circumcenters, incenters, and Euler's Line using a graphing calculator, software such as *The Geometric Supposer, The Geometer's Sketchpad,* or *Cabri Geometry* (which also is on the TI-92) could be used.

Extension

Use the tools that you learned in coordinate geometry (midpoint, slope, and distance formulas), to prove any or all of the conjectures regarding the centers of triangles that were explored on the graphing calculator. A partial list would contain the following ideas:

- Concurrence of the lines or line segments in each of the following groups: medians, altitudes, perpendicular bisectors, and angle bisectors.
- The centroid, circumcenter, and orthocenter are collinear, but the incenter is not.
- The distance from the incenter to each side is equal.
- The distance from the circumcenter to each vertex is equal.
- The centroid is located 2/3 of the distance from the vertex to the midpoint.

Try to show that the above statements are true with a triangle with specific coordinates for each of the vertices. Lastly, use variables (see Figure 5.48.) to show that the statements are true for all triangles. This exercise in generalization would help students gain a feel for the power of mathematics.

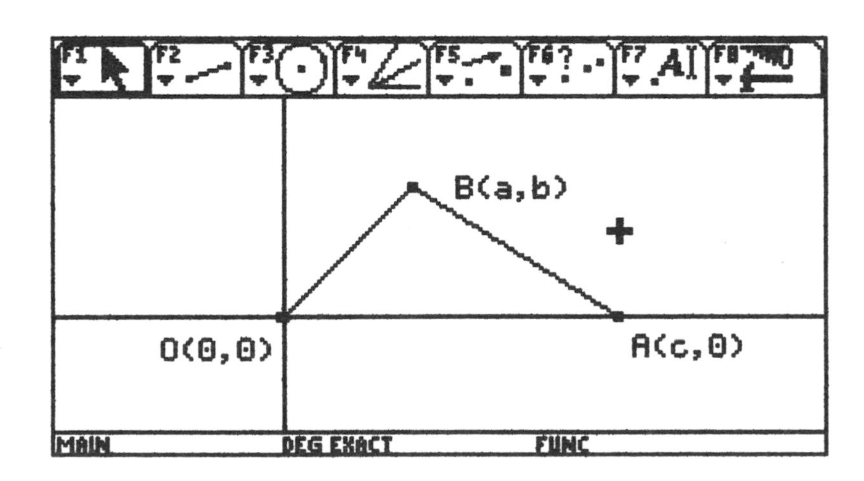

FIGURE 5.48 **Triangle with variable coordinates**

Standards Addressed

C.12.1 Identify, describe, and analyze properties of figures, relationships among figures, and relationships among their parts.

C.12.2 Use geometric models to solve mathematical and real-world problems.

The interplay between different standards (in this case, Geometry, Measurement, and Algebraic Relationships) is well illustrated by Sample Problem 3 and its extension. The transition from the more intuitive level, using technology, to the more abstract level, using deductive reasoning, is also apparent.

Assessment

Assessment of a project like the centers of a triangle can be done well in student groups, while continuing to hold the individual student accountable for his or her own learning. One way to assess individual learning is to have the individual do only a small portion of the problem. Another way to assess the transfer of knowledge from the group to the individual is to have the individual solve a mini-version of the problem that the group did together. Some preliminary work is done individually and in groups. Then, in groups, the class addresses the larger problem. After completing the group problem, each individual student's transfer is assessed by a smaller version of the problem or only a portion of the larger problem.

Sample Problem 4 with Commentary

Example Involving Transformational Geometry

Towns X and Y are on opposite sides of a river (see Figure 5.49). The towns are to be connected with a bridge, $\overline{JK}$, perpendicular to the river, so that the distance $\overline{XJ} + \overline{JK} + \overline{KY}$ is as small as possible. Where should the bridge be located?

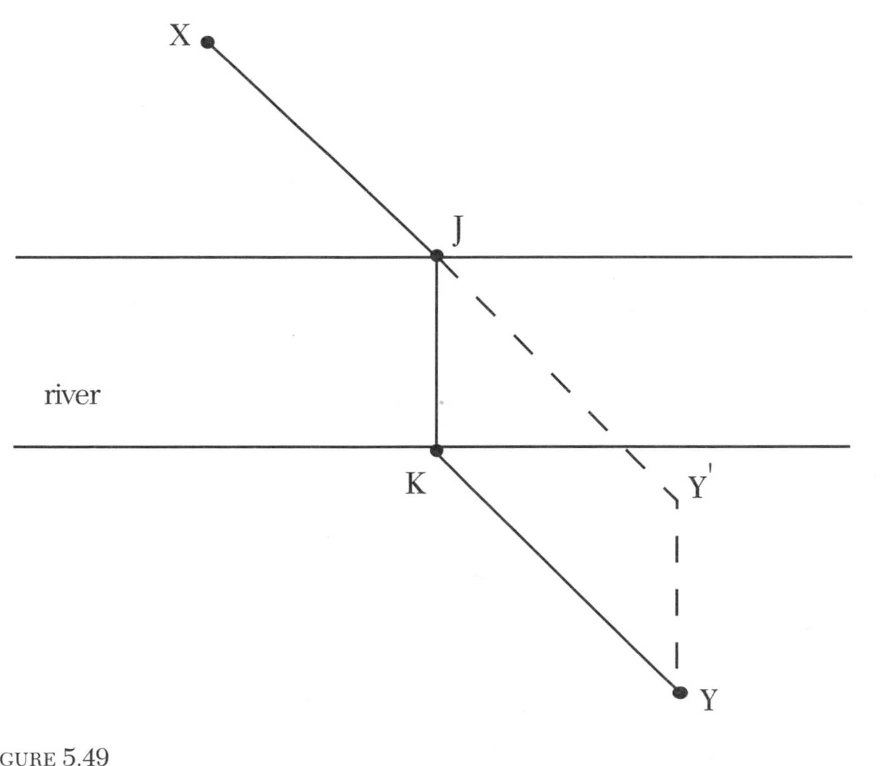

FIGURE 5.49

The solver needs to translate point Y up a distance of $\overline{JK}$ to point Y′, then the shortest distance from X to Y′ is a straight line. Where that line hits the side of the river is where the bridge should be built. Because the length of the bridge is fixed, only the distance between X and Y′ needs to be minimized. If Y′ is the point to which Y translates, then the fact that JKYY′ forms a parallelogram should also foster some interesting discussion.

Standards Addressed

C.12.1 Identify, describe, and analyze properties of figures, relationships among figures, and relationships among their parts.

C.12.2 Use geometric models to solve mathematical and real-world problems.

Extension Problem 1

The amusement park has a miniature golf course. The green around each hole is bounded by a wooden frame. You are playing a game and are now at hole #3. Point B represents the location of the ball; point H, the position of the hole. (See Figure 5.50.) Can you make this hole in one? Show the path the ball will take to make a hole in one.

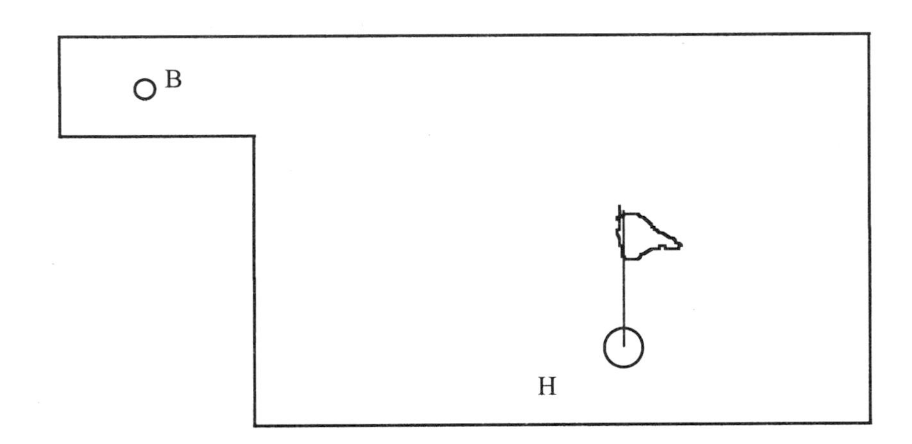

FIGURE 5.50

This problem can be solved by reflecting this diagram in the right border, as shown in Figure 5.51.

Then draw a line from B to the reflection H′. The part of this line that lies in the reflection should then be reflected back into the original part of the diagram to get the actual path of the ball after it bounces off the right border.

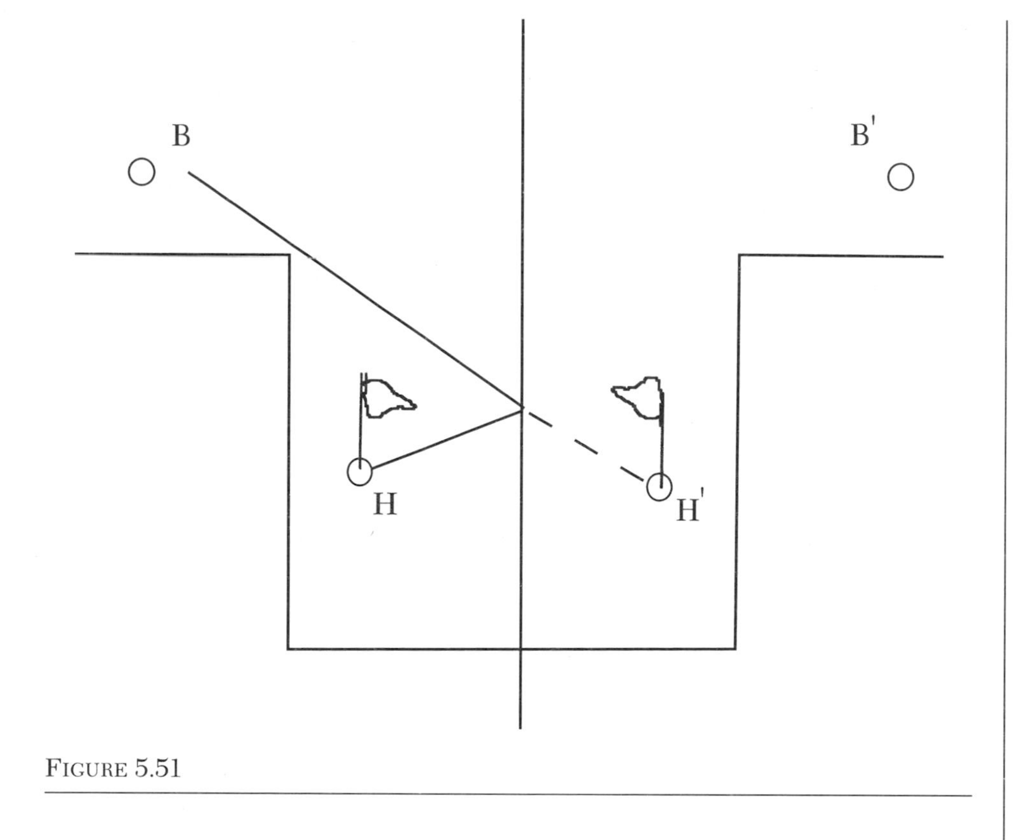

FIGURE 5.51

Extension Problem 2

A farmer standing at a chicken coop at point A wishes to drive over to check the condition of the fence on Independence Avenue and then drive to Freedom Lane to check the condition of the fence there. (See figure 5.52.) After the fences are checked on these two streets, the farmer must go to the barn located at point B. What is the shortest route that can be taken?

To solve this problem, one can reflect point B in the line of Freedom Lane to get point B′, and then reflect this point in the line of Independence Avenue to get point B″, as shown in Figure 5.53.

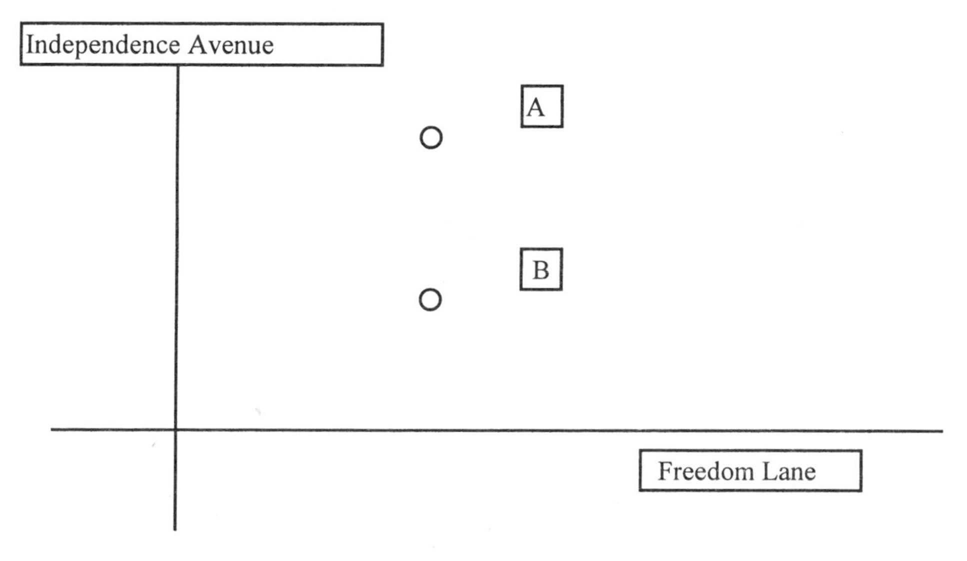

FIGURE 5.52

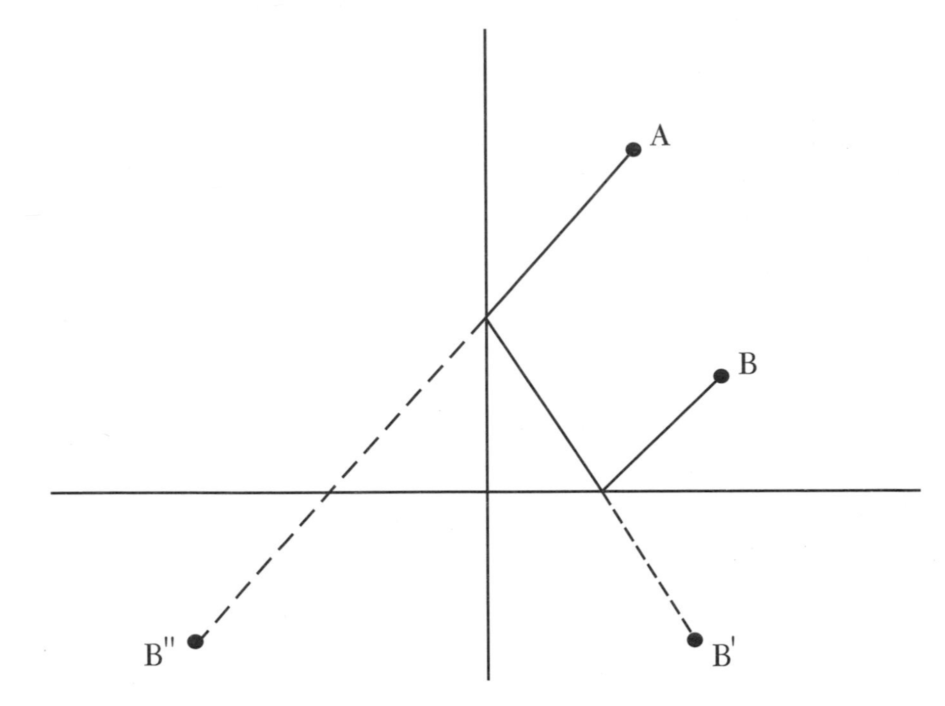

FIGURE 5.53

Drawing the line from A to B″, reflect the line off Freedom Lane and head to B′. Reflect in the line off Freedom Lane and head for A. This gives the shortest route from A to B.

The last two examples are similar to billiards and pool tables in that they heavily involve reflections, which leads directly to the fact that the incoming angle is always equal to the outgoing angle. This is an extremely important concept in physics, involving such things as rays of light reflecting off a surface.

Additional Resources

Additional rich problems such as these can be found is sources such as:

1. The 1989 NCTM Standards and Addenda Series books.
2. Professional Journals:
   - *Wisconsin Teacher of Mathematics* published by the Wisconsin Mathematics Council.
   - *Mathematics Teacher* published by NCTM.
3. Problems found in the curricula developed under the auspices of the National Science Foundation.

Sample Problem 4 and its extensions were contributed by James F. Marty.

Learning Measurement in Secondary School

Measurement is the principle foundation upon which data collection rests. By measuring lengths, weights, volume of liquids, and time intervals, students develop a feeling for measurement and the inherent problem of the inaccuracy of the measurements. Later students encounter indirect measurements such as area, density, or velocity. Students quickly realize that measurement, by its nature, is not exact. The degree of accuracy depends on the situation. Although measuring the length of a room with an inaccuracy 1/4 of an inch may mean very little, certain machining operations require measurement within a thousandth of an inch. Only in the idealized setting of geometry is the student dealing with precise quantities.

In high school, students work more with the concept of indirect measurement. For such purposes, a suggested major project might consist of doing a scale drawing of the school or some building project that the school district is contemplating or completing.

With international economies interacting with one another on a daily basis, students must learn to convert between different measuring systems such as between metric and U.S. Customary units. Indirect measurement concepts such as knowledge of functions, right-triangle trigonometric relationships, and application of algebraic techniques such as using proportional reasoning and solving equations is also important.

The close relationship between geometry and measurement is apparent. Measurement is more than just reading a ruler and involves concepts of speed, velocity, acceleration, ohms, temperature, watts, and rates (such as miles/hour). Assessments at this level should ask students to measure and then calculate derived attributes from measured attributes.

Such technological instruments as the Calculator-Based Lab (CBL) and the Calculator-Based Ranger (CBR) make the collection of data relatively easy and available to all students. The use of these instruments, along with such things as probes and calipers, make real-time collection of data possible. Knowing the difference between analog and digital measuring devices is also essential. Students must be well aware of the limitations of technology by asking questions such as these: How reliable are the results provided by technology? What is it telling us? Is the information reasonable?

Interpretations of results varies. Questions such as these must be addressed: If one multiplies together three measurements that are accurate to the nearest tenth, can one obtain an answer that is accurate to the nearest thousandth? How are accuracies combined?

Students in high school are primarily concerned with refining their basic measurement skills and learning more about indirect measurement and derived attributes. Additionally, students at this level must be prepared to logically justify the reasonableness of their answers.

Students in high school are primarily concerned with refining their basic measurement skills and learning more about indirect measurement and derived attributes.

National Council of Teachers of Mathematics Measurement Standard, *Principles and Standards for School Mathematics* (Standards 2000 Project 2000)

Mathematics instructional programs should include attention to measurement so that all students:

- Understand measurable attributes of objects and the units, systems, and processes of measurement.
- Apply appropriate techniques, tools, and formulas to determine measurements.

Wisconsin's Model Academic Standards for Mathematics, Standard D: Measurement

Students in Wisconsin will select and use appropriate tools (including technology) and techniques to measure things to a specified degree of accuracy. They will use measurements in problem-solving situations. (Wisconsin Department of Public Instruction 1998, 10–11.)

TABLE 5.21 **Wisconsin's Model Academic Standards for Measurement**

Wisconsin's performance standards	Elaboration
By the end of grade 12 students will:	
D.12.1 Identify, describe, and use derived attributes (e.g., density, speed, acceleration, pressure) to represent and solve problem situations.	It is highly likely that the content described in this standard will be covered in physical science courses.
D.12.2 Select and use tools with appropriate degree of precision to determine measurements directly within specified degrees of accuracy and error (tolerance).	While part of this standard might be achieved in mathematics class, it is probably more appropriately covered in science and technology courses where the actual tools are more available.
D.12.3 Determine measurements indirectly, using ■ estimation ■ proportional reasoning, including those involving squaring and cubing (e.g., reasoning that areas of circles are proportional to the squares of their radii) ■ techniques of algebra, geometry, and right triangle trigonometry ■ formulas in applications (e.g., for compound interest, distance formula ■ geometric formulas to derive lengths, areas, or volumes of shapes and objects, (e.g., cones, parallelograms, cylinders, pyramids) ■ geometric relationships and properties of circles and polygons (e.g., size of central angles, area of a sector of a circle) ■ conversion constants to relate measures in one system to another (e.g., meters to feet, dollars to Deutschmarks)	While there is a strong emphasis on geometric techniques (as was the case also in standard D.8.4), there is no requirement that the requisite geometric instruction be presented only in a traditional, formal course of study. Similarly, the mention of use of trigonometric techniques does not necessarily imply a rigorous, formal study of trigonometry for all students. And, finally, the value of using estimation to obtain useful measurements must be asserted. Yet, this desirable skill does not come automatically for all students. It must be encouraged and practiced systematically.

Additional Standards Information

Additional support information can be found in the NCTM *Curriculum and Evaluation Standards*, 157–166, and in the NCTM PSSM, 320–323.

Sample Problem 1

A horse is secured by a 50-foot-long rope that is tied to a stake. The stake is placed 10 feet from the corner of a 20-foot-by-40-foot barn. A line from the stake to the corner makes a 135-degree angle with the sides of the barn.

Under these conditions, how much area does the horse have to graze in?

This problem is explained in more detail under the geometry standard and illustrates the close relationship between measurement and geometry. Standards addressed are listed in that section of this document.

Sample Problem 1 can be found in the *New Standards Project* 1995, pp. 49–51.

Sample Problem 2 with Commentary

This example illustrates that real-world applications involving area consist of the shapes that are not most commonly studied.

When draining a swamp so the area can be used for some other purpose, one needs to know how much soil is needed to fill the remaining hole. By using a measuring stick and small raft, enough random samples can be obtained to estimate the average depth of the swamp to be about four feet. It is still necessary to know the area that the swamp covers.

Discussion

If the swamp is relatively small, a rope with foot markers could be used with two people on opposite sides of the swamp to measure the width across at regular intervals, say twenty-five feet (see Figure 5.54).

If the swamp is too large to use a rope to measure across, then right-triangle trigonometry can be used to find the length across the swamp at 25-foot intervals (see Figure 5.55).

By either the rope or right-triangle trigonometry method, dimensions such as shown in Figure 5.56 can be determined. To find an approximation for the area using rectangular regions, use the top length as a side of the rectangle (see Figure 5.57), adding the areas of the rectangles to get an approximation of the required area.

This requires the following calculation for the area: $25(110) + 25(120) + 25(136) + 25(130) + 25(124) + 25(138) + 25(146) + 25(140) + 25(138) + 25(142) = 25(110 + 120 + 136 + 130 + 124 + 138 + 146 + 140 + 138 + 142) = 25(1{,}324) = 33{,}100$ square feet. Thus, the volume of soil needed would be approximately $(33{,}100)(4) = 132{,}400$ cubic feet or approximately 4,900 cubic yards (because 27 cubic feet = 1 cubic yard and soil is usually sold in cubic yards).

Using the same reasoning, we could have used the bottom as the length of the rectangles to get an approximation for the required area (see Figure 5.58).

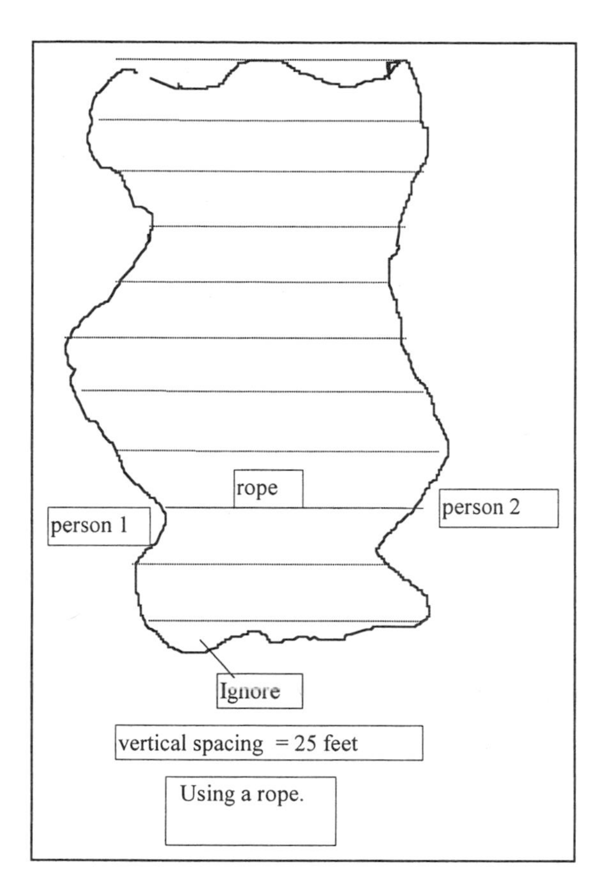

FIGURE 5.54

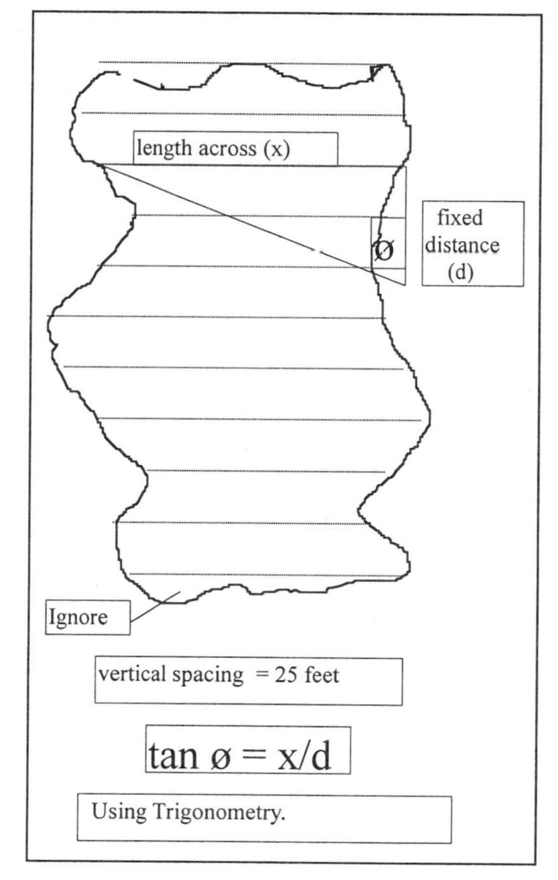

FIGURE 5.55

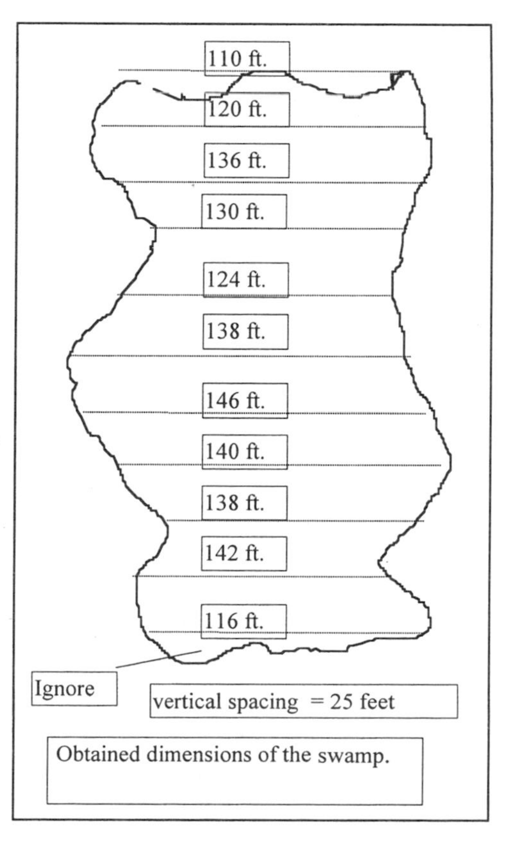

FIGURE 5.56

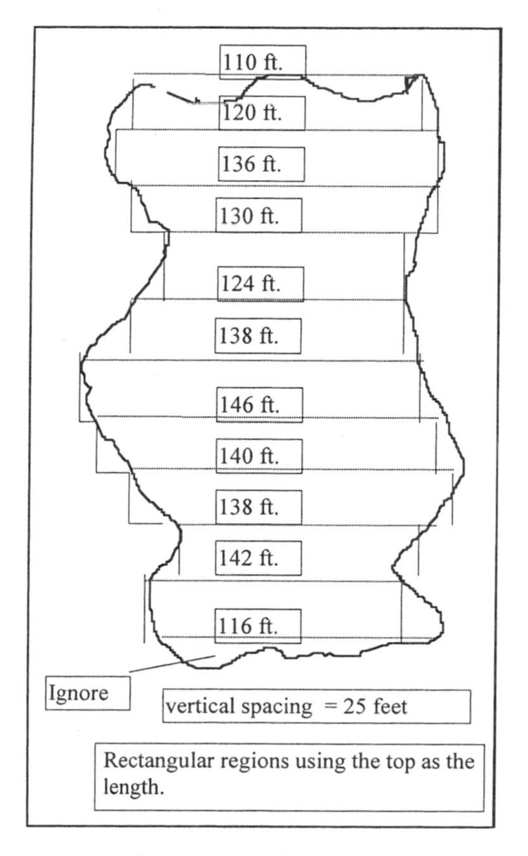

FIGURE 5.57

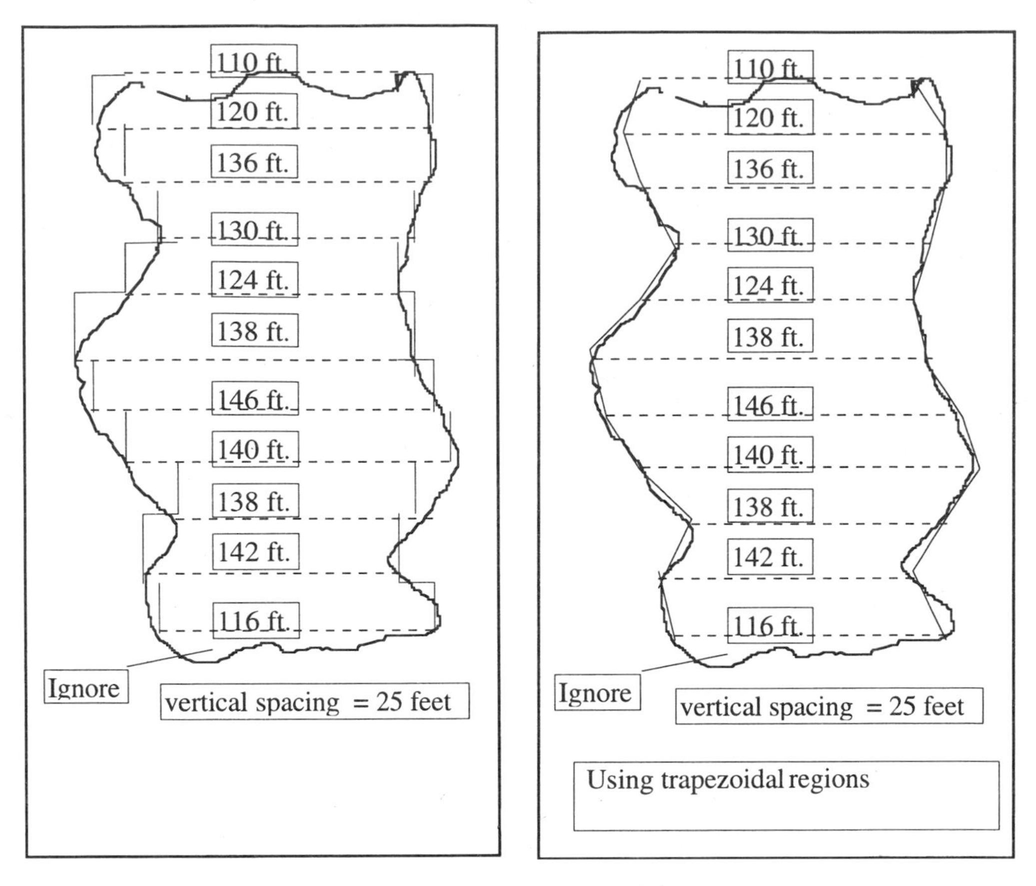

FIGURE 5.58 FIGURE 5.59

The result would require the following calculation for the area:

25(120) + 25(136) + 25(130) + 25(124) + 25(138) + 25(146) + 25(140) + 25(138) + 25(142) + 25(116) = 25(120 + 136 + 130 + 124 + 138 + 146 + 140 + 138 + 142 + 116) = 25(1330) = 33,250 square feet. Thus, the volume of soil needed would be approximately (33,250)(4) = 133,000 cubic feet, or 4,925 cubic yards.

The approximations are quite close, only about 25 cubic yards apart, but some of that closeness could be attributed to the fact that the lengths across the swamp vary by only 36 feet (146–110) and the choice of spacing, namely 25 feet. A closer approximation of the actual area may be obtained if we used trapezoidal rather than rectangular areas to calculate the area (see Figure 5.59).

Using the trapezoid area formula

$$A = \frac{1}{2}(b_1 + b_2)h$$

where A is the area of the trapezoid, b_1 and b_2 are the lengths of the bases, and h is the height of the trapezoid, the calculation for the area would then be: 1/2(110 + 120)25 + 1/2(120 + 136)25 + 1/2(136 + 130)25 + 1/2(130 + 124)25 + 1/2(124 + 138)25 + 1/2(138 + 146)25 + 1/2(146 + 140)25 + 1/2(140 + 138)25 + 1/2(138 + 142)25 + 1/2(142 + 116)25 = 12.5(110 + 120 + 120 + 136 + 136 + 130 + 130 +124 + 124 + 138 + 138 + 146 + 146

+ 140 + 140 + 138 + 138 +142 + 142 + 116) = 12.5(2,654) = 33,175 square feet. Thus, the volume of the soil needed would be approximately (33,175)(4) = 132,700 cubic feet, or 4,916 cubic yards.

The results of the calculations seem to indicate that the trapezoidal method may yield an approximation which is closer to the actual area of the swamp. This is supported by the drawings contained in Figures 5.57, 5.58, and 5.59. Calculations necessary to perform the trapezoidal method (including the area formula) are much more complex. It is not feasible to calculate the exact area. Some questions to be considered are these: How close an estimate do we need to get to the actual area for our results to be useful? In what situations is it better to use the trapezoid method rather than the rectangle method? How could one modify the technique used in order to get an approximation closer to the actual area?

Some of these ideas will be pursued further when studying such topics as sequences and series, allowing the number of rectangles to become infinite. In calculus, the concepts presented lead the student to an understanding of how to arrive at the area under a curve and ultimately a better understanding, by using these ideas, and the concept of a limit, of the definite integral. Perhaps more importantly, it also illustrates that some functions are not capable of being integrated and that techniques such as the trapezoidal method are absolutely necessary to get an approximation. Instead of the area of a swamp, try to estimate the area of the region bounded by $x = 1$, $x = 2$, the x-axis, and the graph of the function $y = x^2$. Students need to realize that the real world works predominantly with irregular shapes such as the shape of the swamp. This type of problem not only illustrates the need for estimation or approximation but also gives every student an opportunity to see examples of where one cannot get exact answers. It also provides them with ideas as to what strategies they might employ to obtain such an approximation.

Sample Problem 2 can be found in "Areas of 'different shapes,'" by J. F. Marty. *Wisconsin Teacher of Mathematics,* vol. 48, no. 3 (1997): 23–25.

Sample Problem 3 with Commentary

To determine absorption rate of different brand names of paper towels, conduct an experiment. The following items are required: a paper plate, pieces of paper towels (of several different brand names), an eye dropper, some food coloring, and a transparency with a graph grid on it. For each brand of paper towels, put the piece over the paper plate so that the back is not touching the bottom of the plate. Using the eye dropper, place a drop of food coloring on the paper towel and let the towel absorb the food coloring. Then use the transparency grid to estimate the area it took on the towel to absorb the food color. The towel that contains the smallest area stained by the food coloring absorbed the best. This is another excellent example of indirect measurement. Like the swamp problem, direct measurement is not possible. By using one brand name of paper towels and applying various numbers of drops, one can plot the area of the moisture versus number of drops applied.

The slope will represent absorbency per drop. To determine the line of best fit, the student should try to get the equation of the line by using a ruler and approximating. Then technology such as using a graphic calculator or software such as *Data Insights* can be used to verify the validity of the approximation.

Sample Problem 3 was contributed by John Janty, Waunakee High School.

Standards Addressed

This activity addresses Standard D.12.3—determine measurements indirectly. It also addresses two statistics and probability standards E.12.2—organize and display data from statistical investigations and E.12.3—interpret and analyze information from organized and displayed data.

Finally, it addresses three Algebraic Relationships Standards. They are F.12.1 analyze and generalize patterns of change and then represent them with an algebraic equation, F.12.2—use linear functions to describe the relationship among variable quantities in a problem, and F.12.4—model and solve real-world problems using algebraic equations.

Additional Resources

Additional rich problems such as these can be found in sources such as:

1. The 1989 NCTM Standards and Addenda Series books.
2. Professional Journals:
   - *Wisconsin Teacher of Mathematics* published by the Wisconsin Mathematics Council.
   - *Mathematics Teacher* published by NCTM.
3. Problems found in the curricula developed under the auspices of the National Science Foundation.

Learning Statistics and Probability in Secondary School

The average citizen today needs to understand statistics and probability to stay informed, properly interpret reports using statistics, and participate in our society because more and more of this information will be used to sell a product or to advance a cause.

In today's society newspapers and television bombard us steadily with data of all kinds, such as employment statistics, medical data from clinical trials, and school demographics. The average citizen today needs to understand statistics and probability to stay informed, properly interpret reports using statistics, and participate in our society because more and more of this information will be used to sell a product or to advance a cause. More and more jobs require employees to use data analysis. Modern technology has made it much easier to analyze data. As a result the amount of statistical information that individuals must interpret will continue to increase.

Data analysis is a relatively new subject in schools, but students today must be educated about data interpretation. Data analysis is best learned by starting with a hands-on approach, then gradually moving toward more abstract and theoretical approaches.

Probability and statistics are not new ideas, but they are being used more to describe the world around us. Thus, a literate person today needs to be able to understand and do statistics and probability. New and exciting classroom materials are available for instruction in data analysis. Owing to advances in technology, a student can now generate data and then do a regression analysis on that data to find the particular mathematical model that best fits or describes what is happening with the data. The student is engaged from the very beginning with the collection of raw data and thus has a vested interest throughout the process.

National Council of Teachers of Mathematics Data Analysis and Probability Standard, *Principles and Standards for School Mathematics* (Standards 2000 Project 2000)

Mathematics instructional programs should include attention to data analysis, statistics, and probability so that all students

- formulate questions that can be addressed with data and collect, organize, and display relevant data to answer them.
- select and use appropriate statistical methods to analyze data.
- Develop and evaluate inferences, and predictions that are based on data.
- Understand and apply basic concepts of chance and probability.

Wisconsin's Model Academic Standards for Mathematics, Standard E: Statistics and Probability

Students in Wisconsin will use data collection and analysis, statistics and probability in problem-solving situations, employing technology where appropriate. (Wisconsin Department of Public Instruction 1998, 12–13.)

TABLE 5.22 **Wisconsin's Model Academic Standards for Statistics and Probability**

Wisconsin's performance standards	Elaboration
By the end of grade 12 students will:	
.12.1 Work with data in the context of real-world situations.	This standard is the end of a continuum begun in Standard E.4.1 and continued with Standard E.8.1. Here, the work is more sophisticated, dealing for the first time with bi-variate data. Also, the notion of using random samples and control groups is included.
E.12.2 Organize and display data from statistical investigations.	The skills and concepts of visual presentation of data are much more sophisticated and complex than those seen in the middle grades. The emphasis should not be on mechanical computation. Rather, technology should be used where applicable. It is the ideas that are most important!
E.12.3 Interpret and analyze information from organized and displayed data.	As just mentioned above, it is the concepts which are most important. Technology should be used rather than memorization of formulas.
E.12.4 Analyze, evaluate, and critique the methods and conclusions of statistical experiments reported in journals, magazines, news media, advertising, etc.	This standard speaks to the real-world application of data analysis principles. Students must become intelligent consumers and interpreters of data, regardless of the form or medium in which it is presented.
E.12.5 Determine the likelihood of occurrence of complex events.	Students expand their techniques of determining probability of events. Complex events can be compound, consisting of two or more linked events one of which may or may not be dependent upon the outcome of the other(s). While theory plays an important part in determining such probabilities, one should not lose sight of the practicality (and accessibility to students) of experiments and simulations.

Additional Standards Information

Additional support information can be found in the NCTM *Curriculum and Evaluation Standards,* 167–175 and in the NCTM PSSM, 324–333.

Sample Problem 1

From a child's family history, the probability that the child has a rare recessive gene is 0.02. The child has two siblings. Based on this information, calculate the following probabilities:

1. The child has the rare gene.
2. The child does not have the rare gene.
3. Exactly one of the three children has the rare gene.
4. All the children have the rare gene.
5. At the most one child has the rare gene.

Some strategies that may employed to solve this problem include making a table, making a tree diagram, or using Pascal's Triangle.

Answers to the problem:

1. 0.02
2. 0.98
3. 0.0576
4. 0.000008
5. 0.9988

Standards Addressed

E.12.5 Determine the likelihood of occurrence of complex events.

Extension

From a child's family history, the probability that the child has a rare recessive gene is 0.02. The child has three siblings. Based on this information calculate the following probabilities:

1. The child has the rare gene.
2. The child does not have the rare gene.
3. None of the children have the rare gene.
4. Exactly one of the children have the rare gene.
5. Exactly two of the children have the rare gene.
6. At least two of the children have the rare gene.
7. At the most two of the children have the rare gene.
8. All of the children have the rare gene.

Discussion

Because there are only two possible outcomes (either you have the rare gene or you do not) and there are four trials (four siblings), this could be solved by using a formula and knowing that it is a binomial distribution with n = 4, p = 0.02, and 1–p = 0.98.

The formula for finding the probability that x of the four children have the rare gene is

$$_{4}C_{x}\ (0.02)^{x}\ \ (0.98)^{4-x} \qquad \text{where } _{n}C_{x} = \frac{n!}{x!\ (n-x)!}.$$

The correct answers are:

1. 0.02
2. 0.98
3. ${}_4C_0\,(0.02)^0\,(0.98)^4 = 0.9224$
4. ${}_4C_1\,(0.02)^1\,(0.98)^3 = 0.0753$
5. ${}_4C_2\,(0.02)^2\,(0.98)^2 = 0.0023$
6. ${}_4C_2\,(0.02)^2\,(0.98)^2 +$
   ${}_4C_3\,(0.02)^3\,(0.98)^1 +$
   ${}_4C_4\,(0.02)^4\,(0.98)^0 = 0.00233648$
7. ${}_4C_0\,(0.02)^0\,(0.98)^4 +$
   ${}_4C_1\,(0.02)^1\,(0.98)^3 +$
   ${}_4C_2\,(0.02)^2\,(0.98)^2 = 0.99996848$
8. ${}_4C_4\,(0.02)^4\,(0.98)^0 = 0.00000016$

Sample Problem 1 and its extension was an original contributed by James F. Marty.

Standards Addressed

E.12.1 Work with the context of real-world situations.

E.12.2 Organize and display data from statistical investigations.

This problem illustrates the powerful use of technology to analyze data.

Sample Problem 2 with Commentary

The following dosage chart, see table 5.23, was prepared by a drug company for doctors who prescribed Tobramycin, a drug that combats serious bacterial infections such as those in the central nervous system, for life-threatening situations.

1. Plot (weight, usual dosage) and draw a best-fit line (figure 5.60).
2. Plot (weight, maximum dosage) on the same axes. Draw a best-fit line.
3. Find the slope for each line. What do they mean, and how do they compare? ____________________
4. Write the equations of the two lines. ____________________
5. Are the lines parallel? Why or why not? ____________________
6. Use a graphing calculator to plot (usual dosage, maximum dosage). Use the calculator to construct a regression line for this data set. How does this line compare to the two lines found in 1 and 2? ____________________

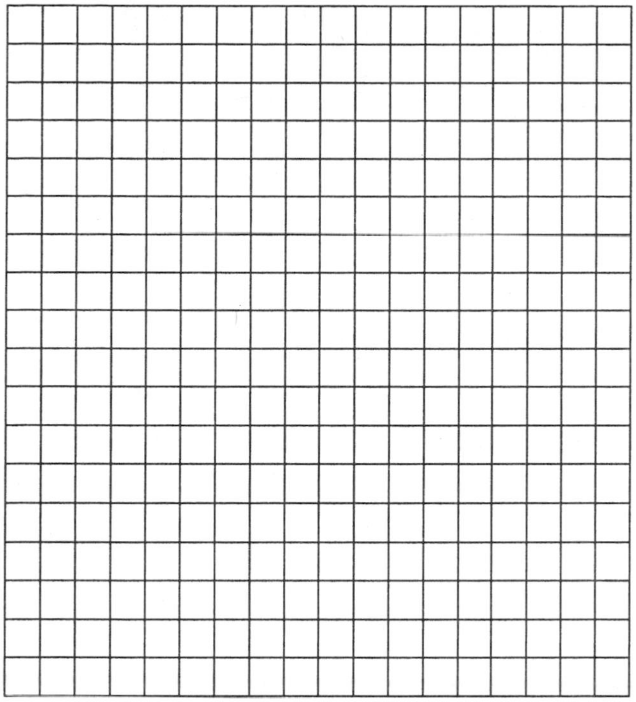

FIGURE 5.60

TABLE 5.23

Weight (pounds)	Usual dosage (mg)	Maximum dosage (mg)
88	40	66
99	45	75
110	50	83
121	55	91
132	60	100
143	65	108
154	70	116
165	75	125
176	80	133
187	85	141
198	90	150
209	95	158

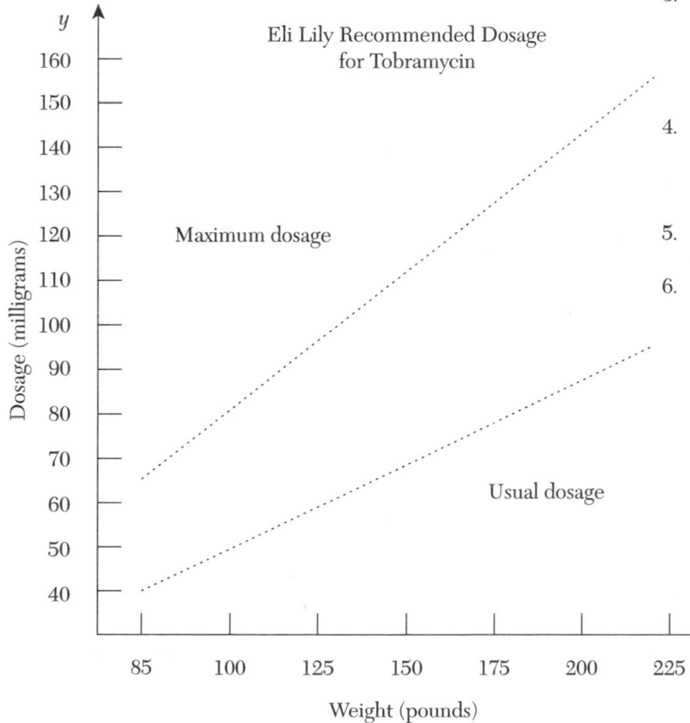

FIGURE 5.61

1. and 2. See figure 5.61.

3. The slope for usual dosage is about 0.45, and for maximum dosage it is about 0.76. For every pound increase in weight, you can increase the usual dosage by .45 milligrams, compared to .76 milligrams increase per pound for maximum dosage.

4. The (weight, usual dosage) equation is approximately $y = 0.46x$. The (weight, maximum dosage) equation is approximately $y = 0.76x - 0.05$.

5. The lines are not parallel because they have different slopes.

6. $y = 1.67x - 0.46$. The ratio of the slopes, the change in maximum dosage to weight to the change in usual dosage to weight (0.76/0.46), is the slope of the new line. (Weight factors out of the ratio.)

This problem about prescription medicine illustrates the importance of slope and reinforces the notion of rate of change. Students draw median-fit lines, eyeball lines, or regression lines, depending on the background of the class.

Sample Problem 2 can be found in "Weights and Drug Doses," by G. Burrill, J. C. Burrill, P. Coffield, G. Davis, J. de Lange, D. Resnick, and M. Siegel. In *Data analysis and statistics across the curriculum: Addenda series, Grades 9–12* (Reston, VA: National Council of Teachers of Mathematics, 1992) p. 39.

Sample Problem 3

You are working as an actuary for a major auto insurance company. Your company plans to offer a senior citizen's accident policy, and you must predict the likelihood of an accident as a function of the driver's age. From previous accident records, you find the following information:

TABLE 5.24

Age (years)	Accidents/100 Million Kilometers Driven
20	440
30	280
40	200
50	200
60	280
70	440
80	680

From the data, you notice that the accident rate falls with age, but then begins to rise again. Based on this observation, you decide the data is not linear and assume that a quadratic function may serve as a better model.

1. Use your graphing calculator to perform a quadratic regression on this set of data. Write the equation expressing accidents per 100 million kilometers in terms of age.
2. How many accidents per 100 million kilometers would you expect for a 75-year-old driver?
3. How many accidents per 100 million kilometers would you expect for an 18-year-old driver?
4. Based on your model, who is safer: a 16-year-old driver or a 72-year-old driver?
5. What age driver appears to be the safest?
6. Your company decides to insure licensed drivers up to the age where the accident rate reaches 830 per million kilometers. What, then, is the domain of this quadratic function?

Answers to Sample Problem 3:

1. if x = age and y = accidents per 100 million kilometers driven, then $y = 0.4x^2 - 36x + 1000$.
2. 550 accidents/100 million km.
3. 482 accidents/100 million km.
4. 72-year-olds appear to be safer.
5. 45-year-olds appear to be the safest.
6. $\{x{:}16 \leq x \leq 85\}$.

Sample Problem 3 can originally be found in *Algebra and trigonometry functions and applications*, 2nd ed., by P. A. Foerster (Menlo Park, CA: Addison-Wesley Publishing Company, 1990) p. 211. It has been modified by James F. Marty. Extension 1 can be found in "Try This," by G. Burrill, J. C. Burrill, P. Coffield, G. Davis, J. de Lange, D. Resnick, and M. Siegel. In *Data analysis and statistics across the curriculum: Addenda series, Grades 9–12* (Reston, VA: National Council of Teachers of Mathematics, 1992) p. 41. Extension 2 is an original problem contributed by James F. Marty.

Standards Addressed

E.12.1 Work with the context of real-world situations.

E.12.2 Organize and display data from statistical investigations.

This problem illustrates the powerful use of technology to analyze data.

Extension 1

Most states have a driver's manual that gives a table similar to the following one, expressing the number of feet required for a car to stop and the speed at which it is traveling prior to applying the brakes. The manuals also give the "rule of thumb" of following two car lengths for every 10 miles per hour. Does the linear "rule of thumb" fit the data in the chart?

TABLE 5.25

Speed (mph)	Stopping distance (ft.)
10	15
20	40
30	75
40	120
50	175
60	240
70	315
80	400

Extension 2

If we sketch a graph of the raw data (y versus x) and think it looks like a power function ($y = a \cdot x^m$) we can very easily verify if it truly is a power function by plotting $\log y$ versus $\log x$. If it is a power function then the graph $\log y$ versus $\log x$ should be linear.

For example, given the data points in Table 5.26, the graph of y versus x is shown in Graph 5.1. The graph of $\log y$ versus $\log x$ is shown next. Notice that graph 5.2 is linear. This will always be true if the original data forms a power function.

Table 5.26

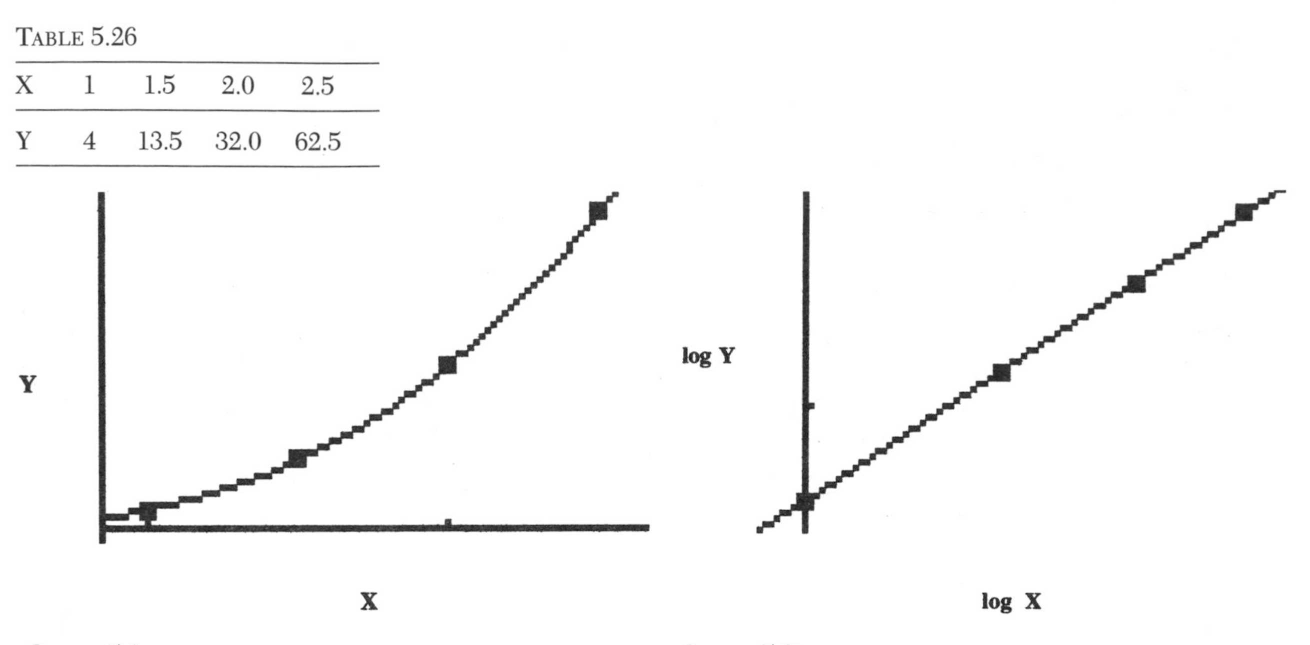

X	1	1.5	2.0	2.5
Y	4	13.5	32.0	62.5

Graph 5.1

Graph 5.2

To properly address the spirit of the standards, algebraic ideas need to be implemented throughout the K-12 mathematics curriculum.

From a graph of the original data (y versus x) you suspect that it can be modeled using an exponential function, ($y = a \cdot b^x$). The function can be tested using logarithms. If the function that models the given data is indeed an exponential function, then the graph log y versus x is linear.

For example, given the data points in Table 5.27, the graph of y versus x is shown on the left below. The graph of log y versus x is shown on the right below. Notice that the graph on the right is linear. This will always be true if the original data forms an exponential function.

The first set of data is modeled by the function $y = 4 \cdot x^3$.
The second set of data is modeled by the function $y = 3 \cdot 2^x$.

Table 5.27

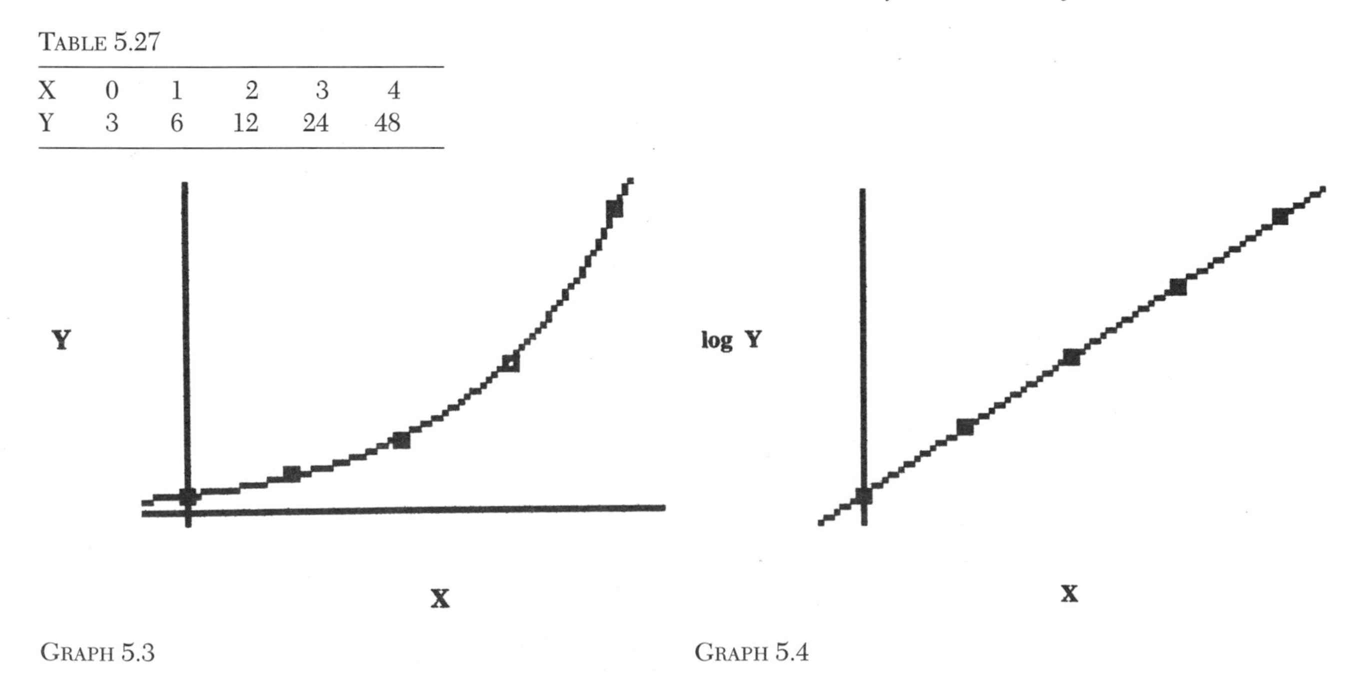

X	0	1	2	3	4
Y	3	6	12	24	48

Graph 5.3

Graph 5.4

Engineers and other professionals who frequently need to do data analysis quickly, often use this process to determine the appropriate mathematical model to use. The need to convert to logarithms can be avoided if they use what is known as log-log paper or semi-log paper where both or one of the scales are logarithmic rather than uniform.

Standards Addressed

This is another problem that involves more than one standard. The three standards used here are: Number Operations and Relationships, Statistics and Probability, and Algebraic Relationships. The specific standards addressed are: B.12.2—compare real numbers using the logarithmic scale, E.12.3—interpret and analyze information from organized and displayed data, and F.12.2—use linear, exponential, and power functions in a variety of ways. More information on this type of data analysis can be found under the Number Operations and Relationships Standard.

Sample Problem 4 with Commentary

A company that produces frozen pizzas has decided to capitalize on the Super Bowl victory by the Green Bay Packers by including with each one of its pizzas a trading card of a player from the championship team. They have produced six different cards. The company has ordered equal amounts of each card and has randomly placed them into each type and size of pizza that it produces so that a customer's chance of getting a certain card are equal with each pizza purchased.

Q. How many pizzas, on average, would a customer be expected to buy to get a complete set of all six cards?
Q. Why would the company want to know this?
Q. Why would a model be a more practical way to answer the question (than for example keeping track of how many pizzas a customer actually purchased)?

Step #1: Identify the Model

1. Explain why a standard die would make an ideal model.
2. Write a calculator statement which would generate the appropriate random numbers.

Step #2: Define a Trial

1. How do you know when the trial is complete?

Step #3/4: Repeat and Record 50 Trials.

Step #5: Find the Expected Value.

Q. How many pizzas would a customer need to buy (on average) to collect all six of the cards?
A. The average number of rolls of a standard, fair die needed to obtain all six outcomes is 14.7. How is this situation related to this problem AND how close were you to this number?

Sample Problem 4 uses the Monte Carlo Method of finding the probability of an event. Using computers has made it much easier to create simulations of such things as natural disasters. Predicting the effects of radioactivity can be computer simulated without requiring an accident such as Chernobyl to provide an opportunity for real data collection.

Sample Problem 4 was contributed by John Janty.

Standards Addressed

E.12.5 Determine the likelihood of occurrences of complex events.

Additional Resources

Additional rich problems such as these can be found in sources such as:

1. The 1989 NCTM Standards and Addenda Series books.
2. Professional Journals:
   - *Wisconsin Teacher of Mathematics* published by the Wisconsin Mathematics Council.
   - *Mathematics Teacher* published by NCTM.
3. Problems found in the curricula developed under the auspices of the National Science Foundation.

Learning Algebraic Relationships in Secondary School

Algebra is often called the language of mathematics.

Algebra is often called the language of mathematics. Patterns that express a relationship between two or more quantities can be represented using an algebraic equation or inequality. The concepts and terminology of algebra can be useful in attacking many different problems, and skill with algebraic thinking can enable students to be creative problem solvers.

Students need to start experiencing algebraic concepts at a very young age. In elementary and middle school students learn to represent quantities with symbols and use equations to relate the symbols (for example, $A = lw$, the area of a rectangle is equal to the product of the length and the width). They also learn the elementary properties of equality (for example, because $2 + 3 = 5$, it is also true that $5 = 2 + 3$; or as another example, because $2 + 5 = 7$ and $3 + 4 = 7$, then $2 + 5 = 3 + 4$). Algebra taught more formally in high school then builds on this base of knowledge.

Students need to be able to translate a real-world problem into a mathematical model, solve the mathematical problem, and translate the result back into the real-world situation.

Algebraic development in high school becomes much more formalized through the use of an axiomatic system. The applications to the real world should emphasize modeling, problem solving, and reasoning. Formalization of vocabulary to describe algebraic relationships is an emphasis of the secondary program. Students need to be able to translate a real-world problem into a mathematical model, solve the mathematical problem, and translate the result back into the real-world situation. See Figure 5.62. Students can give real-world interpretations of their mathematical results.

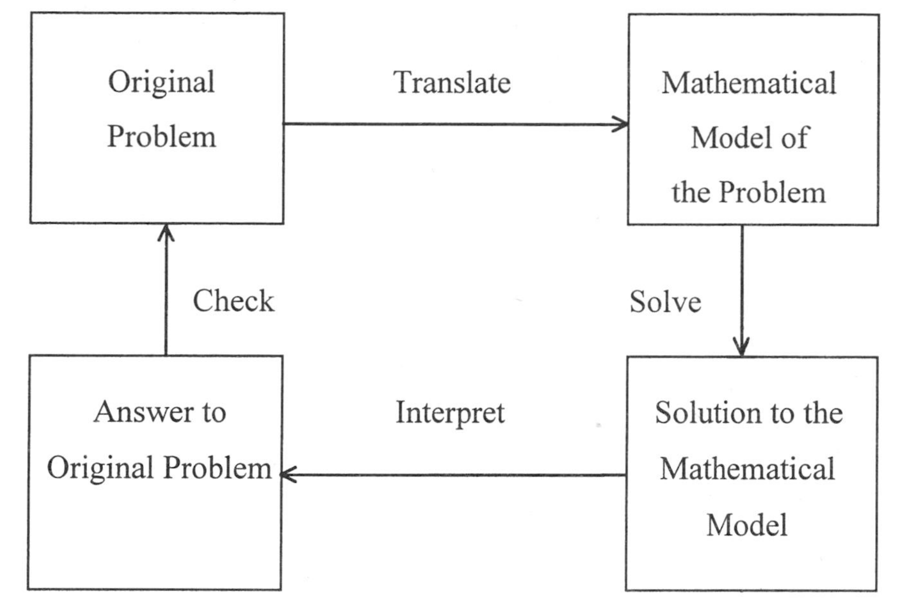

FIGURE 5.62

There should be a strong emphasis on being able to work with multiple representations of problems. Those representations should include: numeric, geometric, and algebraic. Many of the algebra problems that students encounter have strong geometric connections. Additionally, problems could relate to number theory and lead to generalizations. Although students may still be introduced to new concepts at a concrete level, progress must be made to a more abstract level. Students should feel comfortable with using symbols to represent quantities and view algebra as the language of mathematics. The language (syntax) and theorems of algebra need to be used effectively. When appropriate, discovery learning should be fused with recent technology that assists in collecting, analyzing, and presenting data. Application of algebraic skills to real-world problems naturally leads to connections between mathematics and other disciplines (such as business, science, and fine arts) and should be part of the curriculum. *Wisconsin's Model Academic Standards* are written to reflect the mandated two-year high school requirement. Secondary instruction should challenge many students to advance beyond those standards.

Wisconsin's Model Academic Standards *are written to reflect the mandated two-year high school requirement; secondary instruction should challenge many students to advance beyond those standards.*

Assessment must reflect instructional practices and should include mastery of algebraic skills and their use in an applications and problem-solving context. A fine balance should be sought between assessment that employs technology and assessment that does not involve the use of technology. The student must understand when the use of technology is appropriate for simplifying expressions or solving a problem and when it is not.

Many concepts that seem quite elementary in algebra, such as slope and rate of change, appear in a different context in real-life experiences and advanced mathematical topics. The study of linear functions should involve the gathering and analyzing of data using a variety of techniques. Instructors are encouraged to search out rich data-collecting activities that model these important concepts.

To properly address the spirit of the standards, algebraic ideas need to be implemented throughout the K–12 mathematics curriculum.

```
  83
x 42
----
   6
 160
 120
3200
----
3486
```

To properly address the spirit of the standards, algebraic ideas need to be implemented throughout the K–12 mathematics curriculum. All secondary teachers should be knowledgeable about the algebra standards outlined for grades K–4 and 5–8 in both the WMAS and the NCTM *Principles and Standards for School Mathematics.* Changing curricula at the elementary and middle levels incorporate more algebraic thinking into the curriculum than ever before. Often the intermediate algorithm for multiplication of whole numbers illustrated at the left is taught. In expanded form, the problem looks like this: $(80 + 3)(40 + 2) = 3200 + 160 + 120 + 6 = 3486$. There are no variables involved in any of these expressions, yet this is algebra! This process can be used as a springboard to begin the study of multiplication of binomials: $(x + 2)(x + 1) = x^2 + x + 2x + 2 = x^2 + 3x + 2$. Students could use algebra tiles to build and enhance their understanding of this multiplicative process.

To successfully function in the Twenty-first Century, everyone will find a need to be capable of algebraic thinking! All students do not need all the topics that have been traditionally taught in high school algebra (such as less work on rational expressions), but they do need a wide variety of experiences that generate and enhance algebraic thinking.

National Council of Teachers of Mathematics Standard for Algebra (Standards 2000 Project 2000)

Mathematical instructional programs should include attention to patterns, functions, symbols, and model so that all students

- Understand patterns, relations, and functions.
- Represent and analyze mathematical situations and structures using algebraic symbols.
- Use mathematical models to represent and understand quantitative relationships.
- Analyze change in various contexts.

Wisconsin's Model Academic Standards for Mathematics, Standard F: Algebraic Relationships

Students in Wisconsin will discover, describe, and generalize simple and complex patterns and relationships. In the context of real-world problem situations, the student will use algebraic techniques to define and describe the problem to determine and justify appropriate solutions. (Wisconsin Department of Public Instruction 1998, 14–15).

TABLE 5.28 **Wisconsin's Model Academic Standards for Algebraic Relatonships**

Wisconsin's performance standards	Elaboration
By the end of grade 12 students will:	
F.12.1 Analyze and generalize patterns of change (e.g., direct and inverse variation) and numerical sequences, and then represent them with algebraic expressions and equations.	This standard continues the work described in standards F.4.3 and F.8.2. Here the major difference is the expectation of deeper analysis of patterns of change along with the recognition that some patterns of change are nonroutine in nature.
F.12.2 Use mathematical functions (e.g., linear, exponential, quadratic, power) in a variety of ways.	As with standard F.12.1, this standard marks a continuation of conceptual and skill development begun in earlier grades. At this level the student works in a more sophisticated way with a wider range of functions, including a strong emphasis on use of technology, particularly graphing calculators. It is important to maintain the link between algebraic skills and concepts and problems from the real-world.
F.12.3 Solve linear and quadratic equations, linear inequalities, and systems of linear equations and inequalities.	This standard encompasses skills and processes that have appeared in traditional first-and-second year algebra courses. However, technology greatly facilitates the integration of numerical and graphical techniques that produce finer and finer intervals that "approximate" actual solutions.
F.12.4 Model and solve a variety of mathematical and real-world problems by using algebraic expressions, equations, and inequalities.	This statement provides a nice summarization of the rationale for the study of algebraic relationships. Students who have successfully completed a K-12 program of instruction in Algebra will certainly have achieved this standard.

Additional Standards Information

Additional support information can be found in the NCTM *Curriculum and Evaluation Standards*, 150–156, and in the NCTM PSSM, 296–306.

Sample Problem 1

Stacy wants to buy a cell phone. The Wisconsin Cell Phone Company has three plans from which to choose:

Plan 1: $9.95/month with no free minutes and $.25/minute
Plan 2: $19.95/month with 30 free minutes and $.10/minute after that
Plan 3: $29.95/month with 120 free minutes and $.05/minute

When is Plan 1 the cheapest?
When is Plan 2 the cheapest?
When is Plan 3 the cheapest?

Discussion

The problem can be approached by using a variety of methods, including algebra, graphing, or tables. The cost for a specific number of minutes can be calculated using tables (i.e., every 30 minutes for each of the plans). When it can be determined which plan is the most economical, an approximation method can be used to discover the exact number of minutes where the plan becomes less expensive. Students could also graph the table values and look for points of intersection. As a result they would have both a numerical and geometric representation to aid in understanding the problem.

Algebraically the following equations are valid if the point of intersection is beyond the number of free minutes:

Plan 1: $\$9.95 + .25x = Y1; x \geqslant 0$
Plan 2: $\$19.95 + .10\,(x - 30) = Y2; x \geqslant 30$
Plan 3: $\$29.95 + .05\,(x - 120) = Y3; x \geqslant 120$

Setting $Y1 = Y2$, and solving for x will find the number of minutes where Plan 2 becomes more economical than Plan 1 (> 46 2/3 min.). Students can determine where Plan 3 becomes cheaper (> 120 minutes) in a similar manner.

In determining when Plan 3 is less expensive than Plan 1, the algebraic solution is 70. But 70 is not in the domain for the formula for Plan 3 so this answer is not valid. To get the true solution, use the formula $Y3 = \$29.95$ rather than the formula for Y3 given previously.

Sample Problem 1 was contributed by Ann Krause, Blackhawk Technical College, and James F. Marty, Waukesha North High School.

Extension

Realistically, phone companies charge customers using a modified step-function (i.e., greatest integer function) rather than a linear function. Solve this problem using step-functions. Include graphs, a written explanation, and write equations for the graphs.

Sample Problem 2 with Commentary

Drug Dose Problem: A patient receiving a dose of medicine can be modeled by adding 16 milliliters of food coloring to a clear glass bowl containing four cups of water. The assumption that the kidneys remove one quarter of the medicine from the body in a four-hour period is modeled by removing one cup of water from the bowl and replacing it with a cup of clear water. How much medicine remains after eight hours?

Discussion

The procedure described above simulates the patient actually taking medicine, and needs to be followed and repeated. Students need to experience the problem by working cooperative groups to add the food coloring, scoop the "blood," replace with "fresh blood," recording the data, calculate concentration levels, and observe the color changes. Follow-up discussion helps students to communicate thinking and formalize their observations into mathematical expressions, functions, and graphical representations. By constructing tables or graphs, using a calculator or pencil/paper, students can visualize how much medicine is in the blood after any multiple of four hours. This problem can be used to introduce recursive notation. If $a(n)$ denotes the amount of medicine in the blood at the end of n four hour periods, then $a(n) = 0.75\, a(n-1)$.

Sample Problem 2 can be found in "Drugs and pollution in the algebra class," by J. T. Sandefur. *Mathematics Teacher,* vol. 85, no. 2 (1992): 139–145. Reston, VA: National Council of Teachers of Mathematics. There the problem was discussed in more detail and suggestions on how to teach it were given.

Extension 1

Challenge students to find out how many four-hour periods are necessary before the medicine in the blood is less than 0.0001 times of the original dose. Students will discover and use the formula $a(n) = (0.75)^n a(0)$. A geometric approach to the problem will introduce the concept of limit.

Extension 2

Let the student assume that a dose of medicine is given every four hours. The new question becomes: "How high does the level of medicine eventually get?" Steps analogous to those in the original problem can be used. This problem can allow an introduction to matrices.

Sample Problem 3 with Commentary

Sketch a reasonable graph for five of the following relationships:

1. The distance you have gone depends on how long you have been going (at a constant speed).
2. The number of used aluminum cans you collect and the number of dollars refunded to you are related.
3. The distance required to stop your car depends on how fast you are going when you apply the brakes.
4. The mass of a person of average build depends on his or her height.
5. Your car is standing on a long, level highway. You start the motor and floorboard the gas pedal. The speed you are going depends on the number of seconds that have passed since you stepped on the gas pedal.
6. The altitude of a punted football depends on the number of seconds since it was kicked.
7. The maximum speed your car will go depends on how steep a hill you are going up or down.
8. Dan Druff's age and the number of hairs he has growing on his head are related.
9. The distance you are from the band and how loud it sounds to you are related.

10. You fill up your car's gas tank and start driving. The amount of gas you have left in the tank depends on how far you have driven.
11. Your age and your height are related to one another.
12. You pull the plug out of the bathtub. The amount of water remaining in the tub and the number of seconds since you pulled the plug are related to each other.
13. The price you pay for a carton of milk depends on how much milk the carton holds.
14. Calvin Butterball desires to lose some weight, so he reduces his food intake from 8,000 calories per day to 2,000 calories per day. His weight depends on the number of days that have elapsed since he reduced his food intake.
15. The price you pay for a pizza depends on the diameter of the pizza.
16. The distance you are from the reading lamp and the amount of light it shines on your book are related.
17. You climb to the top of the 190-meter tall Tower of Americas and drop your algebra book. The distance the book is above the ground depends on the number of seconds that have passed since you dropped it.
18. As you blow up a balloon, its diameter and the number of breaths you have blown into it are related.
19. The temperature of your cup of coffee is related to how long it has been cooling.
20. You turn on the hot water faucet. As the water runs, its temperature depends on the number of seconds it has been since you turned on the faucet.
21. The time the sun rises depends on the day of the year.
22. As you breathe, the volume of air in your lungs depends upon time.
23. You start running and go as fast as you can for a long period of time. The number of minutes you have been running and the speed you are going are related to each other.
24. As you play with a yo-yo, the number of seconds that have passed and the yo-yo's distance from the floor are related.
25. You run a mile once each day. The length of time it takes you to run it depends on the number of times you have run it in practice.
26. When you dive off the three-meter diving board, time and your position in relation to the water's surface are related to each other.
27. The rate at which you are breathing depends on how long it has been since you finished running a race.
28. Taryn Feathers catches the bus to work each morning. Busses depart every 10 minutes. The time she gets to work depends on the time she leaves home.
29. Milt Famey pitches his famous fast ball to Stan Dupp, who hits it for a home run. The number of seconds that have elapsed since Milt released the ball and its distance from the ground are related.
30. The amount of postage you must put on a first-class letter depends on the weight of the letter.
31. The amount of water you have put on your lawn depends on how long the sprinkler has been running.

32. You plant an acorn. The height of the resulting oak tree depends on the number of years that have elapsed since the planting.
33. A leading soft drink company comes out with a new product, Ms. Phizz. They figure that there is a relationship between how much of the stuff they sell and how many dollars they spend on advertising.
34. Your car stalls, so you get out and push. The speed at which the car goes depends on how hard you push.
35. The number of letters in the corner mailbox depends upon the time of the day.
36. The diameter of a plate and the amount of food you can put on it are related to each other.
37. The number of cents you pay for a long-distance telephone call depends on how long you talk.
38. The grade you could make on a particular test depends upon how long you study for it.
39. You go from the sunlight into a dark room. The diameter of your pupils and the length of time you have been in the room are related.
40. The grade you could make on a particular test depends on how much time elapses between the time you study for it and the time you take the test.
41. You pour some cold water from the refrigerator into a glass but forget to drink it. As the water sits, its temperature depends on the number of minutes that have passed since you poured it.
42. Your efficiency at studying algebra depends on how late at night it is.
43. You pour some popcorn into a popper and turn it on. The number of pops per second depends on how long the popper has been turned on.
44. How well you can concentrate on algebra homework and how late at night it is are related.

This unit on sketching graphs is found in Paul A. Foerster's book entitled *Algebra and Trigonometry Functions and Applications* published by Addison-Wesley in 1990 (62–64).

Discussion

Requiring students to explain their strategies and solutions will result in a very wide variety of answers and explanations. This discussion will prompt further discussion on domain and range and will lead to such discoveries as asymptotes and growth/decay. Students relate to the contexts represented and can find one or more graphed relationships about which they are knowledgeable, thus increasing motivation and involvement. This is a good illustration of how mathematics permeates our lives. Students start to think about using mathematical modeling to describe what they observe in the world immediately around them. Students see some of the rich connections between mathematics and other disciplines and connections within mathematics itself.

When students are introduced to linear functions, they encounter slope-intercept form, $y = mx + b$, or function form, $f(x) = mx + b$, where b represents the y-intercept and m represents the slope of the line. Although most students graph quite well, too many have little or no real understanding of what the slope really means. The many uses of linear functions in mathematics and in every day lives make it imperative that every student have an understanding of exactly what slope really represents.

Slope is referred to as the rate of change. The familiar formula for slope is

$$m = \frac{y_2 - y_1}{x_2 - x_1}$$

or further along in the curriculum,

$$m = \frac{\Delta y}{\Delta x} .$$

Thus, the rate of change is the ratio between how much two quantities change (in our current example the change in y divided by the change in x). In velocity or speed, we are comparing the change in distance to the change in time (miles/hour, meters/second). A demographer (one who studies populations and their characteristics) may compare the change in population to the change in time. A company executive might compare the change in profits to the change in time. A department supervisor in a manufacturing plant may compare the change in costs to the change in the number of items manufactured (total cost divided by the number of units produced). Health field practitioners may analyze the spread of an epidemic by comparing the change in the number of newly reported cases and the change in time (newly reported cases/day). Most of the time, the focus is on how fast a variable we cannot control (the dependent variable) changes compared to how fast a variable that we can control (the independent variable) is changing. Frequently, the quantity that can be controlled is time.

When examining slope there are two items of concern:

1. Is the rate of change changing?
2. Is the rate increasing or decreasing?

On a linear graph, the slope or rate of change is constant. For any two points on a nonvertical line, the ratio between the change in quantities will always be the same. Analysis of the graph of a nonlinear function such as a quadratic function ($y = ax^2 + bx + c$, where $a \neq 0$) shows the ratio between the change in quantities will not always be the same. To facilitate the discussion, an average rate of change can be used. To find the average rate of change, select two points on the graph and calculate the slope of the segment that connects the two points. In order to compare the rates of change on different parts of the curve, let the change in the independent variable be constant. For example, in the case of speed, let the change in time stay constant at one hour or one second and, in the case of population changes, use one year. Keeping the denominator fixed at one allows for easier calculations and facilitates easy comparison.

Examine the function $y = x^2$ *and* $y = -x^2$ on various intervals.

The first thing observed is that the slopes in Figures 5.63 and 5.65 are all positive. In both cases, the values of y are increasing as x increases; therefore, it can be concluded that the function is *increasing*. Likewise, the slopes are all negative in Figures 5.64 and 5.66. In these two cases, the values of the func-

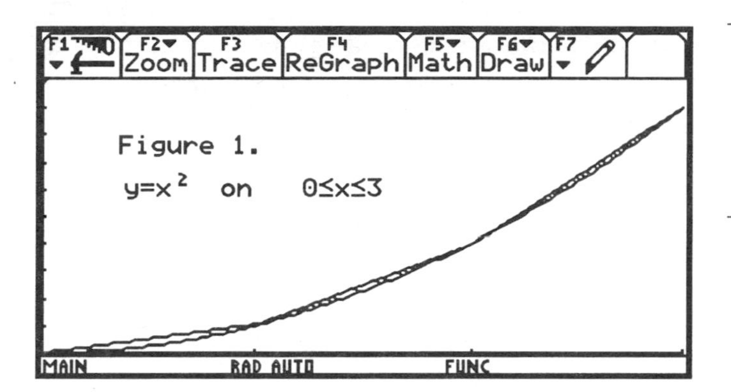

x	y
0	0
1	1
2	4
3	9

average interval	rate of change
0 < x < 1	1
1 < x < 2	3
2 < x < 3	5

FIGURE 5.63

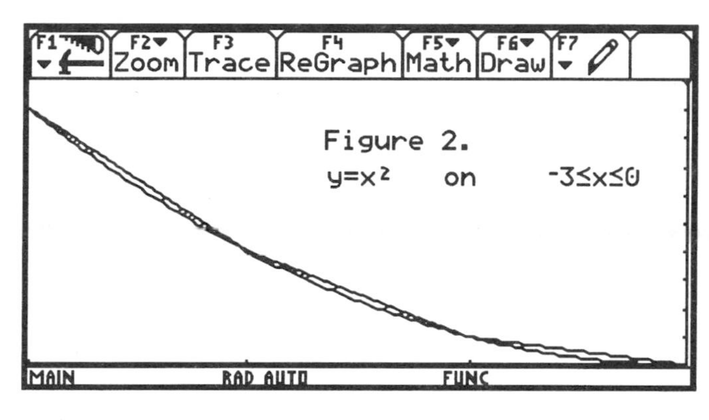

x	y
–3	9
–2	4
–1	1
0	0

average interval	rate of change
–3 < x < –2	–5
–2 < x < –1	–3
–1 < x < 0	–1

FIGURE 5.64

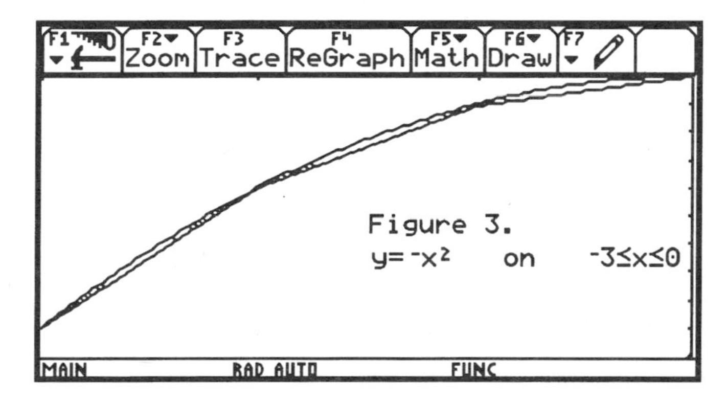

x	y
–3	–9
–2	–4
–1	–1
0	0

average interval	rate of change
–3 < x < –2	5
–2 < x < –1	3
–1 < x < 0	1

FIGURE 5.65

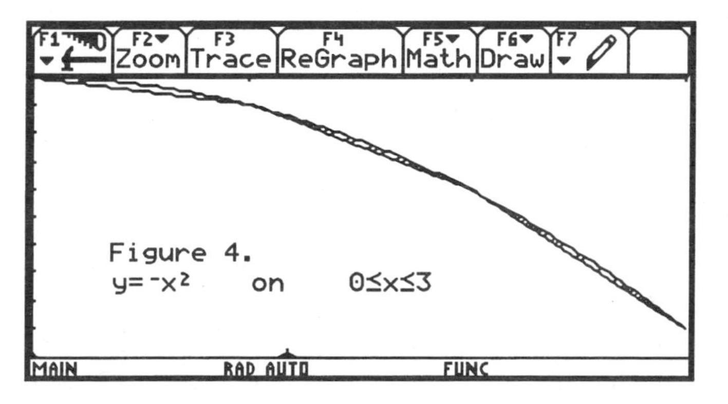

x	y
0	0
1	–1
2	–4
3	–9

average interval	rate of change
0 < x < 1	–1
1 < x < 2	–3
2 < x < 3	–5

FIGURE 5.66

tion (y) are decreasing as x increases; therefore, it can be indicated that the function is *decreasing*. Based on observations made of the figures, it can be conjectured that the sign of the slope can help determine when a function is increasing or decreasing. The second thing that should be observed is that the slopes of the lines are increasing as x increases in Figures 5.63 and 5.64. Both of these curves are part of a curve shaped like the one on the right in Figure 5.67.

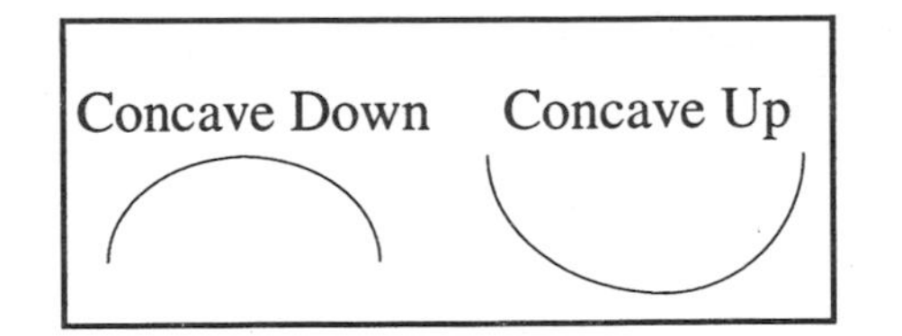

FIGURE 5.67

This curve is a concave up. In Figures 5.65 and 5.66, the slopes are decreasing. Because they are part of a curve such as the one on the left in Figure 5.67, those curves are concave down.

Based on the latter observations, a conjecture could be made that the increasing or decreasing of the slopes has an effect on the concavity of a curve.

Next, the same analysis as above could be done using the functions and intervals given in Table 5.29.

Table 5.30 summarizes possible findings.

TABLE 5.29

Function	Interval
$y = x^3$	$0 \leqslant x \leqslant 3$
$y = x^3$	$-3 \leqslant x \leqslant 0$
$y = x^3$	$-3 \leqslant x \leqslant 0$
$y = -x^3$	$0 \leqslant x \leqslant 3$

TABLE 5.30

Increasing/decreasing	Concavity	Shape of that portion of the curve
Increasing	Up	
Decreasing	Up	
Increasing	Down	
Decreasing	Down	

Between the points where increasing/decreasing changes or a change in concavity occurs, the curve must assume exactly one of the four shapes given in Table 5.30.

When a function switches from increasing to decreasing, it has reached a local maximum (or peak). If the function switches from decreasing to increasing, we say it has reached a local minimum (or valley) (See Figure 5.68). Such a point is called a local maximum/minimum because it may not be the largest/smallest value of function (y) on the entire graph; but, it will be the largest/smallest value of the function in a small interval around that value of x.

Any point where the concavity changes is known as a point of inflection (IP) (See Figure 5.69).

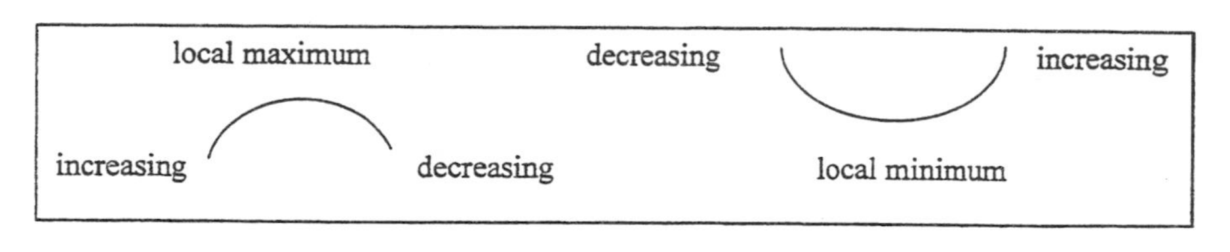

FIGURE 5.68

The following is an anecdote about Sam Walton, the founder of Wal-Mart. The origination and validity of this story has not been fully documented, but when you drive by any Wal-Mart store, you can observe a satellite dish on the roof. Each store sends its sales and inventory reports to the main office in Arkansas every night via the satellite dish where Walton and his staff kept track of the sales of every item in their stores. When the sales of an item fell one week, it was considered an accident. When the sales of an item fell two consecutive weeks, it was not considered an accident, and Walton had his warehouses quit ordering that item. Walton was intelligent enough to realize that sales would continue to increase, but at a slower rate. (See part c of Figure 5.69). Maintaining less inventory with fewer items in storage undoubtedly contributed greatly to making Sam Walton the billionaire he became. Numerically, Sam Walton was searching for the point of inflection. Mathematicians call this graph a logistics curve.

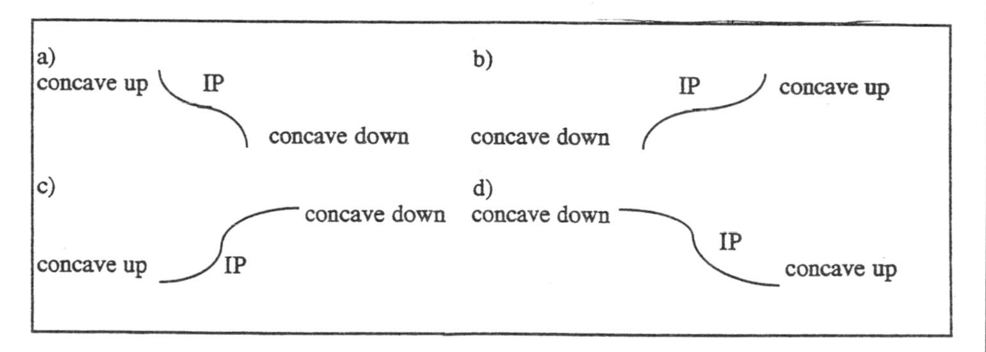

FIGURE 5.69

These examples directly address some of the ideas expressed in Wisconsin's Standard F.12.2. They reflect two major themes brought out in all of the sets of standards: first, the use of technology should be encouraged when it is appropriate; and, second, key algebraic ideas can be taught early in the curriculum, but at a more intuitive level.

Sample Problem 3 can be found in *Algebra and trigonometry functions and applications,* 2nd ed., by P. A. Foerster (Menlo Park, CA: Addison-Wesley Publishing Company, 1990) pp. 62–64. The discussion and follow-up can be found in "Rate of change," by J. F. Marty. *Wisconsin Teacher of Mathematics,* vol. 49, no. 1 (1998): 24–26.

Standards Addressed

F.12.2 Use mathematical functions (e.g., linear, exponential, quadratic, power) in a variety of ways, including recognizing that a variety of mathematical and real-world phenomena can be modeled by the same type of function; translating different forms of representations; describing the relationships among variable quantities in a problem; and using appropriate technology to interpret properties of their graphical representations (e.g., changes in rates of change, maximum and minimum).

Additional Resources

Additional rich problems such as these can be found in sources such as:

1. The 1989 NCTM Standards and Addenda Series books.
2. Professional Journals:
   - *Wisconsin Teacher of Mathematics* published by the Wisconsin Mathematics Council.
   - *Mathematics Teacher* published by NCTM.
3. Problems found in the curricula developed under the auspices of the National Science Foundation.

Performance Assessment Task

Assessment of conceptual understanding calls for opportunities for students to take in information, to weigh it, and to respond with explanation. Performance tasks allow students to "produce an answer from their own knowledge and in their own words" (Wisconsin Department of Public Instruction, 1995, 2). The Wisconsin Center for Education Research (WCER) at the University of Wisconsin–Madison, in cooperation with the Department of Public Instruction, developed several performance assessment items from 1994–1998. The "Rent-a-Car" task, complete with scoring rubric and student work, follows as an example of such assessment. It is the kind of item that could be used to prepare students for the constructed response component of the *Wisconsin Knowledge and Concepts Examination* (WKCE). The item is used with permission from the Office for Educational Accountability, Wisconsin Department of Public Instruction.

Rent-a-Car

Sally is on vacation and plans to rent a car. Here are her options:

Rent-a-Wreck: $15.95 a day plus 19¢ a mile
Lemons for Hire: $18.50 a day plus 16¢ a mile

1. Describe in detail the solution method(s) you would use to solve the following problem:

How many miles would Sally have to drive to make the Lemons for Hire the more economical choice?

[N.B.: No credit is given for solving the problem. REPEAT! Just describe the method(s) you would use to solve it.]

Suppose the rent-a-car conditions were changed as follows:

Rent-a-Wreck: \$29.95 a day plus 19¢ a mile, but the first 100 miles are free.
Lemons for Hire: \$28.50 a day plus 16¢ a mile, but the first 50 miles are free.

2. Would the method(s) you described for the first problem have to be changed in order to solve this problem?

   If you answer NO, then give the reasons why you think that the method(s) would apply.

   If you answer YES, then describe in detail what those changes would be.

Scoring

Potential Elements of a Proficient Response

Question 1

Some possible solutions:

- Construct a table that compares the prices of the two options based on the number of miles driven.
- Difference in base price is \$2.55. Difference in price per mile is \$.03. So, Sally would have to drive more than \$2.55/\$.03 = 85 miles for the Lemons for Hire option to be more economical.
- Wreck: cost = \$25.95(number of days) + \$.19 (total number of miles)
- Lemon: cost = \$28.50(number of days) + \$.16 (total number of miles) Let cost equations be equal, holding the number of days constant at one, and solve for the total number of miles. If Sally drives more than this solution, then the Lemons for Hire option would be more economical.
- Graphical representation of the two linear equations or a set of equations that are representative of the two options. The student must indicate how to interpret the graphs to recognize when the Lemons for Hire option becomes more economical.

Question 2

- Modify one of the four possible solutions specified above (or alternate solution) by increasing the price of the two options by the constant amounts given.

Major Flaws in Response

- Does not argue that after a certain number of miles, the Lemons for Hire plan becomes more economical.
- Does not sufficiently describe the method to solve the problem.
- Does not specify changes to the original method or does not justify reason for not giving changes.

Minor Flaws in Response

- Does not adequately consider the characteristics that the particular method chosen highlight (e.g., does not demonstrate understanding that the intersection point of the two linear equations in the point at which the two options cost the same).

- Makes errors in calculations.

Potential Elements of an Advanced Response

- Demonstrates more than one method to solve the problem.
- Discusses linear nature of the two models, and specifies how to solve system of linear equations (e.g., substitution method).

Student Responses

See Figures 5.70 and 5.71 for an example of an advanced student performance.

RENT-A-CAR

Sally is on vacation and plans to rent a car for one day.
Here are her options:

Rent-a-Wreck: $25.95 a day plus 19¢ a mile
Lemons for Hire: $28.50 a day plus 16¢ a mile

1. Describe in detail the solution method(s) you would use to solve the following problem:

    How many miles would Sally have to drive to make the **Lemons for Hire** the more economical choice?

    No credit is given for solving the problem.
    REPEAT! Just describe the method(s) you would use to solve it.

~~Figure out how much it costs~~
Find an equation that would get the two options to equal each other, Then put a larger ammount of miles in each equation to see if you get a better deal at Lemons for Hire.

You could also just figure it out by trial-and-error and put different ammounts of miles in the problem to figure out how many miles you would have to drive to get better milage.

You could also put each on a coordinate graph and graph the costs until rent-a-wrecks line crosses over Lemons for Hire.

FIGURE 5.70

FORM 372

Suppose the rent-a-car conditions were changed as follows:

Rent-a-Wreck: $25.95 a day plus 19¢ a mile, but the first 100 miles are free.
Lemons for Hire: $28.50 a day plus 16¢ a mile, but the first 50 miles are free.

2. Would the method(s) you described for the first problem have to be changed in order to solve this problem?

If you answer NO, then give the reasons why you think that the method(s) would apply.

If you answer YES, then describe in detail what those changes would be.

No you could use the same methods, but you will have to add the free miles into the equation or chart.

FIGURE 5.71

Pertinent Periodicals from the National Council of Teachers of Mathematics

Mathematics Teacher

Published monthly from September through May. Emphasizes practical ways of helping teachers in grades 8–12 to teach mathematics effectively. Offers special features such as "Activities" for duplication and classroom use: "Sharing Teaching Ideas," "Technology Tips," "Monthly Calendar," and "Media Clips." Reviews of instructional materials include computer software.

Special Focus Issues:

"Algebraic Thinking"	February 1997
"Connections"	November 1993
"Data Analysis"	February 1990
"Emerging Programs"	November 1995

Mathematics Teaching in the Middle School

Published eight times per year—September, October, November–December, January, February, March, April, and May. NCTM's newest journal for grades 5–9 premiered April 1994. Addresses the learning needs of all middle school students, the demands these needs place on their teachers, and the issues that capture the vitality of mathematics and the characteristics of the middle-grades student. Focuses on technology in the classroom, assessment, the use of mathematics in the world around us, professional development, and product reviews; materials for bulletin boards; and three sections for students.

Special Focus Issues:

"Algebraic Thinking"	February 1997
"Geometry"	March–April 1998
"Mathematical Connections"	March–April 1996

Pertinent Periodicals from the Wisconsin Mathematics Council

Wisconsin Mathematics Teacher

Published three times yearly, this publication contains articles on current topics in mathematics education along with classroom ideas for PreK–2, 3–5, 6–8, and 9–12.

Organizations Supporting Mathematics Teachers in Wisconsin

National Council of Teachers of Mathematics (www.nctm.org)
Wisconsin Mathematics Council (www.wismath.org)
Milwaukee Area Mathematics Council

References

Chapter 5 Introduction

Commission on Standards for School Mathematics. 1989. *Curriculum and Evaluation Standards for School Mathematics.* Reston, VA: National Council of Teachers of Mathematics.

Cuoco, A. A., and F. R. Curico, eds. 2001. *The Roles of Representation in School Mathematics: 2001 Yearbook*. Reston, VA: National Council of Teachers of Mathematics.

Friel, S. N., and R. B. Corwin. 1990. The Statistics Standards in K–8 Mathematics. *Arithmetic Teacher* 38(2): 73–77.

Harris, A. van S., and L. M. Waldo. 1911. *First Journeys in Numberland.* Chicago: Scott, Foresman and Company.

Standards 2000 Project. 2000. *Principles and Standards for School Mathematics.* Reston, VA: National Council of Teachers of Mathematics.

Reinhart, S. C. 2000. Never Say Anything a Kid Can Say! *Mathematics Teaching in the Middle School* 5(8): 478–83.

U.S. Department of Education. 1997. *Introduction to TIMSS: The Third International Mathematics and Science Study.* Washington, DC: U.S. Department of Education.

Wisconsin Department of Public Instruction. 1998. *Wisconsin's Model Academic Standards.* Madison, WI: Wisconsin Department of Public Instruction.

Early Beginnings in Mathematics

Baroody, A. J. 2000. Does Mathematics Instruction for Three- to Five-Year-Olds Really Make Sense? *Young Children* 55(4): 61–67.

Bredekamp, S., and C. Copple. 1997. *Developmentally Appropriate Practice in Early Childhood Programs.* Rev. ed. Washington, DC: National Association for the Education of Young Children.

Clements, D. H. 1999. Subitizing: What Is It? Why Teach It? *Teaching Children Mathematics* 5(7): 400–405.

Ginsburg, H. P., and J. Baron. 1993. Cognition: Young Children's Construction of Mathematics. In *Research Ideas for the Classroom: Early Childhood Mathematics*, edited by R. J. Jensen. Reston, VA: National Council of Teachers of Mathematics.

Greenes, C. 1999. Ready to Learn: Developing Young Children's Mathematical powers. In *Mathematics in the Early Years*, edited by J. V. Copley. Reston, VA: National Council of Teachers of Mathematics.

Jensen, E. 1998. *Teaching with the Brain in Mind.* Alexandria, VA: Association for Supervision and Curriculum Development.

Miller, L. H., and H. Weber. 2000. *Early Beginnings.* Unpublished paper developed for the Wisconsin Department of Public Instruction Mathematics Guidelines Task Force. Sheboygan, WI: Sheboygan Area School District.

Mokros, J. 1996. *Beyond Facts and Flashcards: Exploring Math with Your Kids.* Portsmouth, NH: Heinemann.

National Association for the Education of Young Children. 1990. *Guidelines for Appropriate Curriculum Content and Assessment in Programs Serving Children Ages 3 through 8.* A position statement of the National Association of Young Children and the National Association of Early Childhood Specialists in State Departments of Education. Washington, DC: National Association for the Education of Young Children.

Payne, J. N., and D. M. Huinker. 1993. Early Number and Numeration. In *Research Ideas for the Classroom: Early Childhood Mathematics*, edited by R. J. Jensen. Reston, VA: National Council of Teachers of Mathematics.

Perry, G., and M. S. Duru. 2000. *Resources for Developmentally Appropriate Practice: Recommendations from the Profession.* Washington, DC: National Association for the Education of Young Children.

Richardson, K., and L. Salkeld. 1995. Transforming Mathematics Curriculum. In *Reaching Potentials: Transforming Early Childhood Curriculum and Assessment*, edited by S. Bredekamp and T. Rosegrant. Washington, DC: National Association for the Education of Young Children.

Learning Mathematics in the Elementary Grades

Banchoff, T. F. 2000. The Mathematician as a Child and Children as Mathematicians. *Teaching Children Mathematics* 6(6): 350–356.

Burns, M. 1996. *50 Problem Solving Lessons Grades 1–6.* Sausalito, CA: Math Solutions Publications.

Burton, G. 1993. *Addenda Series, Grades K–6: Number Sense and Operations.* Reston, VA: National Council of Teachers of Mathematics.

Caliandro, C. K. 2000. Children's Inventions for Multidigit Multiplication and Division. *Teaching Children Mathematics* 6(6): 420–24.

Coburn, T. 1993. *Addenda Series, Grades K–6: Patterns.* Reston, VA: National Council of Teachers of Mathematics.

Dacey, L. S., and R. Eston. 1999. *Growing Mathematical Ideas in Kindergarten.* Sausalito, CA: Math Solutions Publications.

Del Grande, J., and L. Morrow. 1993. *Addenda Series, Grades K–6: Geometry and Spatial Sense.* Reston, VA: National Council of Teachers of Mathematics.

Economopoulos, K., C. Tierney, and S. J. Russell. 1998. *Arrays and Shares: Multiplication and Division.* Menlo Park, CA: Dale Seymour Publications.

Falkner, K. L., and T. P. Carpenter. 1999. Children's Understanding of Equality: A Foundation for Algebra. *Teaching Children Mathematics* 6(4): 232–36.

Fox, T. B. 2000. Implications of Research on Children's Understanding of Geometry. *Teaching Children Mathematics* 6(9): 572–76.

Geringer, Laura. 1985. *A Three Hat Day*. New York: HarperCollins Publishers.

Isaacs, A. C., and W. M. Carroll. 1999. Strategies for Basic Facts Instruction. *Teaching Children Mathematics* 5(9): 508–15.

Lindquist, M. 1992. *Addenda Series, Grades K–6: Making Sense of Data.* Reston, VA: National Council of Teachers of Mathematics.

McIntosh, A., B. Reys, and R. Reys. 1999. *Number Sense: Simple Effective Number Sense Experiences.* White Plains, NY: Dale Seymour Publications.

Mokros, J., S. J. Russell, and K. Economopoulos. 1995. *Beyond Arithmetic: Changing Mathematics in the Elementary Classroom.* White Plains, NY: Dale Seymour Publications.

National Council of Teachers of Mathematics. 1997. Algebraic Thinking Focus Issue. *Teaching Children Math.*, 3(6): 261–356. Reston, VA: National Council of Teachers of Mathematics.

———. 1999. Geometry and Geometric Thinking Focus Issue of *Teaching Children Math.*, 5(6): 307–386. Reston, VA: National Council of Teachers of Mathematics.

———. 1991. *Professional Standards for Teaching Mathematics.* Reston, VA: National Council of Teachers of Mathematics.

Reed, K. M. 2000. How Many Spots Does a Cheetah Have? *Teaching Children Mathematics* 6(6): 346–49.

Sheffield, L. J. 2000. Creating and Developing Promising Young Mathematicians. *Teaching Children Mathematics* 6(6): 416–19.

Teppo, A. R., ed. 1999. *Reflecting on Practice in Elementary School Mathematics.* Reston, VA: National Council of Teachers of Mathematics.

Tierney, C., M. Berle-Carman, and J. Akers. 1998. *Things that Come in Groups: Multiplication and Division.* White Plains, NY: Dale Seymour Publications.

Trafton, P. R., and D. Thiessen. 1999. *Learning through Problems: Number Sense and Computational Strategies.* Portsmouth, NH: Heinemann.

Whitin, D. J., and S. Wilde. 1992. *Children's Mathematics: Cognitively Guided Instruction.* Portsmouth, NH: Heinemann.

Wickett, M. S. 2000. Amanda Bean and the Gator Girls: Writing and Solving Multiplication Stories. *Teaching Children Mathematics* 6(5): 282–85, 303.

The Young Adolescent Learner of Mathematics

Armstrong, Thomas. 1994. *Multiple Intelligences in the Classroom.* Alexandria, VA: Association for Supervision and Curriculum Development.

Ball, D. L. 1996. Teacher Learning and the Mathematics Reforms: What We Think We Know and What We Need to Learn. *Phi Delta Kappan,* March: 500–508.

Caine, R. N., and G. Caine. 1994. *Making Connections: Teaching and the Human Brain.* Menlo Park, CA: Addison-Wesley.

Carnegie Council on Adolescent Development. October 1995. *Great Transitions: Preparing Adolescents for a New Century.* Washington, DC: Carnegie Council on Adolescent Development.

Clewell, B. C., B. T. Anderson, and M. E. Thorpe. 1992. *Breaking the Barriers.* San Francisco, CA: Jossey-Bass.

Flanders, J. R. 1987. How Much of the Content in Mathematics Textbooks Is New? *Arithmetic Teacher,* 18–23.

Gardner, H. 1983. *Frames of Mind: The Theory of Multiple Intelligences.* New York: Basic Books.

———. 1993. *Multiple Intelligences: The Theory in Practice.* New York: Basic Books.

Lampert, M. 1991. Connecting Mathematical Teaching and Learning. In *Integrating Research on Teaching and Learning Mathematics,* edited by E. Fennema, T. P. Carpenter, and S. J. Lamon. New York: State University of New York Press, 121–152.

Lappan, G., J. T. Fey, W. M. Fitzgerald, S. N. Friel, and E. D. Phillips. 1998. Comparing and Scaling. *Connected Mathematics Project.* Menlo Park, CA: Dale Seymour.

National Council of Teachers of Mathematics. 1991. *Addenda Series, Grades 5–8: Dealing with Data and Chance.* Reston, VA: National Council of Teachers of Mathematics.

———. 1991. *Addenda Series, Grades 5–8: Developing Number Sense in the Middle Grades.* Reston, VA: National Council of Teachers of Mathematics.

———. 1992. *Addenda Series, Grades 5–8: Geometry in the Middle Grades.* Reston, VA: National Council of Teachers of Mathematics.

———. 1994. *Addenda Series, rades 5–8: Measurement in the Middle Grades.* Reston, VA: National Council of Teachers of Mathematics.

———. 1991. *Addenda Series, Grades 5–8: Patterns and Functions.* Reston, VA: National Council of Teachers of Mathematics.

———. 1994. *Addenda Series, Grades 5–8: Understanding Rational Numbers and Proportions.* Reston, VA: National Council of Teachers of Mathematics.

———. 1995. *Assessment Standards for School Mathematics.* Reston, VA: National Council of Teachers of Mathematics.

———. 1989. *Curriculum and Evaluation Standards for School Mathematics.* Reston, VA: National Council of Teachers of Mathematics.

———. 1991. *Professional Standards for Teaching Mathematics.* Reston, VA: National Council of Teachers of Mathematics.

Perkins, D. N. 1981. *The Mind's Best Work.* Cambridge, MA: Harvard University Press.

Romberg, T. A., et. al. 1998. Sampler. *Mathematics in Context.* Chicago, IL: Encyclopedia Britannica Educational Corporation.

Stenmark, J. K., ed. 1991. *Mathematics Assessment: Myths, Models, Good Questions, and Practical Suggestions.* Reston, VA: National Council of Teachers of Mathematics.

Task Force on Education of Young Adolescents. 1989. *Turning Points: Preparing American Youth for the 21st Century.* Washington, DC: Carnegie Council of Adolescent Development.

Wisconsin Department of Public Instruction. 1986. *A Guide to Curriculum Planning in Mathematics.* Madison, WI: Wisconsin Department of Public Instruction.

———. January 1995. *Wisconsin Student Assessment System: Performance Assessment Sampler.* Madison, WI: Wisconsin Department of Public Instruction.

———. 1998. *Wisconsin's Model Academic Standards for Mathematics.* Madison, WI: Wisconsin Department of Public Instruction.

The Secondary Learner of Mathematics

Aichele, D. B., ed. 1994. *Professional Development for Teachers of Mathematics.* Reston, VA: National Council of Teachers of Mathematics.

Assessment Standards Working Groups. May, 1995. *Assessment Standards for School Mathematics.* Reston, VA: National Council of Teachers of Mathematics.

Bezuszka, S. J. and M. Kenney. 1982. *Number Treasury: A Sourcebook of Problems for Calculators and Computers.* Palo Alto, CA: Dale Seymour Publications.

Buerk, D., ed. 1994. *Empowering Students by Promoting Active Learning in Mathematics: Teachers Speak to Teachers.* Reston, VA: National Council of Teachers of Mathematics.

Burke, M. J., ed. 2000. *Learning Mathematics for a New Century.* Reston, VA: National Council of Teachers of Mathematics.

Burrill, G., J. C. Burrill, P. Coffield, G. Davis, J. de Lange, D. Resnick, and M. Siegel. 1992. *Addenda Series, Grades 9–12: Data Analysis and Statistics Across the Curriculum.* Reston, VA: National Council of Teachers of Mathematics.

Commission on Standards for School Mathematics. May, 1989. *Curriculum and Evaluation Standards for School Mathematics.* Reston, VA: National Council of Teachers of Mathematics.

Commission on Teaching Standards for School Mathematics. March, 1991. *Professional Standards for Teaching Mathematics.* Reston, VA: National Council of Teachers of Mathematics.

Commission on the Future of the Standards. April, 2000. *Principles and Standards for School Mathematics.* Reston, VA: National Council of Teachers of Mathematics. (With the purchase of this well written publication, one receives a card that can be sent to NCTM entitling the purchaser to receive a superb CD with electronic examples of activities which emulate the ideas presented in the PSSM for the various grade levels. The E-Standards Task Force is considering ways to expand and improve future electronic versions (both Web and CD) of the PSSM, and the Illuminations project is providing Web-base resources to "illuminate" the messages of the document with funding provided by MCI WorldCom. For current information on these and other efforts and for other information surrounding about the document, visit www.nctm.org.)

Cooney, T. J., ed. 1990. *Teaching and Learning Mathematics in the 1990's.* Reston, VA: National Council of Teachers of Mathematics.

Coxford, F. C., ed. 1988. *The Ideas of Algebra, K–12.* Reston, VA: National Council of Teachers of Mathematics.

Coxford Jr., A. F., L. Burks, C. Giamati, and J. Jonik. 1991. *Addenda Series, Grades 9–12: Geometry from Multiple Perspectives.* Reston, VA: National Council of Teachers of Mathematics.

Curcio, F. R., ed. 1987. *Teaching and Learning: A Problem-Solving Focus.* Reston, VA: National Council of Teachers of Mathematics.

Davis, B. D., C. A. Maher, and N. Noddings. 1990. *Constructivist Views on the Teaching and Learning of Mathematics.* Reston, VA: National Council of Teachers of Mathematics.

Driscoll, M. 1999. *Fostering Algebraic Thinking: A Guide for Teachers, Grades 6–10.* Portsmouth, NH: Heinemann.

Edwards, E. L., ed. 1990. *Algebra for Everyone.* Reston, VA: National Council of Teachers of Mathematics.

Elliott, P. C., ed. 1996. *Communication in Mathematics K–12 and Beyond.* Reston, VA: National Council of Teachers of Mathematics.

Fey, J., ed. 1992. *Calculators in Mathematics Education.* Reston, VA: National Council of Teachers of Mathematics.

Froelich, G., K. G. Bartkovich, and P. A. Foerster. 1991. *Addenda Series, Grades 9–12: Connecting Mathematics.* Reston, VA: National Council of Teachers of Mathematics.

Fuys, D., D. Geddes, and R. Tischler. 1988. *The van Hiele Model of Thinking in Geometry Among Adolescents.* Reston, VA: National Council of Teachers of Mathematics.

Grouws, D. A. and T. J. Cooney, eds. 1988. *Effective Mathematics Teaching.* Reston, VA: National Council of Teachers of Mathematics.

Hansen, V. P., ed. 1984. *Computers in Mathematics Education.* Reston, VA: National Council of Teachers of Mathematics.

Heid, M. K., J. Choate, C. Sheets, and R. M. Zbiek. 1995. *Addenda Series, Grades 9–12: Algebra in a Technological World.* Reston, VA: National Council of Teachers of Mathematics.

Hirsch, C. R., ed. 1985. *The Secondary School Mathematics Curriculum.* Reston, VA: National Council of Teachers of Mathematics.

House, P. A., ed. 1995. *Connecting Mathematics Across the Curriculum.* Reston, VA: National Council of Teachers of Mathematics.

Joint Committee of the Mathematical Association of America and the National Council of Teachers of Mathematics. 1980. *A Sourcebook of Applications of School Mathematics.* Reston, VA: National Council of Teachers of Mathematics.

Kenney, M. J., ed. 1991. *Discrete Mathematics Across the Curriculum, K–12.* Reston, VA: National Council of Teachers of Mathematics.

Krulik, S., ed. 1980. *Problem Solving in School Mathematics.* Reston, VA: National Council of Teachers of Mathematics.

Lindquist, M. M., ed. 1987. *Learning and Teaching Geometry, K–12.* Reston, VA: National Council of Teachers of Mathematics.

Meiring, S. P., R. N. Rubenstein, J. E. Schultz, J. de Lange, and D. L. Chambers. *Addenda Series, Grades 9–12: A Core Curriculum—Making Mathematics Count for Everyone.* Reston, VA: National Council of Teachers of Mathematics.

Morrow, L. J., ed. 1998. *The Teaching and Learning of Algorithms in School Mathematics.* Reston, VA: National Council of Teachers of Mathematics.

Schoen, H. L., ed. 1986. *Estimation and Mental Computation.* Reston, VA: National Council of Teachers of Mathematics.

Schulte, A. P., ed. 1981. *Teaching Statistics and Probability.* Reston, VA: National Council of Teachers of Mathematics.

Seymour, D. 1986. *Visual Patterns in Pascal's Triangle.* Palo Alto, CA: Dale Seymour Publications.

Sharron, S., ed. 1979. *Applications in School Mathematics.* Reston, VA: National Council of Teachers of Mathematics.

Shufelt, G., ed. 1983. *The Agenda in Action.* Reston, VA: National Council of Teachers of Mathematics.

Silvey, L., ed. 1983. *Mathematics for the Middle Grades (5–9).* Reston, VA: National Council of Teachers of Mathematics.

Stenmark, J. K., ed. 1991. *Mathematics Assessment: Myths, Models, Good Questions, and Practical Suggestions.* Reston, VA: National Council of Teachers of Mathematics.

Stiff, L. V., ed. 1999. *Developing Mathematical Reasoning in Grades K–12.* Reston, VA: National Council of Teachers of Mathematics.

Trafton, P. R., ed. 1989. *New Directions for Elementary School Mathematics.* Reston, VA: National Council of Teachers of Mathematics. (This is an excellent source to discover what students are taught prior to grades 9–12.)

Trentacosta, J., ed. 1997. *Multicultural and Gender Equity in the Mathematics Classroom: The Gift of Diversity.* Reston, VA: National Council of Teachers of Mathematics.

Wagner, S. and C. Kieran, eds. 1989. *Research Issues in the Learning and Teaching of Algebra.* Reston, VA: National Council of Teachers of Mathematics.

Webb, N. L., ed. 1993. *Assessment in the Mathematics Classroom.* Reston, VA: National Council of Teachers of Mathematics.

Planning Curriculum
in

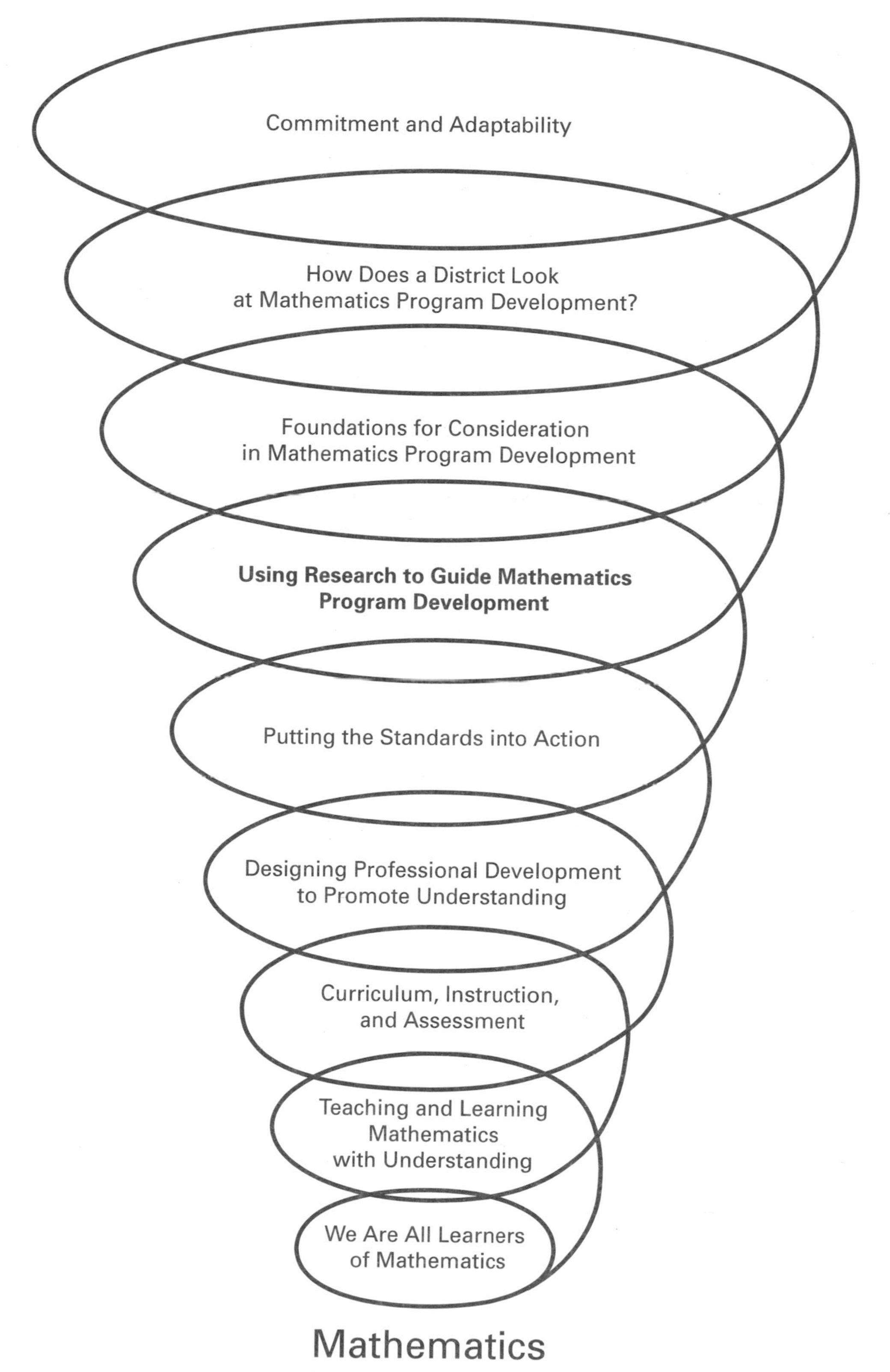

Mathematics

Spring (spring) *n.* 1. a source; origin; or beginning: *a spring into action.* 2. elasticity; resilience: *a spring forward with commitment.* (Morris, W., ed., 1971. *The American Heritage Dictionary of the English Language.* New York: American Heritage Publishing Company, Inc. and Houghton Mifflin Company, 1250, 2nd definition.)

Using Research to Guide Mathematics Program Development

6

John C. Moyer
Marquette University

Learning Research

In the late 1960s the complexity of understanding how humans learn mathematics became apparent to mathematics education researchers. Mathematics had begun to play an increasingly important role in society, yet mathematics was widely and erroneously considered to consist only of isolated facts and skills, rules and procedures, memorization and practice. Mathematical understanding was neither widely valued nor strongly desired. What use was there in trying to understand a subject that consisted only of isolated facts and skills? Mathematics educators were seeking a new perspective on teaching and learning that would reduce the widespread acceptance of mathematics learning without understanding. So mathematics educators turned for direction to the newly created field of cognitive science. Today, after some 30 years of research, there exists a growing body of evidence from cognitive science researchers suggesting there are ways to help almost all students develop a moderately deep understanding of the principles and structure of mathematics.

After some 30 years of research, there exists a growing body of evidence from cognitive science researchers suggesting there are ways to help almost all students develop a moderately deep understanding of the principles and structure of mathematics.

The recommendations of this guide are based upon this growing body of evidence, which has given rise to a cognitive conception of the learning process and the development of competent performance. As compared to the behaviorist perspective that has shaped teaching practices in mathematics for the past 90 years, the cognitive perspective on learning implies very different principles for teaching, designing curriculum, and conducting assessment than is typically found in U.S. schools today. These principles form the basis for the reform effort currently underway in U.S. mathematics education.

The cognitive perspective on learning implies very different principles for teaching, designing curriculum, and conducting assessment.

In this section I present an overview of these principles and the learning research that spawned them. Learning research, however, is not the same as instructional research; it does not prescribe how teachers should design and deliver instruction. Rather, it can only suggest principles of instruction that appear consistent with the way people learn and thus have the potential to foster learning. Hence, I also present some results from the growing body of research investigating the effectiveness of reform-based principles of instruction.

Learning research is not the same as instructional research; it does not prescribe how teachers should design and deliver instruction.

In a general sense, the view of mathematics learning embodied by the past 30 years of research is that mathematics students must construct for

"Effective teachers are those who can stimulate students to learn mathematics."

—National Research Council 1989, 58

themselves new knowledge and understandings based on what they already know and believe (National Research Council 1999). From this point of view, "no one can teach mathematics. Effective teachers are those who can stimulate students to learn mathematics" (National Research Council 1989, 58). In cognitive learning theory, a common view of understanding is that it occurs as the result of actively constructing new knowledge that is connected to existing knowledge (Hiebert and Carpenter 1992). The more numerous the connections to existing knowledge, the greater the understanding.

The cognitive view of learning carries with it many consequent principles for instruction. For example, one teaching principle that has emerged from the view that new knowledge must be constructed from existing knowledge is that teachers need to determine the incomplete understandings and false beliefs that their students have brought with them to the learning of mathematics. Having done so, the teacher must then help students simultaneously clarify their existing knowledge and build new knowledge based upon it. The result is a more mature knowledge structure that is a modification and expansion of the student's previous one.

This new cognitive view of mathematics learning is in marked contrast to the older behaviorist view, which assumes that students' minds are blank slates upon which mathematical knowledge can be imprinted. Because behaviorism values only overtly measurable objectives, there is a tendency for behaviorist-based instruction to emphasize computations and procedures rather than understanding. Hence, behaviorist-based instruction's primary method for imprinting mathematical knowledge is through drill and practice. Partly because it is so difficult to measure, understanding as previously defined is not typically found among the (behavioral) objectives of behaviorist-based instruction. To the extent that behaviorist theories address the question of understanding, it is assumed to occur as a by-product of the largely procedural knowledge imprinted by drill and practice.

In the spirit of the reform efforts underway in mathematics education, the underlying theme of this guide is that understanding should be the most fundamental goal of mathematics instruction, the goal upon which all others depend. Mathematics education researchers (e.g., Hiebert et al. 1997; Carpenter and Lehrer 1999) have hypothesized principles of mathematics instruction that promote understanding. Although it is true that these instructional principles derive logically and intuitively from learning research, existing instructional research does not provide unassailable proof that implementing these derived principles will promote mathematics understanding in the majority of students. In fact, within the complex social environment of schools, so many intervening variables occur simultaneously that some researchers (see Killion 1999) have suggested that it may not be possible to prove the validity of any instructional principles. However, most researchers and practitioners agree that it is possible to compile a body of evidence that is persuasive beyond a reasonable doubt.

The more numerous the connections to existing knowledge, the greater the understanding.

In spite of the fact that the existing body of evidence is incomplete and that there are many unanswered questions related to the large-scale implementation of cognitive-based instructional principles, emerging research strongly suggests that conscientious implementation of cognitive-based instructional principles can help all students learn and understand mathematics. Because these principles hold extraordinary promise for success, they form the basis for the instructional practices advocated in this guide.

A number of researchers and practitioners have identified instructional variables that are critical for facilitating students' understanding. Hiebert et al. (1997) have identified five dimensions of instruction that should be considered when designing mathematics instruction that promotes understanding: (1) nature of classroom tasks, (2) role of the teacher, (3) social culture of the classroom, (4) mathematical tools as learning supports, and (5) equity and accessibility. Although there is some variation in the way various sources categorize the critical variables for instruction, the variables themselves tend to be fairly consistent. That is, some sources group certain variables together in a single dimension, whereas others separate them. For example, the National Council of Teachers of Mathematics (NCTM) (2000) proposes six principles for school mathematics, corresponding to six dimensions of high-quality mathematics instruction: equity, curriculum, teaching, learning, assessment, and technology. These six dimensions are consonant with Hiebert et al.'s five dimensions and with the three dimensions enumerated by Carpenter and Lehrer (1999) (reproduced elsewhere in this guide). It has been somewhat arbitrarily decided to use the Hiebert categorization to organize presentation of the research and principles upon which the recommendations of this guide are based.

Emerging research strongly suggests that conscientious implementation of cognitive-based instructional principles can help all students learn and understand mathematics.

Nature of Classroom Tasks

Choosing worthwhile instructional tasks is one of the most important decisions a teacher makes. "No other decision that teachers make has greater impact on students' opportunity to learn and on their perceptions about what mathematics is than the selection or creation of the tasks with which the teacher engages the students in studying mathematics" (Lappan 1993, 524). Worthwhile mathematical tasks engage students in problem solving using sound and significant mathematics, elicit a variety of solution methods, and require mathematical reasoning; such problems also prompt responses that are rich enough to reveal mathematical understanding (Blume et al. 1998). The type of tasks students are asked to perform determines the type of learning that results (Doyle 1983, 1988). For example, drill and practice tasks teach students how to execute procedures quickly and accurately; listening to lectures teaches students how to reproduce their teachers' words and actions. On the other hand, solving problems, reflecting on the solutions, and discussing them is more likely to promote understanding of mathematical ideas because it forces students to build relationships between new situations and their existing knowledge. Cognitive principles derived from cognitive research imply that students develop mathematical understanding as they solve mathematics problems and reflect on their solutions. Mathematics understanding is amplified when the instructional tasks posed by the teachers require students to use previously learned concepts and techniques in the process of discovering new ones (Lappan and Briars 1995). Wood and Sellers

Choosing worthwhile instructional tasks is one of the most important decisions a teacher makes; the type of tasks students are asked to perform determines the type of learning that results.

The emphasis on understanding should not be construed as advocating the elimination of procedural learning from the curriculum.

A challenge for the design of instructional programs is to strike the appropriate balance between activities desinged to promote understanding and those designed to promote an "automaticity" of skills.

(1997) have found that on tests of standardized achievement and conceptual understanding, students receiving two years of problem-centered instruction outperformed students who lacked such instruction. Similar results on a larger scale have been reported for the public school students in Puerto Rico (Coeyman 2000).

The emphasis on understanding should not be construed as advocating the elimination of procedural learning from the curriculum. Both conceptual knowledge and procedural knowledge are required for mathematical expertise. Well-rehearsed procedures allow some aspects of mathematical tasks to be completed with relatively little mental effort, thus freeing the students' mental resources to concentrate on other, potentially more important aspects (Case 1978). A challenge for the design of instructional programs is to strike the appropriate balance between activities designed to promote understanding and those designed to promote an "automaticity" of skills that allows students to perform mathematical tasks without devoting undue attention to procedural detail. Furthermore, relating procedures to conceptual knowledge enhances students' flexibility in applying and adjusting procedures to fit a variety of problem-solving situations (Davis and McKnight 1980; Hatano 1988; Hiebert and Lefevere 1986).

Until recently, teachers did not have easy access to a rich set of worthwhile mathematical tasks to use in their daily instruction. The Third International Mathematics and Science Study (TIMSS) (Schmidt, McKnight, and Raizen 1997) criticized many curricula implemented in the United States as being "a mile wide and an inch deep" (122). According to the National Council of Teachers of Mathematics (2000, 14), "A curriculum is more than a collection of activities [no matter how worthwhile]: it must be coherent, focused on important mathematics, and well articulated across the grades." In recognition of the shortcomings of U.S. curricula, beginning in the 1990s the National Science Foundation (NSF) sponsored the development of 13 problem-centered "reform" curricula designed to foster understanding in students. At the elementary level these are *Everyday Mathematics; Investigations in Number, Data, and Space;* and *Math Trail Blazers;* at the middle level they are *Connected Mathematics Project, Mathematics in Context, MathScape, Middle Grades Math Thematics, and Middle Mathematics Through Applications;* at the secondary level they are *ARISE Program, Core-Plus Mathematics, Interactive Mathematics Program, MATH Connections,* and *SIMMS Integrated Mathematics.* The American Association for the Advancement of Science (AAAS) evaluated the four NSF-sponsored middle-level curricula, together with nine popular commercial middle-level curricula, using 24 instructional criteria derived from research on learning and teaching (American Association for the Advancement of Science 2000). The results of the evaluation are aligned with the findings of Schmidt, McKnight, and Raizen (1997) in that the AAAS gave satisfactory ratings to the four NSF-sponsored reform curricula but not to any of the nine popular commercial curricula.

The AAAS evaluation does not address the issue of how well students who use curricula that implement the principles derived from research on learning and teaching actually perform in the classroom. The issue is being investigated, however, by a variety of groups and researchers. In one noteworthy study, the U.S. Department of Education's Mathematics and Science Expert Panel (1999) reviewed 61 mathematics programs; 5 of the programs were rated exemplary, and 5 were rated promising. All 10 top-rated programs were judged to exemplify the mathematics called for in the NCTM's curriculum and evaluation standards (1989) and the Project 2061's benchmarks for science literacy (American Association for the Advancement of Science 1993). In addition, the five exemplary programs (*Cognitive Tutor Algebra, College Preparatory Mathematics, Connected Mathematics, Core-Plus Mathematics Project,* and *Interactive Mathematics Program*) provided convincing evidence of effectiveness in multiple sites with multiple populations. The five promising programs (*Everyday Mathematics, MathLand, Middle-School Mathematics through Applications Project, Number Power,* and *The University of Chicago School Mathematics Project*) provided preliminary evidence of effectiveness in one or more sites. Other research is now becoming available (e.g., Zawojewski, Robinson, and Hoover 1999; Huntley et al. 2000) that documents the superiority of the problem-centered approach of reform curricula, as compared with more traditional curricula, in fostering development of understanding and problem-solving ability. Furthermore, emerging research supports the conclusion that students using reform curricula perform at least as well on computation and skill tasks as do students using traditional curricula (e.g., Fuson, Carroll, and Drueck 2000; Carroll 1997).

Emerging research supports the conclusion that students using reform curricula perform at least as well on computation and skill tasks as do students using traditional curricula.

—e.g., Fuson, Carroll, and Drueck 2000; Carroll 1997

Role of the Teacher and the Social Culture of the Classroom

Good curricula need good teachers to implement them effectively. Even if teachers use good curricula that provide effective sequences of mathematics topics and educational tasks, their students will not automatically learn with understanding. The NCTM's teaching principle states that "effective mathematics teaching requires understanding what students know and need to learn and then challenging and supporting them to learn it well" (2000, 16). This principle implies that effective teaching begins with teacher knowledge. That is, teachers must themselves have a "profound understanding of fundamental mathematics (PUFM)" (Ma 1999; Schifter 1999), and they must also understand their students' thinking (Carpenter and Lehrer 1999; Driscoll 1999; Bell, O'Brien, and Shiu 1980). Both types of understanding are needed because effective instruction requires that teachers strike a balance between telling students the mathematics they need to learn and allowing them the leeway to construct their own mathematical knowledge (Ball 1993; Lampert 1991). Without PUFM, teachers will not be able to direct students' sense-making activities in productive directions or to help them make connections to important mathematics they already know. Without understanding their students' thinking, teachers will not be able to build on what the students already know or to recognize learning when it occurs.

Teachers must themselves have a "profound understanding of fundamental mathematics (PUFM)" (Ma 1999; Schifter 1999), and they must also understand their students' thinking.

—Carpenter and Lehrer 1999; Driscoll 1999; Bell, O'Brien, and Shiu 1980

Beyond mathematics knowledge and knowledge of student thinking, teachers need yet another type of knowledge. In the terminology of Shulman (1987), effective teachers need pedagogical content knowledge (knowledge about how the organizational and idiosyncratic aspects of a particular academic discipline affect the way it should be taught). In mathematics, pedagogical content knowledge arises from the interactions between teachers' mathematical knowledge and their knowledge of general teaching methods, but it is shaped by classroom practice. Mathematics teachers draw upon their pedagogical content knowledge when they identify goals, choose tasks, anticipate difficulties, ask questions, represent ideas, assess student understanding, and generally implement instructional strategies designed to help their students understand mathematics concepts.

A major role for the teacher is to establish a classroom environment that supports active construction of knowledge.

A reform view of teaching begins with the premise that students are not passive absorbers of information but rather must take an active part in constructing knowledge and understanding. Therefore, a major role for the teacher is to establish a classroom environment that supports active construction of knowledge. This role differs dramatically from the more traditional role in which the teacher's responsibility is to impart information, demonstrate procedures, oversee student practice, and act as the sole arbiter of mathematical correctness. This more traditional role is consonant with the view that understanding can occur by listening and imitating, not with the view that students must construct understanding for themselves. Recent research has shown that a problem-centered environment that incorporates and values students' intuitive solution methods and provides opportunities for student interaction and discussion can foster learning with understanding (Fennema and Romberg 1999).

The effective teacher makes assessment an integral part of the classroom environment.

The effective teacher makes assessment an integral part of the classroom environment. Two major uses of assessment need to be distinguished. The first, formative assessment, involves giving and receiving feedback during a learning activity for the purpose of improving teaching and learning. The second, summative assessment, involves measuring what students have learned at the end of a learning activity, often for purposes of grading or accountability. Opportunities for formative assessment and feedback should occur continuously as part of instruction. Effective teachers continually attempt to learn about their students' thinking and understanding (Black and William 1998). In addition, they help students learn and perform self-assessment that includes seeking and using feedback on one's progress (Schoenfeld 1985; Chi et al. 1989, 1994). A key principle of assessment is that what is assessed must be congruent with one's learning goals. Many summative assessments overly emphasize memory for procedures and facts (Porter et al. 1993) and thus are not good indicators of the type of learning that is the object of reform classrooms (Schoenfeld 1988).

The type of questions that teachers ask is important.
—Driscoll 1999

Neither continuous formative assessment nor the development of understanding can be achieved without teachers asking questions that challenge students' thinking. The type of questions that teachers ask is important (Driscoll 1999). Prichard and Bingaman (1993) found that achievement improves when teachers ask students high-level questions, as compared with asking low-level questions. Furthermore, when teachers ask students to explain their answers, they are providing support for student learning at the same time that they are assessing student understanding. In addition, when done skillfully, teacher questioning can prompt discussion among students.

In traditional classrooms, students are often required to work alone. The reform view is that doing mathematics is a collaborative activity that depends on communication and social interaction. Accordingly, effective teachers encourage classroom discourse and social interaction, and they use it to help students make connections among ideas and implicit ideas explicit (Lampert 1986; Mack 1990). Research indicates that communication among students positively influences learning (Dessart and Suydam 1983; Lampert 1989).

Mathematics is a collaborative activity that depends on communication and social interaction.

Mathematical Tools as Learning Supports

Mathematical tools are used to support learning by representing mathematical ideas and problem situations. They include such things as manipulative materials, pencil and paper, calculators, and computers. Using a somewhat broadened definition, mathematical tools can even include written symbols and oral language.

Mathematical tools are used to support learning by representing mathematical ideas and problem situations.

There is general consensus that concrete materials are valuable instructional tools. Many research studies show that the use of concrete materials can be effective in increasing student understanding (Suydam and Higgins 1977; Sowell 1989). Lesh (1979) has found that the use of manipulative materials, when connected to other representations (pictorial, verbal, symbolic, and real world) lead to a robust understanding of difficult mathematical concepts. However, there are inconsistencies in the body of research on the use of manipulative materials. This might be expected considering the wide variety of materials available and the uses to which they can be put. However, results vary even among similar, carefully controlled studies using the same concrete materials. For example, studies by Resnick and Omanson (1987) and Labinowicz (1985) show that the use of base-10 blocks had little impact on student learning. However, studies by Fuson and Briars (1990) and Hiebert and Wearne (1992) have yielded results that confirmed the value of these materials in helping students learn place value concepts. These contradictory results raise a red flag for teachers, warning that manipulative materials must be used circumspectly. To be effective, students must construct meaning for all the tools they use, including manipulative materials. This implies that teachers must structure manipulative-based experiences so that students are encouraged to construct meaning for the mathematical ideas the materials represent. Conversely, teachers must not structure manipulative-based tasks in ways that inadvertently encourage rote manipulation and memorization to the detriment of thinking and sense making.

Students must construct meaning for all the tools they use, including manipulative materials.

The NCTM technology principle states that "technology is essential in teaching and learning mathematics; it influences the mathematics that is taught and enhances students' learning" (2000, 24). This principle is in accord with general research findings that calculators and computers are tools that amplify problem solving and understanding. Hembree and Dessart (1986) have conducted a meta-analysis of 79 calculator studies (none of which analyzed graphing calculator usage). They found (Hembree and Dessart 1986, 1992) that the use of calculators improved student understanding, problem solving, and attitudes toward mathematics without affecting computational

Calculators and computers are tools that amplify problem solving and understanding.

—Hembree and Dessart 1986

The use of graphing calculators can foster problem solving and understanding of function and graphing concepts.

skills. Indeed, there is evidence that overemphasis on manual skills can obscure students' recognition of problem situations in which to employ those same skills (Mathematical Sciences Education Board 1990). Dunham and Dick (1994) have conducted a survey of graphing calculator research and concluded that the use of graphing calculators can foster problem solving and understanding of function and graphing concepts. Generally speaking, research shows that the use of computers in instruction can help students learn mathematics more deeply than otherwise (Sheets 1993; Boers-van Oosterum 1990; Rojano 1996; Mathematical Sciences Education Board 1990).

References

American Association for the Advancement of Science. 1993. *Benchmarks for Science Literacy.* New York: Oxford University Press.

———. 2000. *Middle Grades Mathematics Textbooks: A Benchmarks-Based Evaluation.* Washington, D.C.: American Association for the Advancement of Science.

Ball, D. L. (1993). With an Eye on the Mathematical Horizon: Dilemmas of Teaching Elementary School Mathematics. *Elementary School Journal* 93 (4) (March): 373–97.

Bell, A. W., D. O'Brien, and C. Shiu. 1980. Designing Teaching in the Light of Research on Understanding. In *Proceedings of the Fourth International Conference for the Psychology of Mathematics Education,* edited by R. Karplus. Berkeley, CA: The International Group for the Psychology of Mathematics Education.

Black, P., and D. William. 1998. Inside the Black Box: Raising Standards Through Classroom Assessment. *Phi Delta Kappan* 80 (2) (October): 139–48.

Blume, G. W., J. S. Zawojewski, E. A. Silver, and P. A. Kenney. 1998. Focusing in Worthwhile Mathematical Tasks in Professional Development: Using a Task from the National Assessment of Educational Progress. *Mathematics Teacher* 91 (2) (February): 156–61.

Boers-van Oosterum, M. A. M. 1990. Understanding of Variables and Their Uses Acquired by Students in Traditional and Computer-Intensive Algebra. Ph.D. diss., University of Maryland.

Carpenter, T. P., and R. Lehrer. 1999. Teaching and Learning Mathematics with Understanding. In *Mathematics Classrooms That Promote Understanding,* edited by E. F. Fennema and T. A. Romberg. Mahwah, NJ: Lawrence Erlbaum.

Carroll, W. M. 1997. Results of Third-Grade Students in a Reform Curriculum on the Illinois State Mathematics Test. *Journal for Research in Mathematics Education* 28 (2) (March): 237–42.

Case, R. 1978. Implications of Developmental Psychology for the Design of Effective Instruction. In *Cognitive Psychology and Instruction,* edited by A. M. Lesgold, J. W. Pellegrino, S. D. Fokkema, and R. Glaser. New York: Plenum.

Chi, M. T. H., M. Bassok, M. W. Lewis, P. Reimann, and R. Glaser. 1989. Self-Explanations: How Students Study and Use Examples in Learning to Solve Problems. *Cognitive Science* 13 (2) (April–June): 145–82.

Chi, M. T. H., N. deLeeuw, M. Chiu, and C. LaVancher. 1994. Eliciting Self-Explanations Improves Understanding. *Cognitive Science* 18 (3) (July–September): 439–77.

Coeyman, M. 2000. Puerto Rico Closes the Gap. *Christian Science Monitor,* National Section, May 30.

Davis, R. B., and C. McKnight. 1980. The Influence of Semantic Content on Algorithmic Behavior. *Journal of Mathematical Behavior* 3: 39–87.

Dessart, D., and M. Suydam. 1983. *Classroom Ideas from Research on Secondary School Mathematics.* Reston, VA: National Council of Teachers of Mathematics.

Doyle, W. 1983. Academic Work. *Review of Educational Research* 53 (2) (summer): 159–99.

———. 1988. Work in Mathematics Classes: The Context of Students' Thinking During Instruction. *Educational Psychologist* 23 (2) (spring): 167–80.

Driscoll, M. 1999. *Fostering Algebraic Thinking: A Guide for Teachers in Grades 6–10.* Portsmouth, NH: Heinemann.

Dunham, P. H., and T. P. Dick. 1994. Research on Graphing Calculators. *Mathematics Teacher* 87 (6) (September): 440–45.

Fennema, E. F., and T. A. Romberg, eds. 1999. *Mathematics Classrooms that Promote Understanding.* Mahwah, NJ: Lawrence Erlbaum.

Fuson, K. C., and D. J. Briars. 1990. Using a Base-Ten Blocks Learning/Teaching Approach for First- and Second-Grade Place-Value and Multidigit Addition and Subtraction. *Journal for Research in Mathematics Education* 21 (3) (May): 180–206.

Fuson, K. C., W. M. Carroll, and J. V. Drueck. 2000. Achievement Results for Second and Third Graders Using the Standards-Based Curriculum Everyday Mathematics. *Journal for Research in Mathematics Education* 31 (3) (May): 277–95.

Hatano, G. 1988. Social and Motivational Bases for Mathematical Understanding. In *Children's Mathematics,* edited by G.B. Saxe and M. Gearheart. San Francisco: Jossey-Bass.

Hembree, R., and D. J. Dessart. 1986. Effects of Hand-Held Calculators in Pre-College Mathematics Education: A Meta-Analysis. *Journal for Research in Mathematics Education* 17 (2) (March): 83–99.

———. 1992. Research on Calculators in Mathematics Education. In *Calculators in Mathematics Education: 1992 Yearbook,* edited by J. G. Fey and C. R. Hirsch. Reston, VA: National Council of Teachers of Mathematics.

Hiebert, J., and T. P. Carpenter. 1992. Learning and Teaching with Understanding. In *Handbook of Research on Mathematics Teaching and Learning,* edited by D. A. Grouws. New York: Macmillan.

Hiebert, J., T. P. Carpenter, E. Fennema, K. C. Fuson, D. Wearne, H. Murray, A. Olivier, and P. Human. 1997. *Making Sense: Teaching and Learning Mathematics with Understanding.* Portsmouth, NH: Heinemann.

Hiebert, J., and P. Lefevre. 1986. Conceptual and Procedural Knowledge in Mathematics: An Introductory Analysis. In *Conceptual and Procedural Knowledge: The Case of Mathematics,* edited by J. Hiebert. Hillsdale, NJ: Lawrence Erlbaum.

Hiebert, J., and D. Wearne. 1992. Links Between Teaching and Learning Place Value with Understanding in First Grade. *Journal for Research in Mathematics Education* 22 (2) (March): 98–122.

Huntley, M. A., C. L. Rasmussen, R. S. Villarubi, J. Sangtong, and J. T. Fey. 2000. Effects of Standards-Based Mathematics Education: A Study of the Core-Plus Mathematics Project Algebra and Functions Strand. *Journal for Research in Mathematics Education* 31 (3) (May): 328–61.

Killion, J. 1999. *What Works in the Middle: Results-Based Staff Development.* Oxford, OH: National Staff Development Council.

Labinowicz, E. 1985. *Learning from Students: New Beginnings for Teaching Numerical Thinking.* Menlo Park, CA: Addison-Wesley.

Lampert, M. 1989. Arithmetic as Problem Solving. *Arithmetic Teacher,* 36:34–36

———. 1986. Teaching Multiplication. *Journal of Mathematical Behavior* 5: 241–80.

———. 1991. Connecting Mathematical Teaching and Learning. In *Integrating Research on Teaching and Learning Mathematics,* edited by E. Fennema, T. P. Carpenter, and S. J. Lamon. Albany: State University of New York Press.

Lappan, G. 1993. What Do We Have and Where Do We Go from Here? *Arithmetic Teacher* 40 (9) (May): 524–26.

Lappan, G., and D. Briars. 1995. How Should Mathematics Be Taught? In *Prospects for School Mathematics,* edited by I. M. Carl. Reston, VA: National Council of Teachers of Mathematics.

Lesh, R. 1979. Mathematical Learning Disabilities: Consideration for Identification, Diagnosis, and Remediation. *Applied Mathematical Problem Solving,* edited by R. Lesh, D. Mierkiewicz, and M. G. Kantowski. Columbus, OH: ERIC/SMEAC.

Ma, L. 1999. *Knowing and Teaching Elementary Mathematics: Teachers' Understanding of Fundamental Mathematics in China and the United States.* Mahwah, NJ: Lawrence Erlbaum.

Mack, N. K. 1990. Learning Fractions with Understanding: Building on Informal Knowledge. *Journal for Research in Mathematics Education* 21 (1) (January): 16–32.

Mathematical Sciences Education Board. 1990. *Reshaping School Mathematics: A Philosophy and Framework for Curriculum.* Washington, D.C.: National Academy Press.

National Council of Teachers of Mathematics. 1989. *Curriculum and Evaluation Standards for School Mathematics*. Reston, VA: National Council of Teachers of Mathematics.

———. 2000. *Principles and Standards for School Mathematics*. Reston, VA: National Council of Teachers of Mathematics.

National Research Council. 1989. *Everybody Counts: A Report to the Nation on the Future of Mathematics Education*. Washington, D.C.: National Academy Press.

———. 1999. *How People Learn: Brain, Mind, Experience, and School*. Washington, D.C.: National Academy Press.

Porter, A. C., M. W. Kirst, E. J. Osthoff, J. S. Smithson, and S. A. Schneider. 1993. "Reform Up Close: A Classroom Analysis." *Final Report to the National Science Foundation on Grant no. SPA-8953446 to the Consortium for Policy Research in Education Research*. Madison: Wisconsin Center for Education Research, University of Wisconsin–Madison.

Prichard, M. K., and S. Bingaman. 1993. "Instructional Activities and Decisions." In *Research Ideas for the Classroom: High School Mathematics*. Edited by P. S. Wilson. Reston, VA: NCTM.

Resnick, L. B., and S. F. Omanson. 1987. Learning to Understand Arithmetic. In *Advances in Instructional Psychology*, vol. 3, edited by R. Glaser. Hillsdale, NJ: Lawrence Erlbaum.

Rojano, T. 1996. Developing Algebraic Aspects of Problem Solving Within a Spreadsheet Environment. In *Approaches to Algebra: Perspectives for Research and Teaching*. Edited by N. Bednarz, C. Kieran, and L. Lee. Boston: Kluwer.

Schifter, D. 1999. Reasoning About Operations: Early Algebraic Thinking in Grades K–6. In *Developing Mathematical Reasoning in Grades K–12, 1999 Yearbook*, edited by L. V. Stiff. Reston, VA: National Council of Teachers of Mathematics.

Schmidt, W. H., C. C. McKnight, and S. Raizen. 1997. *A Splintered Vision: An Investigation of U.S. Science and Mathematics Education*. Boston: Kluwer.

Schoenfeld, A. H. 1985. Mathematical Problem Solving. Orlando, FL: Academic Press.

———. 1988. When Good Teaching Leads to Bad Results: The Disasters of Well Taught Mathematics Classes. *Educational Psychologist* 23 (2) (spring): 145–66.

Sheets, C. 1993. *Effects of Computer Learning and Problem-Solving on the Development of Secondary School Students' Understanding of Mathematical Functions*. Ph.D. diss., University of Maryland.

Shulman, L. 1987. Knowledge and Teaching: Foundations of the New Reform. *Harvard Educational Review* 57 (1) (February): 1–22.

Sowell, E. J. 1989. Effects of Manipulative Materials in Mathematics Instruction. *Journal for Research in Mathematics Education* 20 (5) (November): 498–505.

Suydam, M. N., and J. L. Higgins. 1977. *Activity-Based Learning in Elementary School Mathematics: Recommendations from Research*. Columbus, OH: ERIC Center for Science, Mathematics, and Environmental Education.

United States Department of Education Mathematics and Science Expert Panel. 1999. *Exemplary Promising Mathematics Programs*. Washington DC: United States Department of Education.

Wood, T., and P. Sellers. 1997. Deepening the Analysis: Longitudinal Assessment of a Problem-Centered Mathematics Program. *Journal for Research in Mathematics Education* 28 (2) (March): 163–86.

Zawojewski, J. S., M. Robinson, and M. Hoover. 1999. Reflections on Developing Formal Mathematics and the Connected Mathematics Project. *Mathematics Teaching in the Middle School* 4 (5) (February): 324–30.

Planning Curriculum in Mathematics

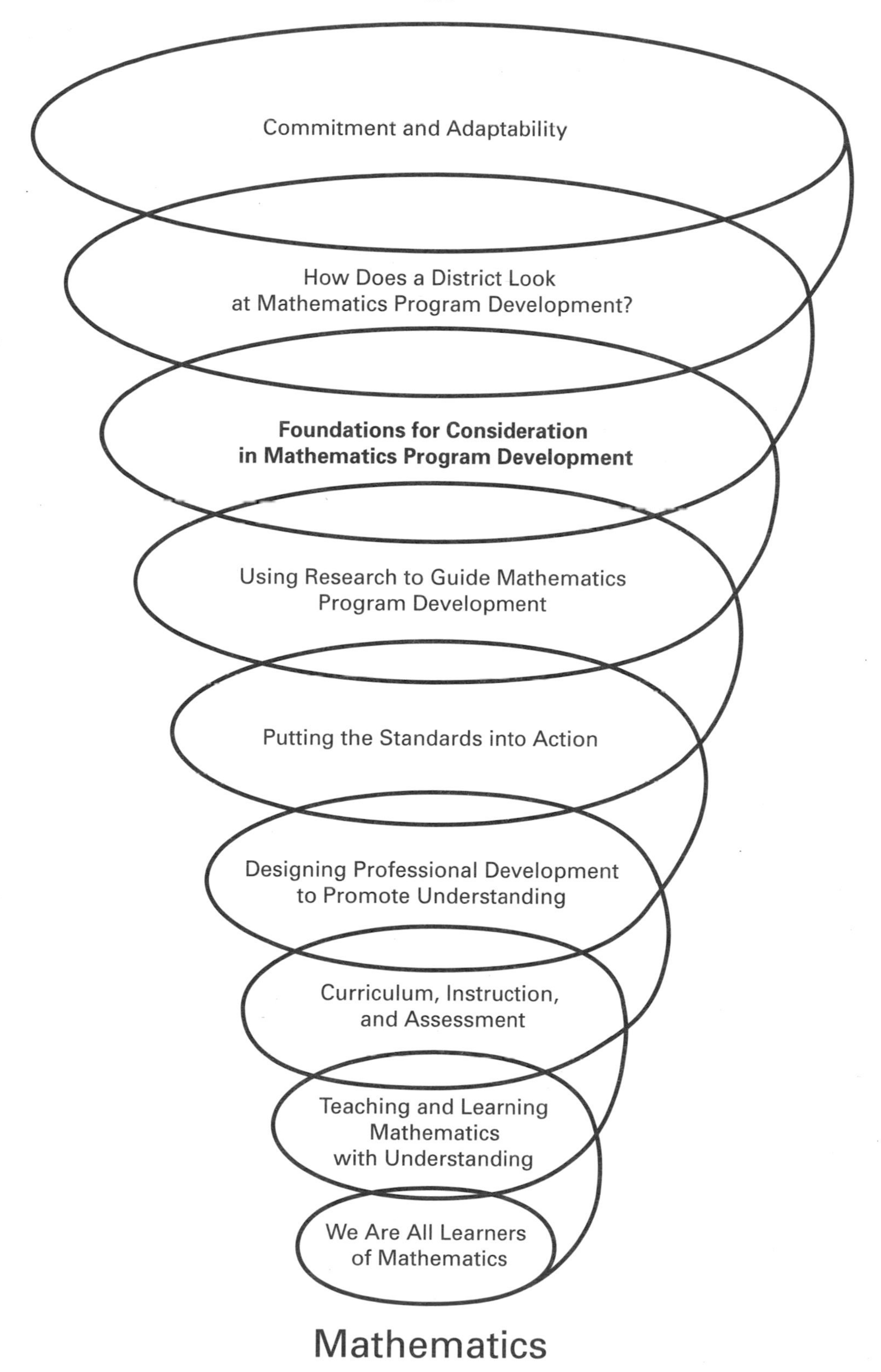

Spring (spring) *n.* 1. a source; origin; or beginning: *a spring into action.* 2. elasticity; resilience: *a spring forward with commitment.* (Morris, W., ed., 1971. *The American Heritage Dictionary of the English Language.* New York: American Heritage Publishing Company, Inc. and Houghton Mifflin Company, 1250, 2nd definition.)

Foundations for Consideration in Mathematics Program Development

7

When the agency decided, in response to school district requests, to produce guides for planning curriculum certain principles were identified that transcended discipline lines and served as foundations for the task force efforts. Among them were: teaching and learning with understanding; and curriculum, instruction, assessment, and professional development and the interrelationships between and among them (in previous chapters). In addition, tenets upon which agency efforts rested were identified: equity, enlightened use of technology, and a specific agency thrust, service learning (all discussed in this chapter). The mathematics guide task force felt that foundations of pertinence to mathematics at this time were: openness and time (in this chapter), and commitment and adaptability (final chapter in this document). Collectively, these foundations undergrid all of the endeavors undertaken in planning curriculum; indeed, they often weigh heavily in the success or failure of viable efforts.

Foundations often weigh heavily in the success or failure of viable efforts.

Openness

> Openness is a term that has gained acceptance to designate a willingness to learn, a desire to capitalize on the findings of research, and a genuine concern for improvement. So often in the preparation of these materials, previous efforts of the department, of national and state groups, and of researchers to effect improvements similar to those being called for in the guide, surfaced. Yet there still seems to be, in some circles, uncertainty, ambivalence, and lack of conviction regarding what needs to be done mathematically as we go forward into a new millennium. With mushrooming knowledge, instantaneous communication, and hard evidence, it would seem that improvement would come more readily. It is openness to the realm of possibilities that will make this happen.

Openness is a willingness to learn, a desire to capitalize on the findings of research, and a genuine concern for improvement.

Our world has become smaller because of instant communication. Knowledge acquisition has become easier because we can access information online, download it, and instantly put it to use. We have literally at our fingertips research on how mathematics is best taught and learned (Schoenfeld, Studies in Mathematical Thinking and Learning Series), on what kind of mathematics is worthy of understanding (Romberg and Kaput 1999), on mathematical endeavors that yield results in other countries (Wilson and Blank 1999; Ma 1999), on "Learning Mathematics for a New Century" (Burke and Curcio 2000), and so on.

"The classrooms described... provide existence proofs that classrooms that promote understanding can and do exist... Students who studied in them did come to understand mathematics."

—Fennema, Sowder, and Carpenter 1999, 185

Findings from research have been put to use in classrooms (Fennema and Romberg 1999) and have been proven: "The classrooms described . . . provide existence proofs that classrooms that promote understanding can and do exist. . . . Students who studied in them did come to understand mathematics [worthy of understanding]." Continuing, the authors suggest that "a critical question remains: In what ways can many more classrooms be developed so that all students have the opportunity to learn with understanding?" (Fennema, Sowder, and Carpenter 1999, 185). (See Impact of National Science Foundation-funded programs in Appendix A.)

It is clear in an age of continual learning that hose who adapt and learn readily and efficiently are advantaged.

It is clear in an age of continual learning that those who adapt and learn readily and efficiently are advantaged. The growth of technological endeavors, of multinational industries, of e-commerce, and so on are excellent examples of knowledge assimilated quickly and used innovatively. Education, as a rule, does not move quickly to learn or even to capitalize on its own fruits. The effect of this reluctance is that education is often several years behind where the "real world" is. Never has this been more true. A current cartoon suggests, "I don't get it! They make us learn reading, writing, and arithmetic to prepare us for a world of videotapes, computer terminals, and calculators!"

In the United States, a curriculum that has focused on arithmetic, followed by one year of algebra, followed by one year of geometry, often taught using behaviorist methodology has resulted in a fragmented mathematics and has often perpetuated content and methodology without regard to current knowledge. United States mathematics programs are unique in the world in teaching mathematics as separate entities. Other countries teach "maths," using whatever mathematics is needed to solve "realistic" (Fruedenthal 1983) problems that lead students toward building understanding.

A focus on memorized facts and routine procedures, on a curriculum that "design[s] each course primarily to meet the prerequisites of the next course" (Romberg and Kaput 1999, 4); on routine functions that isolate mathematics as a discipline from other "thinking" disciplines; on workbooks, lists of problems, paper-and-pencil manipulations, and so on, has resulted in "a tedious, uninteresting path." There are many hurdles to clear, such as the layer-cake approach described by Steen—a few strands (e.g., arithmetic, algebra, geometry, calculus) arranged horizontally with little opportunity to informally develop intuition along the multiple roots of mathematics (1990, 4). There is unacceptably high attrition from the mathematics program, and there is "little resemblance to what a mathematician or user of mathematics does" (Romberg and Kaput 1999, 5). Nationally there has been concern about our general mathematical performance. Mathematics does not "work" for many citizens. It is not unusual to hear comments such as "Mathematics ... I hated it ... never did understand it ... thank goodness for calculators!" "Innumeracy, an inability to deal comfortably with the fundamental notions of number and chance, plagues far too many otherwise knowledgeable citizens" (Paulos 1988, 3).

Yet when there is talk about changing the mathematics that is taught to address identified needs, conversations often devolve into a skills-versus-problem-solving argument. Wu calls this a "bogus dichotomy in mathematics education. There is not 'conceptual understanding' and 'problem-solving skill' on the one hand and 'basic skills' on the other nor can one acquire the former without the latter." He continues with an illustration that pictures a "violinist who still worries about fingering positions" and "an opera singer without the requisite high notes." (Wu 1999, 14).

"There is not 'conceptual understanding' and 'problem-solving skill' on the one hand and 'basic skills' on the other nor can one acquire the former without the latter."

—Wu 1999, 14

Schoenfeld, in *Mathematical Problem Solving*, talks about his revised expectations for problem solving based on in-depth research via videotaped analysis of college mathematics students. Students' resources were far weaker than their performance on standard tests, they had little or no awareness of mathematical heuristics (general problem-solving techniques), they did not know how to deploy the resources (basic mathematical knowledge) they had at their disposal, they were unable to tackle nonstandard problems efficiently, many had serious misunderstandings about mathematics, and they "did not perceive their mathematical knowledge as being useful to them, and consequently did not call upon it" (1985, 13). The disconnect between skills and problem solving was clearly evident.

It is not then a case of teaching skills and procedures or of teaching problem solving. It is a case of teaching both, of making connections, and of building conceptual understanding so that retention and transfer are facilitated. "Learning [with understanding] is generative" (Carpenter and Lehrer 1999, 19). Skills and procedures can be best built when problem-solving situations call for them. Problem solving can best be accomplished when there is automaticity with skills and procedures. Understanding truly surfaces when nonroutine situations are encountered.

Mathematical progress today is dependent upon using what we know about mathematics teaching and learning. It requires programs that are predicated on this knowledge. It calls for teachers who continually strive to know current mathematics techniques congruent with student teaching and learning, and who possess pedagogical content knowledge in sync with such understandings. Assessments need to reflect the content and the manner in which mathematics is taught. Support for such endeavors needs to be enlightened and pervasive.

Mathematical progress today is dependent upon using what we know about mathematics teaching and learning.

Much of the knowledge that exists today did not exist at the time that teachers, parents, and community members attended school. Consequently, "those who are charged with the reform are often those who were most successful with 'the old ways' " (Sparks 1999). If we are to truly make a difference in mathematical understanding, an openness to using what we know about mathematics in this time and place is imperative. This often means changing belief systems ... a Herculean task but one of great importance if progress is to be made.

"I have a son who is a doctor. As I visited with him the other evening, he was researching something on his computer. Looking up, he noted, 'You know, Mother, yesterday's knowledge is not sufficient to treat today's patient.'

"Today's children need to benefit from today's knowledge."

—Grunow 2000

I could not help but think, these are children of a new millennium. We are in possession of 'how to' knowledge that has not existed before. Today's children need to benefit from today's knowledge" (Grunow 2000).

Time

> All areas of educational endeavor lament the lack of it. Everyone wishes they had more. Entire studies and books are written about the management of it. We try to cram more and more into less and less of it. We know that time to learn, cogitate, and reflect is an important component of the assimilation of knowledge. We understand that the amount of time needed to learn complex material is roughly proportionate to the complexity of the information. We know that learning time needs to be more than time on task. But we tend to ignore all we know and continue to overschedule to the detriment of real learning.

We tend to ignore all we know [about the time required for complex learning] and continue to overschedule to the detriment of real learning.

Meaningful learning requires time. Students who are building understanding need time to investigate problems in-depth, discuss conjectures with classmates, reflect on their learning, and summarize mathematical understanding. To realistically accomplish these tasks, schools need to examine the amount of time allocated to mathematics classes, ensuring that the time is uninterrupted, and to look at extended timeframe learning opportunities. The "typical" American schedule assigns "an impartial" 51 minutes for each class period (no matter how well or poorly students comprehend the material), spends "only 41 percent of secondary school time on core academic subjects," offers about 5.6 hours of classroom time a day, and runs a 180-day schedule. "Our timebound mentality has fooled us all into believing that schools can educate all of the people all of the time in a school year of 180 six-hour days . . . we ask the impossible of our students. We expect them to learn as much as their counterparts abroad in only half the time" (National Education Commission on Time and Learning, 7).

The standards movement in education has called for more demanding subject matter for all students in all curricular areas. "It is important to be realistic about the amount of time it takes to learn complex subject matter" (Bransford, Brown, and Cocking 1999, 46). Citing a study of algebra students in which regular algebra students received an average of 65 hours of instruction and homework during the year and students in honors algebra received approximately 250 hours of instruction and homework, Singley and Anderson conclude, "in all domains of learning, the development of expertise occurs only with major investments of time, and the amount of time it takes to learn materials is roughly proportional to the amount of material being learned" (Singley and Anderson 1989, 46).

"The amount of time it takes to learn materials is roughly proportional to the amount of materials being learned."

—Singley and Anderson 1989, 46

School learning has other unique ramifications. Klausmeier (1985) notes that learners in school settings are often faced with tasks that do not immediately have meaning or logic. It takes time to explore underlying concepts and to contemplate connections to previous learning before learning with understanding occurs. Pacing is extremely important. There needs to be time for processing information. Pezdek and Miceli (1982) found that third graders could not mentally integrate information delivered in pictorial and verbal formats in 8 seconds; it took 15 seconds to do so. Coverage of topics requires time. If topics are covered too quickly, learning and transfer are hindered. Some students learn only an isolated set of facts and others cannot grasp principles because they do not have enough specific knowledge to make them meaningful. "The implication is that learning cannot be rushed; the complex cognitive activity of information integration requires time" (Bransford, Brown, and Cocking 1999, 46).

There is also a difference in learning and transfer in relation to the way instructional time is used. Monitoring or using feedback to assess one's progress is an important component of learning. "Feedback that signals progress in memorizing facts and formulas is different from feedback that signals the state of students' understanding" (Chi et al., 1989, 1994). Additionally, students need to know about the degree to which they know "when, where, and how to use the knowledge they are learning" (Bransford, Brown, and Cocking 1999, 47). Mere chapter and problem completion can give erroneous feedback regarding understanding. Contrasting cases can provide good feedback about learning for both perceptual and conceptual learning (Gibson and Gibson 1955; Bransford et al. 1989). Helping students see potential transfer implications is also useful (Anderson, Reder, and Simon 1996). "While time on task is necessary for learning, it is not sufficient for effective learning. Time spent learning for understanding has different consequences for transfer than time spent simply memorizing facts or procedures from textbooks or lectures. In order for learners to gain insight into their learning and their understanding, frequent feedback is critical: students need to monitor their learning and actively evaluate their strategies and their current levels of understanding" (Bransford, Brown, and Cocking 1999, 66).

Monitoring or using feedback to assess one's progress is an important component of learning.

Time spent learning for understanding has different consequences for transfer than time spent simply memorizing facts or procedures from textbooks or lectures.

Though volumes have been written about implementing the standards, there is notably little said about the time required to reap the kind of results that are sought. In a report called *Prisoners of Time* (1994), the National Education Commission on Time and Learning calls time "the missing element in the school reform debate" and the "overlooked solution to the standards problem" (9). The commission found five unresolved issues that present insurmountable barriers to efforts to improve learning that have to be addressed before adequate progress can be made.

Though volumes have been written about implementing the standards, there is notably little said about the time required to reap the kind or results that are sought.

- The fixed clock and calendar is a fundamental design flaw that must be changed.
- Academic time has been stolen to make room for a host of nonacademic activities.
- Today's school schedule must be modified to respond to the great changes that have reshaped American life outside the school.
- Educators do not have the time they need to do their job properly.
- Mastering world-class standards will require more time for almost all students.

The commission went on to contend that:

- High-ability students are forced to spend more time than they need on a curriculum developed for students of moderate ability. Many become bored, unmotivated, and frustrated. They become prisoners of time.
- Struggling students are forced to move with the class and receive less time than they need to master the materials. They are pushed on before they are ready, they fall further and further behind, they are penalized with poor grades, and they often ultimately drop out. They, too, are also prisoners of time.
- "Average" students also get caught as teachers try to motivate the gifted and help those in difficulty. Average students are robbed of quality focused time. Typical students are also prisoners of time.

The paradox is that the more the school tries to be fair in allocating time, the more unfair the consequences.

"The paradox is that the more the school tries to be fair in allocating time, the more unfair the consequences. Providing equal time for students who need more time guarantees unequal results. If we genuinely intend to give every student an equal opportunity to reach high academic standards, we must understand that some students will require unequal amounts of time, i.e., they will need additional time" (National Education Commission on Time and Learning 1994, 15).

"There is a conviction in American schools that the only valid use of teachers' time is 'in front of the class,' that 'reading, planning, collaboration with other teachers and professional development are somehow a waste of time.'"

—NECTL 1994, 17

The commission then contends that the other prisoners of time are the staff. There is a conviction in American schools that the only valid use of teachers' time is "in front of the class," that "reading, planning, collaboration with other teachers and professional development are somehow a waste of time" (17). The National Education Association states, "resolution of the time issue remains one of the most critical problems confronting educators today" (National Education Commission on Time and Learning 1994, 19). A representative of the American Federation of Teachers at a GOALS 2000 U.S. Department of Education Teacher Forum noted that teachers, principals, and administrators need time—to come up to speed as academic standards are overhauled, to come to grips with new assessment systems, to make productive and effective use of greater professional autonomy. If time is not allotted, "it sends a powerful message: don't take reform too seriously. Squeeze it in on your own time" (19).

Some findings regarding teacher changes further illuminate the issue. According to a RAND study (National Education Commission on Teaching and Learning 1994, 17), new teaching strategies can require as much as 50 hours of instruction, practice, and coaching before teachers become comfortable with them. A study of successful urban schools indicates they needed up to 50 days of external technical assistance for coaching and strengthening of staff skills through professional development. Recent professional development studies are finding success when there is a combination of introductory study, learning institutes, and ongoing online follow-up (see exemplary Wisconsin professional development models in Appendix F of this document).

"We cannot continue to abandon teachers at every critical stage of their development and then send them into the classroom with a mandate to 'teach for understanding'; this is dishonest and irresponsible."

—Askey 1999, 49

In an article on knowing and teaching elementary mathematics, Richard Askey discusses the grounding that is necessary for teachers to facilitate the development of understanding: "We cannot continue to abandon teachers at every critical stage of their development and then send them into the classroom with a mandate to 'teach for understanding.' This is dishonest and irresponsible. As things stand now, we are asking teachers to do the impossible" (Askey 1999, 49).

Though we must view international studies cautiously because of the many differences involved, there may be some things we can consider as we look at international practices. "For teachers across all countries, time is both a resource and a constraint" (Continuing to Learn from TIMSS Committee 1999, 65). Cursory examination of "instructional time" places the United States in the top half of nine countries examined, but core academic time comparisons show that in the final four years of secondary school, French, German, and Japanese students receive "more than twice as much core academic instruction as American students" (24). Additionally, "among U.S. twelfth-grade advanced mathematics students who were currently taking a mathematics course, a much lower percentage reported receiving five or more hours of mathematics instruction per week than the international average" (National Center for Education Statistics, February 1998, 74). Interestingly enough, other countries do not ignore the co-curricular and extracurricular activities that fill the rest of the school schedule; they offer them at school or at other sites after the academic day.

Research on homework indicates that it does yield results.

—Walberg, Paschal, and Weinstein 1985

Out-of-school learning is also a factor. Japanese students (15 percent in grade four, rising to 50 percent in grade nine) attend *jukus*—private tutorial services that "enrich instruction, provide remedial help, and prepare students for university examinations" (National Education Commission on Time and Learning 1994, 25). *Jukus* exist in conjunction with the Japanese belief that hard work, not ability or aptitude, is the key to meeting high standards. If more time is required for mastery, diligence and application are necessary. German and European students are accustomed to homework, spending two or more hours on it daily. In the United States, only 29 percent report spending as much time on homework. Interestingly, however, more U.S. twelfth-grade advanced mathematics students report being assigned homework three or more times per week than the international average (90 percent as contrasted with 65 percent) (National Center for Education Statistics 1998, 74).

Research on homework indicates that it does yield results. Walberg, Paschal, and Weinstein, in a synthesis of 15 empirical studies on the effects of homework on learning by elementary and secondary students, found the results to be "large and consistent. . . . When homework is merely assigned without feedback from teachers it appears to raise, on average, the typical student at the 50th percentile to the 60th percentile. But, when it is graded or commented upon, homework appears to raise learning from the 50th to the 79th percentile. This graded-homework effect is among the largest ones discovered in educational research literature" (Walberg, Paschal, and Weinstein 1985, 76). In an analysis of instructional practices that contribute to achievement done by the First in the World Consortium, it was stated, "the data also indicate that differences exist in how homework is assigned and used. Few students are more likely than U.S. students to have homework assigned every day and to discuss their completed homework in class" (National Institute on Student Achievement, Curriculum, and Assessment 1999, 31).

Teachers work varies from country to country, but teachers everywhere are very busy and spend much time outside of school... preparing.

An analysis of teacher time is also of interest. Teachers' work differs from country to country, but teachers everywhere are very busy and spend much time outside of school preparing and grading tests, reading and grading student work, planning lessons, meeting with students and parents, engaging in professional development, reading, keeping records, and completing administrative tasks. Studies done in the 1990s are showing that U.S. teachers spent 2.2 hours per week outside of the school day preparing or grading tests, 2.5 hours planning lessons, and 3.5 hours recordkeeping and doing administrative tasks. Japanese teachers spent 2.4 hours on tests, 2.7 hours on lesson plans, and 4 hours on recordkeeping and administrative tasks outside the school day. Japanese teachers have an eight- to nine-hour official workday and a 240-day school year. U.S. teachers have a seven- to eight-hour workday and an approximately 180-day school year. German teachers have a 5- to 5.5-hour school day and an approximately 184-day school year. Japanese teachers have a broad range of in-school responsibilities, "including cleaning of a portion of the school each day" (Stevenson and Nerison-Low 1997, 127–133).

Japanese teachers generally deal with more students in each classroom, but teach fewer classes—usually four hours a day. The rest of the time is spent in reviewing the day's lesson, preparing for the next day, and collaborating with colleagues. Japanese teachers prepare their lessons and teach them to one another, asking for critiques to improve the lessons. German teachers teach only 21 to 24 hours a week. Non-classroom time is spent on preparation, grading, in-service education, and consulting with colleagues, though German teachers teach in the morning and return to their homes shortly after midday. In Japan and Germany, "teachers are held to high standards" (National Education Commission on Time and Learning 1994, 26).

The National Education Commission on Time and Learning offers eight recommendations to help put time at the top of the nation's reform agenda:

- Reinvent schools around learning, not time.
- Fix the design flaw: Use time in new and better ways.
- Establish an academic day.
- Keep schools open longer to meet the needs of children and communities.
- Give teachers the time they need.
- Invest in technology.
- Develop local action plans to transform schools.
- Share the responsibility: Finger pointing and evasion must end. (29)

Elementary students should study mathematics for at least an hour a day; middle-grades and high school students should "be required to study the equivalent of a full year of mathematics in each grade."

—Standards 2000 Project 2000, 371

The National Council of Teachers of Mathematics states in its *Principles and Standards for School Mathematics*, "Learning mathematics with understanding requires consistent access to high-quality mathematics instruction." Elementary students should study mathematics for at least an hour a day "under the guidance of teachers who enjoy mathematics and are prepared to teach it well." Middle-grades and high school students should "be required to study the equivalent of a full year of mathematics in each grade" (four years of high school mathematics). "All middle-grades and high school students should be expected to spend a substantial amount of time every day working on mathematics outside of class, in activities ranging from typical homework assignments and projects to problem solving in the workplace" (Standards 2000 Project 2000, 371).

The challenge is, of course, for the mathematics to be worthwhile and challenging, for the delivery to facilitate development of understanding and be motivating, and for assessment of mathematical endeavors to be aligned with content and delivery and gleaned from multiple, authentic measures.

Equity

> Mathematical equity is provision of a mathematics education for *all* students that: empowers them to function mathematically in our society; helps them become confident in their abilities to "do" mathematics; leads them to become problem solvers of problems of all natures; and fits them for further mathematical endeavors. In an era of achievement, "because the great masses of students need to be educated for thinking for work rather than for low-skilled tasks, and educational success is a necessity rather than a luxury for a chosen few . . . schools are now expected to ensure that all students learn and perform at high levels" (Darling-Hammond, Wise, and Klein 1999, 2). Yet, mathematics as a discipline has been one of the least equitable. There is more tracking in mathematics than any subject area. Mathematics, more than any discipline, has been used to sort and separate students, often removing life choices in accordance with the results. No discipline needs examine equity more than mathematics!

Mathematical equity is provision of a mathematical education for* all *students that empowers them to function mathematically in our society.

In 1989, the NCTM *Curriculum and Evaluation Standards* stated that "equity has become an economic necessity." Making an impassioned plea for mathematical literacy for all, the standards characterized mathematics as a "critical filter for employment and full participation in our society. We cannot afford to have the majority of our population mathematically illiterate" (Commission on Standards for School Mathematics 1989, 4). The NCTM *Professional Standards for Teaching Mathematics* reemphasized that the "most compelling goal" is "the comprehensive mathematics education of every child . . . *all students*." "Every child" was then identified more specifically—"students who have been denied access in any way to educational opportunities as well as those who have not; students who are African American, Hispanic, American Indian, and other minorities as well as those who are considered to be a part of the majority; students who are female as well as those who are male; and students who have not been successful in school and in mathematics as well as those who have been successful." It was concluded that "it is essential that schools and communities accept the goal of mathematical education for *every* child" (Commission on Teaching Standards for School Mathematics 1991, 4). Continuing with the theme, the NCTM *Assessment Standards for School Mathematics* specified an equity standard. "In the past we wanted all students to learn some mathematics, but we differentiated

"We cannot afford to have the majority of our population mathematically illiterate."

—Commission on Standards for School Mathematics 1989, 4

among the types of mathematics education different groups of students received. Now we have high expectations for all students, envisioning a mathematics education that develops each student's mathematical power to the fullest." The standards then elucidated, "In an equitable assessment, each student has an opportunity to demonstrate his or her mathematical power. Because different students show what they know and can do in different ways, assessments should allow for multiple approaches" (Assessment Standards Working Groups 1995, 15).

The NCTM *Principles and Standards for School Mathematics* offer an equity principle: "Excellence in mathematics education requires equity—high expectations and strong support for all students" (Standards 2000 Project 2000, 11). Contending that "the vision of equity in mathematics education challenges a pervasive societal belief in North America that only some students are capable of learning mathematics," (12) the principles continued:

- Equity requires high expectations and worthwhile opportunities for all.
- Equity requires accommodating differences to help everyone learn mathematics.
- Equity requires resources and support for all classrooms and all students (12–13).

"Equity does not mean that every student should receive identical instruction."

—Standards 2000 Project 2000, 12

It was carefully explained that "equity does not mean that every student should receive identical instruction; instead, it demands that reasonable and appropriate accommodations be made as needed to promote access and attainment for all students" (12).

L. A. Steen, in *Why Numbers Count: Quantitative Literacy for Tomorrow's America*, compellingly points out, "The relentless quantification of society continues unabated. The tendency to reduce complex information to a few numbers is overwhelming—in health care, in social policy, in political analysis, in education. . . . Literacy is about . . . solving problems and using technology; it is about practices as well as knowledge, procedures as well as concepts. Numbers count because ideas count" (1997, xxvii). The thrust of the plea is that to survive successfully in today's world, all citizens need to possess "numeracy."

To acquire mathematical proficiency, all students obviously need access to opportunities that will help them build understanding.

To acquire mathematical proficiency, all students obviously need access to opportunities that will help them build understanding. In the United States it is common practice to "label" and "sort" students. "The practice of dividing students into instructional groups on the criterion of assumed similarity in ability or attainment is widespread . . . from assigning them temporarily to separate groups within a single classroom to setting up classes in differentiated streams or tracks. Students may be streamed only for various subjects or for their entire range of school-based learning" (Oakes 1985, ix). Whether this practice is done to ensure continued challenge for the best and brightest, to facilitate a reduction in student variability for the management and teaching of a group, or to bring together students of like dispositions, research continues to show that the practice is extremely inequitable. Oakes looked at the results from John Goodland's University of California at Los Angeles study of thirty-eight schools, "A Study of Schooling," and found that "there were clear differences between upper and lower tracks in regard to the content and quality of instruction, teacher-student and student-student relationships, the expectations of teachers for their students, the affective climate of classrooms, and other elements of the educational enterprise. It appears that those students for whom the most nurturant learning would appear to be appropriate received the least. Not only do individual schools differ widely in the quality of education they provide, but also, it appears, quality varies substantially from track level to track level within individual schools" (1985, xi).

There [are] clear differences between upper and lower tracks in regard to the content and quality of instruction.

—Oakes 1985, xi

Additionally, minority students were overrepresented in the lower tracks, as were white children of low-income families. Vocational programs served as forms of tracking in which poor and minority children were overrepresented. Surprisingly, gifted and talented students often were "not well provided for in upper tracks, just as slower students are not well provided for in the lower tracks" (Oakes 1985, xii). Indeed, gifted mathematics students frequently handle memorization and procedures so well that it is assumed that they know the mathematics concepts represented. They are not given time to "muck about" with the mathematics (Phillips 1987) and to build understandings. They suffer perhaps more than students who are given time and with whom innovative approaches are used.

"Evidence suggests that teachers themselves are tracked, with those judged to be the most competent and experienced assigned to the top tracks" (Darling-Hammond 1997, 269). In a study done in Chicago, "Closing the Divide," Dreeben (1987) found that outcomes among students were explained not by socioeconomic status, race, or ability levels but by the quality of instruction ... having enough time, covering a substantial amount of rich curricular materials, matching instruction appropriately to the ability levels of groups. Across the board, experimental studies offer strong evidence that "what students learn is substantially a function of the opportunities they are provided." Likewise, it has also become clear that "differences in teacher expertise are a major reason for the differences in learning opportunities across schools and classrooms" (Darling-Hammond 1997, 271).

Additionally, resources matter. In a large-scale study of more than one thousand school districts, Ferguson (1991) found that money matters in education. Student achievement increases with expenditure levels. The single most important determinant of student achievement is teacher expertise and experience. Small student–teacher ratios are statistically significant determinants of students outcomes. "Knowledgeable teachers working in personalized settings are the most important key to learning" (Ferguson 1991, 490). Where teachers are poorly trained and class sizes are large, students suffer from worksheets, rote drill, superficial texts, sitting for long periods of the day, working on boring tasks of low cognitive level, and seldom talking about what they know or constructing and solving problems that involve higher thinking skills (Metz 1978). Equalization of resources for school funding is needed to ensure equity.

Resources matter.

Fennema and Leder, in *Mathematics and Gender,* define *equity* as equal opportunity to learn mathematics, equal educational treatment, and equal educational outcomes (1990, 3–4). They contend that although justice for both females and males has been an underlying belief in mathematics education,

Fennema and Leder define equity *as equal opportunity to learn mathematics, equal educational treatment, and equal educational outcomes.*

studies have been based on the use of a male standard, females' achievement and beliefs about themselves in comparison with those of males. Female contributions to mathematics are minimized because of attitudes toward females as mathematicians. Classrooms perpetuate these notions. When students construct knowledge, the mental activity that takes place is as important as overt behavior. Knowledge that is brought to the activity, confidence in approaching the task at hand, control felt in doing the activity, interest in the task—all affect the quality of participation. These autonomous learning behaviors are important in working with high-cognitive-level activities. Fennema and Leder suggest that females have less opportunity to develop autonomous learning behaviors (ALBs): "Teachers often interact differently with their males and female students, with males attracting more and qualitatively different interactions" (1990, 17). Females express lower confidence in their ability to do mathematics, often suffer negative consequences such as unpopularity when they achieve mathematical success (Clarkson and Leder 1984), and are uncertain about the appropriateness of doing mathematics (Boswell 1985); females feel they have to work harder to attain success in mathematical endeavors (Fennema 1985), and these "attributions of success" often lead to cognitive, motivational, and/or emotional deficits (Weiner 1980). The "practices, values, expectations, and beliefs of both individuals and society must be examined if the currently existing gender differences in mathematics participation and performance are to be understood and changed" (Fennema and Leder 1990, 21).

Autonomous learning behaviors are important in working with high-cognitive-level activities.

Secada and Berman (1999) suggest interestingly that the present emphasis on teaching for understanding raises some questions. Context is important for the connections that develop in building understanding. It is important that those contexts do connect with student backgrounds and match the diversity of the students who are learners of mathematics. Extension of mathematics into different domains also needs to reflect domains with which diverse students are familiar. In classrooms that promote understanding, students are encouraged to talk about thinking used in solving problems. This opens student thinking to analysis; differences become clearer. Equitable classrooms work with those differences to position them as strengths rather than as unenlightened positions. Sharing of solution strategies also opens students to scrutiny. In classrooms where the substance of the solution is valued, widespread participation is valued, appreciation for diversity in thought is expressed, and the merit of the strategy rests on the mathematics, not the doer of the mathematics, students feel free to offer their findings and to open them to discussion.

Equitable classrooms work with differences to position them as strengths rather than as unenlightened positions.

Students are likewise encouraged to question, to assume responsibility for their own learning, and to share in the "mathematical authority" (Secada and Berman 1999, 37) afforded by such discourse. If this classroom norm differs considerably from home behavior expectations, it can create an uncomfortable dichotomy for students. Classrooms that honor equity work with autonomous components to make them work for the good of understanding and the students. Hailing the emphasis on building understanding for each student as a noble goal, Secada and Berman suggest that care in implementation must be taken to make sure that all students do have an opportunity to construct knowledge and that those who facilitate such knowledge building be cognizant of the possible inequities in application that could arise.

NCTM carefully points out that rich mathematics for all does not mean "sameness" for all. Cognitively Guided Instruction, a philosophy of mathematical knowledge building, is built on listening to student cognitions to inform teacher decision making and curriculum implementation (Fennema, Carpenter, and Peterson 1989). Some of the outstanding premises of the *Professional Standards for Teaching Mathematics* (Commission on Teaching Standards for School Mathematics 1991) are the affirmations of teachers as curriculum decision makers, as selectors of worthwhile mathematical tasks, as facilitators of discourse for building understanding and for listening to students' cognitions, and as creators of environments that foster construction of knowledge. Thus, teachers listen to students and develop the curriculum in conjunction with classroom and student readiness, picking up where students are and moving forward appropriately. What could be more equitable than continual progress based on expressed understanding?

What could be more equitable than continual progress based on expressed understanding?

Building on student understanding also implies differentiation. "Teachers must be ready to engage students in instruction through different learning modalities, by appealing to differing interests, and by using varied rates of instruction along with varied degrees of complexity" (Tomlinson 1999, 2). Acknowledgement of student learning diversities has prompted *Teaching with the Brain in Mind* (Jensen 1998); *Frames of Mind: The Theory of Multiple Intelligences* (Gardner 1983); *A Different Kind of Classroom: Teaching with Dimensions of Learning* (Marzano 1992); *Cognitive Type Theory and Learning Style* (Mamchur 1996); *Concept-Based Curriculum and Instruction: Teaching beyond the Facts* (Erickson 1998); and *Discovering and Exploring Habits of Mind* (Costa and Kallick 2000). If knowledge is to be fed into complex weblike structures and multiple connections, then it would seem logical to address knowledge building through many "representations" and translations between and among them (Post et al. 1993). Selection and use of representations was thought so important that NCTM added it as a fifth process standard PSSM: "By listening carefully to students' ideas and helping them select and organize representations that will show their thinking, teachers can help students develop the inclination and skills to model problems effectively, to clarify their own understanding of a problem, and to use representations to communicate effectively with one another" (Standards 2000 Project 2000, 209).

"Teachers must be ready to engage students in instruction through different learning modalities, by appealing to differing interests, and by using varied rates of instruction along with varied degrees of complexity."

—Tomlinson 1999, 2

Equity is a goal of all mathematics programs. Consequently, disparity is cause for concern. In the Council of Chief State School Officers publication *State Indicators of Science and Mathematics Education 1999*, this analysis is given: "All states have a significant disparity between the percent of white students at or above the Basic level and the percentage for the largest minority group" (Blank and Langesen 1999, 16). In mathematics disparity, the Wisconsin state difference at the eighth grade level between white and black students in percent at or above Mathematics Basic level is 63 percentage points; between white and Hispanic students, 37 points. The disparity between white and minority students increased from 1992 to 1996 by 19 points. Still, it needs to be noted that the trend for all three populations from 1982 to 1994 has been upward; all populations are making proress.

Attention to closing gaps and to disaggregating data to determine strengths and weaknesses does seem to be indicating progress on many fronts. On the new main National Assessment of Educational Progress (NAEP), gender differences on math scores have almost disappeared in elementary and middle school and, though high school girls continued to lag behind boys on the Third International Mathematics and Science Study (TIMSS), there is less difference in Grade 12 scores between males and females than there is on the long-term trend (*Education Week on the Web*, 21 June 2000; www.edweek.org/ew/ewstory.cfm?slug=41tl.h19). Math scores are rising across the country at a national average rate of about one percentile point per year, "a pace outstripping that of the previous two decades and suggesting that public education reforms are taking hold . . . [though] "progress is far from uniform" (Rand News Release, 25 July 2000; www.rand.org/publications/mr/mr92k). Finding Wisconsin to be second in achievement progress by students from similar families, the factors that seemed to contribute to the differences were lower pupil–teacher ratios, higher participation in public prekindergarten programs, and a higher percentage of teachers who are satisfied with the resources they are provided for teaching. Low teacher turnover also seemed to contribute.

Math scores are rising across the country at a national average rate of about one percentile point per year, "a pace outstripping that of the previous two decades and suggesting that public education reforms are taking hold."

—Rand New Release, July 2000, 25

Another analysis of NAEP data found that rural students show "highly comparable levels of achievement relative to their nonrural counterparts" (Lee and McIntire 2000, 3) and, in fact, "outperform nonrural students" on the most recent testing in 1996. Given that many rural students are poor and attend schools whose instructional resources and course offerings are limited, the level of their academic performance relative to their nonrural counterparts is encouraging. Indeed, the rural schools can provide a model of strength worth studying and emulating. The factors that seem to give the competitive edge are the better social/organizational context that includes teacher training, safe and orderly climate, and collective support.

Another contributing factor is that more students are taking "tough" math (Viadero 2000, 8; www.edweek.org; Cole 2000, 5; www.ccsso.org). In math, 63 percent of students took three years of high school mathematics, compared with 49 percent in 1990. The proportion of students taking tough math courses such as advanced placement calculus, third-level algebra, and analytical geometry increased from 25.2 percent in 1982 to 41.4 percent in 1998. Nationally, 5 percent of students took advanced placement math exams. "We're coming back to what is important for students to know," according to Christopher T. Cross, president of the Council for Basic Education (Viadero 2000, 8).

More students are taking 'tough' math.

—Viadero 2000, 8

Noting that "equity is a critical factor in the nation's economic viability," Croom, in discussing mathematics for all students, suggests that "eliminating the social injustices of past schooling practices will require the support of policymakers, administrators, teachers, parents, and others concerned about excellence and equity in mathematics education." He continues: "All children can learn challenging mathematics with appropriate support and an equitable learning environment" (Croom 1997, 7). This concern and belief have been evidenced in current legislation. The Improving America's Schools Act (1994), a reauthorization of the Elementary and Secondary Education Act (ESEA) of 1965, called for challenging content for all students, for identification of students who are not proficient, and for efforts to move those students through adequate yearly progress measures to proficiency. Current state assessment measures, the *Wisconsin Student Assessment System*, has responded to this national call for such identification, and coupled with state legislation, 115.38, that asked for identification of schools in need of improvement and assistance to those schools to facilitate growth, Wisconsin continues to try to assist its school districts in providing an education of excellence for *all* Wisconsin students.

Technology

> In a discussion of technology-enriched learning of mathematics, Wattenberg and Zia state, "It is easy to feel like kids in a candy store as each new year brings new and more exciting technology for doing and learning science and mathematics—personal computers, scientific and graphing calculators, calculators and computers with computer algebra systems linked to devices for collecting scientific data, the World Wide Web, Java applets, and so much more" (Wattenberg and Zia 2000, 67). There is increasing access to primary resources, massive and real-time data sets, and museum-quality collections. "Mathematics with its tools for analysis and visualization is at the very center of the best technology-enriched, inquiry-driven learning, and mathematical modeling is especially important to enable our students to understand and build the sophisticated and compelling simulations that are becoming so common and important. Yet, questions remain: What tools have been shown to improve our students' learning? When and how do we use them?" (68).

"Mathematics with its tools for analysis and visualization is at the very center of the best technology-enriched, inquiry-driven learning."

—Wattenberg and Zia 2000, 68

In the foreword to *Standards for Technological Literacy: Content for the Study of Technology* (Technology for All Americans Project 2000), the contention is made that "we are a nation increasingly dependent on technology. Yet, in spite of this dependence, U.S. society is largely ignorant of the history and fundamental nature of the technology that sustains it" (v). Technology is defined (broadly speaking) as "how people modify the natural world to suit their own purposes." An analysis of the technology content standards indicates that study of the various components could certainly enhance understanding of the role of educational technology in mathematical learning and could enlighten debate regarding the use of technological tools in mathematics.

Technology is defined (broadly speaking) as "how people modify the natural world to suit their own puposes."

—Technology for All Americans Project 2000, v

The technology content standards are:

- Students will develop an understanding of The Nature of Technology. This includes acquiring knowledge of: (1) the characteristics and scope of technology; (2) the core concepts of technology; and (3) the relationships among technologies and the connections between technology and other fields.
- Students will develop an understanding of Technology and Society. This includes learning about: (4) the cultural, social, economic, and political effects of technology; (5) the effects of technology on the environment; (6) the role of society in the development and use of technology; and (7) the influence of technology on history.

- Students will develop an understanding of Design. This includes knowing about: (8) the attributes of design; (9) engineering design; and (10) the role of troubleshooting, research and development, invention and innovation, and experimentation in problem solving.
- Students will develop Abilities for a Technological World. This includes being able to: (11) apply the design process; (12) use and maintain technological products and systems; and (13) assess the impact of products and systems.
- Students will develop an understanding of The Designed World. This includes selecting and using: (14) medical technologies; (15) agricultural and related biotechnologies; (16) energy and power technologies; (17) information and communication technologies; (18) transportation technologies; (19) manufacturing technologies; and (20) construction technologies.

Technology allows students to visualize and experience mathematics in heretofore impossible ways.

Mathematics and technology are closely intertwined. Many of the technological inventions of today rest heavily on mathematical inception, mathematics for development, and mathematics for implementation. But the interaction is reciprocal. Much of the mathematical change that has taken place has occurred in response to technological advances. "Technology is an important resource for teaching and learning mathematics. Calculators, computers, and the World Wide Web are invaluable for students and teachers in the classroom. Technology allows students to visualize and experience mathematics in heretofore impossible ways, engage in real-world (rather than contrived) problem solving, perform rapid and complex computations, and generate their own representations of their own learning. Furthermore, technology allows students to undertake projects that connect with global communities, integrate mathematics with other subjects, and fit students' individual needs and interests" (International Society for Technology in Education, 2000, p. 96).

Technology enhances mathematics learning; technology supports effective mathematics teaching; and technology influences what mathematics is taught.

—Standards 2000 Project 2000, 24–27

Principles and Standards for School Mathematics calls electronic technologies—calculators and computers—essential tools for teaching, learning, and doing mathematics. Research on the appropriate use of technology shows that students can expand mathematical knowledge and conceptual understandings with the use of technology (Boers-van Oosterum 1990; Sheets 1993; Groves 1994; Dunham and Dick 1994). Noting that "technology should not be used as a replacement for basic understanding and intuitions," but to "foster those understandings and intuitions," the *standards* urge "responsible" use of technology to further mathematical understanding. Three points are then made: (1) technology enhances mathematics learning, (2) technology supports effective mathematics teaching, and (3) technology influences what mathematics is taught (Standards 2000 Project 2000, 24–27).

"Students using calculators tended to have better attitudes toward mathematics... better self-concepts... and there was no loss in student ability to perform paper-and-pencil computational skills."

—Hembree and Dessart 1992, 129

Hembree and Dessart, in a meta-analysis of 79 non-graphing calculator studies, found "improvement in students' understanding of arithmetical concepts and in their problem-solving skills . . . students using calculators tended to have better attitudes toward mathematics . . . better self-concepts . . . and there was no loss in student ability to perform paper-and-pencil computational skills when calculators were used as part of mathematics instruction" (Hembree and Dessart 1992, 129). Research on the use of scientific calculators with graphing capabilities "also has shown positive effects on student achievement." The positive effects were many: improvements in graphing ability, conceptual understanding of graphs, the ability to relate graphical representations to other representations, function concepts, spatial visualization, problem solving, flexibility in thinking, perseverance, and focus on conceptual understanding. Students who work with graphing calculators are more likely to solve problems graphically than use other methods such as algebra. "Studies of graphing calculators have found no negative effect on basic skills, factual knowledge, or computational skills" (Hembree and Dessart 1992, 129).

Research on the use of scientific calculators with graphing capabilities "also has shown positive effects on students achievement."

—Hembree and Dessart 1992, 129

In a report of a study done by the Policy Information Center, Research Division, for the Educational Testing Service, some caveats regarding the positive results of technological use are offered. The national study of the relationship between different uses of educational technology and various educational outcomes drew data from the 1996 National Assessment of Educational Progress (NAEP) in mathematics and used samples from 6,227 fourth-graders and 7,146 eighth-graders. The study examined information on frequency of computer use for mathematics in the school, access to computers and frequency of computer use in the home, professional development of mathematics teachers in computer use, and kinds of instructional uses of computers by mathematics teachers and their students. The results found that technology is a factor in academic achievement, but that results depend on how it is used. "When computers are used to perform certain tasks, namely applying higher order concepts, and when teachers are proficient enough in computer use to direct students toward productive uses more generally, computers do seem to be associated with significant gains in mathematics achievement, as well as an improved social environment in the school" (Policy Information Center, Research Division, Educational Testing Service 1998, 32).

Technology is a factor in academic achievement, but the results depend on how it is used.

The technology standards contend that technology is not always understood by an accessing public. Waits and Demana, who have pioneered the use of technology in the classroom, share some of their reflections regarding the use of calculators in mathematics. They learned: (1) change can occur if we put the potential for change in the hands of everyone (the handheld calculator); (2) it takes practiced teachers to change the practice of teachers (teachers need to learn how to use technology); (3) calculators cause changes in the mathematics we teach; and (4) calculators cause changes in the way we teach and in the way students learn (Waits and Demana 2000). Elaborating, Henry Pollak, regarding the effect of technology, is quoted:

- Some mathematics becomes less important (paper-and-pencil calculation and symbol-manipulation techniques).
- Some mathematics becomes more important (discrete mathematics, data analysis, parametric representations, and nonlinear mathematics).
- Some mathematics becomes possible (fractal geometry, predictive statistics) (Pollak 1986).

Additionally, Waits and Demana found:

- Calculators reduce the drudgery of applying arithmetic and algebraic procedures when those procedures are not the focus of the lesson.
- Calculators with computer interactive geometry allow for investigations that lead to a much better understanding of geometry (Vonder Embse and Engebretsen 1996).
- Calculators help students see that mathematics has value, interest, and excitement (Bruneningsen and Krawiec 1998).
- Calculators make possible a "linked multiple-representation" approach to instruction.
- "Before calculators we studied calculus to learn how to obtain accurate graphs. Today we use accurate graphs to help us study the concepts of calculus" (Waits and Demana 2000, 57).

The best use of technology balances the use of paper-and-pencil techniques and technology.

—Waits and Demana 2000, 59

Waits and Demana believe that the best use of technology balances the use of paper-and-pencil techniques and technology. There are "appropriate uses" for both; the two can complement one another. A good balance can be effected by having students:

1. Solve problems using paper and pencil and then support the results using technology.
2. Solve problems using technology and then confirm the results using paper-and-pencil techniques.
3. Solve problems for which they choose whether it is most appropriate to use paper-and-pencil techniques, technology, or a combination of both.

"These approaches help students understand the proper use of technology" (Waits and Demana 2000, 59). (Appropriate use of manipulatives was also urged by Waits and Demana.)

The National Council of Teachers of Mathematics has offered two position statements titled "Calculators and the Education of Youth" and "The Use of Technology in the Learning and Teaching of Mathematics" (see Appendix G).

"What has not yet been fully understood is that computer-based technologies can be powerful pedagogical tools... extensions of human capabilities and contexts for social interactions supporting learning.

—Bransford, Brown, and Cocking 1999, 218

Bransford, Brown, and Cocking believe that technology has a broad role to play in education. "What has not yet been fully understood is that computer-based technologies can be powerful pedagogical tools . . . extensions of human capabilities and contexts for social interactions supporting learning" (Bransford, Brown, and Cocking 1999, 218). As such, technology resources for education function in a social environment, "mediated by learning conversations with peers and teachers." Kozma and Schank (1998) concur, "emphasis in U.S. schools [has been] on individual learning and performance—what students can do by themselves" (Kozma and Schank 1998, 3). The twenty-first century, they continue, will make much different demands. "Symbolic analysts (Reich 1991)—problem identifiers, problem solvers, strategic brokers" will be needed. They will need a much different set of skills. "They will need to be able to use a variety of tools to search and sort vast amounts of information, generate new data, analyze them, interpret their meaning, and transform them into something new." Communities of learning will be needed to support development of such skills. "In our vision of communities of understanding, digital technologies are used to interweave schools, homes, workplaces, libraries, museums, and social services to integrate education into the fabric of the community" (Kozma and Schank 1998, 5).

"In our vision of communities of understanding, digital technologies are used to interweave schools, homes, workplaces, libraries, museums, and social services to integrate education into the fabric of the community."

—Kozma and Schank 1998, 5

Service Learning

> Service Learning is a method by which young people learn and develop through active participation in thoughtfully organized service experiences. These services meet actual community needs; are coordinated in collaboration with the school and community; are integratd into each young person's academic curriculum; provide structured time for young people to think, talk, and write about what they did and said during the service project; provide young people with opportunities to use newly acquired academic skills and knowledge in real life situations in their own communities; enhance what is taught in the school by extending student learning beyond the classroom; and help foster the development of a sense of caring for others (Alliance for Service-Learning in Education Reform (ASLER) 1998). Mathematics is best learned in meaningful contexts. Certainly the situations proposed in service learning have potential for meaningful mathematics explorations.

The situations proposed in service learning have potential for meaningful mathematics explorations.

The Wisconsin Department of Public Instruction Strategic Plan (2000) identifies among its goals:

GOAL 3. HELP ALL STUDENTS BECOME CARING, CONTRIBUTING, RESPONSIBLE CITIZENS.

3.5 Classroom instruction is enhanced by offering students a variety of opportunities for meaningful participation such as student clubs, service learning, music, athletics and drama.

3.6 Schools will infuse citizenship knowledge and skills across the curriculum.

The U.S. Department of Labor (Wisconsin Department of Public Instruction 1998a, 4) lists the following characteristics that employers look for in teens: listening-to-learn skills; listening and communication; **adaptability**: creative thinking and **problem solving**, especially in response to barriers/obstacles; personal management: self-esteem, goal-setting/self-motivation, personal career development/goals, pride in work accomplished; group effectiveness: interpersonal skills, negotiation, teamwork; organizational effectiveness and leadership; making a contribution; and competence in reading, writing, and **computation**.

The Wisconsin Department of Public Instruction citizenship guide, *Citizenship Building a World of Good, A Tool Kit for Schools and Communities* lists seven characteristics of successful schools in developing caring, contributing, productive, and responsible citizens:

- Core values
- Safe and orderly places
- **Family and community involvement**
- Addressing of societal values
- Positive relationships
- **Engagement of students' minds**
- **High expectations**

In an attempt to capitalize on one of the most valuable resources of schools—its own students—service learning has become an important thrust. Kielsmeier defines service learning as "a way of teaching and learning that engages students in active service tied to curriculum" (Kielsmeier 2000, 652). Service learning, he continues, "can transform the idealism of youth into a powerful force for educational change and democratic renewal." Fitting well with Wisconsin's identified goal of caring, contributing, and responsible citizens, service learning is a consideration for all disciplines. How, then, can mathematics capitalize on service learning? How can service learning use mathematics as a resource?

Service learning is "a way of teaching and learning that engages students in active service tied to curriculum."
—Kielsmeier 2000, 652

The National Service-Learning Cooperative (1998) lists Essential Elements of Service-Learning. It is interesting to note that these elements carefully parallel many of the components necessary for the construction of mathematical knowledge.

The Essential Elements of Service-Learning (National Service-Learning Cooperative 1998) parallel many of the components necessary for the construction of mathematical knowledge.

1. In effective service learning, there are clear educational goals that require the application of concepts, content, and skills from the academic disciplines and involve students in the construction of their own knowledge.

   Principles and Standards for School Mathematics (Standards 2000 Project 2000, 20) states, "research on the learning of complex subjects such as mathematics has solidly established the important role of conceptual understanding in the knowledge and activity of persons who are proficient ... along with factual knowledge and procedural facility (Bransford, Brown, and Cocking 1999).

   Effective mathematics learning is facilitated by identification of concepts that need to be developed at specific stages and by alignment of curricula with the envisioned goals. Conceptual development is a focus of effective mathematics programs (Erickson 1998). Students need to be able to identify the mathematics content that is being addressed in explorations in which they engage. Skills need to be developed in conjunction with meaningful problem solving. Construction of mathematical understanding is the thrust of the mathematics program.

In effective service-learning, students are involved in construction of their own knowledge. Construction of mathematical understanding is the thrust of the mathematics program.

2. In effective service learning, students are engaged in tasks that challenge and stretch them cognitively and developmentally.

Mathematics is best learned when the context in which a problem is situated is meaningful and pertinent, when the task at hand is engaging and motivating, when the cognitive level is demanding, when the task calls for the use of previous knowledge in building new understandings, and when the problem moves the student to new, deeper, or broader understandings.

In effective service learning, students are engaged in tasks that challenge and stretch them cognitively and developmentally. Mathematics is best learned when the cognitive level is demanding.

At a National Service-Learning Leader School, Malcolm Shabbaz, an alternative high school located in Madison, Wisconsin, more than half of the 2,600 students participate in activities that combine academic learning and service (Riley and Wofford 2000, 670). Working for local service projects such as raising funds for a respite center for children caught in situations of family violence, students use their talents and skills to benefit others. Students also travel to communities in Appalachia, the Mississippi Delta, and Native American reservations to gain in-depth knowledge of the history and culture of the peoples living there, along with opportunities to repair the homes of senior citizens and work with Head Start classes in those locales.

Noting that "many children from all backgrounds do not understand mathematics enough to use it or cannot even do many tasks accurately," *Toward a Mathematics Equity Pedagogy* (TMEP), an approach based on research on children's thinking and the best from traditional pedagogy and powerful elements of reformed teaching, was developed, has been used, and is reaping successes (Fuson, LaCruz, Smith, LoCicero, Hudson, Ron, and Steeby 2000). There are six aspects of TMEP:

1. Start where children are and keep learning meaningful. Use many meaning-focused classroom activities.
2. Set high-level mathematical goals and expectations for all. Bring children up to the higher mathematics.
3. Develop a collaborative math talk culture of understanding, explaining, and helping.
4. Build on the best of traditional instruction.
5. Facilitate the learning of general school competencies. Increase self-regulatory actions so that children become more organized, understand what they do not understand, are involved in setting some learning goals, and are helped to reflect on their own progress.
6. Mobilize learning help in the home.

3. In effective service learning, assessment is used as a way to enhance student learning as well as to document and evaluate how well students have met content and skill standards.

In effective service learning, assessment is used as a way to enhance student learning. In good mathematics programs, assessment is used to determine where students are in understanding of the topic at hand in order to make curricular decisions regarding the next mathematical tasks to be introduced.

The diverse purposes for which mathematics assessments are used are characterized in the *Assessment Standards for School Mathematics* (Commission on Teaching Standards for School Mathematics 1995, 25) as monitoring students' progress to promote growth, making instructional decisions to improve instruction, evaluating students' achievement to recognize accomplishments, and evaluating programs to modify them.

In good mathematics programs, assessment is used to determine where students are in understanding of the topic at hand in order to make curricular

In effective service-learning, students are engaged in service tasks that have clear goals and significant consequences for themselves and others. Worthwhile Mathematical Tasks, tenets of worthy mathematics programs, represent mathematics as an ongoing human activity.

decisions regarding the next mathematical tasks to be introduced. Good mathematics assessments identify what students know (as opposed to what they don't know!) individually and as a class. "Assessment occurs at the intersection of the important mathematics that is taught with how it is taught, what is learned and how it is learned" (Assessment Standards Presentation Package 1995).

4. In effective service learning, students are engaged in service tasks that have clear goals, meet genuine needs in the school or community, and have significant consequences for themselves and others.

The *Professional Standards for Teaching Mathematics* (1991, 25) offer as Standard 1, Worthwhile Mathematical Tasks. Those tasks should be based on:

- sound and significant mathematics;
- knowledge of students' understandings, interests, and experiences;
- knowledge of the range of ways that diverse students learn mathematics;

and should

- engage students' intellect;
- develop students' mathematical understandings and skills;
- stimulate students to make connections and develop a coherent framework for mathematical ideas;
- call for problem formulation, problem solving, and mathematical reasoning;
- promote communication about mathematics;
- represent mathematics as an ongoing human activity;
- display sensitivity to, and draw on, students' diverse background experiences and dispositions;
- promote the development of all students' dispositions to do mathematics.

Service-learning tasks, because of the rich contexts in which they are situated, challenge students to employ unique thinking, first of all, in recognizing important problems, and then in formulating solution plans and in analyzing how solution steps should be taken. The mathematics that is demanded in service-learning tasks often involves higher-level thinking skills such as statistical analyses and projections. When solution strategies have been employed, reflection on results is important. Often, service-learning tasks require another component—the need to convince others of a possible best solution. Such tasks are indeed worthwhile mathematical tasks.

In effective service learning, formative and summative evaluation is employed. Rich mathematical tasks ask students to formulate a plan and to check solutions for reasonableness of results.

5. In effective service learning, formative and summative evaluation are employed in a systematic evaluation of the service effort and its outcome.

The *Assessment Standards for School Mathematics* (1995, 4) suggests that the assessment process can be thought of as four interrelated phases that highlight principal points at which critical decisions need to be made: plan assessment, gather evidence, interpret evidence, and use results. The *standards* suggest that these phases are not necessarily sequential and that each phase has decisions and actions that occur within the phase.

Rich mathematical tasks ask students to formulate a plan for solving the problem. Execution then often involves reconsidering the approach to the problem, a sifting and winnowing so as to speak of solution strategies. When the problem is solved, checking of the solution for reasonableness against expected results is necessary. The impact of the solution is evaluated. If there is a discrepancy in the result and the expected outcome, reexamination of approach and of the solution is in order. Likewise, the result often requires explanation and defense of result.

6. In effective service learning, student voice is maximized in selecting, designing, implementing, and evaluating the service project.

Standard 3 of the *Professional Teaching Standards* (1991, 45) is the Student's Role in Discourse. It proposes that classroom discourse should be promoted so that students:

- listen to, respond to, and question the teacher and one another;
- use a variety of tools to reason, make connections, solve problems, and communicate;
- initiate problems and questions;
- make conjectures and present solutions;
- explore examples and counterexamples to investigate a conjecture;
- try to convince themselves and one another of the validity of particular representations, solutions, conjectures, and answers; and
- rely on mathematical evidence and argument to determine validity.

The *Wisconsin Model Academic Standards,* (1998, 5) in "Standard A. Mathematical Processes" ask students to "develop effective oral and written presentations employing correct mathematical terminology, notation, symbols, and conventions for mathematical arguments and display of data... [and] to organize work and present mathematical procedures and results clearly, systematically, succinctly, and correctly."

7. In effective service-learning, diversity is valued as demonstrated by its participants, its practice, and its outcomes.

In a discussion of developing mathematical reasoning, Malloy states, "effective mathematics instruction of students with diverse cultural backgrounds challenges the notion that one culture or one method of reasoning dominates the learning process. In order for all children to become mathematically literate, mathematics educators must be poised to use varied instructional strategies that address the needs of the diverse populations of children who will live and become productive in the twenty-first century" (Stiff and Curcio 1999, 13).

A meta-analysis of research on the learning of mathematics (Grouws and Cebulla 1999, 124) indicates that the use of small groups of students to work on tasks that deal with important mathematical concepts and ideas can increase student mathematics achievement. There is a richness and diversity of input of solution strategies, there is clarification of the problem itself, individual members can work on various parts of the solution, and evaluation of solutions is quite automatic; students check conclusions with one another. Conclusions are more valid because of the diversity of input. By the same token, focus on the task minimizes differentness of approach.

In effective service learning, student voice is maximized. In the mathematics classroom, student discourse is used to socially construct mathematical understanding.

In effective service learning, diversity is valued. Mathematics educators need to be poised to use varied instructional strategies that address the needs of the diverse populations of children in today's classrooms.

In effective service-learning, communciation and interaction with the community are promoted. The learning of mathematics is a social process; seeking and giving community input can serve as a performance measure of effectiveness of classroom efforts.

8. In effective service-learning, communication and interaction with the community are promoted and partnerships and collaboration are encouraged.

Children are constructors of their own mathematical knowledge (von Glasersfeld 1995); learning is a social process (Vygotsky 1978), and writing and talking are tools for reflecting thought as well as generating new thoughts (Murray 1968; Vygotsky 1978). Mathematics classrooms that engage students in formulating and testing hypotheses, discussing various approaches to problem-solving, keeping explorations open-ended so students feel free to contribute a diversity of responses, sharing personal mathematical connections so they develop mathematical ownership, and questioning solutions so students can learn to consider and defend their work (Whitin and Whitin 2000, 213–214), help students gain confidence in their mathematical abilities.

Certainly, taking those learnings one step further by seeking and giving community input can only strengthen such efforts. It can also serve as a performance measure of effectiveness of such classroom efforts. It is important that many classroom experiences precede actual community interaction so that confidence levels run high and tactics are many and varied.

In effective service-learning, students are prepared for all aspects of service work. Classrooms in which problem-solving discourse is usual build mathematics understanding and norms for social interaction.

9. In effective service-learning, students are prepared for all aspects of their service work. They possess a clear understanding of tasks and roles, as well as the skills and information required by the tasks; awareness of safety precautions; and knowledge about and sensitivity to the people with whom they will be working.

In a discussion of perspectives on mathematics education, Willoughby (2000, 9–10) analyzed that all people should have the following:

- A solid understanding of the significance and use of numbers
- Proficiency in the basic operations
- Ability to use thinking to solve problems (as opposed to tedious algorithms)
- Ability to decide which operations, arguments, or other mathematical thinking are appropriate to a real solution
- Ability to decide when approximations or estimates are appropriate
- Ability to use technology intelligently, including when to use technology
- Ability to collect, organize, and interpret data intelligently, to extrapolate or interpolate from the data, and to recognize unsound statistical procedures
- An understanding of probability concepts
- An understanding of the role of functions in modeling the real world
- A knowledge of the various geometric relationships
- A knowledge of and ability to use simple useful trigonometric information
- An intuitive understanding of the foundations of calculus

The ability to recognize situations in which mathematical thinking is likely to be helpful, to formulate problems in mathematical terms, and to interpret the results of mathematical analysis so that others can understand them, is useful in service-learning situations.

Classrooms in which problem-solving discourse is usual, where mathematical conjectures are posed and checked as a matter of course, and where solutions are explained and defended help students become in-depth thinkers. Not only do the classroom endeavors build mathematical understanding, but they also build norms for social interaction, problem investigation, and solution interpretation. With practice, these techniques would lend themselves well to broader venues.

10. In effective service learning, student reflection takes place before, during, and after service; uses multiple methods that encourage critical thinking; and is a central force in the design and fulfillment of curricular objectives.

In effective service learning, student reflection takes place. Well-taught mathematics carries students forward in knowing how to evaluate results.

When students become aware of themselves as learners, they can develop the ability to monitor their learning strategies and resources and can assess their readiness for particular performances (Bransford, Brown, and Cocking 1999, 55).

1) Metacognition can help students "analyze what they did and why . . . highlight the generalizable feature of the critical decisions and actions and focus on strategic levels rather than on specific solutions.
2) Building on previous experiences and identifying connections between the old learning and the new helps students increase transfer.
3) Students need to move pieces of knowledge to the conceptual level to increase understanding. "Conceptual understanding requires a higher-level, integrative thinking ability that needs to be taught systematically through all levels of schooling. Drawing patterns and connections between related facts, ideas, and examples leads to synthesizing information at a conceptual level" (Erickson 1998, 8). Critical content needs to correlate to disciplinary concepts and conceptual ideas.

If we can produce students who think critically, they will be able to handle not only academic tasks but life challenges. Well-taught mathematics carries students forward in knowing how to identify critical components of tasks, formulate solutions and execute them step-by-step, and evaluate results. These pieces lend themselves well to the broader curriculum and ultimately to the community.

11. In effective service learning, multiple methods are designed to acknowledge, celebrate, and further validate students' service work.

In effective service learning, multiple methods are designed to validate students' service work. "Humans are motivated to develop competence and to solve problems."

—White, 1959, in Bransford, Brown, and Cocking 1999, 48

"Learners of all ages are more motivated when they can see the usefulness of what they are learning and when they can use that information to do something that has an impact on others—especially their local community" (McCombs, 1996; Pintrich and Schunk, 1996). Social opportunities affect motivation.

"Although extrinsic rewards and punishments clearly affect behavior, people work hard for intrinsic reasons as well" (Bransford, Brown, and Cocking, 1999, p. 48). "Humans are motivated to develop competence and to solve problems" (White, 1959, in Bransford, Brown, and Cocking, 1999, p. 48).

And, does service learning work? A decade of research on service learning is yielding interesting results. Though much of the research used self-reports

Service learning provides an avenue for students to become active, positive contributors to society.

or information from surveys administered before and after a service experience, some used qualitative methods and case studies. The major findings of the research are (Billig 2000, 658–664):

1) Service learning has a positive effect on the personal development of public school youths.
2) Students who participate in service learning are less likely to engage in "risk" behaviors.
3) Service learning has a positive effect on students' interpersonal development and the ability to relate to culturally diverse groups.
4) Service learning helps develop students' sense of civic and social responsibility and their citizenship skills.
5) Service learning provides an avenue for students to become active, positive contributors to society.
6) Service learning helps students acquire academic skills and knowledge.
7) Students who participate in service learning are more engaged in their studies and more motivated to learn.
8) Service learning is associated with increased student attendance.
9) Service learning helps students to become more knowledgeable and realistic about careers.
10) Service learning results in greater mutual respect between teachers and students.
11) Service learning improves the overall school climate.
12) Engaging in service learning leads to discussions of teaching and learning and of the best ways for students to learn.
13) Service learning leads to more positive perceptions of schools and youths on the part of community members.

References

Openness

Burke, M. J., and F. R. Curcio. 2000. *Learning Mathematics for a New Century: 2000 Yearbook.* Reston, VA: National Council of Teachers of Mathematics.

Carpenter, T. A., and R. Lehrer. 1999. "Teaching and Learning Mathematics with Understanding." In *Mathematics Classrooms That Promote Understanding,* edited by E. Fennema and T. A. Romberg. Mahwah, NJ: Lawrence Erlbaum Associates.

Fennema, E., and T. A. Romberg. 1999. *Mathematics Classrooms That Promote Understanding.* Mahwah, NJ: Lawrence Erlbaum Associates.

Fennema, E., J. Sowder, and T. A. Carpenter. 1999. "Creating Classrooms That Promote Understanding." In *Mathematics Classrooms That Promote Understanding,* edited by E. Fennema and T.A. Romberg. Mahwah, NJ: Lawrence Erlbaum Associates.

Freudenthal, H. 1983. *Didactical Phenomenology of Mathematical Structures.* Dordrecht, The Netherlands: Reidel.

Grunow, J. 1999. *Mathematics for the New Millennium Professional Development Sessions.* June 28–July 1. Stevens Point, Wisconsin.

Ma, L. 1999. *Knowing and Teaching Elementary Mathematics.* Mahwah, NJ: Lawrence Erlbaum Associates.

Paulos, J. A. 1988. *Innumeracy: Mathematical Illiteracy and Its Consequences.* New York: Vintage Books, A Division of Random House, Inc.

Romberg, T. A. 1999. Opening Keynote Speech for Mathematics for the New Millennium Institute, June 28–July 1. Stevens Point, Wisconsin.

Romberg, T. A., and J. J. Kaput. 1999. "Mathematics Worth Teaching, Mathematics Worth Understanding." In *Mathematics Classrooms That Promote Understanding,* edited by E. Fennema and T. A. Romberg. Mahwah, NJ: Lawrence Erlbaum Associates.

Schoenfeld, A. H. 1995. *Mathematical Problem Solving.* New York: Academic Press, Inc.

Sparks, B. E. 1999. Closing Keynote Address for Mathematics for the New Millennium Institute, June 28–July 1. Stevens Point, Wisconsin.

Steen, L., ed. 1990. *On the Shoulders of Giants: New Approaches to Numeracy.* Washington, DC: National Academy Press.

Wilson, L. D., and R. K. Blank. 1999. *Improving Mathematics Education Using Results from NAEP and TIMSS.* Washington, DC: Council of Chief State School Officers, State Education Assessment Center.

Wu, H. 1999. "Basic Skills versus Conceptual Understanding: A Bogus Dichotomy in Mathematics Education." *American Educator, 23* (3) (Fall): 14–20.

Time

Anderson, J. R., L. M. Reder, and H. A. Simon. 1996. "Situated Learning and Education." *Educational Researcher 25* (4) (May): 5–96.

Askey, R. 1999. "Knowing and Teaching Elementary Mathematics." *American Educator, 23* (3): 6–3, 49.

Bransford, J. D., A. L. Brown, and R. R. Cocking, eds. 1999. *How People Learn: Brain, Mind, Experience, and School.* Washington, DC: National Academy Press.

Bransford, J. D., J. J. Franks, N. J. Vye, and R. D. Sherwood. 1989. "New Approaches to Instruction: Because Wisdom Can't be Told." In *Similarity and Analogical Reasoning,* edited by S. Vosniadou and A. Ortony. Cambridge, U.K.: Cambridge University Press.

Chi, M. T. H., M. Bassok, M. W. Lewis, P. Reimann, and R. Glaser. 1989. "Self-explanations: How Students Study and Use Examples in Learning to Solve Problems." *Cognitive Psychology 1:* 145–182.

Chi, M. T. H., N. deLeeuw, M. Chiu, and C. LaVancher. 1994. "Eliciting Self-Explanations Improves Understanding." *Cognitive Science 18:* 439–477.

Continuing to Learn from TIMSS Committee. 1999. *Global Perspectives for Local Action: Using TIMSS to Improve U.S. Mathematics and Science Education.* Washington, DC: National Academy Press.

Gibson, J. J., and E. J. Gibson. 1955. "Perceptual Learning: Differentiation or Enrichment." *Psychological Review 62:* 32–51.

Klausmeier, H. J. 1985. *Educational Psychology* 5th ed. New York: Harper and Row.

National Center for Education Statistics. 1998. *Pursuing Excellence: A Study of U.S. Twelfth-Grade Mathematics and Science Achievement in International Context.* Washington, DC: U.S. Department of Education.

National Education Commission on Time and Learning. 1994. *Prisoners of Time.* Washington, DC: U.S. Government Printing Office.

National Institute on Student Achievement, Curriculum, and Assessment. 1999. *A First Look at What We Can Learn from High Performing School Districts: An Analysis of TIMSS Data from the First in the World Consortium.* Jessup, MD: U.S. Department of Education ED Pubs.

Pezdek, K., and L. Miceli. 1982. "Life Span Differences in Memory Integration as a Function of Processing Time." *Developmental Psychology 18* (3): 485–490.

Singley, K, and J. R. Anderson. 1989. *The Transfer of Cognitive Skill.* Cambridge, MA: Harvard University Press.

Standards 2000 Project. 2000. *Principle and Standards for School Mathematics.* Reston, VA: National Council of Teachers of Mathematics.

Stevenson, H. W., and R. Nerison-Low. 1997. *To Sum It Up: Case Studies of Education in Germany, Japan and the United States.* Washington, DC: U.S. Government Printing Office.

Walberg, J. J., R. A. Paschal, and T. Weinstein. 1985. "Homework's powerful effects on learning." *Educational Leadership,* April: 76–79.

Equity

Assessment Standards Working Groups. 1995. *Assessment Standards for School Mathematics.* Reston, VA: National Council of Teachers of Mathematics.

Blank, R. K., and D. Langesen. 1999. *State Indicators of Science and Mathematics Education 1999: State-by-State Trends and New Indicators from the 1997–98 School Year.* Washington, DC: Council of Chief State School Officers.

Boswell, S. L. 1985. "The Influence of Sex-Role Stereotyping on Women's Attitudes and Achievement in Mathematics." In *Women and Mathematics: Balancing the Equation,* edited by S. F. Chipman, L. R. Brush, and D. M. Wilson. Hillsdale, NJ: Lawrence Erlbaum.

Clarkson, P., and G. C. Leder. 1984. "Causal Attributions for Success and Failure in Mathematics: A Cross-Cultural Perspective." *Educational Studies in Mathematics, 15:* 413–422.

Cole, A. 2000. "More Kids Take Tough Courses." *Impact: Eisenhower Professional Development Evaluation Publication*. May–July: 5.

Commission on Standards for School Mathematics. 1989. *Curriculum and Evaluation Standards for School Mathematics.* Reston, VA: National Council of Teachers of Mathematics.

Commission on Teaching Standards for School Mathematics. 1991. *Professional Standards for Teaching Mathematics.* Reston, VA: National Council of Teachers of Mathematics.

Costa, A. L., and B. Kallick, eds. 2000. *Discovering and Exploring Habits of Mind.* Alexandria, VA: Association for Supervision and Curriculum Development.

Croom, L. 1997. "Mathematics for All Students: Access, Excellence, and Equity." In *Multicultural and Gender Equity in the Mathematics Classroom: The Gift of Diversity (1997 Yearbook),* edited by J. Trentacosta and M. J. Kenney. Reston, VA: National Council of Teachers of Mathematics.

Darling-Hammond, L. 1997. *The Right to Learn: A Blueprint for Creating Schools That Work.* San Francisco: Jossey-Bass Publishers.

Dreeben, R. 1987. "Crossing the Divide: What Teachers and Administrators Can Do to Help Black Students Reach Their Reading Potential." *American Educator* (Winter): 28–35.

Erickson, H. L. 1998. *Concept-Based Curriculum and Instruction: Teaching Beyond the Facts.* Thousand Oaks, CA: Corwin Press, Inc.

Fennema, E. 1985. "Attribution Theory and Achievement in Mathematics. In *The Development of Reflections,* edited by S. R. Yussen. New York: Academic Press.

Fennema, E., T. P. Carpenter, and P. Peterson. 1989. "Teachers' Decision Making and Cognitively Guided Instruction: A New Paradigm for Curriculum Development." In *School Mathematics: The Challenge to Change,* edited by N. F. Ellerton and M. A. (Ken) Clements eds. Geelong, Victoria, Australia: Deaking University Press.

Fennema, E., and G. C. Leder. 1990. *Mathematics and Gender.* New York: Teachers College Press.

Ferguson, R. F. 1991. "Paying for Public Education: New Evidence on How and Why Money Matters." *Harvard Journal on Legislation, 28* (2) (Summer): 465–498.

Gardner, H. 1983. *Frames of Mind: The Theory of Multiple Intelligences.* New York: Basic Books.

Jensen, E. 1998. *Teaching with the Brain in Mind.* Alexandria, VA: Association for Supervision and Curriculum Development.

Lee, J., and W. G. McIntire. 2000. *Understanding Rural Student Achievement: Identifying Instructional and Organizational Differences Between Rural and Nonrural Students.* Orono, ME: University of Maine.

Mamchur, C. 1996. *A Teacher's Guide to Cognitive Type Theory and Learning Style.* Alexandria, VA: Association for Supervision and Curriculum Development.

Marzano, R. J. 1992. *A Different Kind of Classroom: Teaching with Dimensions of Learning.* Alexandria, VA: Association for Supervision and Curriculum Development.

Metz, M. H. 1978. *Classrooms and Corridors: The Crisis of Authority in Desegregated Secondary Schools.* Berkeley: University of California Press.

National Council of Teachers of Mathematics. 2000. *Principles and Standards for School Mathematics.* Reston, VA: National Council of Teachers of Mathematics.

Oakes, J. 1985. *Keeping Track: How Schools Structure Inequality.* New Haven: Yale University Press.

Phillips, E. 1987. Middle Grades Mathematics Project (Summer Training Institute).

Post, T. R., K. A. Cramer, M. Behr, R. Lesh, and G. Harel. 1993. "Curriculum Implications of Research on the Learning, Teaching, and Assessing of Rational Number Concepts." In *Rational Numbers: An Integration of Research,* edited by T. P. Carpenter, E. Fennema, and T. A. Romberg. Mahwah, NJ: Erlbaum.

Secada, W. G., and P. W. Berman. 1999. "Equity As a Value-Added Dimension in Teaching for Understanding in School Mathematics." In *Mathematics Classrooms That Promote Understanding,* edited by E. Fennema and T. A. Romberg. Mahwah, NJ: Lawrence Erlbaum Associates.

Steen, L. A. 1997. *Why Numbers Count: Quantitative Literacy for Tomorrow's America.* New York: College Entrance Examination Board.

Tomlinson, C. A. 1999. *The Differentiated Classroom: Responding to the Needs of All Learners.* Alexandria, VA: Association for Supervision and Curriculum Development.

Viadero, D. 2000. "Study: More Students Taking Tough Math." *Education Week,* June 14: 8.

Wheelock, A. 1992. *Crossing the Tracks.* New York: New Press.

Weiner, B. 1980. "The Order of Affect in Rational (Attributions) Approaches to Human Motivation." *Educational Research, 19:* 4–11.

Technology

Boers-van Oosterum, M. A. M. 1990. Understanding of Variables and Their Uses Acquired by Students in Traditional and Computer-Intensive Algebra. Ph.D. diss., University of Maryland College Park.

Bransford, J. D., A. L. Brown, and R. R. Cocking, eds. 1999. *How People Learn: Brain, Mind, Experience, and School.* Washington, DC: National Academy Press.

Bruneningsen, C., and W. Krawiec. 1998. *Exploring Physics and Mathematics with the CBL System.* Dallas, TX: Texas Instruments.

Dunham, P. H., and T. P. Dick. 1994. "Research on Graphing Calculators." *Mathematics Teacher* 87 (September): 440–445.

Grouws, D. A., and K. J. Cebulla. 1999. "Mathematics." In *Handbook of Research on Improving Student Achievement,* edited by G. Cawelti. Arlington, VA: Educational Research Service.

Groves, S. 1994. "Calculators: A Learning Environment to Promote Numbers Sense." Paper presented at the annual meeting of the American Educational Research Association, April, New Orleans.

Hembree, R., and D. Dessart. 1992. "Research on Calculators in Mathematics Education." In *Calculators in Mathematics Education: 1992 Yearbook,* edited by J. T. Fey. Reston, VA: National Council of Teachers of Mathematics.

International Society for Technology in Education. 2000. *National Educational Technology Standards for Students: Connecting Curriculum and Technology.* Eugene, OR: International Society for Technology in Education (in collaboration with the U.S. Department of Education).

Kozma, R., and P. Schank. 1998. "Connecting with the 21st Century: Technology in Support of Educational Reform." In *ASCD Yearbook 1998: Learning with Technology,* edited by C. Dede. Alexandria, VA: Association for Supervision and Curriculum Development.

National Council of Teachers of Mathematics. 1999. Position statements. Available on the Web site: www.nctm.org. March.

Policy Information Center, Research Division, Educational Testing Service. 1998. *Does it Compute?: The Relationship Between Educational Technology and Student Achievement in Mathematics.* Princeton, NJ: Educational Testing Service.

Pollak, H. O. 1986. "The Effects of Technology on the Mathematics Curriculum." In *Proceedings of the Fifth International Congress on Mathematical Education,* edited by Marjorie Canss. Boston: Birkhauser Press.

Reich, R. 1991. *The Work of Nations: Preparing Ourselves for 21st-Century Capitalism.* New York: Alfred Knopf.

Sheets, C. 1993. Effects of Computer Learning and Problem-Solving Tools on the Development of Secondary School Students' Understanding of Mathematical Functions. Ph.D. diss., University of Maryland College Park.

Standards 2000 Project. 2000. *Principles and Standards for School Mathematics.* Reston, VA: National Council of Teachers of Mathematics.

Technology for All Americans Project. 2000. *Standards for Technological Literacy: Content for the Study of Technology.* Reston, VA: International Technology Education Association.

Vonder Embse, C., and A. Engebretsen. 1996. "Using Interactive Geometry Software for Right-Angle Trigonometry." *Mathematics Teacher* 89 (October): 602–605.

Waits, B. K., and F. Demana. 2000. "Calculators in Mathematics Teaching and Learning: Past, Present, and Future." In *Learning Mathematics for a New Century: 2000 Yearbook,* edited by M. J. Burke and F. R. Curcio. Reston, VA: National Council of Teachers of Mathematics.

Wattenberg, F., and L. L. Zia. 2000. "Technology-Enriched Learning of Mathematics: Opportunities and Challenges." In *Learning Mathematics for a New Century: 2000 Yearbook,* edited by M. J. Burke and F. R. Curcio. Reston, VA: National Council of Teachers of Mathematics.

Service Learning

Alliance for Service Learning in Education Reform. 1998. *Essential Elements of Service Learning for Effective Practice for Organizational Support.* Minneapolis, MN: National Service Learning Cooperative.

Assessment Standards Working Groups. 1995. *Assessment Standards for School Mathematics.* Reston, VA: National Council of Teachers of Mathematics.

Billig, S. H. 2000. "Research on K–12 School-Based Service-Learning: The Evidence Builds." *Phi Delta Kappan, 81* (9) (May): 658–664.

Bransford, J. D., A. L. Brown, and R. R. Cocking., eds. 1999. *How People Learn: Brain, Mind, Experience, and School.* Washington, DC: National Academy Press.

Commission on Teaching Standards. 1991. *Professional Standards for Teaching Mathematics.* Reston, VA: National Council of Teachers of Mathematics.

Erickson, H. L. 1998. *Concept-Based Curriculum and Instruction: Teaching Beyond the Facts.* Thousand Oaks, CA: Corwin Press, Inc.

Fuson, K. C., Y. D. LaCruz, S. T. Smith, A. M. LoCicero, K. Hudson, P. Ron, and R. Steeby. 2000. "Blending the Best of the Twentieth Century to Achieve a Mathematics Equity Pedagogy in the Twenty-First Century." In *Learning Mathematics for a New Century,* edited by M. J. Burke and F. R. Curcio. Reston, VA: National Council of Teachers of Mathematics.

Grouws, D. A., and K. J. Cebulla. 1999. "Mathematics." In *Handbook of Research on Improving Student Achievement,* edited by G. Cawelti. Arlington, VA: Educational Research Service.

Kielsmeier, J. C. 2000. "A Time To Serve, a Time To Learn: Service-Learning and the Promise of Democracy." *Phi Delta Kappan, 81* (9) (May): 652–657.

Malloy, C. E. 1999. "Developing Mathematical Reasoning in the Middle Grades: Recognizing Diversity." In *Developing Mathematical Reasoning in Grades K–12,* edited by L. V. Stiff and F. R. Curcio. Reston, VA: National Council of Teachers of Mathematics.

McCombs, B. L. 1996. "Alternative Perspectives for Motivation." In *Developing Engaged Readers in School and Home Communities,* edited by L. Baker, P. Afflerback, and D. Reinking. Mahwah, NJ: Erlbaum.

Murray, D. 1968. *A Writer Teaches Writing: A Practical Method of Teaching Composition.* Boston: Houghton Mifflin.

National Service-Learning Cooperative. 1998. *Essential Elements of Service-Learning.* St. Paul, MN: National Youth Leadership Council.

Pintrich, P. R., and D. Schunk. 1996. *Motivation in Education: Theory, Research and Application.* Columbus, OH: Merrill Prentice-Hall.

Riley, R. W., and H. Wofford. 2000. "The Reaffirmation of *The Declaration of Principles.*" *Phi Delta Kappan, 81* (9) (May): 670–672.

Standards 2000 Project. 2000. *Principles and Standards for School Mathematics.* Reston, VA: National Council of Teachers of Mathematics.

Student Services/Prevention and Wellness Team. 1998. *Citizenship: Building a World of Good.* Madison: Wisconsin Department of Public Instruction.

Trentacosta, J., and M. J. Kenney, eds. 1997. *Multicultural and Gender Equity in the Mathematics Classroom: The Gift of Diversity: 1997 Yearbook.* Reston, VA: National Council of Teachers of Mathematics.

Von Glasersfeld, E. 1995. *Radical Constructivism: A Way Of Knowing And Learning.* London: Falmer.

Vygotsky, L. S. 1978. *Mind in Society: The Development of Higher Psychological Processes.* Cambridge, MA: Harvard University Press.

Whitin D. J., and P. Whitin. 2000. "Exploring Mathematics Through Talking and Writing." In *Learning Mathematics for a New Century: 2000 Yearbook,* edited by M. J. Burke and F. R. Curcio. Reston, VA: National Council of Teachers of Mathematics.

Willoughby, S. S. 2000. "Perspectives on Mathematics Education." In *Learning Mathematics for a New Century,* edited by M. J. Burke and F. R. Curcio. Reston, VA: National Council of Teachers of Mathematics.

Wisconsin Department of Public Instruction. 1998. *Citizenship Building a World of Good: A Tool Kit for Schools and Communities.* Madison: Wisconsin Department of Public Instruction.

———. 2001. *Wisconsin Department of Public Instruction Strategic Plan.* www.dpi.state.wi.us/dpi/dpi-mission.html.

———. 1998. *Wisconsin's Model Academic Standards for Mathematics.* Madison: Wisconsin Department of Public Instruction.

Planning Curriculum

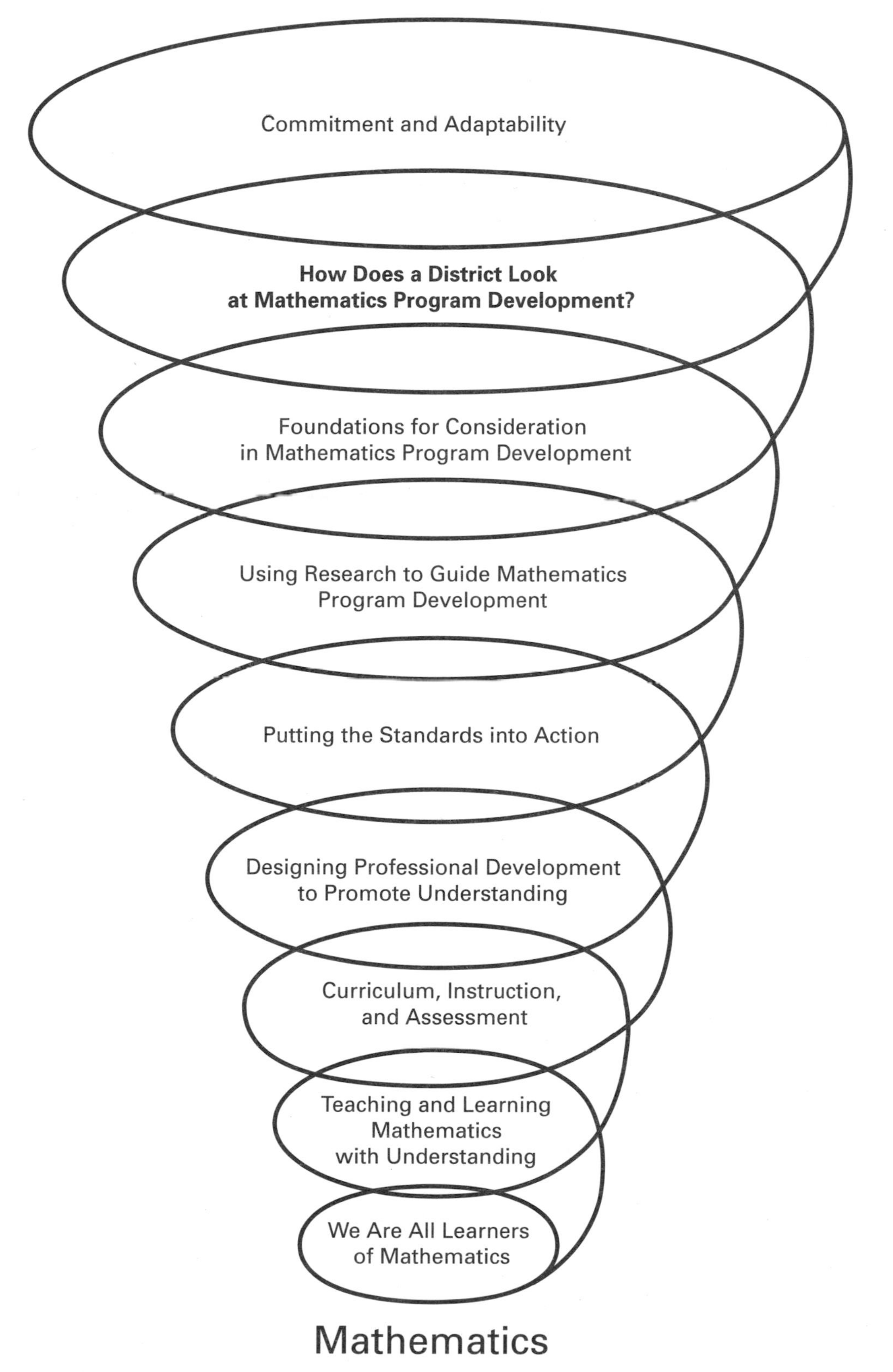

Mathematics

Spring (spring) *n.* 1. a source; origin; or beginning: *a spring into action.* 2. elasticity; resilience: *a spring forward with commitment.* (Morris, W., ed., 1971. *The American Heritage Dictionary of the English Language.* New York: American Heritage Publishing Company, Inc. and Houghton Mifflin Company, 1250, 2nd definition.)

How Does a District Look at Mathematics Program Development?

8

A rapidly changing society precipitates curricular change. The change in mathematics that has occurred as a result of the technological explosion, research on teaching and learning for understanding, international comparisons, the standards movement, and the emphasis on assessment and accountability has impacted mathematics teachers and curriculum planners heavily. Existing programs must be evaluated to see if they are meeting students' present requirements and anticipating future potentialities—and everything is operating under time constraints. The *Wisconsin State Journal*, September 6, 2000 reads, "The internet has become a much greater tool than was ever anticipated. Internet access is now available in about half of the homes in the United States and is used daily by half of the workforce." Teachers who have just retooled to teach AP computer science find that another shift in language is anticipated. Schools have to develop mechanisms to address continual change. A curriculum review every five years is no longer sufficient. Continual evaluation, adaptation, refitting, and retooling need to be part of the school mechanism. This is particularly true for mathematics. Because of technology, there is probably no other discipline that is changing so much.

Continual evaluation, adaptation, refitting, and retooling need to be part of the school mechanism.

We need to assess what is needed in mathematics programs. "Better schooling will result in the future—as it has in the past and does now—chiefly from the steady, reflective efforts of the practitioners who work in schools and from the contributions of the parents and citizens who support . . . public education" (Tyack and Cuban 1995, 135). Contending that a new millennium offers educational opportunities to rethink the way our schools will work in the twenty-first century, Marsh (1999) suggests that we need to redefine the common purposes of schools: to help students participate in a democratic society, experience a productive work life, and engage in lifelong learning. In an information age, schools for the democratic society will be expected "for the first time in history" (Marsh, 2) to educate all of our youth to a high level, to help students understand the social problems of a world "characterized by increasing specialization of knowledge and complexity of effects," (Marsh, 2) and to address the complexities of increasing diversity. Participation in the world of work will be challenging; "if thinking for a living is the new reality in the world of work, will thinking for an education become the norm in our schools?" (Marsh, 3). The skills of self-monitoring and teamwork will

"If thinking for a living is the new reality in the world of work, will thinking for an education become the norm in our schools?"

—Marsh, 3

If lifelong learning is to occur, then skills in research, critical analysis, and selection of task become extremely important.

become paramount with the work of the future being done in small and rapidly shifting organizations, under tight deadlines, using strong teamwork, with new attention to quality and results indicators, and in frequent partnership with external groups. If lifelong learning is to occur, then skills in research, critical analysis, and selection of task become extremely important. In all instances, problem solving—the stuff of which mathematics is made—rises to the fore.

> [Yet], many adults believe they have a clear sense of what mathematics is and why they despise it. These thoughts and feelings have been slam-dunked into them by their years in school, struggling with textbooks and teachers living a kind of mathematical mythology of memorizing math facts (orally and with flashcards); computing page after page of sums, differences, products, and quotients (with and without remainders); going to the chalkboard to work out the answer; trying to remember the rules and procedures such as "write down the 2 and carry the 1" and "invert and multiply"; and developing careful, step-by-step proofs (beyond a shadow of a doubt). Common myths about mathematics include "math is only for a few good men; most mere mortals are not good at it" [and] 1) doing math means getting the one right answer; 2) mathematics is a collection of rules, theorems, and procedures to be memorized; 3) mathematics is really just arithmetic, working with bigger and more complicated numbers. Teaching math involves steadily working through the textbook, page by page, assigning sheets of drill exercises from the workbooks or worksheets for practice.
>
> Perhaps the saddest outcome of this situation is that each year a significant percentage of children come to believe that they are incapable of doing math. (Zemelman, Daniels, and Hyde 1998, 83)

"Perhaps the saddest outcome of this situation is that each year a signficant percentage of children come to believe that they are incapable of doing math."

—Zemelman, Daniels, and Hyde 1998, 83

These comments from a book called *Best Practice: New Standards for Teaching and Learning in America's Schools (2nd Ed.),* in a chapter called, "The Way It Used to Be," unfortunately do not describe only mathematics classrooms that used to be, but some classrooms that still are. When we come to an examination of mathematics programs, we need to be very aware of what is, what needs to be, and the difference between the two. There is a great deal of interest in improving mathematics programs in the schools. This concern has been prompted by a change in mathematics itself, by a society that is highly mathematized, and by an economy that demands mathematical understanding. It has also been fueled by international comparisons of mathematical expertise that are not necessarily flattering, by high attrition rates of students from mathematics courses, and by state assessments that find a high proportion of our students not sufficiently proficient in mathematics.

When we come to an examination of mathematics programs, we need to be very aware of what is, what needs to be, and the difference between the two.

How Does a School District Go About Examining Its Mathematics Program?

First Steps

Select a Broad-Based Committee

Because mathematics is a K–12 venture situated in a broad school culture and because mathematics, possibly more than any other discipline, is undergoing epistemological shifts in expectations, it is necessary to involve representatives of the concerned stakeholders in the study. The study group should include mathematics teacher representatives from PK–2, 3–5, 6–8, and 9–12; a principal from each level, the curriculum coordinator, a member of the school board, a parent, and members of other equally concerned constituencies.

Envision the Place of Mathematics in Society and Research Its Mission in the Schools

In order to examine the mathematics program, it is necessary for all participants to be familiar with where mathematics has been and where it is today. They need to recognize the contributions that mathematics makes to society and think about the needed inputs of mathematics in this era. They should contemplate what will develop in the new millennium and anticipate the kind of mathematics that will contribute positively to continued growth. There is a strong demand on the schools from the public—parents, professional organizations, employers, governments, and so forth—for teaching and learning with understanding. Research has produced an ever-growing body of knowledge about teaching and learning mathematics that can provide direction for addressing the building of understanding.

In order to examine the mathematics program, it is necessary for all participants to be familiar with where mathematics has been and where it is today.

Examination of the school mathematics program must rest on a shared vision of what effective mathematics education is. An analysis of mathematics education belief structures should be part of the initial efforts. Then, knowledge of exemplary mathematics programs can be built by examining research and literature. Romberg and Kaput (1999), discussing mathematics worth teaching and mathematics worth understanding, suggest that a shift in expectations of school mathematics has occurred. They identify areas to be examined: "(a) traditional school mathematics, to clarify what the shift is away from; (b) mathematics as a human activity, to portray the direction the shift is toward; (c) mathematics worth teaching, to provide an overview to the . . . initial question 'What is it that students are expected to understand?'; and (d) speculations about school mathematics in the future" (3–17).

Examination of the school mathematics program must rest on a shared vision of what effective mathematics education is.

Additionally, those participating in the mathematics evaluation need to contemplate the implications of current brain research, recent economic developments, and international comparisons; the effect of the standards, assessments, and accountability components; the impact of diversity and accessibility; the contributions of school financial structure; and the ramifications of state and national interventions. Cook (1995, 7–37) offers an excellent listing of considerations for Strategic Planning for America's Schools that could provide a basis for "big picture" discussions, including:

Demographics

The aging population
The restructuring of the family
The emergence of minorities

Redefining "Public" Education

Economic transitions

Beyond Economic Transitions
The management of information
The polarizing of society
System obsolescence
The transformation of mainstream values as we move from an age of information age of biogenetics

Competition
When participants have dealt with these foundations, then contemplation of the mathematics program can ensue.

Internal Analysis

Identify Strengths in the Existing Program

In considering internal strengths, "only those strengths that directly relate to the stated mission should be considered here."
—Cook 1995, 53

Defining strengths as "those internal qualities, circumstances, or conditions that contribute to the [program's] ability to achieve its mission," Cook continues, "only those strengths that directly relate to the stated mission should be considered here" (Cook 1995, 53).

An excellent starting tool for districts is a joint publication from the National Council of Teachers of Mathematics and the Association for Supervision and Curriculum Development called *A Guide for Reviewing School Mathematics Programs* (Blume and Nicely 1991). The guide reflects critical elements in the areas of goals, curriculum, instruction, evaluation, and teacher and administrator responsibility. The authors of the guide suggest it can be effectively used in several formats, such as with committees of teachers and administrators at grade levels K–4, 5–8, and 9–12 who meet together, respond to the questions together, and then analyze responses collectively. While the questions could well provide good information to individual teachers, the authors encourage group dialogue and consensus building.

To focus appropriately on the "strengths" of the current program, actual data needs to be assembled.

To focus appropriately on the "strengths" of the current program, actual data needs to be assembled (Council of Chief State School Officers May 2000; Wisconsin Department of Public Instruction 1999). That data needs to include: (1) the number of students in mathematics courses as a percentage of the total school population; (2) the number of students who continue in mathematics programs throughout the entire grade school and high school experience; (3) test data, both standardized and classroom-based; (4) longitudinal data regarding individual achievement over time and program measurement over time (the fourth grade, two years running, etc.); and (5) student success in mathematics programs at technical schools, colleges, and universities and on the job. Affective measures need to be gathered also: (1) the motivation and enthusiasm of mathematics students based on student questionnaires; (2) the attendance of students in mathematics classes as weighed against attendance in other classes; (3) teacher attitude measures; and (4) parent and community evaluations as expressed in documents seeking input.

Affective measures need to be gathered also.

Analysis of data and inputs should yield indicators of programs that seem to be especially effective and of practices that seem to yield results. It is then important to study those programs and practices in depth to determine what works and pinpoint effective elements. It is important that those items be defined in relation to the context in which they are generated because the context may contribute to their effectiveness. The next step is to decontextualize the effective components. With those items isolated, it might then be possible to conceive of a study of those pieces to determine generalizability.

Pinpoint Needed Improvements

Defining "weaknesses" as "internal characteristics, conditions, or circumstances that restrict or prevent the realization of the mission," Cook notes, "all [programs] have them; the trick is to distinguish between those weaknesses that are tolerable and those that are critical" (Cook 1995, 54).

In considering weaknesses, "the trick is to distinguish between those weaknesses that are tolerable and those that are critical."

—Cook 1995, 54

In identifying strengths, areas in need of improvement will undoubtedly surface. It is important that those elements be identified and then decontextualized so they can be studied impartially and intensely to determine underlying implications. Just as with strengths, it is useful to have the weaknesses backed by quantifiable data, if possible.

Prioritization of needs is a next step. The identified needed improvements should be considered in light of greatest effect and importance of impact. Cook's suggestion might be helpful: Identify the tolerable and the critical.

External Analysis

Characterizing the external analysis as the "most exciting part of the planning discipline because it is futuristic," Cook states, "The purpose of the external analysis is to prevent surprises that may negatively affect the [program's] ability or opportunity to accomplish its mission" (Cook, 57).

"The purpose of the external analysis is to prevent surprises that may negatively affect the [program's] ability or opportunity to accomplish its mission."

—Cook, 57

Gather Information

Once needs have been identified, isolated, and prioritized, information must be gathered about all aspects of the pinpointed improvements. Outside assistance is often helpful at this stage. In the first place, knowledgeable outside evaluation of self-study data may isolate additional needs, help in analyzing priorities, or negate an identified need (it may not be a problem at all). Also, the outside "expert" may be able to help identify additional resources.

It is at this point that districts often choose to visit other schools to see what takes place in other classrooms. They may decide they need to look at different materials. They may realize that professional development is needed. Again, much information needs to be gleaned and considered. Various subgroups could be assigned to research each of the components to be addressed. Expert assistance may be sought in building the information base. The subcommittees should then report back to the whole study group with the information they have gathered.

Formulate an Action Plan

Cook continues, "each strategy will be developed by several such plans, all containing step-by-step directions, time lines, assignments of responsibilities, and cost-benefit analyses" (Cook, 65). He cautions, "action plans that are long on process and short on getting things done are nothing more than a means of postponing dedicated effort and, hence, a denial of accountability[!]."

Prioritization of the next steps is based on the information gathered. Action plans need to be formulated, as well as timelines for the execution of the plans.

Implement

Once the action plan has been formulated and the timeline designed, the first step needs to be taken. *It is extremely important that this first step is taken*, that resources are allocated so the first step is effective, and that analysis of that first step is undertaken to determine effectiveness. So many times all of the above procedures are performed, the study group meets and talks and researches and plans, but *nothing* happens. Inaction is deadly. The study group becomes discouraged because they wasted their time, the other staff loses faith in the group because they were expecting results, and the process loses any credibility for future endeavors.

It is extremely important that this first step is taken, that resources are allocated so the first step is effective, and that analysis of that first step is undertaken to determine effectiveness.

It is also important that the first step be successful. Often, too many objectives are identified, efforts are fragmented, resources are inadequate, participants don't know which investigation to pursue, and results are few or inconclusive. Instead, as the action plan is formulated, the objective must be clearly identified, commitment must be sought from each of the participants, specific tasks must be identified and assigned, sufficient resources must be allocated, and unity of purpose must be asserted. If the first step yields results, the group gains confidence to move forward. If the task undertaken is limited in scope, it stands a greater chance of success.

It is also important that the first step be successful.

The first step having been taken, the next step needs to be actualized in the same manner. It is vital that each step be clearly identified, that resources be sufficient to support it, that a commitment be made to making it work, and that results be carefully analyzed. Success with each subsequent step will build confidence for the next endeavors. Though various subgroups may pursue various avenues, it is important that the process be small enough to be controlled so that meaningful results can be obtained. Fragmentation of purpose can sabotage the best efforts.

Some Sample "First Steps"

Wisconsin's Model Academic Standards *"specify what students should know and be able to do, what they might be asked to do to give evidence of having met the intent of the standards, and how well they must perform. They include content, performance and proficiency standards."*

—Wisconsin Department of Public Instruction 1998, ix

One of the first steps that many Wisconsin school districts are taking is the alignment of their curricula with *Wisconsin's Model Academic Standards* (WMAS) (1998). Since WMAS "reflects the collective values of the citizens," [is] "tailored to prepare young people for economic opportunities that exist in Wisconsin, the nation, and the world," (ix), and is in alignment with the National Council of Teachers of Mathematics *Principals and Standards for School Mathematics*, the document reflects mathematics worth teaching and mathematics worth understanding (Romberg and Kaput 1999). *Wisconsin's Model Academic Standards* "specify what students should know and be able to do, what they might be asked to do to give evidence of having met the intent of the standards, and how well they must perform. They include content, performance, and proficiency standards" (ix). Because the WMAS are "a model to be met or exceeded," because "in the future, mastery of subject matter will be objectively measured against these new standards at grades four, eight, and twelve," and because "assessment, including high school graduation exams, [will be] based on standards" (v), it is to a district's advantage to have its curriculum in alignment with the WMAS.

Districts that have undertaken this task have used several methods: concept mapping, the standards-based model, curriculum mapping, curriculum evaluation, examination of curricular materials, implementation of selected materials, bringing all components on board, and ongoing program support.

Because "assessment, including high school graduation exams [will be] based on standards" (v), it is to a district's advantage to have its curriculum in alignment with the WMAS.

Unpacking the Standards through Concept Mapping

One of the first steps in doing an alignment is to understand the intent of the standard. An effective identified method for this is the use of a concept map (Grunow 1999).

One of the first steps in doing an alignment is to understand the intent of the standard.

> In working with a group of elementary teachers one afternoon, a question was raised about the meaning of Performance Standard F.4.4, Algebraic Reasoning: Recognize variability in simple functional relationships by describing how a change in one quantity can produce a change in another (e.g., number of bicycles and the total number of wheels) (14). I suggested use of a concept map to help develop meaning. Our first task was to put in circles the important components. The group identified one quantity and another quantity to be put in circles. Then it was felt that change (in another circle) needed to be placed above the first quantity and that a downward arrow needed to show that change was affecting the first quantity. Then, an arrow from that quantity to the second quantity needed to be drawn. A new arrow then needed to emanate from the second quantity to change (in another circle) below the second quantity. Having seen what was really happening, the group then decided to identify the process (in another circle above the whole group) as functional relationships (see Figure 8.1). An interesting discussion ensued regarding whether the arrows could also be reversed . . . would a change in the second quantity impact the first in a reciprocal fashion? To illuminate the discussion, one of the teachers suggested consideration of the "barnyard problem" that deals with the number of animals and number of feet for the animals. Based on that problem, it was decided that the relationships probably were reciprocal . . . that a change in one would affect the other no matter which change came first.
>
> Use of the concept map proved to be a revelation. In identifying the important components, the participants found the meaning of the statement. In visualizing relationships between and among the components, they developed an understanding of the function process. In selecting a task that fit the picture, they discovered how to teach the standard. Discussion of direction of functionality helped participants develop understanding of the concept (Erickson 1998). The participants began discussion of additional examples to teach to the standard . . . extensions. Finally, they talked about how they could assess student understanding of the concept! This one activity had unpacked the standard, helped teachers identify how to teach to develop understanding of the standard, led to consideration of extension activities, and

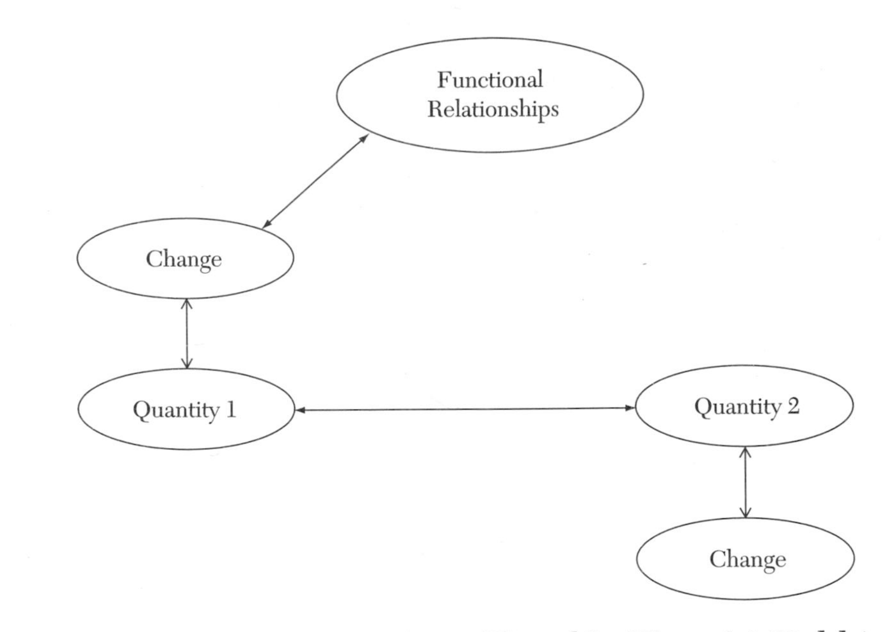

FIGURE 8.1 **Using Concept Mapping to "Unpack" a *Wisconsin's Model Academic Standards* Item**

had culminated in discussions of how to informally and formally assess understanding! One of the participants concluded triumphantly, "Now I know how to go back and help my colleagues as we address curriculum alignment. Understanding the standard is a first step!"

The Standards-Based Model

The *Standards-Based Model* (Grunow, Kniess, and Mortel 1998) is another piece developed to help teachers and school districts ensure that the standards are being addressed in the curriculum. (See Appendix D for examples of use.) Developed for the Mathematics for the New Millennium professional development project, the complete model was much sought, has been made available on disc, and is posted on the DPI mathematics Web site (www.dpi.state.wi.us/dpi/dlsis/cal/caltmtm2.html).

Recognizing that the "opportunity to learn" has been provided and verification that material has been taught are major components of the assessment process.

Emphasizing connectedness, the model could be used in many ways. Recognizing that the "opportunity to learn" has been provided and verification that material has been taught are major components of the assessment process, teachers use the model to keep track of each major investigation done by their classes. In some instances, the teacher records the activity upon its completion. In other cases, the teacher uses the Standards-Based Model for planning for an activity, a unit, and so forth. When the task is age-appropriate, some teachers use the form with their classes to reflect on an investigation, with students and teachers filling in the form as the reflective discussion ensues. Some teachers use the form as a record of ongoing activities addressing a specific standard.

A compendium of completed Standards-Based Models (or similar forms) has several advantages: (1) it urges consideration of curriculum/instruction/assessment as the performance standard is contemplated; (2) it looks at connections within the discipline and with other disciplines; (3) it facilitates recordkeeping; and (4) faithfully kept, it gives a good picture of coverage of the standards. Whether the form is used to plan or record completed activities, the amount of attention devoted to the standard can be determined. If a standard is being neglected, that omission can also be identified.

Curriculum Mapping

As it was originally conceived, "mapping is a technique for recording time on task data and then analyzing this data to determine the 'fit' to the officially adopted curriculum and the assessment/testing programs" (English 1983, 13). Jacobs suggests that today, with computers available to all, "it is possible for teachers to easily complete maps, compile them, share them with colleagues, and communicate with educators in other buildings" (Jacobs 1997, 8). She then describes the process in a series of phases.

"Mapping is a technique for recording time on task data and then analyzing this data to determine the 'fit' to the officially adopted curriculum and the assessment/testing programs."

—English 1983, 13

Phase 1: Collecting the Data

Each teacher describes three major elements that comprise the curriculum on the curriculum map and the approximate time spent on each of the concepts and topics:

1. The processes and skills emphasized
2. The content in terms of essential concepts and topics, or the content as examined in essential questions (and/or the content as reflected in *Wisconsin's Model Academic Standards*)
3. The products and performances that are the assessments of learning (and/or the Wisconsin State Assessment System results)

Phase 2: The First Read-Through

Teachers read through one another's maps for a grade level or grade-level span. In consensus groups, they identify areas in which they gained information, places where they located gaps or repetitions, potential areas for integration, mismatches between outcomes and curriculum, and meaningful and nonmeaningful assessments. "They are red-flagging areas that need attention—not rewriting the curriculum" (Jacobs 1997, 12).

Phase 3: Mixed-Group Review Session

Groups of teachers that do not work together in grade level teams or as partners share results from small-team synthesis. The collection of information by a facilitator at this stage is helpful.

Phase 4: Large-Group Review

The entire mathematics staff meets. A compendium of results from the Mixed-Group Review Session is offered. The group studies the results to identify emerging patterns. The process is still nonjudgmental.

Phase 5: Determine Those Points That Can Be Revised Immediately

Things such as repetitions can often be easily negotiated in this session.

Phase 6: Determine Those Points That Will Require Long-Term Research and Development

Task forces with specific thrusts can be appointed to pursue each of the identified components. Because of the process used to isolate the points, "people like these groups. They feel that they are real and purposeful" (15).

Phase 7: The Review Cycle Continues

"Curriculum review should be active and ongoing" (15). It is no longer sufficient to review the curriculum every five years. Too much is happening from year to year. The curriculum document should be a continual work in progress.

Representative Next Steps

When all sought information is gathered, the information is conveyed to the group. It is then up to the group to prioritize once again. Next steps need to be determined. Resources and personnel should be sought to actualize the plan.

Often at this point it is decided that curricula that reflect teaching for understanding should be found and examined, examples of teaching for understanding should be seen and experienced, and ways of assessing for understanding should be investigated.

Curriculum Evaluation

The Education Development Center (Goldsmith, Mark, and Kantrov 1998) offers Choosing a Standards-Based Mathematics Curriculum, *an excellent resource to serve as a basis for curricula evaluation.*

In an article on the role of curriculum materials in teacher learning and instructional reform, Ball and Cohen (December 1996) make the case that "the enacted curriculum" is actually jointly constructed by teachers, students, and materials in particular contexts. The curriculum is important, but so are the teacher's decision making based on listening to student cognitions, the development of understanding facilitated by the teacher and articulated by the students in classroom discourse, the environment in the classroom that fosters mathematical thinking, pacing, and so forth. Curriculum evaluation needs to proceed in light of those understandings.

The Association of State Supervisors of Mathematics and the National Council of Supervisors of Mathematics published a *Guide to Selecting Instructional Materials for Mathematics Education* (Fall 1993), which offers succinct criteria for materials evaluation. More recently, the Education Development Center (Goldsmith, Mark, and Kantrov 1998) offered *Choosing a Standards-Based Mathematics Curriculum,* an excellent resource to serve as a basis for curricula evaluation.

Following the publication of the national standards and the development of the state standards for mathematics, the question was often asked, "Where are the curricula that build the understanding called for in the standards?"

Following the publication of the national standards and the development of the state standards for mathematics, the question was often asked, "Where are the curricula that build the understanding called for in the standards?" In response to those queries, textbook companies instituted massive curricular changes to reflect the standards.

Also, the National Science Foundation offered funding to curriculum consortiums interested in developing total mathematics curricula at the elementary, middle school, and high school levels. Several such projects were funded, with the stipulation that the developed materials reflect the vision of the standards and help students build the conceptual understanding called for in the standards. Consequently, the projects had to have student research components and had to demonstrate gains in student understanding resulting from the use of the curricula. The resulting curricula are now published and are finding use in classrooms (see NSF-funded curricula in Appendix A . The strong research component continues; documentation of student gains in sites where the materials are used is continually sought (See Appendix A for Impact Data).

For the first time in history, scientific evaluations of curricula are being done to facilitate materials selection. The American Association for the Advancement of Science (AAAS), through its Project 2061, has developed extensive, well-researched criteria for evaluation and has used them to examine middle school mathematics, algebra, and biology materials. The results of their analyses can be seen on their Web site (www.project2061.org/matheval). The U.S. government established an expert panel to identify exemplary and promising mathematics programs. The results of that examination can be ordered through a toll-free number—(877) 433–7827—and can be found on their Web site (www.enc.org).

The projects had to have student research components and had to demonstrate gains in student understanding resulting from the use of the curricula. The resulting curricula are now published and are finding use in classrooms.

Examination of Curricular Materials

After the informational groundwork has been laid and curricular alignment has been pursued, districts often decide that the district or grade level components need to update curricula. Sample materials are ordered from publishers and from the National Science Foundation-funded curricula for examination, the goal being to narrow the field to two or three selections to be examined in-depth. Reviewing curricular materials, many of which are packed in glitzy packages, is a difficult task. The following process has worked well with many school districts.

Teachers identify the unit that is their greatest strength. They then examine a series in its entirety, looking at that unit to see if it does as good a job as they do and to see if it is something with which they feel they could work. Next, the teachers identify a unit that is a weakness at their grade level, a unit that they, as teachers, are not fond of, or with which students seem to experience difficulty. The teachers then look through the series, focusing on that unit to see how it is addressed. This is a good unit to pilot teach because it may prove to offer a better way to develop understanding of a difficult topic or it may introduce an approach that is preferred by the teachers.

After the initial examination phase, many schools opt to pilot two or three of the curricula that appear promising. Pilot programs, when carefully planned and executed, can yield good information about curricula. There are many variables to be considered in developing a pilot program:

Pilot programs, when carefully planned and executed, can yield good information about curricula.

1. *Funding.* Enough investment needs to be made in materials, manipulatives, and so forth, to result in a viable picture regarding the effectiveness of the curricula. But since an adoption will ultimately ensue, huge investments in series that may not be adopted would be counterproductive.
2. *Participants.* Some members of the staff may be interested in working with a pilot project; others may not be. Piloting is not easy and needs to be carefully done from an action research perspective. While interested

parties will probably provide the best trial, total staff participation will strengthen ownership and give everyone a better understanding of the issues being addressed. On the other hand, mandatory participation may not give materials a fair trial on all counts.

3. *Scope.* The traditional pilot program of a few years ago was a full year's trial of a single curriculum (often with the materials provided by the company). A more current approach is trial of a replacement unit or two from a couple of sources (often from sample materials). Regardless of approach, it is important that the concepts identified to be taught at the grade level are addressed so that students do not lose out in the process. If a full year is instituted, it is important that the concepts embodied in the series reflect the district curriculum for that grade level. If infusion units are chosen, it is again important that those units mesh with the grade-level curriculum in a meaningful sequence.

A fair trial of the materials will find teachers trying one or two of the sets and analyzing the strengths and weaknesses of each.

If there are two or three sets of materials under consideration, pilot teachers should try components of each set. After investing time in the trial of a curriculum unit, it is natural for teachers to become attached to that set of materials. A fair trial of the materials will find teachers trying one or two of the sets and analyzing the strengths and weaknesses of each. When the pilot is finished, the pilot teachers then need to meet either with each other or with the entire staff who will be using the materials to discuss their findings. A very effective method of sharing ideas is to actually teach a segment to the group in the manner in which the materials indicate—a sample lesson (see Appendix D for some forms developed for reflection during the pilot process in the Muskego-Norway School District).

At this point, it often helpful to bring in representatives from each of the publishing companies to discuss what support the companies will provide to users. It is also helpful to hear from teachers who are using the materials. During the pilot process, it is wise to have pilot teachers travel to school districts where the materials are being used to see them in action. All of this information should be shared and evaluated. Many districts have textbook selection criteria. Final selections need to be weighed against these standards. Financial considerations regarding contrasting costs of the selected series may be part of the selection equation also. All potential users need to weigh in on the final decision for ownership purposes.

All potential users need to weigh in on the final decision for ownership purposes.

Implementation of Selected Materials

A committee may be necessary to present the selection and the rationale for the selection to the board of education for their approval. An implementation plan should be drawn up including timelines, costs, and accompanying professional development experiences. Teaching for understanding requires a facilitative posture rather than a "stand and deliver" approach. Teachers need to see methodologies for building understanding in action, to experience knowledge development under those conditions, and to collaborate on strategies for effectively teaching the new curriculum. Materials, manipulatives, software, hardware, professional development opportunities, and assessment plans should be part of the presentation for the board and must be part of the implementation plan. It is imperative that all parties understand that adequate resources need to be allocated to properly implement the curriculum. Effective results are dependent on teachers having the necessary training, good materials and accompanying resources, and district support!

Materials, manipulatives, software, hardware, professional development opportunities, and assessment plans must be part of the implementation plan.

Another extremely important component of the plan is a strategy for the assessment of the program results. The implementation should begin with good baseline data, obtained on an identified assessment instrument that will be used at the beginning of the year, perhaps mid-year, and at the end of the year (preferably with different forms). This is an extremely important component. Many districts get into a curriculum implementation, realize that they want to evaluate the results they are obtaining, and have no baseline data from which to work. Data on affective responses also needs to be accumulated through attitude surveys, student journals, teacher reports, etc.

Another extremely important component of the plan is a strategy for the assessment of the program results.

Bringing All Components on Board

If the group process as outlined in this piece has taken place, there has been continuous involvement of all interested stakeholders in the process. When curricular change has been identified, it is important to continue to work with the focus group to develop a plan for the dissemination of information. The rest of the staff, the parents, and the community need to be aware of the curricular change and its ramifications. They need to be prepared to discuss:

- Why were the new materials chosen?
- What are the hoped-for results from adoption of the series?
- What do the various components of the series look like?
- Will there be homework?
- What will the homework look like?
- If the mathematics looks different, how can parents help their children with their homework?
- If a child does not appear to understand, how can parents help the child formulate a good question to ask the next day to help clarify thinking?
- How will the materials address the mathematical needs of *all* children?
- What will the mathematics classroom look like?
- How can parents make arrangements to come see the materials in use in the classroom?
- How will student progress be reported?
- What can parents do to support the implementation?
- How can the teacher be contacted when questions arise?

Several extremely effective vehicles have been developed by school districts to facilitate the curricular change process.

A classroom newsletter that tells about the material that will be covered in an upcoming unit, that shares especially meaningful mathematical developments, and that addresses questions can be an excellent communication device. A tear-off component the size of a postcard that can be returned to the classroom with comments, kudos, suggestions, or questions is a helpful addition to a classroom newsletter.

Sharing resources with parents is an appreciated gesture. An informative publication from the U.S. Department of Education, *Helping Your Child Learn Math* (Kanter and Darby 1999), can be reproduced and distributed to parents. *Family Math* (Equals 1989) has good problems and suggestions for doing mathematics at home. The following Web sites for parents can also be quite useful: www.ed.gov/pubs/parents/Math; www.enc.org/focus/topics/family/index.htm; and www.forum.swarthmore.edu/parents.citizens.html.

Scheduling parent visits so that they can come and participate in an actual mathematics lesson has been well received. Parents have an opportunity to experience the building of understanding themselves, they have advanced notice so the visit can be scheduled, and they have an opportunity to talk with the teacher and the students on a one-to-one basis.

Placing assignments and other pertinent information on a classroom Web site not only allows parents access to what is happening in the classroom, but also helps absent students keep up. Such features as problems of the week and challenges of the day are also useful on a classroom Web site.

One school offered night sessions for interested parents to come in and actually work with the same materials that their children were using. Results were overwhelmingly positive, inspiring such remarks as, "Boy, I wish I'd learned math that way!"

Parent math nights where general information is conveyed and where students share their experiences with an investigation are well received. Parents are pleased to hear about the program. They especially like hearing their children tell about the mathematics they are learning.

Open-door classrooms, where standing invitations to parents are extended, let parents know that their presence is desired, that their opinions are valued, and that the mathematical learning taking place in the classroom is worthy.

Open-door classrooms, where standing invitations to parents are extended, let parents know that their presence is desired, that their opinions are valued, and that the mathematical learning taking place in the classroom is worthy.

Ongoing Program Support

"Schools need critical financial, technical, and political support from external sources" (Newmann and Wehlage 1995). Those external sources help schools focus on student learning and enhance organizational strategies by setting high standards for learning, supporting extensive sustained, schoolwide staff development, and increasing school autonomy. In studies of successful schools, the authors found progress when there is a concentration on the intellectual quality of student work, when schoolwide capacity is built for authentic pedagogy, and when there is support from the external environment. The Department of Public Instruction has an excellent resource to help direct district support efforts. Targeting high academic standards, leadership, vision, evidence of success, professional development, family, school, and community partnerships, and standards of the heart (see Figure 8.2.), the resource is available from:

In studies of successful schools, the authors found progress when there is a concentration on the intellectual quality of student work, when schoolwide capacity is built for authentic pedagogy, and when there is support from the external environment.

Goals 2000
Wisconsin Department of Public Instruction
125 South Webster Street
Madison, WI 53707-7841
(800) 243-8782
www.dpi.state.wi.us/dpi/dlsis/edop/g2.html
Bulletin No. 01001

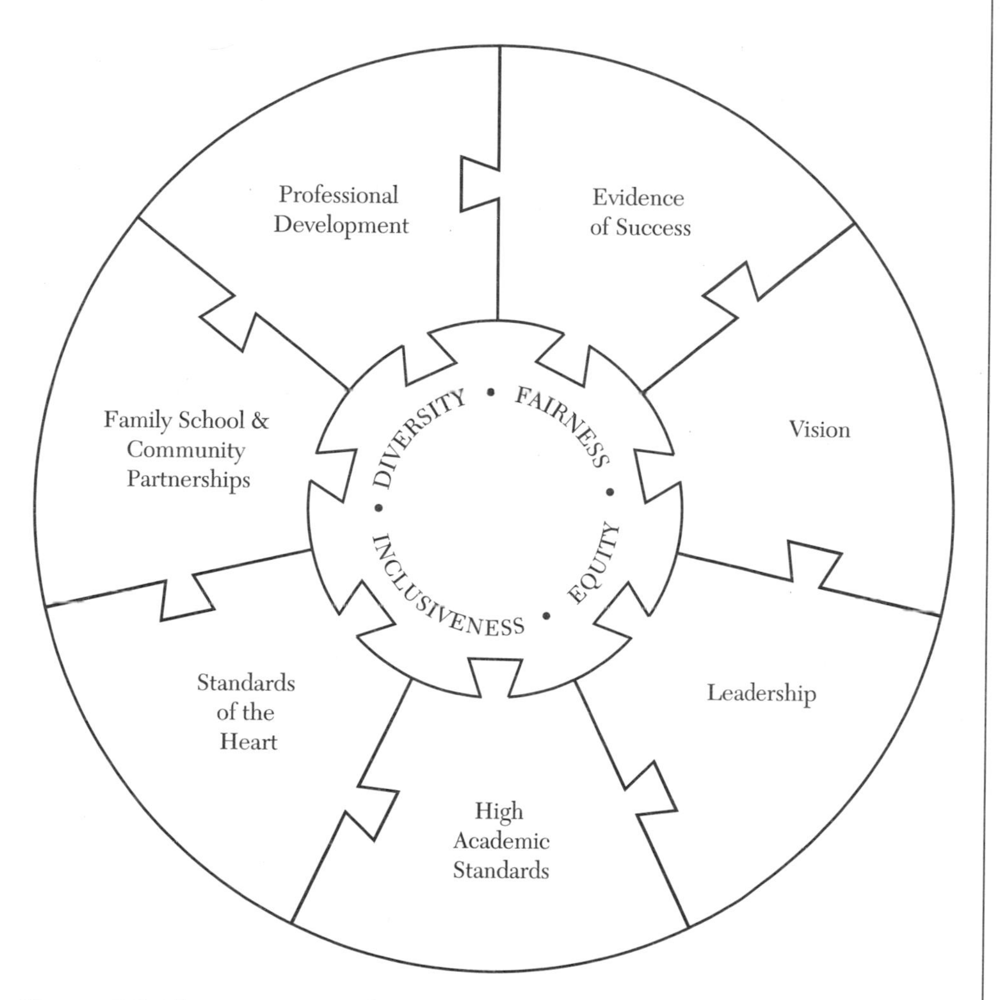

FIGURE 8.2 **Characteristics of Successful Schools**
From Wisconsin Department of Public Instruction (July, 2000). (www.dpi.state.wi.us/dpi/dlsis/edop/g2.html)

Improvement: District Level

District improvement is, of course, an ongoing story. Please find "Getting Started at Improving Your Math Education Program: One District's Ongoing Story" in Appendix E: Readings. This district has done an outstanding job of evolving a process for improving their mathematics program, a task not easily accomplished. Our sincere thanks go to the district and to the curriculum coordinator for this interesting account.

District improvement is, of course, an ongoing story.

References

How Does a District Look at Mathematics Program Development?

American Association for the Advancement of Science (AAAS). *Project 2061: Textbook Evaluation.* Washington, DC: AAAS (project2061.aaas.org). 1999.

Association of State Supervisors of Mathematics and National Council of Supervisors of Mathematics. Fall 1993. *Guide to Selecting Instructional Materials for Mathematics Education.* Golden, CO: Association of State Supervisors of Mathematics and National Council of Supervisors of Mathematics.

Ball, D. L., and D. K. Cohen. 1996. "What Is—Or Might Be—the Role of Curriculum Materials in Teacher Learning and Instructional Reform?" In *Educational Researcher, 25 (9)*, 6–8, 14.

Blume, G. W., and R. F. Nicely Jr. 1991. *A Guide for Reviewing School Mathematics Programs.* Reston, VA: National Council of Teachers of Mathematics and Association for Supervision and Curriculum Development.

Cook, W. J. Jr. 1995. *Strategic Planning for America's Schools.* Arlington, VA: American Association of School Administrators.

Council of Chief State School Officers. May 2000. *Using Data on Enacted Curriculum in Mathematics and Science.* Washington, DC: Council of Chief State School Officers.

English, F. W., ed. 1983. "Contemporary Curriculum Circumstances." In *Fundamental Curriculum Discussions.* Alexandria, VA: Association for Supervision and Curriculum Development.

Erickson, H. L. 1998. *Concept-Based Curriculum and Instruction: Teaching Beyond the Facts.* Thousand Oaks, CA: Corwin Press, Inc.

Equals. 1989. *Family Math.* Berkeley, CA: Lawrence Hall of Science, University of California.

Goldsmith, L. T., J. Mark, and I. Kantrov. 1998. *Choosing a Standards-Based Mathematics Curriculum.* Newton, MA: Education Development Center, Inc. (mcc@edc/org).

Grunow, J. E. 1999. *Using Concept Maps in a Professional Development Program to Assess and Enhance Teachers' Understanding of Rational Number.* Ann Arbor, MI: UMI (UMI Microform 9910421).

Gurnow, J., B. Kneiss, and J. Mortel. 1998. *Standards-Based Model.* Madison, WI: Department of Public Instruction.

Jacobs, H. H. 1997. *Mapping the Big Picture: Integrating Curriculum and Assessment K–12.* Alexandria, VA: Association for Supervision and Curriculum Development.

K–12 Mathematics Curriculum Center. 1999. *Curriculum Summaries.* Newton, MA: Education Development Center, Inc. (www.edc.org/mcc).

Kanter, P. F., and L. B. Darby. June 1999. *Helping Your Child Learn Math.* Jessup, MD: Education Publications Center, U.S. Department of Education. (www.ed.gov/pubs/edpubs.html).

National Council of Supervisors of Mathematics. 1997. *Supporting Improvement in Mathematics Education.* Golden, CO: National Council of Supervisors of Mathematics (mo@enet.net).

Newmann, F. M., and G. G. Wehlage. 1995. *Successful School Restructuring: A Report to the Public and Educators by the Center on Organization and Restructuring Schools.* Madison, WI: Wisconsin Center for Education Research.

Romberg, T. A., and J. J. Kaput. 1999. "Mathematics Worth Teaching, Mathematics Worth Understanding." In *Mathematics Classrooms That Promote Understanding,* edited by E. Fennema and T. A. Romberg. Mahwah, NJ: Lawrence Erlbaum Associates.

Standards 2000 Project. 2000. *Principles and Standards for School Mathematics.* Reston, VA: National Council of Teachers of Mathematics.

Tyack, D., and L. Cuban. 1995. *Tinkering Toward Utopia: A Century of Public School Reform.* Cambridge, MA: Harvard University Press.

U.S. Department of Education Math and Science Education Expert Panel. 1999. *Exemplary and Promising Mathematics Programs.* Portsmouth, NH: U.S. Department of Education Math and Science Education Expert Panel (www.enc.org).

Wisconsin Department of Public Instruction. July 2000. *Characteristics of Successful Schools.* Madison: Wisconsin Department of Public Instruction (www.dpi.state.wi.us/dpi/dlsis/edop/g2.html).

———. 1999. *Data-Driven Decision-Making Workshops.* Madison: Wisconsin Department of Public Instruction.

———. 1998. *Wisconsin's Model Academic Standards.* Madison: Wisconsin Department of Public Instruction.

Wisconsin State Journal. 2000. Report: 37% Use Internet on the Job. A1, 6 September.

Zemelman, S., H. Daniels, and A. Hyde. 1998. *Best Practice: New Standards for Teaching and Learning in America's Schools (2nd Ed.).* Portsmouth, NH: Heinemann.

Getting Started at Improving Your Math Education Program: One District's Ongoing Story

Covey, S. R. 1989. *The 7 Habits of Highly Effective People: Powerful Lessons in Personal Change.* New York: A Fireside Book.

Wisconsin Department of Public Instruction. 1998. *Wisconsin's Model Academic Standards for Mathematics.* Madison: Wisconsin Department of Public Instruction.

Planning Curriculum in

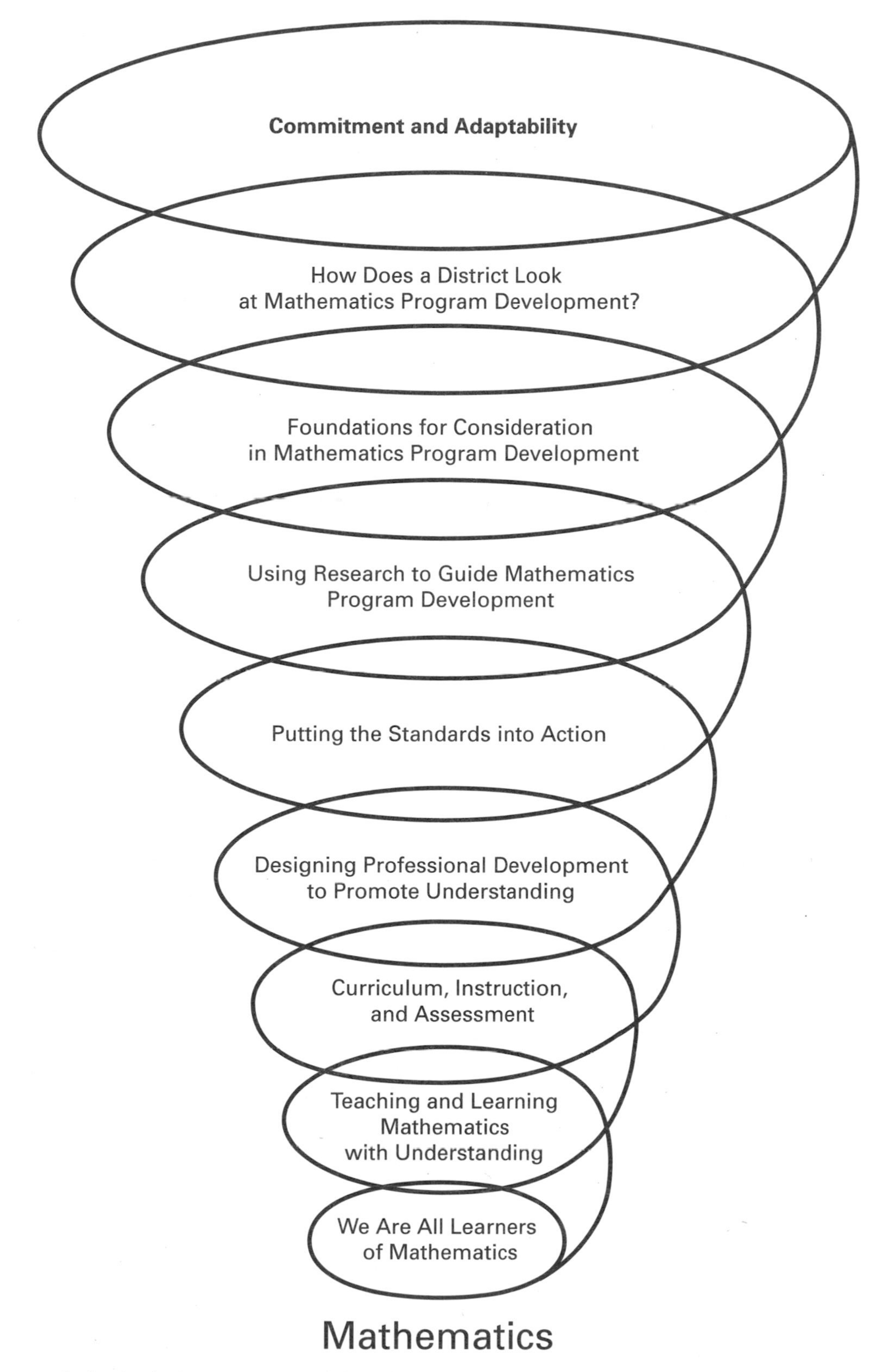

Spring (spring) *n.* 1. a source; origin; or beginning: *a spring into action.* 2. elasticity; resilience: *a spring forward with commitment.* (Morris, W., ed., 1971. *The American Heritage Dictionary of the English Language.* New York: American Heritage Publishing Company, Inc. and Houghton Mifflin Company, 1250, 2nd definition.)

Commitment and Adaptability

9

"The more things change, the more they stay the same." We live in a moving and changing society, yet we often hear the same old educational arguments that have persisted for years. We have extensive research regarding the teaching and learning of mathematics for understanding, but that research is slow to permeate mathematics programs at the classroom level. Excellent curricula have been developed that reflect that research and that show effectiveness with students, but there is hesitation in adopting them. It is imperative that students know how to use technology in the work-a-day world, but a fear persists, despite research to the contrary, that skills will erode. It is not a time when ambivalence can prevail. We need to use what we know. We need to commit to programs that reflect that knowledge. We need to continually assess to inform next moves. Living in an age of change, we need especially to be adaptable.

Public support for mathematics instruction must be raised to a level commensurate with the importance of mathematical understanding to individuals and society.

—National Council of Teachers of Mathematics 1980

In 1980, in *An Agenda for Action: Recommendations for School Mathematics of the 1980s,* the National Council of Teachers of Mathematics offered Recommendation 8: Public support for mathematics instruction must be raised to a level commensurate with the importance of mathematical understanding to individuals and society. It offered recommended actions:

8.1 Society must provide the incentives that will attract and retain competent, fully prepared, qualified mathematics teachers.

- School districts must provide compensation commensurate with the professionalism and qualifications necessary to achieve our educational goals in mathematics.
- School districts should investigate a variety of incentives and conditions that will stop the drain of qualified mathematics teachers to other, more highly compensated fields of work.
- School districts should assure mathematics teachers a classroom environment and conditions conducive to effective teaching, including a reasonable class size.
- School districts should provide teaching conditions and incentives that will attract dedicated and competent people into the mathematics teaching profession.

8.2 Parents, teachers, and school administrators must establish new and higher standards of cooperation and teamwork toward the common goal of educating each student to his or her highest potential.

- Professionals must respond to calls for maintaining educational standards and must work cooperatively with parents to specify these standards.
- Parents must support the maintenance of agreed on standards of achievement and discipline.
- Programs in mathematics that take full advantage of home and school cooperation should be systematically developed.
- Parents should enter the process of determining educational goals as partners, with shared participation and responsibility for the accomplishment of those goals.
- Parents should support the teacher's assignment of homework when it is reasonable and clearly related to the educational objectives.
- Parents should help guide students to an understanding of their critical need to learn mathematics.
- Parents should help students understand that they have the ultimate responsibility for their own learning and must be active and cooperative participants in the process.
- Parents and teachers should cooperate in a mutually supportive attack on the erosion of respect for authority in the classroom.
- Parents and teachers should cooperate in a mutually supportive attack on the erosion of motivation toward academic achievement.

Parents should help guide students to an understanding of their critical need to learn mathematics.

8.3 Government at all levels should operate to facilitate, not dictate, the attainment of goals agreed on cooperatively by the public's representatives and the professionals.

- Government funding agencies should support an emphasis on research and development in applying mathematics to problem solving.
- Teachers should have advisory roles in all decisions of policy and support.
- When legislation affecting education is required, it should be entered into with caution and only after the involvement of educational professionals in formulating and reviewing the mathematical and pedagogical aspects of such legislation.
- Legislation concerning accountability should take into account the multiple factors that determine school achievement.
- Legislators should avoid tendencies to mandate testing as the sole criterion for the evaluation of educational success.
- Mandates for the achievement of minimal competencies should not limit the school mathematics program in its broader range of essential goals.
- Legislation should not determine educational goals but when necessary should facilitate the achievement of cooperatively agreed on objectives.
- Dual respect and effective articulation must become commonplace between the civic leaders who appropriately call for educational accountability and the educational professionals who must formulate responses.

Legislation concerning accountability should take into account the multiple factors that determine school achievement.

These recommendations have been recounted in their entirety because it is interesting to note, first of all, that we have known for a long time—20 years!—what needs to be done to improve mathematics education, and, secondly, because it is edifying to see how many of the suggestions have come to fruition in those 20 years. Additionally, it is enlightening to see what was suggested in 1980 and to reflect on the several struggles that have taken place in recent years regarding the very components that NCTM addressed. On the other hand, it is disturbing to think how long it might take for some of the suggestions to be recognized and implemented!

Through its several landmark standards documents, NCTM has continued to call for actions that will facilitate mathematical progress. Stating that "deciding on the content of school mathematics is an initial step in the necessary change process" (251), the *Curriculum and Evaluation Standards* (1989) presents those content standards and suggests that necessary next steps are curriculum development, the production of standards-based teaching materials and textbooks, development of standards-based tests, instruction consistent with the underlying philosophy of the standards, standards-based teacher preservice and inservice programs, incorporation of technology into mathematics instruction, a focus on students with different needs and interests, equity in the study of mathematics, working conditions that facilitate standards-based explorations, and research on not only what is, but what ought to be, in school mathematics.

"Deciding on the content of school mathematics is an initial step in the necessary change process."

—Commission on School Mathematics 1989, 251

The *Professional Standards for Teaching Mathematics* (1991) offers an entire section of standards for the support and development of mathematics teachers and teaching, including, responsibilities of policymakers in government, business, and industry as well as responsibilities of schools, colleges, universities, and professional organizations. "Next Steps" addresses professionalism, school structure, entry into the profession, school mathematics as part of mathematics, the collegiate curriculum, collaboration between schools and universities, expectations, accrediting and certificating agencies, networking with other disciplines, and research about the "much we need to know that cannot be determined from current practice" (193).

Suggesting that the *Assessment Standards for School Mathematics* (1995) could be used by classroom teachers, curriculum directors, and department chairs to guide classroom, school, and district assessment practices; by policymakers and administrators to generate and support needed improvements in mathematical assessments; and by state or provincial and commercial test publishers to align their assessment systems with those of schools or districts in order to provide more useful data to students, teachers, parents, and the public at large, the standards reflected the learning standard, the equity standard, the openness standard, the inferences standard, and the coherence standard. In looking at next steps, it was suggested that the ideas in the standards needed to be "discussed, thought about, and assimilated" (81), that they needed to be personalized for each involved party, that users needed to reflect on the vision of alternative assessment practices, and that the assessment process is a shared responsibility in which everyone needs to participate.

The assessment process is a shared responsibility in which everyone needs to participate.

The *Principles and Standards for School Mathematics* (2000) asks readers to envision a classroom in which the principles and standards have come to life and asks questions important to mathematics education:

How can all students have access to high-quality mathematics education?
Are good instructional materials chosen, used, and accepted?
How can teachers learn what they need to know?
Do all students have time and the opportunity to learn?
Are assessments aligned with instructional goals?
Is technology supporting learning?

At this time, under the guidance of the Mathematical Association of America, mathematicians around the country are conducting a major review of the undergraduate curriculum.

Responsibilities are then assigned to mathematics teachers, students, mathematics teacher-leaders, school, district, and state or province administrators, higher-education faculty, families, other caregivers, community members, and professional organizations and policymakers.

At this time, under the guidance of the Mathematical Association of America, mathematicians around the country are conducting a major review of the undergraduate curriculum. Their recommendations, which are scheduled to emerge in 2001, "could form the basis for the biggest change in mathematics since calculus reform" (www.chronicle.com). Contributing to the effort are the Mathematical Association of America, the American Mathematical Society, and the Society of Industrial and Applied Mathematics. A consortium of 12 universities that have a $2 million grant from the National Science Foundation to make stronger connections between math and other disciplines, Mathematical Sciences and Their Applications Throughout the Curriculum, is actively evolving a new look at college mathematics that sees mathematical modeling and inquiry, a focus on the application of mathematics to solve real-life problems, as predominate over calculus serving as the "umbrella" under which college-level mathematics is organized and much more significant than the ability to differentiate formulas (Wilson 2000).

There are many efforts afoot to address facilitation of change. There is a huge body of knowledge regarding educational change (see, Senge 1990; Fullan 1993, 1999; Fullan with Stiegelbauer 1991, Newmann and Wehlage 1995; Schmoker 1996; Hargreaves 1997; Caine and Caine 1997; Marsh 1999). Best practices have been identified (Zemelman, Daniels, and Hyde 1998). "Bold plans" for school restructuring have been developed, implemented, and assessed (Stringfield, Ross, and Smith 1996). Successful schools' "winning strategies" (Richardson 2000) have been pinpointed. Successful school reform efforts have been studied and documented (Downs 2000). From these studies, we know that change is difficult but "do-able," that there are plans and effective practices that work, and what the characteristics of "winning schools" are.

Mathematically speaking, we have well-conceived national and state standards.

Mathematically speaking, we have well-conceived national and state standards. Standards-based curricula have been developed, implemented, and evaluated, with very promising results. Standardized assessments are beginning to address understanding and the concepts represented in the standards (WINSS 2000, www.dpi.state.wi.us). There are some outstanding professional development efforts that have been developed to help teachers learn how to teach for understanding (see Professional Development, chapter four of this document). Teacher evaluation systems are beginning to reflect knowledge of standards-based teaching (Danielson and McGreal 2000). Attention to performance and proficiency-based preservice education and teacher advancement systems is being given as a result of new legislation (Chapter PI 34, legislatively approved 2/17/00 to take effect 7/1/00, tcert@dpi.state.wi.us). The revamping of college mathematics programs is under way (Wilson 2000). Various technologies are being made available to schools (TEACH, www.dpi.state.wi.us/dltcl/pld/teratk12.html). Our educational systems are gradually awakening to the richness of diversity and the value of equitable approaches. Data is being used to drive decision making (CESA Standards and Assessment Center 2000). So, where do we go from here?

Schmoker suggests that *results* are the key to continuous school improvement (1996). We need to continue to set goals, work together toward those goals, look at performance data, assess findings, research and find research, redefine aims, collaborate, and lead based on our findings. Minneapolis (Pascopella 2000) has developed an encompassing set of measurement indicators worthy of consideration:

> **Results** ***are the key to continuous school improvement.***
>
> —Schmoker 1996

What was the growth in test scores from year to year?
What are the graduation rates at the school?
What are the students' attendance rates? What is the change in attendance rates from one year to the next?
How safe did both students and staff report feeling in their school and classroom?
Do students and staff report feeling respected by one another?
How many students were suspended for behavior that threatened the safety of others? What is the change in the number of suspensions of this behavior from one year to the next?
How many students of color were enrolled in advanced courses? What is the change in the number of students enrolled in those courses from one year to the next school year?
How are gifted and talented services delivered?

Newmann and Wehlage (1995) offer a model for school restructuring that includes, first of all, student learning and authentic pedagogy, but is then supported by strong school organizational capacity and positive external support (see Figure 9.1). In *Education on the Edge of Possibility* (1997), Caine and Caine offer a theory to address the paradigm shift that an information society on the verge of a new millennium requires:

1. Disequilibrium is everywhere, and we need to understand that.
2. The brain is equipped to deal with a turbulent world. But to understand this, we first need to come to terms with how the brain learns and to see how this knowledge translates into our everyday lives.
3. The change process is intrinsically transformational.
4. To function best in this new environment, we need to embrace a fundamentally different world view or perceptual orientation (10–11).

In a wonderful book entitled *Who Moved My Cheese*" Johnson (1998) discusses the times in which we live in terms of continual change. "Living in constant white water with the changes occurring all the time at work or in life

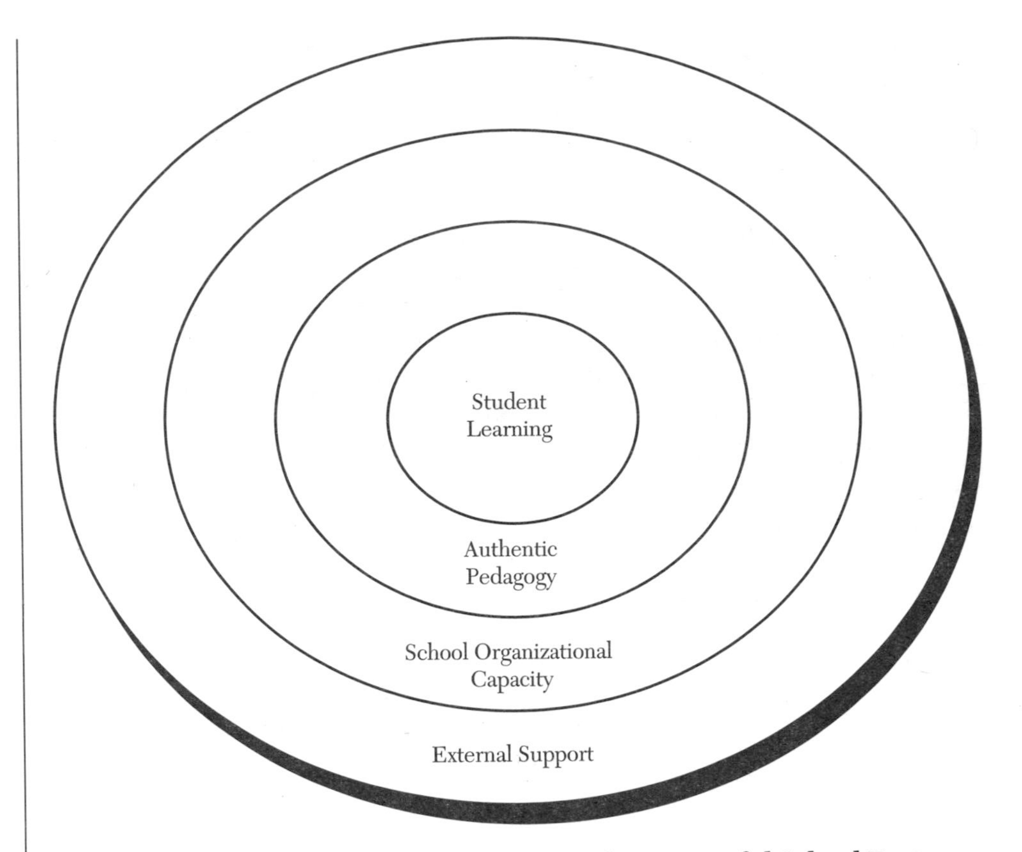

FIGURE 9.1 **Circles of Support—The Context for Successful School Restructuring**
From F. M. Newmann and G. G. Wehlage, 1995, *Successful School Restructuring*. Copyright by Wisconsin Center for Education Research. Used with permission of the Center.

"While in the past we may have wanted loyal employees, today we need flexible people who are not possessive about 'the way things are done around here.'"
—Blanchard in Johnson 1998, 16

can be stressful, unless people have a way of looking at change that helps them understand it." Using the metaphor of cheese as life's goals and mice as "parts of all of us," "the simple and the complex," he presents "the little people." "Sniff" sniffs out change early. "Scurry" scurries into action. "Hem" denies and resists change as he fears it will lead to something worse. "Haw" learns to adapt in time when he sees changing leads to something better. "Whatever parts of us we choose to use, we all share something in common; a need to find our way in the maze and find success in changing times." Ken Blanchard, who co-authored with Johnson *The One Minute Manager*, suggests, "They keep moving our cheese [our goals]. While in the past we may have wanted loyal employees, today we need flexible people who are not possessive about 'the way things are done around here,' because, like every company that wants to not only survive in the future, but stay competitive," [we need to envision and move toward the future]. The chapter titles help point the way:

The Handwriting on the Wall
Change Happens
They Keep Moving Our Cheese
Anticipate Change
Get Ready for the Cheese to Move

Monitor Change
Smell the Cheese Often So You Know When It Is Getting Old
Adapt to Change Quickly
The Quicker You Let Go of Old Cheese, the Sooner You Can Enjoy New Cheese
Change
Move with the Cheese
Enjoy Change!
Savor the Adventure and the Taste of New Cheese!
Be Ready to Quickly Change Again and Again
They Keep Moving the Cheese

Adaptability

The American Heritage Dictionary of the English Language defines *adaptability* as "possessing the attribute of being able to adjust or become adjusted to new or different conditions" (1980, 8).

Adaptability in this era is the name of the game.

Adaptability in this era is the name of the game. We have much knowledge about what must be done to keep mathematics education viable and about how to make it a motive force for the future. If we Hem and Haw too long and ignore what we do know, our Cheese will move and we will have even greater gaps to close. We do need to Sniff out what mathematics for the future looks like, we do need to be cognizant of mathematical history, and we must learn about indicators toward which research would point us. We must be willing to make informed change. When we have all parameters in view, then, for the sake of our children, we need to "Scurry" into action. To ignore what we know, to "Hem," is irresponsible. We cheat worthy students of a right to a challenging and meaningful mathematics education. To leap blindly without carefully considering all parameters is also irresponsible. In a Knowledge Age, to be ignorant is unconscionable. Moves that are taken need to be informed and carefully considered, extensively planned for, and carefully executed. The nice thing is that we live in an age where we recognize the strength of group collaboration and decision making. We can work together for the mathematical good.

In a Knowledge Age, to be ignorant is unconscionable.

Commitment

The American Heritage Dictionary of the English Language defines *commitment* as [the state of] pledging oneself to a position; of doing, performing, or perpetuating; of consigning; entrusting (1980, 145).

When a considered action is taken, it is imperative that we stay the course, carefully monitor what is taking place, and continually assess our findings so we can make informed decisions about our "next steps." So many educational endeavors are begun and abandoned as a new innovation appears. Continual change of approach leaves children bewildered and unsure. Keeping the goal of a mathematical education that produces students who are confident in their mathematical abilities, who are able to use their knowledge to

Continual change of approach leaves children bewildered and unsure.

Commitment calls for us to set worthy, viable mathematical goals.

Adaptability allows us to achieve those goals by making whatever changes are necessary to realize those goals.

solve problems of many varieties, and who are mathematical thinkers capable of envisioning mathematics to come, we need to remain committed to our course. We need to have sufficient knowledge to understand why we are doing what we are doing and abundant faith that, if we keep students and their mathematical power in mind, we will make appropriate decisions. Commitment calls for us to set worthy, viable mathematical goals. Adaptability allows us to achieve those goals by making whatever changes are necessary to realize those goals.

References

Assessment Standards Working Groups. May 1995. *Assessment Standards for School Mathematics.* Reston, VA: National Council of Teachers of Mathematics.

Caine, R. N., and G. Caine. 1997. *Education on the Edge of Possibility.* Alexandria, VA: Association for Supervision and Curriculum Development.

CESA Standards and Assessment Center. 2000. *Using Data to Improve Student Achievement: Data Retreat Participant's Guide.* Green Bay, WI: CESA 7.

Commission on Standards for School Mathematics. March 1989. *Curriculum and Evaluation Standards for School Mathematics.* Reston, VA: National Council of Teachers of Mathematics.

Commission on Teaching Standards for School Mathematics. March 1991. *Professional Standards for Teaching Mathematics.* Reston, VA: National Council of Teachers of Mathematics.

Danielson, C., and T. L. McGreal. 2000. *Teacher Evaluation to Enhance Professional Practice.* Alexandria, VA: Association for Supervision and Curriculum Development.

Davies, P., ed. 1980. *The American Heritage Dictionary of the English Language.* New York: Dell Publishing Company, Inc.

Downs, A. 2000. "Successful School Reform Efforts Share Common Features." *Harvard Education Letter, 16* (2) (March/April: 1–5). Cambridge, MA: Harvard Graduate School of Education.

Fullan, M. 1993. *Change Forces: Probing the Depths of Educational Reform.* New York: Falmer Press.

———. 1999. *Change Forces: The Sequel.* Philadelphia, PA: Falmer Press, Taylor and Francis Group.

Fullan, M., and S. Stiegelbauer. 1991. *The New Meaning of Educational Change (2nd Ed.).* New York: Teachers College Press.

Hargreaves, A., ed. 1997. *Rethinking Educational Change with Heart and Mind: 1997 Yearbook.* Alexandria, VA: Association for Supervision and Curriculum Development.

Johnson, S. J. 1998. *Who Moved My Cheese?* New York: G. P. Putnam's Sons.

Marsh, D. D. 1999. *Preparing Our Schools for the 21st Century: 1999 Yearbook.* Alexandria, VA: Association for Supervision and Curriculum Development.

National Council of Teachers of Mathematics. 1980. *An Agenda for Action: Recommendations for School Mathematics of the 1980s.* Reston, VA: National Council of Teachers of Mathematics.

Newmann, F. M., and G. G. Wehlage. 1995. *Successful School Restructuring: A Report to the Public and Educators by the Center on Organization and Restructuring of Schools.* Madison: Wisconsin Center for Education Research.

Pascopella, A. 2000. "How to Best Evaluate Schools." *Curriculum Administrator 36* (6) (August): 47–51.

Richardson, J. 2000. "Successful School Share Winning Strategies." *Results.* (May): 1, 6. Arlington, VA: National Staff Development Council.

Schmoker, M. 1996. *Results: The Key to Continuous School Improvement.* Alexandria, VA: Association for Supervision and Curriculum Development.

Senge, P. 1990. *The Fifth Discipline: The Art and Practice of the Learning Organization.* New York: Doubleday.

Standards 2000 Project. 2000. *Principles and Standards for School Mathematics.* Reston, VA: National Council of Teachers of Mathematics.

Stringfield, S., S. Ross, and L. Smith. 1996. *Bold Plans for School Restructuring: The New American School Designs.* Mahwah, NJ: Lawrence Erlbaum.

Wilson, R. 2000. "The Remaking of Math." *Chronicle of Higher Education,* January 7. (Section: The Faculty, http://chronicle.com).

Wisconsin Informational Network for Successful Schools. 2000. *WINSS.* Madison: Wisconsin Department of Public Instruction. (www.dpi.state.wi.us; E-mail: winss@dpi.state.wi.us)

Zemelman, S., H. Daniels, and A. Hyde. 1998. *Best Practice: New Standards for Teaching and Learning in America's Schools.* Portsmouth, NH: Heinemann.

Appendix A

After release of the *Curriculum and Evaluation Standards for School Mathematics* (Commission on Standards for School Mathematics 1989), there were many requests for curricula to "teach to the vision of the standards." Consequently, the National Science Foundation issued a request for proposals (RFP) for interested researchers to submit proposals to write comprehensive standards and research-based curricula at the elementary, middle, and high school levels. These projects had to reflect the vision of the standards and needed to show success with students. There was a heavy research component involved in development of the materials. When the materials had been developed and student tested, the projects were to find publishers for their materials.

These projects had to reflect the vision of the standards and needed to show success with students.

The resulting curricula are listed in this appendix along with the specific research projects that have addressed their effectiveness. The list is partial. As the materials are implemented in school districts across the nation, additional research studies are being done to continue to assess efficacy.

Additional information is available on several Web sites:

The K–12 Mathematics Curriculum Center: http://www.edc.org/mcc
Alternatives for Rebuilding Curricula (ARC) (Elementary Curricula): http://www.comap.com/elementary/projects/arc/
Show-Me Center (Middle Grades Curricula): http://showmecenter.missouri.edu
Curricular Options in Mathematics Programs for All Secondary Students (COMPASS): http://www.ithaca.edu/compass

NSF-Funded Standards-Based Curriculum Projects

Elementary

Everyday Mathematics, K–6
Everyday Learning Corporation
2 Prudential Plaza, Suite 1200
Chicago, IL 60601
Ph. (800) 382-7670

Investigations in Number, Data, and Space, K–5
Globe Fearon/Prentice Hall
P.O. Box 182819
Columbus, OH 43272-5286
Ph. (800) 848-9500
http://www.globefearon.com
http://www.phschool.com

MathTrailblazers, K–5
Kendall/Hunt Publishing Co.
4050 Westmark Drive
Dubuque, IA 52004-1840
Ph. (800) 542-6657

Middle Grades

Connected Math Project, Grades 6–8
Globe Fearon/Prentice Hall
P.O. Box 182819
Columbus, OH 43272-5286
Ph. (800) 848-9500
http://www.globefearon.com
http://www.phschool.com

Mathematics in Context, Grades 5–8
Encyclopaedia Britannica
310 S. Michigan Avenue
Chicago, IL 60604
Ph. (800) 554-9862

Middle-School Math Through
Applications Project (MMAP)
Most modules available through self-publishing.
Commercial publisher pending.
Ph. (650) 687-7918

MathScape, Grades 6–8
Creative Publications
1300 Villa Street
Mountain View, CA 94041
Ph. (800) 624-0822

Middle Grade MathThematics, Grades 6–8
McDougal Littell
1560 Sherman Avenue
Evanston, IL 60201
Ph. (800) 323-4068, ext. 3206

Secondary

Contemporary Mathematics in Context
Everyday Learning Corporation
P.O. Box 812960
Chicago, IL 60681
Ph. (800) 382-7670

Interactive Mathematics Program
Key Curriculum Press
1150 65th Street
Emeryville, CA 94608
Ph. (800) 995-MATH (6284)

MATH Connections
It's About Time, Inc.
84 Business Park Dr., Suite 307
Armonk, NY 10504
Ph. (888) 698-TIME (8463)

Mathematics: Modeling Our World
South-Western Educational Publishing
7625 Empire Drive
Florence, KY 41042
Ph. (800) 865-5840

SIMMS Integrated Mathematics
Simon & Schuster Custom Publishing
2055 South Gessner, Suite 200
Houston, TX 77063
Ph. (888) 339-0529

Evidence of Impact Related to Student Outcomes

(NOTE: This listing was compiled for the NCTM NSF Implementation Centers session based on available information in 1998. Research listings may have been inadvertently omitted. For up-to-date information on research related to student outcomes for a particular curriculum, contact the appropriate NSF Center or the individual curriculum development Web site at www.nsf.gov/nsb/search.htm.)

Elementary Curricula

Everyday Mathematics

http://www.everydaylearning.uchicago.edu.index.html.
Impact data: http://everydaylearning.com/Pages/results.html.

"Student Achievement Studies." A booklet from Everyday Learning Corporation presenting results of seven studies and reports on districts using Everyday Mathematics.

"Everyday Mathematics Success Stories." Cases provided by Everyday Learning Corporation about districts' results from using Everyday Mathematics.

Carroll, W. M. 1996a. Mental Computation of Students in a Reform-Based Mathematics Curriculum. *School Science and Mathematics* 96(6): 305–311.

———. 1996b. Students in a Reform Mathematics Curriculum: Performance on the 1993 Third-Grade IGAP. *Illinois School Research and Development Journal* 33 (1): 11–14.

———. 1996c. Use of Invented Algorithms by Second Graders in a Reform Mathematics Curriculum. *Journal of Mathematical Behavior* 14(3): 349–362.

———. 1997a. Mental and Written Computation: Abilities of Students in a Reform-Based Curriculum. *The Mathematics Educator* 2(1): 18–32.

———. 1997b. Results of Third-Grade Students in a Reform Curriculum on the Illinois State Mathematics Test. *Journal for Research in Mathematics Education* 28(2): 237–242.

Fuson, K., W. M. Carroll, and J. Landis. 1996. Levels in Conceptualizing and Solving Addition and Subtraction Compare Word Problems. *Cognition and Instruction* 14 (3): 345–371

Porter, D., and W. Carroll. 1995. Invented Algorithms: Some Examples from Primary Classrooms. *Illinois Mathematics Teacher* (April): 6–12.

Woodward, J., and J. Baxter. 1997. The Effects of an Innovative Approach to Mathematics on Academically Low-Achieving Students in Inclusive Settings. *Exceptional Children* 63(3): 373–388.

Math Trail Blazers (TIMS)

Web site: www.math.uic.edu/IMSE.mtb.html

Three studies have been conducted:

1. A study of standardized test scores (Illinois State mathematics assessment—Illinois Goal Assessment Program (IGAP)) of Math Trailblazers (MTB) and comparison group.
2. A study comparing acquisition of math facts.
3. A longitudinal study of six first grade classrooms using MTB.

Investigations in Number, Data, and Space

Mokros, J., M. Berle-Carman, A. Rubin, and K. O'Neil. 1996. Learning Operations: Invented Strategies That Work. Paper presented at the Annual Meeting of the American Educational Research Association, New York, April.

Mokros, J., M. Berle-Carman, A. Rubin, and T. Wright. 1994. *Full-Year Pilot Grades 3 and 4: Investigations in Number, Data, and Space®*. Cambridge, MA: TERC.

Middle Grades Curricula

Connected Mathematics Project (CMP)

http://www.math.msu.edu/cmp

Impact data and information: http://www.Math.msu.edu/cmp/RepStu-Stu-Achieve.htm

Ben-Chaim, D., J. T. Fey, W. M. Fitzgerald, C. Benedetto, and J. Miller. 1997. A Study of Proportional Reasoning Among Seventh and Eighth Grade Students. Unpublished paper. Weizman Institute.

Hoover, M. N., J. S. Zawojewski, and J. Ridgway. 1997. Effects of the Connected Mathematics Project on Student Attainment. Paper presented at the annual meeting of the American Educational Research Association, Chicago, April.

Lappan, R. T., D. E. Barnes, B. J. Reys, and R. E. Reys. 1998. Standards-Based Middle Grade Mathematics Curricula: Impact on Student Achievement. Paper presented at the annual meeting of the American Educational Research Association, San Diego, April.

Mathematics in Context (MiC)

http://www.showmecenter.missouri.edu/showme/mic/shtml

The impact of the mathematics in context curriculum. Impact data, shared with the MiC project staff, from three districts that used the curriculum for at least two years (available from MiC staff).

Romberg, T. A. 1997. Mathematics in Context: Impact on teachers. In *Mathematics Teachers in Transition*, edited by E. Fennema and B. S. Nelson. Mahwah, NJ: Lawrence Erlbaum Associates.

Middle School Mathematics through Applications Program (MMAP)

http://mmap/wested.org.

Institute for Research on Learning (IRL) has impact data—but not compiled in any formal report to date. Web site: http://www.irl.org.

MathScape

Web site: www2.edu.org/MathscapeSTM/

Research report for *MathScape: Seeing and Thinking Mathematically* (available from Education Development Center, Inc. (EDC) http://www.edc.org/mcc and http://www.edc.org/mcc/materials/htm).

MathThematics (STEM)

Web site: http://www.showmecenter.missouri.edu/stem/shtml.

A middle school curriculum for grades 6 through 8, developed by the STEM Project: Six Through Eight Mathematics.

Field test and materials evaluation results are available from McDougall Littell.

Lappan, R. T., D. E. Barnes, B. J. Reys, and R. E. Reys. 1998. Standards-Based Middle Grade Mathematics Curricula: Impact on Student Achievement. Paper presented at the annual meeting of the American Educational Research Association, San Diego.

Secondary Curricula

MATH Connections

The MATH Connections web site (http://www.mathconnections.com) has student results on standardized tests, data about student confidence in learning mathematics, student perception of usefulness of mathematics, and an external evaluation of cognitive discourse in the classroom.

"Evidence of Effectiveness and Success" is available from the MATH Connections Implementation Center.

SIMMS Integrated Mathematics

Hirstein. 1997. "Systemic Initiative for Montana Mathematics and Science (SIMMS) Project: Student Assessment in the Pilot Study." University of Montana. Preprinted in November 1997 from Monograph 4.

"SIMMS Monograph 4: Assessment Report 1." A collection of articles summarizing research pertaining to SIMMS, including the article mentioned above.

SIMMS Monograph 5: Assessment Report II can be found at http://www.ithaca.edu/compass/frames.htm.

Applications Reform In Secondary Education (ARISE) Program

(Contact ARISE staff for information at http://comap.com/highschool/projects/arise.html.)

Interactive Mathematics Program (IMP)

IMP Evaluation Updates. (Available from the IMPlementation Center, Box 2891, Sausalito, CA 95966): No. 1: Spring 1995; No. 2: Fall 1996; No. 3: Fall 1997.

Evaluation reports on IMP. (Available from Dr. Norman Webb, Wisconsin Center for Education Research, 1025 West Johnson Street, Madison, WI 53706):

- *Impact of the Interactive Mathematics Program on the Retention of Underrepresented Students: Cross-School Analysis of Transcripts for the Class of 1993 for Three High Schools*
- *Comparison of IMP Students with Students Enrolled in Traditional Courses on Probability, Statistics, Problem Solving, and Reasoning*
- *Replication Study of the Comparison of IMP Students with Students Enrolled in Traditional Courses on Probability, Statistics, Problem Solving, and Reasoning*

Study of IMP in Philadelphia. (Available from Dr. Edward Wolff, Mathematics Department, Beaver College, 450 S. Easton Road, Glenside, PA 19038):

- Edward Wolff. *Summary of Matched-Sample Analysis Comparing IMP and Traditional Students at Philadelphia High School for Girls on Mathematics Portion of Stanford-9 Test*

- Edward Wolff. *Summary of Matched-Sample Stanford-9 Analysis Comparing IMP and Traditional Students at Central High School, Philadelphia, PA*

Study of NSF Projects. (Available from Dr. Norman Webb, Wisconsin Center for Education Research, 1025 West Johnson Street, Madison, WI 53706):

- Hal Schoen. *Impact Study of Mathematics Education Projects Funded by the National Science Foundation, 1983–91.*

Summary of IMP evaluation. (Available from the IMPlementation Center, P.O. Box 2891, Sausalito, CA 95966):

- Diane Resek. *Evaluation of the Interactive Mathematics Program.* A paper edition presented at the 1998 annual meeting of the American Educational Research Association (AERA). www.mathimp.org/research/AERA-paper.html.

Alper, L., D. Fendel, S. Fraser, and D. Resek. 1997. Designing a High School Mathematics Program for All Students. *American Journal of Education* 106(1): 148–178.

Clark, D., M. Wallbridge, and S. Fraser. 1992. *The Other Consequences of a Problem-Based Mathematics Curriculum.* Research report no. 3. Victoria, Australia: Mathematics Teaching and Learning Centre, Australian Catholic University, Christ Campus.

White, P., A. Gamoran, and J. Smithson. 1995. Math Innovations and Student Achievement in Seven High Schools in California and New York. Consortium for Policy Research in Education and the Wisconsin Center for Education Research, University of Wisconsin–Madison, January.

Core-Plus Mathematics

http://wmich/edu/cpmp.

Coxford, A. F., and C. F. Hirsch. 1996. A Common Core of Math for All. *Educational Leadership* 53(8): 22–25.

Hirsch, C. R., and A. F. Coxford. 1997. Mathematics for All: Perspectives and Promising Practices. *School Science and Mathematics* 97(5): 232–241.

Hirsch, C. R., and M. L. W. Weinhold. 1999. Everybody Counts—Including the Mathematically Promising. In *Developing Mathematically Promising Students,* edited by L. Sheffield. Reston, VA: National Council of Teachers of Mathematics.

Schoen, H. L., D. L. Bean, and S. W. Ziebarth. 1996. Embedding Communication Throughout the Curriculum. In *Communication in Mathematics: K–12 and Beyond,* 1996 Yearbook of the National Council of Teachers of Mathematics, edited by P. C. Elliott and M. J. Kenney. Reston, VA: The National Council of Teachers of Mathematics.

Schoen, H. L., and S. W. Ziebarth. 1997a. A Progress Report on Student Achievement in the Core-Plus Mathematics Project Field Test. Unpublished manuscript, University of Iowa.

———. 1997b. A Progress Report on Student Achievement in the Core-Plus Mathematics Project Field Test. *NCSM Journal of Mathematics Education Leadership* 1(3): 15–23.

———. 1998. High School Mathematics Curriculum Reform: Rationale, Research, and Recent Developments. In *Annual Review of Research for School Leaders,* edited by P. S. Hlebowitsh and W. G. Wraga. New York: Macmillan.

Tyson, V. 1995. An Analysis of the Differential Performance of Girls on Standardized Multiple-Choice Mathematics Achievement Tests Compared to Constructed Response Tests of Reasoning and Problem Solving. Unpublished Ph.D diss., University of Iowa.

Appendix B

B

Change in certification requirements (Chapter 34, July 1, 2000) called for identification of content consistent with the Wisconsin Standards for Teacher Development and Licensure, Standard 1: Teachers know the subjects they are teaching.

PI 34 Standard 1 Mathematics Content Development Committee

Henry Kepner, University of Wisconsin–Milwaukee
J. Marshall Osborn, University of Wisconsin–Madison
Jack Moyer, Marquette University
Billie Sparks, University of Wisconsin–Eau Claire
Phil Makurat, University of Wisconsin–Whitewater
Ann Krause, Blackhawk Technical College
Jim Marty, Waukesha North High School
John Janty, Waunakee High School
Jim Moser, Mathematics Consultant, Department of Public Instruction (Retired)
Jodean Grunow, Mathematics Consultant, Department of Public Instruction
Kathryn Lind, Teacher Education and Licensing Consultant, Department of Public Instruction (Group Advisor)

Proposed Mathematics Program Approval Guidelines for PI 34 Standard 1

Standard 1: Teachers Know the Subjects They Are Teaching.

The teacher understands the central concepts, tools of inquiry, and structures of the disciplines he or she teaches and can create learning experiences that make these aspects of subject matter meaningful for students.

Teachers responsible for mathematics instruction *at any level* understand the key concepts and procedures of mathematics and have a broad understanding of the K–12 mathematics curriculum. They understand the structures within the discipline, the historical roots and evolving nature of

The teacher understands the central concepts, tools of inquiry, and structures of the discipline he or she teaches and can create learning experiences that make these aspects of subject matter meaningful for students.

mathematics, and the interaction between technology and the discipline. Using content-appropriate methodologies, they facilitate the building of student conceptual and procedural understanding. They listen to student cognitions and use their gleanings to inform their instructional decision making. They understand the importance of helping all students build understanding of the discipline, and they help students build confidence in their abilities to utilize mathematical knowledge. They help students build awareness of the usefulness of mathematics and of the economic implications of fine mathematical preparation.

"The very essence of studying mathematics is itself an exercise in exploring, conjecturing, examining and testing—all aspects of problem solving" (Commission on Teaching Standards for School Mathematics 1991, 95). Teachers need to experience "struggling to create mathematical understandings" (Interstate New Teacher Assessment and Support Consortium 2000, 28) so that they can create learning experiences that will help students become problem solvers. Mathematics teachers are able to formulate and pose worthwhile mathematical tasks, solve problems using several strategies, evaluate results, generalize solutions, use problem-solving approaches effectively, and apply mathematical modeling to real-world situations.

Mathematics teachers are able to formulate and pose worthwhile mathematical tasks, solve problems using several strategies, evaluate results, generalize solutions, use problem-solving approaches effectively, and apply mathematical modeling to real-world situations.

Mathematical reasoning is an indicator of the depth of mathematical understanding. Teachers of mathematics need to be able to make convincing mathematical arguments, frame mathematical questions and conjectures, formulate counterexamples, construct and evaluate arguments, and use intuitive, informal exploration and formal proof. Mathematical discourse is critical to learning mathematics and to communicating about what is being learned. Teachers of mathematics need to be able to express ideas orally, in writing, and visually; to be able to use mathematical language, notation, and symbolism; and to be able to translate mathematical ideas between and among contexts. Teachers of mathematics are able to connect the concepts and procedures of mathematics and to draw connections between mathematical strands, between mathematics and other disciplines, and with daily life. Teachers of mathematics are decision makers charged with selecting appropriate representations to facilitate mathematical problem solving and are responsible for translating between and among representations to explicate problem-solving situations.

Teachers of mathematics are able to connect the concepts and procedures of mathematics and to draw connections between mathematical strands, between mathematics and other disciplines, and with daily life.

Mathematical Processes

All teachers of mathematics develop knowledge of mathematics through the following critical processes:

- Problem solving
- Communication
- Reasoning and formal and informal argument
- Mathematical connections
- Representations
- Technology

Number Operations and Relationships

All teachers of mathematics understand number operations and relationships from both abstract and concrete perspectives and are able to identify real-world applications. They know and represent the following mathematical concepts and procedures and the connections among them:

- Number sense
- Set theory
- Number and operation
- Composition and decomposition of numbers, including place value, primes, factors, multiples, inverses, and the extension of these concepts throughout mathematics
- Number systems through the real numbers, their properties and relations
- Computational procedures
- Proportional reasoning
- Number theory

Teachers of upper-level mathematics know and represent the following additional mathematical concepts and procedures and the connections among them:

- Advanced counting procedures, including union and intersection of sets and parenthetical operations
- Algebraic and transcendental numbers
- The complex number system, including polar coordinates
- Approximation techniques as a basis for numerical integration, fractals, and numerical-based proofs
- Situations in which numerical arguments presented in a variety of classroom and real-world situations (e.g., political, economic, scientific, social) can be created and critically evaluated
- Opportunities in which acceptable limits of error can be assessed (e.g., evaluating strategies, testing the reasonableness of results, and using technology to carry out computations)

Geometry and Measurement

All teachers of mathematics understand and represent geometry and measurement from both abstract and concrete perspectives and are able to identify real-world applications. They know the following mathematical concepts and procedures and the connections among them:

- Formal and informal argument
- Names, properties, and relationships of two- and three-dimensional shapes
- Spatial sense
- Spatial reasoning and the use of geometric models to represent, visualize, and solve problems

- Transformations and the ways in which rotation, reflection, and translation of shapes can illustrate concepts, properties, and relationships
- Coordinate geometry systems, including relations between coordinate and synthetic geometry and generalizing geometric principles from a two-dimensional system to a three-dimensional system
- Concepts of measurement, including measurable attributes, standard and nonstandard units, precision and accuracy, and use of appropriate tools
- The structure of systems of measurement, including the development and use of measurement systems and the relationships among different systems
- Measurement, including length, area, volume, size of angles, weight and mass, time, temperature, and money
- Measuring, estimating, and using measurement to describe and compare geometric phenomena
- Indirect measurement and its uses, including developing formulas and procedures for determining measures to solve problems

Teachers of upper-level mathematics know the following additional mathematical concepts and procedures and the connections among them:

- Systems of geometry, including Euclidean, non-Euclidean, coordinate, transformational, and projective geometry
- Transformations, coordinates, and vectors and their use in problem solving
- Three-dimensional geometry and its generalization to other dimensions
- Topology, including topological properties and transformations
- Opportunities to present convincing arguments by means of demonstration, informal proof, counterexamples, or other logical means to show the truth of statements or generalizations

Statistics and Probability

All teachers of mathematics understand statistics and probability from both abstract and concrete perspectives and are able to identify real-world applications. They know the following mathematical concepts and procedures and the connections among them:

- Use of data to explore real-world issues
- The process of investigation, including formulation of a problem, designing a data collection plan, and collecting, recording, and organizing data
- Data representation through graphs, tables, and summary statistics to describe data distributions, central tendency, and variance
- Analysis and interpretation of data
- Randomness, sampling, and inference

- Probability as a way to describe chances or risk in simple and compound events
- Outcome prediction based on experimentation or theoretical probabilities

Teachers of upper-level mathematics must also understand the following additional mathematical concepts and procedures and the connections among them:

- Use of the random variable in the generation and interpretation of probability distributions
- Descriptive and inferential statistics, measures of disbursement, including validity and reliability and correlation
- Probability theory and its link to inferential statistics
- Discrete and continuous probability distributions as bases for inference
- Situations in which students can analyze, evaluate, and critique the methods and conclusions of statistical experiments reported in journals, magazines, news media, advertising, and so on

Functions, Algebra, and Concepts of Calculus

All teachers of mathematics understand functions, algebra, and basic concepts underlying calculus from both abstract and concrete perspectives and are able to provide real-world applications. They know the following mathematical concepts and procedures and the connections among them:

- Patterns
- Functions as used to describe relations and to model real-world situations
- Representations of situations that involve variable quantities with expressions, equations, and inequalities and that include algebraic and geometric relationships
- Multiple representations of relations, the strengths and limitations of each representation, and conversion from one representation to another
- Attributes of polynomial, rational, trigonometric, algebraic, and exponential functions
- Operations on expressions and solution of equations, systems of equations, and inequalities using concrete, informal, and formal methods
- Underlying concepts of calculus, including rate of change, limits, and approximations for irregular areas

All teachers of upper-level mathematics must also understand the following additional mathematical concepts and procedures and the connections among them:

- Concepts of calculus, including limits (epsilon-delta) and tangents, derivatives, integrals, and sequences and series

- Modeling to solve problems
- Calculus techniques, including finding limits, derivatives, and integrals and using special rules
- Calculus applications, including modeling, optimization, velocity and acceleration, area, volume, and center of mass
- Numerical and approximation techniques including Simpson's rule, trapezoidal rule, Newton's approximation, and linerization
- Multivariate calculus
- Differential equations

Discrete Mathematics (the study of mathematical properties of sets and systems that have a finite number of elements)

All teachers of mathematics understand discrete processes from both abstract and concrete perspectives and are able to identify real-world applications. They know the following mathematical concepts and procedures and the connections among them:

- Counting techniques
- Representation and analysis of discrete mathematics problems using sequences, graph theory, arrays, and networks
- Iteration and recursion

Teachers of upper-level mathematics must also understand the following additional mathematical concepts and procedures and the connections among them:

- Topics, including symbolic logic, induction, linear programming, and finite graphs
- Matrices as a mathematical system and matrices and matrix operations as tools for recording information and for solving problems
- Developing and analyzing algorithms

Wisconsin Standards for Teacher Development and Licensure

1. Teachers know the subjects they are teaching.

The teacher understands the central concepts, tools of inquiry, and structures of the discipline(s) he or she teaches and can create learning experiences that make these aspects of subject matter meaningful for students.

2. Teachers know how children grow.

The teacher understands how children learn and develop and can provide learning opportunities that support their intellectual, social, and personal development.

3. Teachers understand that children learn differently.

The teacher understands how students differ in their approaches to learning and creates instructional opportunities that are adapted to diverse learners.

4. Teachers know how to teach.

The teacher understands and uses a variety of instructional strategies to encourage students' development of critical-thinking, problem-solving, and performance skills.

5. Teachers know how to manage a classroom.

The teacher uses an understanding of individual and group motivation and behavior to create a learning environment that encourages positive social interaction, active engagement in learning, and self-motivation.

6. Teachers communicate well.

The teacher uses knowledge of effective verbal, nonverbal, and media communication techniques to foster active inquiry, collaboration, and supportive interaction in the classroom.

7. Teachers are able to plan different kinds of lessons.

The teacher plans instruction based upon knowledge of subject matter, students, the community, and curriculum goals.

8. Teachers know how to test for student progress.

The teacher understands and uses formal and informal assessment strategies to evaluate and ensure the continuous intellectual, social, and physical development of the learner.

9. Teachers are able to evaluate themselves.

The teacher is a reflective practitioner who continually evaluates the effects of his or her choices and actions on others (students, parents, and other professionals in the learning community) and who actively seeks out opportunities to grow professionally.

10. Teachers are connected with other teachers and the community.

The teacher fosters relationships with school colleagues, parents, and agencies in the larger community to support students' learning and well-being.

References

Commission on Standards for School Mathematics. 1989. *Curriculum and Evaluation Standards for School Mathematics.* Reston, VA: National Council of Teachers of Mathematics.

Commission on Teaching Standards for School Mathematics. 1991. *Professional Standards for Teaching Mathematics.* Reston, VA: National Council of Teachers of Mathematics.

Cook, D. M. 1989. *Strategic Learning in the Content Areas.* Madison, WI: Department of Public Instruction.

Interstate New Teacher Assessment and Support Consortium. 2000. *Draft Standards in Mathematics for Licensing Beginning Teachers.* Council of Chief State School Officers: Interstate New Teacher Assessment and Support Consortium.

Mathematics Content Task Force, Governor's Council on Model Academic Standards. 1999. *Wisconsin's Model Academic Standards for Mathematics.* Madison, WI: Department of Public Instruction. (http://www.dpi.state.wi.us/).

National Board for Professional Teaching Standards. 1991. *Toward High and Rigorous Standards for the Teaching Profession: Initial Policies and Perspectives of the NBPTS.* 3d ed. Detroit, MI: National Board for Professional Teaching Standards National Council of Teachers of Mathematics. 1998. *NCATE program standards.* Reston, VA: National Council of Teachers of Mathematics.

Standards 2000 Project. 2000. *Principles and Standards for School Mathematics.* Reston, VA: National Council of Teachers of Mathematics.

National Research Council. 1991. *Moving Beyond Myths: Revitalizing Undergraduate Mathematics Action Plan.* Washington, DC: National Academy Press.

Peppard, J. 1997. *A Guide to Connected Curriculum and Action Research.* Madison, WI: Department of Public Instruction.

Standards 2000 Writing Group. 1998. *Principles and Standards for School Mathematics: Discussion Draft.* Reston, VA: National Council of Teachers of Mathematics.

Additional References

Darling-Hammond, L., A. E. Wise, and S. P. Klein. 1999. *A License to Teach: Raising Standards for Teaching.* San Francisco: Jossey-Bass.

Darling-Hammond, L., and G. Sykes, eds. 1999. *Teaching as the Learning Profession: Handbook of Policy and Practice.* San Francisco: Jossey-Bass.

Eisenhower National Clearinghouse for Mathematics and Science Education. 1999. *Ideas That Work: Mathematics Professional Development.* Columbus, OH: Eisenhower National Clearinghouse for Mathematics and Science Education.

Grunow, J. E. 1999. *Using Concept Maps in a Professional Development Program to Assess and Enhance Teachers' Understanding.* Ann Arbor, MI: UMI (#9910421).

Hargreaves, A. 1994. *Changing Teachers, Changing Times: Teachers' Work and Culture in the Postmodern Age.* New York, NY: Teachers College Press.

Loucks-Horsley, S., P. W. Hewson, N. Love, and K. E. Stiles. 1998. *Designing Professional Development for Teachers of Science and Mathematics.* Thousand Oaks, CA: Corwin Press.

Appendix C

C

Wisconsin's *Model Academic Standards for Mathematics* (1998) were developed to reflect performances to be anticipated at the end of grades 4, 8, and 12. This was done for two reasons. First of all, it was felt that learning occurs over time. There should be target skills, but students will reach them via different paths in different timeframes. Likewise, schools need to have the autonomy to decide when and where they will address specific standards.

As districts have aligned curricula with WMAS, they often do wish to pinpoint specific performance standards at specific grade levels. James M. Moser, Ph.D., DPI mathematics consultant (retired), has assisted several districts in this process. Working with those districts and with the WMAS, he has developed Suggested School District Mathematics Standards Levels to be Compatible with State of Wisconsin Model Academic Standards. The entire list of these expectations is attached.

As districts have aligned curricula with WMAS, they often do wish to pinpoint specific performance standards at specific grade levels.

This list is not meant to be prescriptive but only to provide a possible model for specification of standards. It is important to remember that although grade-specific standards are useful guideposts, it is in development of the entire concept that understanding occurs.

Thanks to Dr. Moser for this extremely useful guide.

Suggested School District Mathematics Standards Levels to be Compatible with State of Wisconsin Model Academic Standards

TABLE C.1 **Number**

WI Model Academic Standards	School District Levels
By the end of Grade 4, students will:	
B.4.1 Represent and explain whole numbers, decimals, and fractions with: ■ Physical materials ■ Number lines and other pictorial models ■ Verbal descriptions ■ Place value concepts and notation ■ Symbolic renaming (e.g., 43 = 40 + 3 = 30 + 13)	***Level B.4.1.1*** ■ Represent whole numbers up to 12 with physical objects and with pictures. ***Level B.4.1.2*** ■ Represent and explain whole numbers up to 100 with: Physical objects Pictures Number lines ■ Represent and explain fractions with denominators of 2, 3, and 4 by using pictorial and physical models of parts of a whole object. ***Level B.4.1.3*** ■ Represent and explain whole numbers up to 1,000 with: Grouped physical objects Pictures and number lines Place value concepts and notation ■ Represent and explain simple fractions by using pictorial and physical models of : Parts of whole objects Parts of sets ***Level B.4.1.4*** ■ Represent and explain whole numbers up to 10,000 with: Grouped physical objects Pictures and number lines Place value concepts and notation ■ Represent and explain fractions by using pictorial and physical models of Parts of whole objects Parts of sets ***Level B.4.1.5*** ■ Represent and explain decimals (to hundredths) with: Physical objects Pictures Place value concepts and notation

Special Note: Particular numerical levels within a given cell are not necessarily of the same degree of difficulty or sophistication as a similarly numbered level in another cell. For example, Level B.4.6.1 is much more sophisticated than Level B.4.1.1; the latter should probably occur much earlier in a program of instruction for all children.

TABLE C.1 **Number (*continued*)**

WI Model Academic Standards By the end of Grade 4, students will	School District Levels
	Level B.4.1.6 ■ Represent and explain whole numbers and decimals (to hundredths) with: Physical and pictorial models Place value concepts and notation Verbal explanations Symbolic renaming ***Level B.4.1.7*** ■ Identify and represent equivalent fractions for halves, fourths, eighths, tenths, and sixteenths.
B.4.2 Determine the number of things in a set by: ■ Grouping and counting (e.g., by 3s, 5s, and 100s) ■ Combining and arranging (e.g., all possible coin combinations amounting to 30 cents) ■ Estimation, including rounding	***Level B.4.2.1*** ■ Count with meaning sets with up to 12 objects. ***Level B.4.2.2*** ■ Determine the number of things in a set with up to 100 objects by: Counting Grouping by 2s, 5s, and 10s ***Level B.4.2.3*** ■ Determine the number of things in a set with up to 1000 objects by: Counting Grouping by 2s, 5s, 10s and 100s Estimating Rounding to the nearest 10 or 100 ***Level B.4.2.4*** ■ Determine the number of things in a set with up to 10,000 objects by: Counting Grouping Estimating Rounding to the nearest 10 or 100 ***Level B.4.2.5*** ■ Determine the number of things in a set by: Counting Grouping, combining, and arranging Estimating Rounding

(continued)

TABLE C.1 **Number (*continued*)**

WI Model Academic Standards **By the end of Grade 4, students will:**	School District Levels
B.4.3 Read, write, and order whole numbers, simple fractions (e.g., halves, fourths, tenths, unit fractions) and commonly used decimals (monetary units). B.4.4 Identify and represent equivalent fractions for halves, fourths, eighths, tenths, and sixteenths.	***Level B.4.3.1*** ■ Read, write, and order whole numbers up to 12. ***Level B.4.3.2*** ■ Read, write, and order whole numbers up to 100. ***Level B.4.3.3*** ■ Read, write, and order whole numbers up to 1,000. ***Level B.4.3.4*** ■ Read, write, and order: Whole numbers up to 10,000 Decimals with one place Fractions with denominators of 2, 3, or 4 ***Level B.4.3.5*** ■ Read, write, and order: Whole numbers Simple fractions (e.g., halves, fourths, tenths, and unit fractions) Commonly used decimals (monetary units)
B.4.5 In problem-solving situations involving whole numbers, select and efficiently use appropriate computational procedures such as: ■ Recalling the basic facts of addition, subtraction, multiplication and division ■ Using mental math (e.g., 37 + 25, 40 × 7) ■ Estimation ■ Selecting and applying algorithms for addition, subtraction, multiplication, and division ■ Using a calculator	***Level B.4.5.1*** ■ Determine solutions for simple addition and subtraction problems involving whole numbers up to 12 by using manipulative materials or various counting strategies. ***Level B.4.5.2*** ■ Recall basic addition facts. ***Level B.4.5.3*** ■ Recall basic addition and subtraction facts. ***Level B.4.5.4*** ■ In problem situations involving whole numbers, select and efficiently use the following appropriate solution procedures: Modeling with objects, tally marks, pictures Counting forward and back from a number different from 1 Using the basic facts for addition and subtraction Estimating Adding and subtracting two-digit numbers without regrouping Using mental math ***Level B.4.5.5*** ■ In problem situations involving whole numbers, select and efficiently use the following appropriate solution procedures: Counting forward and back from a number different from 1 Using the basic facts for addition and subtraction Estimating Adding and subtracting two- and three-digit numbers Using mental math Using a calculator

Table C.1 **Number (*continued*)**

WI Model Academic Standards **By the end of Grade 4, students will:**	**School District Levels**
	Level B.4.5.6 ■ Recall basic multiplication facts. ***Level B.4.5.7*** ■ Recall basic multiplication and division facts. ***Level B.4.5.8*** ■ In problem situations involving whole numbers, select and efficiently use the following appropriate solution procedures: Modeling Using the basic facts for multiplication Estimating Using addition and subtraction algorithms Using mental math, including counting strategies Using a calculator ***Level B.4.5.9*** ■ In problem situations involving whole numbers, select and efficiently use the following appropriate solution procedures: Using the basic facts of addition, subtraction, multiplication, and division Using mental math, including counting strategies Estimating Selecting and applying algorithms for addition, subtraction, multiplication, and division Using a calculator
B.4.6 Add and subtract fractions with like denominators. B.4.7 In problem-solving situations involving money, add and subtract decimals.	***Level B.4.6.1*** ■ Add and subtract fractions with like denominators in problem solving situations. ***Level B.4.6.2*** ■ In problem-solving situations involving money, add and subtract decimals.

(continued)

TABLE C.1 **Number (*continued*)**

WI Model Academic Standards **By the end of Grade 8, students will**	**School District Levels**
B.8.1 Read, represent, and interpret rational numbers (whole numbers, integers, decimals, fractions, and percents) with verbal descriptions, geometric models, and mathematical notation (e.g., expanded, scientific, exponential). B.8.4 Express order relationships among rational numbers using appropriate symbols (>, <, ≥, ≤, ≠). B.8.5 Apply proportional thinking in a variety of problem situations that include but are not limited to: ■ Ratios and proportions (e.g., rates, scale drawings, similarity) ■ Percents, including those greater than 100 and less than 1 (e.g., discounts, rate of increase or decrease, sales tax)	***Level 5*** ■ Represent whole numbers and decimals in expanded notation. ■ Represent directed quantities with positive and negative numbers. ***Level 6*** ■ Give verbal and symbolic explanations for the meaning of percent. ■ Explain the various interpretations for fractions (e.g., part[s] of a whole object or set, indicated division, ratio comparison). ■ Count, read, and write fractions and mixed numbers and decimals. ■ Express order relationships among rational numbers using appropriate symbols (>, <, ≥, ≤, ≠). ■ Use exponential notation to represent repeated use of a given factor. ***Level 7*** ■ Solve simple problem situations involving whole-number percents between 0 and 100. ***Level 8*** ■ Represent any number in scientific notation. ■ Apply proportional thinking in a variety of problem situations that include but are not limited to: Ratios and proportions (e.g., rates, scale drawings, similarity) Percents, including those greater than 100 and less than 1
B.8.2 Perform and explain operations on rational numbers (add, subtract, multiply, divide, raise to a power, extract a root, take opposites and reciprocals, determine absolute value). B.8.3 Generate and explain equivalencies among fractions, decimals, and percents.	***Level 5*** ■ Add, subtract, and multiply decimals and fractions, excluding mixed numbers. ■ Add integers. ■ Generate and explain equivalencies for fractions, excluding mixed numbers. ***Level 6*** ■ Demonstrate proficiency of all four operations on whole numbers, decimals, and fractions, including mixed numbers. ■ Multiply integers. ■ Determine opposites and reciprocals of whole numbers and fractions. ■ Generate and explain equivalencies between decimals and fractions, including mixed numbers. ***Level 7*** ■ Add, subtract, multiply, and divide integers. ■ Generate and explain equivalencies between decimals and percents. ■ Raise any rational number to a given (whole number) power. ■ Determine absolute value of any rational number. ***Level 8*** ■ Add, subtract, multiply, and divide positive and negative rational numbers. ■ Generate and explain equivalencies among percents, decimals, and fractions, including mixed numbers. ■ Determine the square root of any positive whole number.

Table C.1 **Number (*continued*)**

WI Model Academic Standards **By the end of Grade 8, students will:**	School District Levels
B.8.6 Model and solve problems involving number-theory concepts such as ■ Prime and composite numbers ■ Divisibility and remainders ■ Greatest common factors ■ Least common multiples	***Level 5*** ■ Determine factors and divisors of whole numbers. ■ List multiples of a whole number. ***Level 6*** ■ Determine the least common multiple and greatest common factor of two given numbers. ***Level 7*** ■ Give the prime factorization of any number. ■ Determine the least common multiple and greatest common factor of three given numbers. ■ Express the prime factorization of any number using exponential notation.
B.8.7 In problem-solving situations, select and use appropriate computational procedures with rational numbers such as ■ Calculating mentally ■ Estimating ■ Creating, using, and explaining algorithms ■ Using technology (e.g., scientific calculators, spreadsheets)	***Level 5*** ■ In problem-solving situations suitable for this grade level, select and use the following appropriate computational procedures with rational numbers: Modeling Calculating mentally Estimating Creating, using, and explaining algorithms Using technology (e.g., calculators, computers, database) ***Level 6*** ■ In problem-solving situations suitable for this grade level, select and use the following appropriate computational procedures with rational numbers: Modeling Calculating mentally Estimating Creating, using, and explaining algorithms Using technology (e.g., calculators, computers, database) ***Level 7*** ■ In problem-solving situations suitable for this grade level, select and use appropriate computational procedures with rational numbers such as: Calculating mentally Estimating Creating, using, and explaining algorithms Using technology (e.g., scientific calculators, spreadsheets) ***Level 8*** ■ In problem-solving situations, select and use appropriate computational procedures with rational numbers such as: Calculating mentally Estimating Creating, using, and explaining algorithms Using technology (e.g., scientific calculators, spreadsheets)

(continued)

TABLE C.1 **Number (*continued*)**

WI Model Academic Standards **By the end of Grade 12, students will**	School District Levels
B.12.1 Use complex counting procedures such as union and intersection of sets and arrangements (permutations and combinations) to solve problems.	***Level 10*** ■ Use complex counting procedures such as union and intersection of sets and arrangements (permutations and combinations) to solve problems.
B.12.2 Compare real numbers using ■ Order relations ($>$, $<$) and transitivity ■ Ordinal scales including logarithmic (e.g., Richter, pH ratings) ■ Arithmetic differences. ■ Ratios, proportions, percents, rates of change.	***Level 9*** ■ Compare real numbers using: order relations ($>$, $<$) and transitivity arithmetic differences ***Level 10*** ■ Compare real numbers using: ordinal scales including logarithmic (e.g., Richter, pH ratings) ■ Ratios, proportions, percents, rates of change.
B.12.3 Perform and explain operations on real numbers (add, subtract, multiply, divide, raise to a power, extract a root, take opposites and reciprocals, determine absolute value).	***Level 9*** ■ Perform and explain operations on real numbers (add, subtract, multiply, divide, raise to a power, extract a root, take opposites and reciprocals, determine absolute value).
B.12.4 In problem-solving situations involving the applications of different number systems (natural, integers, rational, real), select and use appropriate ■ Computational procedures ■ Properties (e.g., commutativity, associativity, inverses) ■ Modes of representation (e.g., rationals as repeating decimals, indicated roots as fractional exponents)	***Level 10*** ■ Perform and explain operations involving fractional and negative exponents. ■ Explain and use rules for exponents when multiplying or dividing numbers with like bases. ■ Generate and explain equivalencies between rational numbers and repeating decimals. ■ Routinely assess the acceptable limits of error when: Evaluating strategies Testing the reasonableness of results Using technology to carry out computations
B.12.6 Routinely assess the acceptable limits of error when ■ Evaluating strategies ■ Testing the reasonableness of results ■ Using technology to carry out computations	
B.12.5 Create and critically evaluate numerical arguments presented in a variety of classroom and real-world situations (e.g., political, economic, scientific, social).	***Level 10*** ■ Create and critically evaluate numerical arguments presented in a variety of classroom and real-world situations (e.g., political, economic, scientific, social).

TABLE C.2 **Geometry**

WI Model Academic Standards **By the end of Grade 4, students will:**	School District Levels
C.4.1 Describe two- and three-dimensional figures (e.g., circles, polygons, trapezoids, prisms, spheres) by: ■ Naming them ■ Comparing, sorting, and classifying them ■ Drawing and constructing physical models to specifications ■ Identifying their properties (e.g., number of sides or faces, two- or three-dimensionality, equal sides, number of right angles) ■ Predicting the results of combining or subdividing two-dimensional figures ■ Explaining how these figures are related to objects in the environment	***Level C.4.1.1*** ■ Identify and name simple two- and three-dimensional figures in the real world. ***Level C.4.1.2*** ■ Describe simple figures by identifying obvious properties (e.g., number of sides or corners; do they roll, stack, slide?). ***Level C.4.1.3*** ■ Name, compare, sort, and classify two- and three-dimensional figures. ***Level C.4.1.4*** ■ Draw and construct physical models of figures to given specifications. ■ Give real-world examples of two- and three-dimensional figures. ***Level C.4.1.5*** ■ Give the correct mathematical name and describe the properties of relatively complex figures (e.g., trapezoid, cylinder, prism). ■ Predict the results of combining or subdividing two-dimensional figures.
C.4.2 Use physical materials and motion geometry (such as slides, flips, and turns) to identify properties and relationships, including but not limited to: ■ symmetry ■ congruence ■ similarity C.4.3 Identify and use relationships among figures, including but not limited to: ■ location (e.g., between, adjacent to, interior of) ■ position (e.g., parallel, perpendicular) ■ intersection (of two-dimensional figures)	***Level C.4.2.1*** ■ Describe the location of a figure with respect to one or more other figures (e.g., adjacent to, between, interior of). ***Level C.4.2.2*** ■ Identify congruent figures. ***Level C.4.2.3*** ■ Describe the intersection of 2 two-dimensional figures. ***Level C.4.2.4*** ■ Use physical materials and drawings to carry out slides, flips, and turns of simple two-dimensional figures. ■ Identify figures that have line symmetry. ***Level C.4.2.5*** ■ Identify parallel and perpendicular: Lines, segments Sides of simple two-dimensional figures Faces of simple three-dimensional figures ***Level C.4.2.6*** ■ Use physical materials to identify similar figures.

(continued)

Special Note: Particular numerical levels within a given cell are not necessarily of the same degree of difficulty or sophistication as a similarly numbered level in another cell. For example, Level C.4.4.1 is much more sophisticated than Level C.4.1.1; the latter should probably occur much earlier in a program of instruction for all children.

WI Model Academic Standards	School District Levels
By the end of Grade 4, students will:	
C.4.4 Use simple two-dimensional coordinate systems to find locations on maps and to represent points and simple figures	***Level C.4.4.1*** ■ State the coordinates of locations or objects on simple maps and grids. ***Level C.4.4.2*** ■ When given their coordinates, place or locate points on a two-dimensional grid (including a map).
By the end of Grade 8, students will:	
C.8.1 Describe special and complex two- and three-dimensional figures (e.g., rhombus, polyhedron, cylinder) and their component parts (e.g., base, altitude, and slant height) by ■ Naming, defining, and giving examples ■ Comparing, sorting, and classifying them ■ Identifying and contrasting their properties (e.g., symmetrical, isosceles, regular) ■ Drawing and constructing physical models to specifications ■ Explaining how these figures are related to objects in the environment C.8.2 Identify and use relationships among the component parts of special and complex two- and three-dimensional figures (e.g., parallel sides, congruent faces). C.8.3 Identify three-dimensional shapes from two-dimensional perspectives and draw two-dimensional sketches of three-dimensional objects, preserving their significant features. (Hereafter, the basic essence of these three standards is denoted by "Stem" in the school district levels.)	***Level C.8.1.1*** ■ "Stem" with respect to the entire set of triangles and their component parts. ***Level C.8.1.2*** ■ "Stem" with respect to the entire set of quadrilaterals and their component parts. ***Level C.8.1.3*** ■ "Stem" with respect to the entire set of rectangular solids (including cubes) and their component parts. ***Level C.8.1.4*** ■ "Stem" with respect to the entire set of polygons and their component parts. ***Level C.8.1.5*** ■ "Stem" with respect to the entire set of polyhedra, including prisms and pyramids, and their component parts. ***Level C.8.1.6*** ■ "Stem" with respect to the entire set of circle-related figures (circles, spheres, cones, cylinders) and their component parts.

TABLE C.2 **Geometry (*continued*)**

WI Model Academic Standards	School District Levels
By the end of Grade 8, students will:	
C.8.4 Perform transformations on two-dimensional figures and describe and analyze the effects of the transformations on the figures.	***Level C.8.4.1*** ■ Identify figures that have turn symmetry. ***Level C.8.4.2*** ■ Describe the effects of slides, flips, and turns of two-dimensional figures. ***Level C.8.4.3*** ■ Describe the effect of two or more successive transformations of a two-dimensional figure. ***Level C.8.4.4*** ■ Determine coordinates of slide and flip images of figures located on a two-dimensional grid.
C.8.5 Locate objects using the rectangular coordinate system.	***Level C.8.5.1*** ■ Locate objects using the rectangular coordinate system and positive coordinates (i.e., in the first quadrant). ***Level C.8.5.2*** ■ Locate objects using the rectangular coordinate system (i.e., in all four quadrants).
By the end of Grade 12, students will: C.12.1 Identify, describe and analyze properties of figures, relationships among figures, and relationships among their parts by ■ Constructing physical models ■ Drawing precisely with paper and pencil, hand calculators, and computer software ■ Using appropriate transformations (e.g., translations, rotations, reflections, enlargements) ■ Using reason and logic (Hereafter to be denoted by "Stem" in the school district levels.) C.12.2 Use geometric models to solve mathematical and real-world problems.	***At all levels*** ■ Use geometric models to solve mathematical and real-world problems. ***Level C.12.1.1*** ■ "Stem" with respect to: Perpendicular lines Parallel lines, transversals, and resulting pairs of angles ***Level C.12.1.2*** ■ "Stem" with respect to congruent triangles, including the basic triangle congruence theorems. ***Level C.12.1.3*** ■ "Stem" with respect to all quadrilaterals with particular attention to parallelograms. ***Level C.12.1.4*** ■ "Stem" with respect to similar triangles. ***Level C.12.1.5*** ■ "Stem" with respect to circles. ***Level C.12.1.6*** ■ "Stem" with respect to three-dimensional figures.

(continued)

TABLE C.2 **Geometry (*continued*)**

WI Model Academic Standards **By the end of Grade 12, students will:**	School District Levels
C.12.3 Present convincing arguments by means of demonstration, informal proof, counterexamples, or any other logical means to show the truth of ■ Statements (e.g., these two triangles are not congruent). ■ Generalizations (e.g., the Pythagorean theorem holds for all right triangles).	***Level C.12.3.1*** ■ Present valid arguments using elements of symbolic logic, including concepts and techniques such as: Truth tables Negation, converse, inverse, contrapositive Syllogism, chain rule ***At all levels*** ■ Present convincing arguments by means of demonstration, informal proof, counter-examples, or any other logical means to show the truth of: Statements (e.g., these two triangles are not congruent) Generalizations (e.g., the Pythagorean theorem holds for all right triangles)
C.12.4 Use the two-dimensional rectangular coordinate system and algebraic procedures to describe and characterize geometric properties and relationships such as slope, intercepts, parallelism, and perpendicularity. (Hereafter to be denoted by "Stem" in the School District levels.)	***Level C.12.4.1*** ■ "Stem" with respect to lines, including: Given a line, produce its equation, and conversely Two-point, point-slope, slope-intercept forms ***Level C.12.4.2*** ■ "Stem" with respect to parallel and perpendicular lines. ***Level C.12.4.3*** ■ "Stem" with respect to distance and midpoints. ***Level C.12.4.4*** ■ "Stem" with respect to analysis of special figures such as parallelograms, squares, rhombuses.
C.12.5 Identify and demonstrate an understanding of the three ratios used in right-triangle trigonometry (sine, cosine, tangent).	***Level C.12.5.1*** ■ Identify and demonstrate an understanding of the three ratios used in right-triangle trigonometry (sine, cosine, tangent).

Table C.3 **Measurement**

WI Model Academic Standards

By the end of Grade 4, students will:

D.4.1 Recognize and describe measurable attributes, such as length, liquid capacity, time, weight (mass), temperature, volume, monetary value, and angle size, and identify the appropriate units to measure them.

D.4.2 Demonstrate understanding of basic facts, principles, and techniques of measurement, including

- Appropriate use of arbitrary and standard units (metric and U.S.)
- Appropriate use and conversion of units within a system (such as yards, feet, and inches; kilograms and grams; gallons, quarts, pints, and cups)
- Judging the reasonableness of an obtained measurement as it relates to prior experience and familiar benchmarks

D.4.3 Read and interpret measuring instruments (e.g., rulers, clocks, thermometers).

D.4.4 Determine measurements directly by using standard tools to these suggested degrees of accuracy:

- Length, to the nearest half-inch or nearest centimeter
- Weight (mass), to the nearest ounce or nearest five grams
- Temperature, to the nearest five degrees
- Time, to the nearest minute
- Monetary value, to dollars and cents
- Liquid capacity, to the nearest fluid ounce

School District Levels

Level D.4.1.1

- Compare and measure length using arbitrary units.

Level D.4.1.2

- Explain and use appropriate standard units to measure lengths (inches, centimeters).

Level D.4.1.3

- Compare and measure weight (mass) and capacity using arbitrary units.

Level D.4.1.4

- Explain and use appropriate standard units to measure length (inches, feet, yards, centimeters, meters).

Level D.4.1.5

- Appropriately use a ruler to measure length to the nearest 1/4 inch or nearest centimeter.

Level D.4.1.6

- Explain and use appropriate standard units to measure weight (mass) (ounce, pound).

Level D.4.1.7

- Explain and use appropriate standard units to measure weight (mass) (gram, kilogram).

Level D.4.1.8

- Tell time to the nearest five minutes.
- Give monetary value for commonly used coins and bills and know simple equivalences of coins.

Level D.4.1.9

- Explain and use appropriate standard units to measure liquid capacity (cups, pints, gallons).
- Read a thermometer to the nearest 5° C or F.

Level D.4.1.10

- Give the monetary value for any collection of coins or bills up to a value of $10.00.

(continued)

TABLE C.3 **Measurement (*continued*)**

WI Model Academic Standards **By the end of Grade 4, students will:**	School District Levels
D.4.3 Read and interpret measuring instruments (e.g., rulers, clocks, thermometers).	***Level D.4.1.11*** ■ Estimate area of simple polygons by superimposing a square grid. ■ Compare and measure angle size using arbitrary units.
D.4.4 Determine measurements directly by using standard tools to these suggested degrees of accuracy: ■ Length, to the nearest half-inch or nearest centimeter ■ Weight (mass), to the nearest ounce or nearest five grams ■ Temperature, to the nearest five degrees ■ Time, to the nearest minute ■ Monetary value, to dollars and cents ■ Liquid capacity, to the nearest fluid ounce	***Level D.4.1.12*** ■ Measure weight (mass) to the nearest ounce or nearest gram. ■ Measure liquid capacity of a container to the nearest fluid ounce. ■ Use standard and digital clock to tell time to the nearest minute. ***Level D.4.1.13*** ■ Carry out conversion of units within a given system (e.g., pounds to ounces, pints to gallons, centimeters to meters). ***Level D.4.1.14*** ■ Determine perimeter of simple polygons. ■ Determine area of squares and rectangles. ***Level D.4.1.15*** ■ Estimate length, weight, capacity, and duration of events.
D.4.5 Determine measurements by using basic relationships (such as perimeter and area) and approximate measurements by using estimation techniques.	

WI Model Academic Standards **By the end of Grade 8, students will:**	**School District Levels**
D.8.1 Identify and describe attributes in situations where they are not directly or easily measurable (e.g., distance, area of an irregular figure, likelihood of occurrence). D.8.2 Demonstrate understanding of basic measurement facts, principles, and techniques, including the following: ■ Approximate comparisons between metric and U.S. units (e.g., a liter and a quart are about the same; a kilometer is about six-tenths of a mile) ■ Knowledge that direct measurement produces approximate, not exact, measures ■ The use of smaller units to produce more precise measures D.8.3 Determine measurements directly using standard units (metric and U.S.) with these suggested degrees of accuracy: ■ Lengths, to the nearest millimeter or 1/16 inch ■ Weight (mass), to the nearest 0.1 g or 0.5 ounce ■ Liquid capacity, to the nearest milliliter ■ Angles, to the nearest degree ■ Temperature, to the nearest C° or F° ■ Elapsed time, to the nearest second D.8.4 Determine measurements indirectly using estimation: ■ Conversion of units within a system (e.g., quarts to cups, millimeters to centimeters) ■ Ratio and proportion (e.g., similarity, scale drawings) ■ Geometric formulas to derive lengths, areas, volumes of common figures (e.g., perimeter, circumference, surface area) ■ The Pythagorean relationship ■ Geometric relationships and properties for angle size (e.g., parallel lines and transversals; sum of angles of a triangle; vertical angles)	***Level D.8.1.1*** ■ Use a ruler to measure length to the nearest 1/16 inch or nearest millimeter. ■ Measure weight (mass) to the nearest 0.5 ounce or nearest 0.1 gram. ■ Measure liquid capacity of a container to the nearest milliliter. ■ Read a thermometer to the nearest degree Celsius or Fahrenheit. ■ Use a stop watch to measure elapsed time to the nearest second. ■ Use a protractor to measure angles to the nearest degree. ***Level D.8.1.2*** ■ Explain and use appropriate standard units to measure volume (cubic inch, cubic centimeter). ***Level D.8.1.3*** ■ Carry out exact conversion of units within a given system (e.g., pounds to ounces, pints to gallons, centimeters to meters) as well as approximate conversions between systems. ***Level D.8.1.4*** ■ Use geometric formulas to determine perimeter and area of squares, rectangles, parallelograms, and triangles. ***Level D.8.1.5*** ■ Use geometric formulas to determine the volume and surface area of cubes and rectangular solids. ***Level D.8.1.6*** ■ Use geometric formulas and relationships to find linear, area, and angle measures of polygons. ***Level D.8.1.7*** ■ Determine measurements indirectly using geometric relationships and properties for angle size (e.g., parallel lines and transversals; sum of angles of a triangle; vertical angles). ***Level D.8.1.8*** ■ Use ratio and proportion (e.g., similarity) to determine linear measures of geometric figures. ***Level D.8.1.9*** ■ Use geometric formulas and relationships to find linear, arc length, surface area, volume and angle measures of circles and spheres. ***Level D.8.1.10*** ■ Use geometric formulas and relationships to find linear, surface area, volume, and angle measures of prisms, pyramids, cones, and cylinders. ***Level D.8.1.11*** ■ Use the Pythagorean relationship to find linear measures of right triangles.

(continued)

WI Model Academic Standards **By the end of Grade 12, students will:**	School District Levels
D.12.1 Identify, describe, and use derived attributes (e.g., density, speed, acceleration, pressure) to represent and solve problem situations. D.12.2 Select and use tools with appropriate degree of precision to determine measurements directly within specified degrees of accuracy and error (tolerance). D.12.3 Determine measurements indirectly, using ■ Estimation ■ Proportional reasoning, including those involving squaring and cubing (e.g., reasoning that areas of circles are proportional to the squares of their radii) ■ Techniques of algebra, geometry, and right-triangle trigonometry ■ Formulas in applications (e.g., for compound interest, distance formula) ■ Geometric formulas to derive lengths, areas, or volumes of shapes and objects (e.g., cones, parallelograms, cylinders, and pyramids) ■ Geometric relationships and properties of circles and polygons (e.g., size of central angles, area of a sector of a circle) ■ Conversion constants to relate measures in one system to another (e.g., meters to feet, dollars to deutschmarks)	***At All Levels*** ■ Select and use tools with an appropriate degree of precision to determine measurements directly within specified degrees of accuracy and error (tolerance). ■ Determine measurements indirectly using estimation. ***Level D.12.1.1*** ■ Use conversion constants to relate measures in one system to another (e.g., meters to feet, dollars to deutschmarks). ***Level D.12.1.2*** ■ Identify, describe, and use derived attributes (e.g., density, speed, acceleration, pressure) to represent and solve problem situations. ***Level D.12.1.3*** ■ Use right-triangle trigonometry to find linear and angle measures in right triangles. ***Level 12.1.3.4*** ■ Determine measurements indirectly using techniques of algebra and geometry. ***Level 12.1.3.5*** ■ Determine measurements indirectly, using: proportional reasoning, including those involving squaring and cubing (e.g., reasoning that areas of circles are proportional to the squares of their radii) geometric relationships and properties of circles and polygons (e.g., size of central angles, area of a sector of a circle)

TABLE C.4 **Statistics and Probability**

WI Model Academic Standards **By the end of Grade 4, students will:**	**School District Levels**
E.4.1 Work with data in the context of real-world situations by ■ Formulating questions that lead to data collection and analysis ■ Determining what data to collect and when and how to collect them ■ Collecting, organizing, and displaying data ■ Drawing reasonable conclusions based on data	***At All Levels*** ■ Formulate questions that lead to data collection and analysis. ■ In problem-solving situations, read, extract, and use information presented in graphs, tables, or charts. ■ Draw reasonable conclusions based on data. ***Level E.4.1.1*** ■ Collect and organize data from common classroom events. ■ Organize a set of data by tallying and counting.
E.4.2 Describe a set of data using ■ High and low values and range ■ Most frequent value (mode) ■ Middle value of a set of ordered data (median)	***Level E.4.1.2*** ■ Create simple bar and picture graphs using real classroom data. ***Level E.4.1.3*** ■ Plan, collect, and organize data from surveys and experiments.
E.4.3 In problem-solving situations, read, extract, and use information presented in graphs, tables, or charts.	***Level E.4.1.4*** ■ Create bar graphs, picture graphs, and tables to represent data. ■ Read and interpret information from simple bar graphs, picture graphs, and tables. ***Level E.4.1.5*** ■ Create simple line graphs using real classroom data. ■ Read and interpret information from bar, line, picture, and circle graphs. ***Level E.4.1.6*** ■ Describe a set of data using: High and low values, and range Most frequent value (mode) ***Level E.4.1.7*** ■ Determine the median of a set of ordered data.
E.4.4 Determine if future events are more, less, equally likely, impossible, or certain to occur.	***Level E.4.4.1*** ■ Describe an imagined event as being impossible or certain to occur.
E.4.5 Predict outcomes of future events and test predictions using data from a variety of sources.	***Level E.4.4.2*** ■ Compare likelihood of simple events. ■ Use available data to predict the outcome of a future event. ***Level E.4.4.3*** ■ Determine if future events are more, less, equally likely, impossible, or certain to occur. ***Level E.4.4.4*** ■ Predict outcomes of future events and test predictions using data from a variety of sources.

(continued)

Special Note: Particular numerical levels within a given cell are not necessarily of the same degree of difficulty or sophistication as a similarly numbered level in another cell. For example, Level E.8.7.1 is a bit more sophisticated than Level E.8.1.1.; as a result, the latter should probably occur earlier in a program of instruction for all students.

TABLE C.4 **Statistics and Probability** ***(continued)***

WI Model Academic Standards **By the end of Grade 8, students will:**	**School District Levels**
E.8.1 Work with data in the context of real-world situations by ■ Formulating questions that lead to data collection and analysis ■ Designing and conducting a statistical investigation ■ Using technology to generate displays, summary statistics, and presentations	***At All Levels*** ■ Work with data from real-world situations suitable to this grade level by: Formulating questions that lead to data collection and analysis Designing and conducting experiments, investigations, and surveys Generating appropriate displays, summary statistics, and presentations ■ Use the results of data analysis to: Make predictions. Develop convincing arguments. Draw conclusions.
E.8.2 Organize and display data from statistical investigations using ■ Appropriate tables, graphs, or charts (e.g., circle, bar, or line for multiple sets of data) ■ Appropriate plots (e.g., line, stem-and-leaf, box, scatter)	***Level E.8.1.1*** ■ Determine the arithmetic mean of a set of data. ***Level E.8.1.2*** ■ Organize and display data from statistical investigations using line plots.
E.8.3 Extract, interpret, and analyze information from organized and displayed data by using ■ Frequency and distribution, including mode and range ■ Central tendencies of data (mean and median) ■ Indicators of dispersion (e.g., outliers)	***Level E.8.1.3*** ■ Extract, interpret, and analyze information from organized and displayed data by using: Frequency and distribution, including mode and range Central tendencies of data (mean and median) Indicators of dispersion (e.g., outliers)
E.8.4 Use the results of data analysis to ■ Make predictions. ■ Develop convincing arguments. ■ Draw conclusions.	***Level E.8.1.4*** ■ Organize and display data from statistical investigations using circle graphs. ***Level E.8.1.5*** ■ Organize and display data from statistical investigations using stem-and-leaf plots, box plots, and scatter plots.
E.8.5 Compare several sets of data to generate, test, and as the data indicate, confirm or deny hypotheses.	***Level E.8.1.6*** ■ Compare several sets of data to generate, test, and as the data indicate, confirm or deny hypotheses.
E.8.6 Evaluate presentations and statistical analyses from a variety of sources for ■ Credibility of the source ■ Techniques of collection, organization, and presentation of data ■ Missing or incorrect data ■ Inferences ■ Possible sources of bias	***Level E.8.6.1*** ■ Evaluate simple experiments for quality of data collection. ***Level E.8.6.1*** ■ Evaluate statistical reports for missing or incorrect data. ***Level E.8.6.3*** ■ Evaluate statistical reports for techniques of collection, organization, and presentation of data. ***Level E.8.6.4*** ■ Evaluate statistical reports for correctness of inferences and for possible sources of bias.

TABLE C.4 **Statistics and Probability (*continued*)**

WI Model Academic Standards **By the end of Grade 8, students will:**	School District Levels
E.8.7 Determine the likelihood of occurrence of simple events by ■ using a variety of strategies to identify possible outcomes (e.g., lists, tables, tree diagrams) ■ conducting an experiment ■ designing and conducting simulations ■ applying theoretical notions of probability (e.g., that four equally likely events have a 25% chance of happening)	***Level E.8.7.1*** ■ Create a list to identify all possible outcomes of an event. ■ Generate a table to identify all possible outcomes of an event. ***Level E.8.7.2*** ■ Determine the probability of simple events such as getting heads when flipping a coin or having a 5 turn up with one toss of a die. ■ Conduct a simple experiment to determine the probability of an event that cannot be predicted theoretically. ***Level E.8.7.3*** ■ Create tree diagrams to identify all possible outcomes of an event. ***Level E.8.7.4*** ■ Determine the likelihood of occurrence of simple events by applying theoretical notions of probability (e.g., that four equally likely events have a 25 percent chance of happening). ***Level E.8.7.5*** ■ Determine the likelihood of occurrence of simple events by designing and conducting simulations.

(continued)

WI Model Academic Standards By the end of Grade 12, students will:	School District Levels
E.12.1 Work with data in the context of real-world situations by ■ Formulating hypotheses that lead to collection and analysis of one- and two-variable data ■ Designing a data collection plan that considers random sampling, control groups, the role of assumptions, etc. ■ Conducting an investigation based on that plan ■ Using technology to generate displays, summary statistics, and presentations	***Level E.12.1.1*** ■ Work with data in the context of real-world situations by: Formulating hypotheses that lead to collection and analysis of one- and two-variable data Designing a data collection plan that considers random sampling, control groups, and the role of assumptions Conducting an investigation based on that plan Using technology to generate displays, summary statistics, and presentations
E.12.2 Organize and display data from statistical investigations using ■ Frequency distributions ■ Percentiles, quartiles, deciles ■ Lines of best fit (estimated regression line) ■ Matrices	***Level E.12.1.2*** ■ Organize and display data from statistical investigations using: Frequency distributions Percentiles, quartiles, deciles ***Level E.12.1.1*** ■ Organize and display data from statistical investigations using lines of best fit (estimated regression line). ***Level E.12.1.4*** ■ Organize and display data from statistical investigations using matrices.
E.12.3 Interpret and analyze information from organized and displayed data when given ■ Measures of dispersion, including standard deviation and variance ■ Measures of reliability ■ Measures of correlation	***Level E.12.1.5*** ■ Interpret and analyze information from organized and displayed data when given: Measures of dispersion, including standard deviation and variance Measures of reliability Measures of correlation
E.12.4 Analyze, evaluate, and critique the methods and conclusions of statistical experiments reported in journals, magazines, news media, advertising, etc.	***Level E.12.4.1*** ■ Analyze, evaluate, and critique the methods and conclusions of statistical experiments reported in journals, magazines, news media, advertising, etc.
E.12.5 Determine the likelihood of occurrence of complex events by ■ Using a variety of strategies (e.g., combinations) to identify possible outcomes ■ Conducting an experiment ■ Designing and conducting simulations ■ Applying theoretical probability	***Level E.12.5.1*** ■ Determine the likelihood of occurrence of complex events by: Conducting an experiment Designing and conducting simulations ***Level E.12.5.2*** ■ Determine the likelihood of occurrence of complex events by: Using a variety of strategies (e.g., combinations) to identify possible outcomes Applying theoretical probability

TABLE C.5 **Algebra**

WI Model Academic Standards	School District Levels
By the end of Grade 4, students will:	
F.4.1 Use letters, boxes, or other symbols to stand for any number, measured quantity, or object in simple situations (e.g., $N + 0 = N$ is true for any number)	***Level F.4.1.1*** ■ Solve simple open sentences involving addition, including missing addends. ■ Represent problem situations involving addition with simple open sentences (one-step equations).
F.4.2 Use the vocabulary, symbols, and notation of algebra accurately (e.g., correct use of the symbol =; effective use of the associative property of multiplication)	***Level F.4.1.2*** ■ Solve simple open sentences involving subtraction (e.g., $7 - [] = 3$, $[] - 5 = 4$). ■ Represent problem situations involving addition and subtraction with simple open sentences (one-step equations).
F.4.5 Use simple equations and inequalities in a variety of ways, including ■ using them to represent problem situations ■ solving them by different methods (e.g., use of manipulatives, guess and check strategies, recall of number facts) ■ recording and describing solution strategies	***Level F.4.1.3*** ■ Solve simple open sentences involving multiplication, including missing factors. ■ Represent problem situations involving multiplication with simple open sentences. ***Level F.4.1.4*** ■ Solve simple open sentences involving division. ■ Represent problem situations involving multiplication and division with simple open sentences (i.e., one-step equations). ***Level F.4.1.5*** ■ Demonstrate a basic understanding of equality and inequality.
F.4.3 Work with linear patterns and relationships in a variety of ways, including ■ Recognizing and extending number patterns ■ Describing them verbally ■ Representing them with pictures, tables, charts, and graphs ■ Recognizing that different models can represent the same pattern or relationship ■ Using them to describe real-world phenomena	***Level F.4.3.1*** ■ Identify, describe, extend, and create simple AB patterns using concrete materials, sounds, and pictures (e.g., red, blue, red, blue . . .). ***Level F.4.3.2*** ■ Identify, describe, extend, and create a variety of patterns using concrete materials, sounds, pictures, etc. ***Level F.4.3.3*** ■ Identify, describe, extend, and create a variety of number patterns. ***Level F.4.3.4*** ■ Represent patterns using tables, graphs, verbal rules, or equations. ***Level F.4.3.5*** ■ Recognize that different models can represent the same pattern or relationship. ■ Use patterns to describe real-world phenomena.

(continued)

Special Note: Particular numerical levels within a given cell are not necessarily of the same degree of difficulty or sophistication as a similarly numbered level in another cell. For example, Level F.4.4.1 is much more sophisticated than Level F.4.3.1; the latter should probably occur much earlier in a program of instruction for all children.

TABLE C.5 **Algebra (*continued*)**

WI Model Academic Standards	School District Levels
By the end of Grade 4, students will:	
F.4.4 Recognize variability in simple functional relationships by describing how a change in one quantity can produce a change in another (e.g., number of bicycles and the total number of wheels)	***Level F.4.4.1*** ■ Recognize variability in simple functional relationships by describing how a change in one quantity can produce a change in another (e.g., number of bicycles and the total number of wheels). ***Level F.4.4.2*** ■ Recognize and use generalized properties and relationships of arithmetic (e.g., commutativity of addition, inverse relationship of multiplication and division).
F.4.6 Recognize and use generalized properties and relationships of arithmetic (e.g., commutativity of addition, inverse relationship of multiplication and division)	
By the end of Grade 8, students will:	
F.8.1 Work with algebraic expressions in a variety of ways, including ■ Using appropriate symbolism, including exponents and variables ■ Evaluating expressions through numerical substitution ■ Generating equivalent expressions ■ Adding and subtracting expressions	***Level F.8.1.1*** ■ Understand concepts of variable expressions and equations. ***Level F.8.1.2*** ■ Evaluate numerical expressions using correct order of operations. ***Level F.8.1.3*** ■ Evaluate algebraic expressions and formulas through numerical substitution. ***Level F.8.1.4*** ■ Add and subtract algebraic expressions. ***Level F.8.1.5*** ■ Generate equivalent algebraic expressions.
F.8.2 Work with linear and nonlinear patterns and relationships in a variety of ways, including ■ Representing them with tables, with graphs, and with algebraic expressions, equations, and inequalities ■ Describing and interpreting their graphical representations (e.g., slope, rate of change, intercepts) ■ Using them as models of real-world phenomena ■ Describing a real-world phenomenon that a given graph might represent	***Level F.8.2.1*** ■ Represent linear patterns with the use of variable expressions and equations (e.g., 1, 3, 5, 7, ... by $2n - 1$). ***Level F.8.2.2*** ■ Represent patterns graphically and with tables. ***Level F.8.2.3*** ■ Differentiate between linear and nonlinear patterns and relationships. ***Level F.8.2.4*** ■ Describe linear graphs in terms of slope and intercepts. ***Level F.8.2.5*** ■ Use linear and non-linear patterns and relationships as models of real-world phenomena. ■ Describe a real-world phenomenon that a given graph might represent.

TABLE C.5 **Algebra (*continued*)**

WI Model Academic Standards By the end of Grade 8, students will:		School District Levels
F.8.3	Recognize, describe, and analyze functional relationships by generalizing a rule that characterizes the pattern of change among variables. These functional relationships include exponential growth and decay (e.g., cell division, depreciation).	***Level F.8.3.1*** ■ Recognize and describe simple functional relationships (e.g., total no. of inches = 12 × total no. of feet). ***Level F.8.3.2*** ■ Represent simple functional relationships with graphs, tables, and equations. ***Level F.8.3.3*** ■ Recognize, describe, and analyze functional relationships by generalizing a rule that characterizes the pattern of change among variables.
F.8.4	Use linear equations and inequalities in a variety of ways, including ■ Writing them to represent problem situations and to express generalizations ■ Solving them by different methods (e.g., informally, graphically, with formal properties, with technology) ■ Writing and evaluating formulas (including solving for a specified variable) ■ Using them to record and describe solution strategies	***At All Levels*** ■ Represent problem situations with an equation or inequality. ***Level F.8.4.1*** ■ Solve two-step linear equations with one variable with whole number solutions (e.g., $2x + 5 = 13$). ***Level F.8.4.1*** ■ Solve linear equations with one variable involving Whole-number coefficients A variable appearing on either or both sides of the equal sign ***Level F.8.4.2*** ■ Solve simple linear inequalities with one variable. ***Level F.8.4.3*** ■ Solve simple two-step inequalities with one variable. ***Level F.8.4.4*** ■ Solve linear equations and inequalities with one variable involving Integer coefficients Variable appearing on either or both sides of the equal (or inequality) sign Grouping symbols ***Level F.8.4.5*** ■ Solve any linear equation or inequality using formal addition and multiplication properties of equations and inequalities. ***Level F.8.4.6*** ■ Write and evaluate formulas (including solving for a specified variable).

(continued)

WI Model Academic Standards	School District Levels
By the end of Grade 8, students will:	
F.8.5 Recognize and use generalized properties and relations, including ■ additive and multiplicative property of equations and inequalities ■ commutativity and associativity of addition and multiplication ■ distributive property ■ inverses and identities for addition and multiplication ■ transitive property	***Level F.8.5.1*** ■ Recognize and describe the following generalized properties and relationships: Commutativity and associativity of addition and multiplication Inverses and identities for addition and multiplication ***Level F.8.5.2*** ■ Recognize and describe the distributive property. ***Level F.8.5.3*** ■ Recognize, describe, and use the transitive property as it does (or does not) apply in a variety of relationships.
By the end of Grade 12, students will:	
F.12.1 Analyze and generalize patterns of change (e.g., direct and inverse variation) and numerical sequences and then represent them with algebraic expressions and equations. F.12.2 Use mathematical functions (e.g., linear, exponential, quadratic, power) in a variety of ways, including ■ Recognizing that a variety of mathematical and real-world phenomena can be modeled by the same type of function ■ Translating different forms of representing them (e.g., tables, graphs, functional notation, formulas) ■ Describing the relationships among variables in a problem ■ Using appropriate technology to interpret properties of their graphical representations (e.g., intercepts, slopes, rates of change, changes in rates of change, maximum, minimum) F.12.4 Model and solve a variety of mathematical and real-world problems by using algebraic expressions, equations, and inequalities.	***At All Levels*** ■ Model and solve a variety of mathematical and real-world problems by using algebraic expressions, equations, and inequalities. ■ Use mathematical functions in a variety of ways to recognize that a variety of mathematical and real-world phenomena can be modeled by the same type of function. ***Level F.12.1.1*** ■ Use linear functions in a variety of ways, including Translating different forms of representing them (e.g., tables, graphs, functional notation, formulas) Describing the relationships among variables in a problem Using appropriate technology to interpret properties of their graphical representations (e.g., intercepts, slopes) ***Level F.12.1.2*** ■ Use quadratic functions in a variety of ways, including: Translating different forms of representing them (e.g., tables, graphs, functional notation, formulas) Describing the relationships among variables in a problem Using appropriate technology to interpret properties of their graphical representations (e.g., maximum, minimum) ***Level F.12.1.3*** ■ Use exponential and power functions in a variety of ways, including: Translating different forms of representing them (e.g., tables, graphs, functional notation, formulas) Describing the relationships among variables in a problem Using appropriate technology to interpret properties of their graphical representations (e.g., intercepts, slopes, rates of change, changes in rates of change, maximum, minimum) ***Level F.12.1.4*** ■ Analyze and generalize patterns of change (e.g., direct and inverse variation) and numerical sequences and then represent them with algebraic expressions and equations.

Table C.5 **Algebra (*continued*)**

WI Model Academic Standards By the end of Grade 12, students will:	School District Levels
F.12.3 Solve linear and quadratic equations, linear inequalities, and systems of linear equations and inequalities ■ Numerically ■ Graphically, including use of appropriate technology ■ Symbolically, including use of the quadratic formula	***Level F.12.3.1*** ■ Solve linear equations: Numerically Symbolically Graphically, including use of appropriate technology ***Level F.12.3.2*** ■ Solve systems of linear equations: Numerically Symbolically Graphically, including use of appropriate technology ***Level F.12.3.3*** ■ Solve quadratic equations: Numerically Symbolically, including use of the quadratic formula Graphically, including use of appropriate technology ***Level F.12.3.4*** ■ Solve linear inequalities: Numerically Symbolically Graphically, including use of appropriate technology ***Level F.12.3.5*** ■ Solve systems of linear inequalities: Numerically Symbolically Graphically, including use of appropriate technology

Appendix D Assessment Sample Forms

D

This appendix contains several sample pages from various resources useful for classroom assessment. Included are forms for self-assessment, for rubric scoring of several varieties, for portfolios, for assessment of materials, for the Standards-Based Model, and so on.

The offerings are by no means exhaustive, only representative.

Informal Assessment

Classroom Observations, Interviews, Conferences, and Questions

"Assessment must be an interaction between teacher and students, with the teacher continually seeking to understand what a student can do and how a student is able to do it and then using this information to guide instruction" (Webb and Briars 1990, 108). Many teachers jot down classroom observations about student thinking and work during the class and bring those observations to the forefront during a whole-class discussion of the activity. Students' active participation in group activities is also assessed. Although teachers have always assessed classroom interaction, a conscious effort to utilize specific techniques and to focus on results of informal classroom assessments for teacher decision making has emerged as classrooms focus on gathering valid information about student understanding.

The *Cognitively Guided Instruction* (CGI) philosophy (Carpenter and Fennema 1988) requires teachers to make assessment a regular part of instruction by "regularly asking students to explain how they figured an answer" (Webb 1992, 678). Student strategies are referenced against an extensive map of the domain to determine if the task has meaning for the student and to ascertain next steps to be taken in building understanding. The Graded Assessment in Mathematics (GAIM) project (Brown 1988) was a "continuous assessment scheme for recording the mathematical progress" (Webb 1992, 678) of a group of upper-level students. Based on an analysis of a year's anticipated mathematical growth in six topic areas, students were assessed on all aspects of growth—orally, practically, and in writing, and teachers devised their own

"Assessment must be an interaction between teacher and students, with the teacher continually seeking to understand what a student can do and how a student is able to do it and then using this information to guide instruction."

—Webb and Briars 1990, 108

Teachers [need to] make assessment a regular part of instruction by "regularly asking students to explain how they figured an answer."

—Webb 1992, 678

assessment procedures, with supervision from an outside assessor. The focus was on documentation of continual student growth.

An outstanding collection of attributes that can be assessed through observation, of tips for interviews and conferences, of classroom management techniques for informal assessments, of questioning techniques and sample questions, and of ways to record observations and interviews is given in *Mathematics Assessment: Myths, Models, Good Questions, and Practical Suggestions* (Stenmark 1991, 26–34). Approached from the standpoint of aiding classroom assessment, the suggestions are practical and useful.

Class Activities and Projects

Teachers can frequently choose an activity from a unit of study that can be used for assessment purposes.

"Assessment should become a routine part of the ongoing classroom activity rather than an interruption," (Standards 2000 Project, 23). Teachers can frequently choose an activity from a unit of study that can be used for assessment purposes. Students can complete that activity in class, ask questions, work in groups, or use references. Once they have completed the product as a class activity, the teacher can then take the result and assess it as though it were a test. Frequently, a scoring rubric for such an assessment is used. Students should be aware of how their work will be scored in advance. Ideally, students would be participants in development of the scoring rubric. (See Figure D.1.)

The quality of the mathematical tasks in which students engage for construction of knowledge and for assessment of it often dictates the learning and the results. Selection of a component of a task on which to evaluate student learning is an art. Much can be learned easily about student understanding based on task selection.

Student Self-Assessment

"When students can tell themselves if they have performed at a proficient level, then the ultimate goal of assessment has been reached."

—Howell 1999

"When students can tell themselves if they have performed at a proficient level, then the ultimate goal of assessment has been reached" (Howell 1999). We need to work with students to self-assess. The beginnings are simple. Having a student write a comment under his or her name at the top of the paper regarding the assignment is a first step. "Great, let's do some more!" "I didn't get what we said about problem 5; can we discuss it?"

Observations, Interviews, Conferences, and Questions

Teacher observations, interviews, and conferences have always been important tools in assessing student progress. The change in current assessment processes reflects this important aspect of learning, adding documentation and sometimes involving an outside observer or interviewer.

Observations

We might quietly observe students either individually or as they work in groups. The purpose of an observation may be for mathematics (How far can

4 Advanced Responses

Correct Mathematics
Good Explanation
Additional Charts, Graphs, Tables
Mathematical Synthesis/Efficiency

3 Proficient Responses

Correct Mathematics
An Explanation
An Additional Chart, Graph, Table

2 Basic Responses

Minimal Mathematics
Incomplete or Minimal Explanation

1 Minimal Response

Incorrect Mathematics
Incorrect Explanation

0 No Response

No Attempt Made

FIGURE D.1 **Measuring Up: Scoring Rubric**
Jodean Grunow and Students, Dodgeville Middle School, jodean.grunow@dpi.state.wi.us

students count with one-to-one correspondence?) or for affective characteristics (Does this child's behavior help his learning?)

Any observer should look for only a few aspects in a single observation. Be selective.

Some attributes that can be assessed best through observation are presented in the following list. We stress that this is a partial list and may not show the goals you want for your students. Also, any observer should look for only a few aspects in a single observation. Be selective.

MATHEMATICAL CONCEPTS

- Organizing and interpreting data
- Selecting and using appropriate measurement instruments
- Explaining the relationships between inverse operations
- Extending and describing numeric or geometric patterns
- Estimating regularly
- Using visual models and manipulative materials to demonstrate mathematical concepts
- Showing relationships between perimeter, area, and volume
- Making connections among concrete, representational, and abstract ideas

STUDENTS' DISPOSITION TOWARD LEARNING

- Planning before acting; revising plans when necessary
- Sticking to the task without being easily distracted
- Becoming actively involved in the problem
- Using calculators, computers, or other needed tools effectively
- Explaining organizational and mathematical ideas
- Supporting arguments with evidence
- Asking probing mathematical questions
- Completing the task
- Reviewing the process and the results

MATHEMATICS COMMUNICATION

- Discussing for clarification of the student's own ideas and to communicate to others
- Communicating a process so that it is replicable
- Having the confidence to make a report to the whole class
- Representing capably and fairly a group consensus
- Synthesizing and summarizing the student's own or a group's thinking

GROUP WORK

- Dividing the task among the members
- Agreeing on a plan or structure for tackling the task
- Taking time to ensure that all members understand the task
- Using the time in a productive way
- Remembering to record results
- Considering seriously and using the suggestions and ideas of others

From J.K. Stenmark, ed., *Myths, Models, Good Questions, and Practical Suggestions* (Reston, VA: National Council of Teachers of Mathematics, 1991).

Interviews

Interviews and conferences with students are a source of rich information regarding their thoughts, understandings, and feelings about mathematics. Students receive encouragement and are given opportunities to grow mathematically. Teachers gain solid bases for modification of the curriculum or for necessary reporting of students' progress.

An interview includes a planned sequence of questions, whereas a conference implies discussion, with student and teacher sharing ideas.

"Hey, graphing really makes sense now!" "I need HELP!" These simple comments tell the teacher a great deal more than any number grade. Having students write a mathematics autobiography (see Figure D.2), set mathematics goals (see Figure D.3), or do some daily evaluation (see Figure D.4) fosters first efforts at self-assessment in small, doable tasks.

Simple student comments tell the teacher a great deal more than any number grade.

Just asking students to write a few comments on the bottom of a completed investigation also gets at evaluation, especially if directed responses are indicated. "What was your greatest 'Ah-ha' in this investigation? Tell about it" (see Figure D.5). "Why was the second question in the lesson asked?" An extension of that is a requested reflection on learning. "Having worked through this investigation, what would you say about the relationships among and between fractions, decimals, and percentages?" These kinds of questions help students make the mathematics in what they've done come alive. "We cannot retain much of the mathematics we have seen or heard if we have not appropriated it as our own" (Silver, Kilpatrick, and Schlesinger 1990, 21).

"We cannot retain much of the mathematics we have seen or heard if we have not appropriated it as our own."

—Silver, Kilpatrick, and Schlesinger 1990, 21

Reflection

Reflection on tasks done is also important (see attached Figure D.6). A simple notebook checklist can become an evaluation tool. "Have I included all of the requested materials?" "Is my work organized in a meaningful fashion?" "Have I written down my understandings of mathematical terms used in class?" Reflection on classroom contributions is also important. "What was the best contribution to the class that I made today?" "What was the most difficult idea that I had to deal with today? How did I confront it? Would I work with such difficulty the same way tomorrow?" Reflection on contributions to group work is also important (see Figure D.7). "What was my task in the group today? Did I do it well?" "Did our group work together well today? Why or why not?" "How could I help my group increase our learning tomorrow?"

Journals

Students should be encouraged and required to write about what they are learning. The use of journals can provide the teacher with invaluable insights into student thinking, thus helping the teacher provide instruction based on student knowledge. Creating a journal form where a question is posed, students have a space to respond to the question, and including a place for students to write other comments and concerns can be very useful (see Figure D.8). Journaling can help identify students who need help as well as students

The use of journals can provide the teacher with invaluable insights into student thinking, thus helping the teacher provide instruction based on student knowledge.

Name:
Date:
Hour:

A Success Story

My Mathematics Autobiography

FIGURE D.2 **A Success Story**

Jodean Grunow, Dodgeville Middle School, jodean.grunow@dpi.state.wi.us.

Name:
Date:
Hour:

My Mathematics Goal(s)

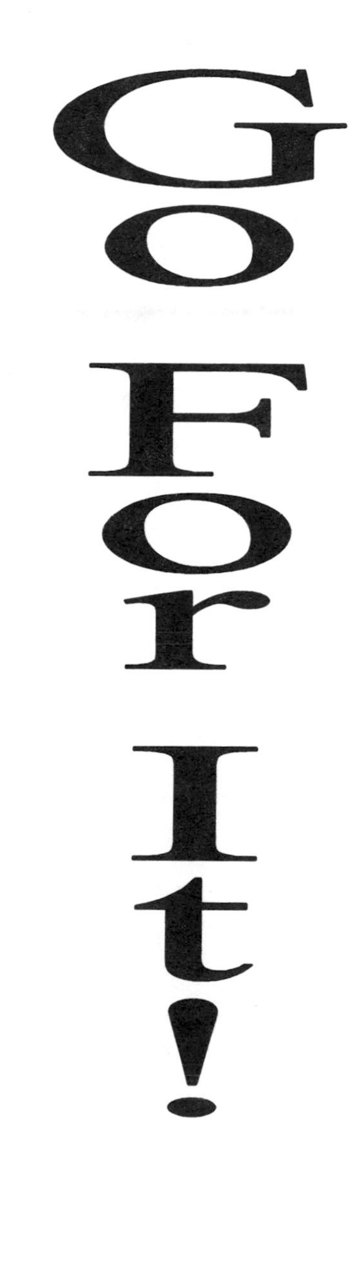

FIGURE D.3 **My Mathematics Goals**
Jodean Grunow, Dodgeville Middle School, jodean.grunow@dpi.state.wi.us.

Name:
Date:
Hour:
Activity:

Individual Daily Evaluation Sheet

Each time I learn something new I feel good.

Today I learned:

We've been trying to implement the Standards in our classroom. Today's lesson used Standard(s):

Special Comments:

FIGURE D.4 **Individual Daily Evaluation Sheet**
Jodean Grunow, Dodgeville Middle School, jodean.grunow@dpi.state.wi.us.

Name:
Period:
Date:

My "Ah-Ha's!"

Date **Ah-Ha!**

FIGURE D.5 **My "Ah-Ha's!"**
Jodean Grunow, Dodgeville Middle School, jodean.grunow@dpi.state.wi.us.

Name:
Hour:
Unit:
Due Date:

Unit Assignment Sheet

Date	Assignment	Grade

Assignment Average:

Test Grade:

Extra Credit Points:

Unit Grade:

FIGURE D.6 **Unit Assignment Sheet**
Jodean Grunow, Dodgeville Middle School, jodean.grunow@dpi.state.wi.us.

Date:
Hour/Group:
Activity:
Director:
Recorder:
Runner:
Checker:

Group Record Sheet

We began promptly and independently: Yes No (Circle one.)

Today we learned:

We worked well when we:

We could improve if we were to:

We would like to know more about:

We worked well throughout the entire hour: Yes No (Circle one.)

Remember:

1. You are responsible for your own behavior.
2. You must be willing to help anyone in your group who asks.
3. You need to discuss the problem thoroughly to see if your group can find a solution. Then ask the teacher for help.

FIGURE D.7 **Group Record Sheet**
Jodean Grunow, Dodgeville Middle School, jodean.grunow@dpi.state.wi.us.

Name:
Period:
Date:

Mathematics Journal

Topic: ______________________________

____

Response: ______________________________

____

____

____

____

____

____

____

Other Comments: ______________________________

____

____

____

FIGURE D.8 **Mathematics Journal**
Jane Howell, Platteville Middle School, howell@mhtc.net.

who understand and are ready to move ahead, and it may raise questions for the teacher that he or /she had not considered. Students can respond to writing prompts such as, "When I worked through this activity . . . "

I learned . . .
I discovered . . .
I was most comfortable with . . .
I was confused about . . .
I predicted . . .
I found that, if I were to teach a younger student about this, I would . . .
I was challenged by . . .

Because of the pertinence of the information, journals should be read as soon as possible by the teacher, and comments should be made and shared with the student. The journals are written for student self-reflection, to convey ideas informally, and for the teacher's edification; no grade needs to be assigned. Care needs to be exercised in the use of journals. If students are asked to journal everyday in every class, sheer overload will defeat the purpose.

Writing in Mathematics

Several alternative assessment strategies require writing. The teacher needs to keep in mind that students *need to be taught* how to write quality responses to mathematical problems. To help students learn what needs to be included in a quality response, the teacher can create a form prompting students to address their understanding of the problem: What was the mathematics in this investigation? How did you address each part of the problem? What extension question would you like to ask after doing this investigation? Students can generate a written response as a group and then evaluate the response to see if all aspects of the investigation are addressed.

Students need to be taught how to write quality responses to mathematical problems.

Students can also, with teacher assistance, score their response on a rubric basis. They can begin with an analytic rubric (see Figure D.9) that assigns value to each part of the response. Moving to holistic scoring is often facilitated by evaluating the response as "acceptable" or "needs revision." Students then need to help devise a holistic rubric that reflects what they believe is embodied in an advanced, a proficient, a basic, or a minimal response. Students, as they progress, need to be given an opportunity to revise or redo their responses before final evaluation.

"It's in the writing that the learning takes place" (Fennema 1998). Students need to explain their reasoning, verbally, with manipulatives, pictorially, graphically, in tables, and in writing. Explaining asks the student to verify original thinking and then to put it in a form that others will understand. In so doing, the student often continues to build on understanding. There is sometimes a debate about asking students to read for information or to explain results. Both are major parts of building the "language of mathematics" and of augmenting knowledge growth. Writing, especially, asks for organization of thought to convey meaning.

"It's in the writing that the learning takes place."

—Fennema 1998

Analytic Scoring for Problem Solving

Understanding the Problem

0 No attempt

1 Completely misinterprets the problem

2 Misinterprets major part of the problem

3 Misinterprets minor part of the problem

4 Complete understanding of the problem

Solving the Problem

0 No attempt

1 Totally inappropriate plan

2 Partially correct procedure but with major fault

3 Substantially correct procedure with minor omission or procedural error

4 A plan that could lead to a correct solution with no arithmetic errors

Answering the Problem

0 No answer or wrong answer based upon an inappropriate plan

1 Copying error; computational error; partial answer for problem with multiple answers; no answer statement; answer labeled incorrectly

2 Correct solution

FIGURE D.9 **Analytic Scoring for Problem Solving**
From: Education Leadership, May 1992. (This complete issue was devoted to alternative forms of assessment.)

Azzolino (1990) offers a list of instructional purposes of writing in mathematics. Writing can:

Demand participation of the student.
Help the student summarize, organize, relate, and associate ideas (connections!).
Provide an opportunity for a student to define, discuss, or describe an idea or concept.
Permit the student to experiment with, create, or discover mathematics independently.
Encourage the personalization, assimilation, and accommodation of the mathematics being taught.
Assist the student in reviewing, refocusing, and reconsidering topics either recently studied or considered long ago.
Assist in recording and retaining mathematical procedures, algorithms, and concepts for future use.
Assist in translating or decoding of mathematical notation.
Assist in symbolizing or coding with proper notation.
Help the teacher diagnose a student's misconceptions and problems.
Provide an appropriate vehicle for the student to express and focus on negative feelings and frustrations as well as to emote and rejoice in the beauty of mathematics.
Assist in the reading, summarizing, or evaluating of texts.
Collect evidence for research.

Being able to explain reasoning is extremely important for constructed response items on formal assessments such as the Wisconsin Student Assessment System Knowledge and Concepts Test. The thrust of the assessment is a read on student knowledge and understanding. This is best obtained through student writing.

Being able to explain reasoning is extremely important for constructed response items on formal assessments such as the Wisconsin Student Assessment System Knowledge and Concepts Test.

Investigations and Open-Ended Questions

As students investigate large mathematical explorations such as comparisons of area and perimeter, surface area and volume, questions can be presented to students for consideration pertinent to the point at hand; in other words, tailored to the understanding of the moment. Students' responses to investigations can indicate if students can

Identify and define a problem and what they already know.
Make a plan; create, modify, and interpret strategies.
Collect needed information.
Organize the information and look for patterns.
Discuss, review, revise and explain results.
Persist, looking for more information, if needed.
Produce a quality product or report (Stenmark 1989, 13).

Camping Investigation is an example of this type of exploration from *The Language of Functions and Graphs, An Examination Module for Secondary Schools* (Shell Centre for Mathematical Education), 14–15.

Camping Investigation

On their arrival at a campsite, a group of campers are given a piece of string 50 meters long and four flag poles with which they have to mark out a rectangular boundary for their tent. They decide to pitch their tent next to a river. This means that the string has to be used for only three sides of the boundary.

1. If they decide to make the width of the boundary 20 meters, what will the length of the boundary be?
2. Describe in words, as fully as possible, how the length of the boundary changes as the width increases through all possible values. (Consider both small and large values of the width.)
3. Find the area enclosed by the boundary for a width of 20 meters and for some other different widths.
4. Draw a sketch graph to show how the area enclosed changes as the width of the boundary increases through all possible values. (Consider both small and large values of the width.)

The campers are interested in finding out what the length and the width of the boundary should be to obtain the greatest possible area.

5. Describe, in words, a method by which you could find this length and width.
6. Use the method you have described in part 5 to find this length and width.

An open-ended question is one in which the student is given a situation and asked to solve the problem and then to explain why or how the solution was reached.

It is important for students to become accustomed to explaining… to emphasize reasoning and to convey knowledge.

An open-ended question is one in which the student is given a situation and asked to solve the problem and then to explain why or how the solution was reached. Consider the illustrative item prepared to reflect items that might appear on the Wisconsin Student Assessment System Knowledge and Concepts Examination, Levels 19–21/22 as shown in Figure D.10.

This item is considered to address two standards: measurement and problem solving. It is a constructed response item. An explanation is given: "Constructed-response items that measure the problem-solving objective generally relate to real-world situations that have elements of several mathematical content areas. The sample item requires the student to use procedures of measurement and algebra to devise and explain a strategy that will lead to the solution of the problem." Not only does the item assess response construction, but it calls for connections and analysis (Shafer and Romberg 1999). The scoring rubric for this item would call for a correct solution and an explanation. Again, it is important for students to become accustomed to explaining … to emphasize reasoning and to convey knowledge.

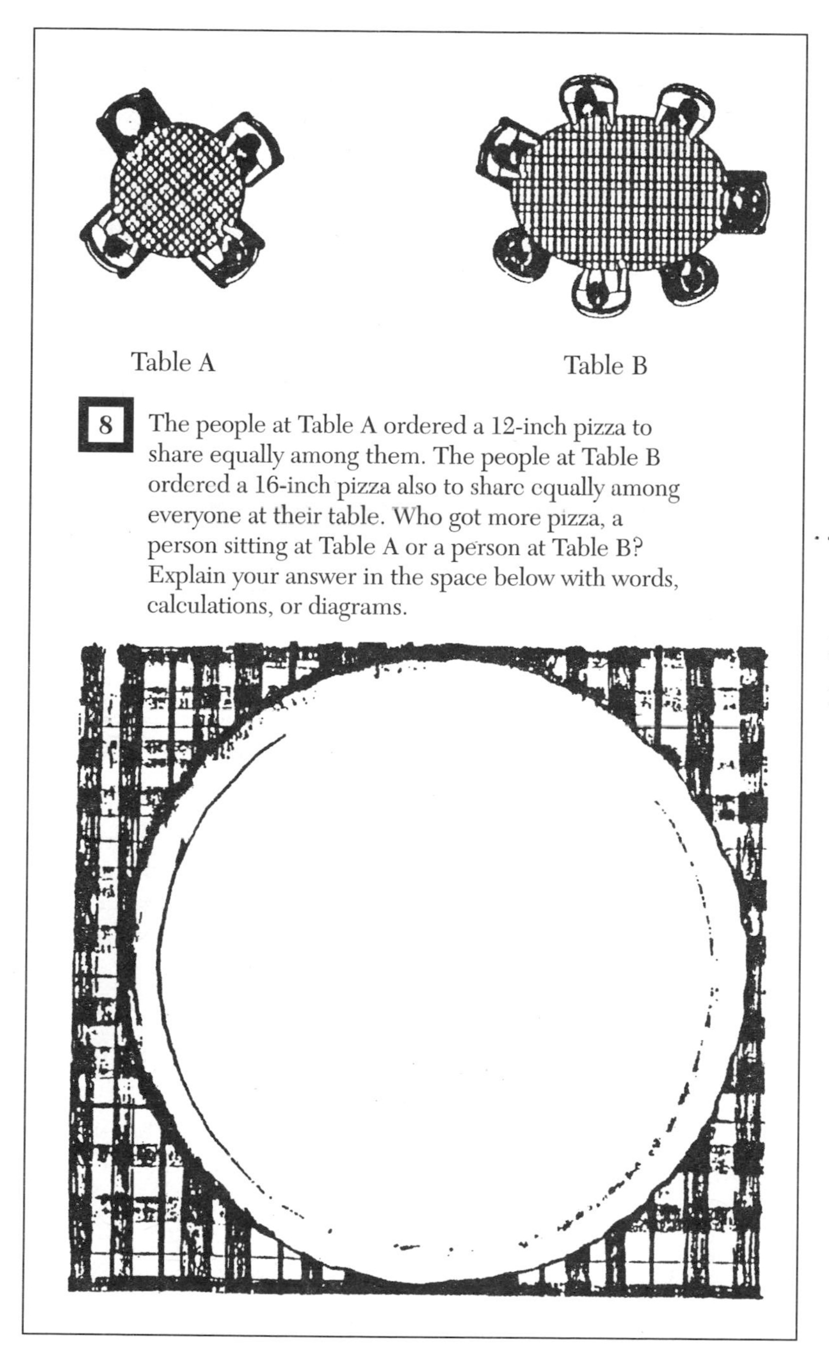

13 Measurement

17 Problem Solving and Reasoning

Constructed-response items that measure the problem-solving objective generally relate to real-world situations that have elements of several mathematical content areas. The sample item requires the student to use procedures of measurement and algebra to devise and explain a strategy that will lead to the solution of the problem.

FIGURE D.10 **Mathematics Illustrative Levels, 19–21/22**

Performance Assessment

The Wisconsin Knowledge and Concepts Test includes constructed-response questions that are a modification of performance assessment tasks and require students to show more than just an answer.

Performance assessment offers students a mathematics investigation. In solving the problem, students demonstrate understanding of the task (particularly if some aspect of it is not clearly defined); describe and explain their approaches to the task and what methods and materials they used; show flexibility in their thinking such as starting with one strategy and recognizing its limitations and switching then to a different strategy; communicate clearly the strategy or strategies used; and show evidence of reflection by assessing if their responses are appropriate for the initial question. The Wisconsin Knowledge and Concepts Test includes constructed-response questions that are a modification of performance assessment tasks and require students to show more than just an answer. Consider the illustrative item prepared to reflect items that might appear on the Wisconsin Knowledge and Concepts Examination, Levels 10–14 as shown in Figure D.11.

This item contributes to the scores in two objectives—measurement and communication. The student demonstrates an understanding of scale by drawing the second tree twice the height of the first. The student can use the ruler provided to determine the desired height, but that is not the only possible strategy. The student then validates the solution by explaining his or her own decision-making process. Concepts in the communication objective include relating daily vocabulary to mathematical terminology and relating models, diagrams, and pictures to mathematical ideas.

In a discussion of mathematical reasoning and the role of assessment tasks, Peressini and Webb (1999, 157) suggest that mathematical reasoning involves "gathering evidence, making conjectures, establishing generalizations, building arguments, and drawing (and validating) logical conclusions about these various ideas and their relationships." With this view of mathematical power, thinking, and reasoning, they contend that selection of "rich tasks that expose students to situations in which they must invoke their mathematical power and demonstrate their ability to think and reason in problem-solving situations" is "one of the most important activities" in which mathematics teachers engage (see Figure D.12).

The assessment Standards Working Groups (1995, 13) suggest that "assessment that enhances mathematics [reasoning] incorporates activities that are consistent with, and sometimes the same as, the activities used in instruction."

The Assessment Standards Working Groups (1995, 13) suggest that "assessment that enhances mathematics [reasoning] incorporates activities that are consistent with, and sometimes the same as, the activities used in instruction." Worthwhile mathematical tasks have the following characteristics:

Based on sound and significant mathematics
Use knowledge of students' understandings, interests, and experiences
Employ knowledge of the range of ways that diverse students learn mathematics
Engage students' intellect
Develop students' mathematical understandings and skills
Stimulate students to make connections and develop a coherent framework for mathematical ideas
Call for problem formulation, problem solving, and mathematical reasoning
Promote communication about mathematics

13 Measurement

18 Communication

This item contributes to the scores in two objective—Measurement and Communication. The student demonstrates an understanding of scale by drawing the second tree twice the height of the first. The student can use the ruler provided to determine the desired height, but that is not the only possible strategy. The student then validates the solution by explaining his or her own decision-making process. Concepts in the Communication objective include relating daily vocabulary to mathematical terminology, and relating models, diagrams, and pictures to mathematical ideas.

FIGURE D.11 **Mathematics Illustrative Items, Levels 10–14**

ESSENTIAL	• The task fits into the core of the curriculum. • It represents a "big idea."	vs.	TANGENTIAL
AUTHENTIC	• The task uses processes appropriate to the discipline. • Students value the outcome of the task.	vs.	CONTRIVED
RICH	• The task leads to other problems. • It raises other questions. • It has many possibilities.	vs.	SUPERFICIAL
ENGAGING	• The task is thought-provoking. • It fosters persistence.	vs.	UNINTERESTING
ACTIVE	• The student is the worker and decision maker. • Students interact with other students. • Students are constructing meaning and deepening. understanding.	vs.	PASSIVE
FEASIBLE	• The task can be done within school and homework time. • It is developmentally appropriate for students. • It is safe.	vs.	INFEASIBLE
EQUITABLE	• The task develops thinking in a variety of styles. • It contributes to positive attitudes.	vs.	INEQUITABLE
OPEN	• The task has more than one right answer. • It has multiple avenues of approach, making it accessible to all students.	vs.	CLOSED

FIGURE D.12 **Criteria for Performance Tasks**

J. K. Stenmark, ed. *Mathematics Assessment: Myths, Models, Good Questions, and Practical Suggestions* (Reston, VA: National Council of Teachers of Mathematics, 1991).

Represent mathematics as an ongoing human activity
Display sensitivity to, and draw on, students' diverse background experiences and dispositions
Promote the development of all students' dispositions to do mathematics (Commission on Teaching Standards 1991, 25).

Shafer and Romberg (1999) note that assessment needs to include attention to levels of reasoning: reproduction, connections, and analysis. Well-selected tasks often embody all components of reasoning. For example, an item could be written that would assess reproduction of understanding regarding factors: "Find all of the factors of 36." The student would need to recall what a factor is, would need to know multiplication facts, and would need to be able to apply both understandings to the item. The item is, however, not highly motivating and is totally recall dependent. If, however, the item became an item based on the game of Taxman (Mathematical Sciences Education Board 1993, 101–14), it would be much more engaging and much more could be learned. The game calls for a listing of numbers, i.e., 1–6. One student selects a number. The partner gets to take all the factors of that number. If 4 is selected, the first player gets 4 and the partner gets 1 and 2, factors of 4, and has a total of 3. The partner can then select 3, 5, or 6. However, there are no remaining factors to be selected, so the game terminates. An assessment item built on the game might give 10 numbers and ask the student to select his first move and explain why he had selected it. The item is interesting, it checks understanding of factors in a manner that allows the student to actually apply his knowledge, and it connects to previous learning (playing the game).

Shafer and Romberg (1999) note that assessment needs to include attention to levels of reasoning: reproduction, connections, and analysis. Well-selected tasks often embody all components of reasoning.

A performance item might take Taxman a step farther and ask for a mathematical analysis of the best pick and the worst pick if the game were to use 50 or 100 (Lappan et al. 1996). This item would then ask the student to conceive of how he/she might convey mathematically what would happen. It would be calling for analysis, certainly connections to the original components, and would check not only factors but also tacit understanding of primes, composites, abundant, deficient, and perfect numbers. (Hopefully, because the student is experiencing all of those components of number as the game is played, the class would discuss the terms and talk about how they arrived at them.)

The Wisconsin Center for Educational Research and the Department of Public Instruction worked on a project designed to produce performance assessments for the state assessment. The *Wisconsin Criteria for Quality Tasks* (Fortier and Moser, 1992) called for authenticity (realistic and relevant), integration (drawing on knowledge from several disciplines, integrating with higher level thinking skills, and using processes such as writing and computing), collaboration (group work), transferability (placed in a meaningful context and requiring transfer of skills from one discipline to another), tools and technology (as used in the classroom), duration (sustained efforts over longer periods of time), sense of time (drawing on past, present, and future orientations), and a sense of place (aesthetics, civility, etc.), sense of culture (foster a feeling for the richness of culture[s]).

Scoring of [performance] tasks required development of assessment rubrics.

Scoring of such tasks required development of assessment rubrics. The rubrics characterized the degree to which students had met the requirements of the task and included the designations of exemplary, competent, satisfactory, inadequate, and no attempt (see Figure D.13). Descriptions of components of each were written in a generic form and also in task-specific forms. Scoring was done by two assessors and were juried if there was a discrepancy. Scoring was highly reliable. With permission from the Department of Public Instruction Office of Educational Accountability, one of these performance items follows (see Figure D.14).

There have been far-reaching efforts designed to aid in use of performance tasks. Some notable examples and excellent sources of performance items are the following:

Clarke, D. 1988. *Assessment Alternatives in Mathematics.* Melbourne, Australia: Jenkin Buxton Printers Pty. Ltd.

Danielson, C. 1997. *A Collection of Performance Tasks and Rubrics.* Larchmont, NY: Eye on Education.

MacDonell, A., and C. Anderson, eds. 1999. *Balanced Assessment for the Mathematics Curriculum: Berkeley, Harvard, Michigan State, Shell Centre.* White Plains, NY: Dale Seymour Publications. (Available for elementary, middle school, and high school, two levels, these materials come with examples of student work and analysis of scoring criteria for the work.)

New Standards. 1997. *Performance Standards.* Pittsburgh: National Center on Education and the Economy and the University of Pittsburgh.

Shell Centre for Mathematical Education. 1984. *Problems with Patterns and Numbers.* University of Nottingham, UK: Joint Matriculation Board.

———. 1985. *The Language of Functions and Graphs.* University of Nottingham, UK: Joint Matriculation Board.

Students, too, can help in evaluation of learning and assessment tasks.

Students, too, can help in evaluation of learning and assessment tasks. At the time that the Wisconsin Center for Educational Research and the Department of Public Instruction were working on items for a performance component of the state assessment, a set of questions was developed to help students evaluate tasks when they had completed them. The set of questions is very easily used with students and helps them assess both the mathematics in the task and the task itself. (See Figure D.15.)

Concept Mapping

Concept mapping is an assessment method that combines an open-response format and the hierarchical thrust of the superitem method of assessing domain understanding.

Concept mapping is an assessment method that combines an open-response format and the hierarchical thrust of the superitem method of assessing domain understanding. Concept maps use nodes (circles) for mathematical concepts; lines, to connect the concepts and form propositions; arrows, to indicate directionality; placement, to show hierarchy. "To construct a concept map, the developer must have complete understanding of the domain to be represented in order to (a) appropriately select the major concepts; (b) hierarchically place those concepts; (c) adequately represent the relationships between and among the concepts; and (d) visually represent cause and effect" (Grunow 1999, 44).

Wisconsin Student Assessment System
Performance Assessment Development Project:
Draft Mathematics Scoring Rubric

(E) EXEMPLARY RESPONSE:

E.1 Complete in every way with clear, coherent, unambiguous and insightful explanation;
E.2 Shows understanding of underlying mathematical concepts, procedures, and structures;
E.3 Examines and satisfies all essential conditions of the problem;
E.4 Presents strong supporting arguments with examples and counterexamples as appropriate;
E.5 Solution and work is efficient and shows evidence of reflection and checking of work;
E.6 Appropriately applies mathematics to the situation.

(D) PROFICIENT RESPONSE:

More than COMPETENT; but not EXEMPLARY.

(C) COMPETENT RESPONSE:

C.1 Gives a fairly complete response, with reasonably clear explanations;
C.2 Exhibits minor flaws in underlying mathematical concepts, procedures, and structures;
C.3 Examines and satisfies most essential conditions of the problem;
C.4 Draws some accurate conclusions, but reasoning may be faulty or incomplete;
C.5 Solution and work show some evidence of reflection and checking of work;
C.6 Applies mathematics to the situation.

(B) MINIMAL RESPONSE:

Needs development; minimally ADEQUATE but not COMPETENT.

(A) INADEQUATE RESPONSE:

A.1 Response is incomplete and explanation is insufficient or not understandable;
A.2 Exhibits major flaws in underlying mathematical concepts, procedures, and structures;
A.3 Fails to address essential conditions of the problem;
A.4 Uses faulty reasoning and draws incorrect conclusions;
A.5 Shows no evidence of reflection and checking of work;
A.6 Fails to apply mathematics to the situation.

(TF) or (AB) NO ATTEMPT:

Provides irrelevant or no response;
Copies part of the problem but does not attempt a solution;
Illegible response.

FIGURE D.13 **Wisconsin Student Assessment System Performance Assessment Development Project: Draft Mathematics Scoring Rubric**

DRAFT: Scoring Institute, April 22–23, 1994

A Mile of Coins

Your class is going to have an overnight field trip to an amusement park. The class has decided to pay for the trip by collecting a mile of coins. You must gather enough coins of one single type that they would stretch for one mile if the coins were laid out side by side. The coins also must provide enough money to pay all of the expenses of the trip.

Here is some needed information:

Diameter of Coins

Penny 3/4 inch   Nickel 7/8 inch   Dime 11/16 inch   Quarter 15/16 inch

Travel Expenses

Bus transportation	$ 900
Lodging in Motel	$2,600
Park Admission	$1,500
Food	Students provide for themselves

Measurement Equivalents

1 mile = 5,280 feet   1 foot = 12 inches

Which type of coin should be used to make your Mile of Coins? ________________

Show all of your work and describe the reasoning you used to determine the answer. If your choice provides more money than is needed to meet expenses, give a reasonable plan of how the excess will be used.

(See attached advanced student response.)

FIGURE D.14A **A Mile of Coins**

A MILE OF COINS

Your class is going to have an overnight field trip to an amusement park. The class has decided to pay for the trip by collecting *A Mile of Coins*. The Mile of Coins must be enough coins of one single type that they would stretch for one mile if the coins were laid out side-by-side. The coins also must provide enough money to pay all of the expenses of the trip.

Here is some needed information:

Diameter of coins

Penny $\frac{3}{4}$ inch Nickel $\frac{7}{8}$ inch Dime $\frac{11}{16}$ inch Quarter $\frac{15}{16}$ inch

Travel Expenses

Bus transportation	$900
Lodging in Motel	$2600
Park Admission	$1500
Food	Students provide for themselves

$4000

2600
1500
900
5000

Measurement Equivalents

1 mile = 5280 feet 1 foot = 12 inches

1. Which type of coin should be used to make your Mile of Coins? Dimes

   Show all work and describe the reasoning you used to determine the answer. If your choice provides more money than is needed to meet expenses, give a reasonable plan of how the excess will be used.

Pennies

P = $\frac{3}{4}$ in

$\frac{12}{3/4}$

16 pennies = 1 foot
16 x 5280 =
84,480 pennies
$844.80
Not enough money

Nickels

N = $\frac{7}{8}$ in 48 nickels = 42 inches

$\frac{12}{7/8}$

42/12 =
3.5 feet = 48 nickels
5280 ÷ 3.5 =
1508 4/7
1508 x 48
72,384 + 28
$3620.60
Not enough

FIGURE D.14B **A Mile of Coins (*continued*)**

Mile of Coins

Dimes

D = $\frac{11}{16}$ in
256 dimes = 176 in
$14\frac{2}{3}$ feet
5280 ÷ $14\frac{2}{3}$
360 times
360 × 256
92,160 dimes
$9,216.00

The best choice for the class is dimes. It will give an excess of money but not as much as the quarters, plus dime come out exactly at a mile. You wouldn't have to round the number of dimes. And People wo[uld] be more ~~likely~~ likely to give dimes becaus they are worth less than quar[ters].

You could take the extra mone[y] and pay for the ~~students~~ class's food and/or stay in a more expens[ive] hotel.

1.

P

FIGURE D.14C **A Mile of Coins (*continued*)**

What did you think about this task?

Now that you have finished this task, would you please take time to respond to the following questions? Your honest comments will be used to improve these tasks. Again, thank you for your help!

Write the name of the task: ______________________________

1. How did the task catch your interest and motivate you to do your best work?

2. What mathematics did this task require you to use? Please explain.

3. What changes do you recommend to make this task better?

4. What materials/supplies did you use, or would you have liked to use, to complete this task?

FIGURE D.15 **What Did You Think About This Task?**

Scoring [of concept maps] is highly reliable.

Concept-based assessment, as envisioned by Ruiz-Primo and Shavelson (1997), is composed of a task, a form for the student's response, and a scoring system. They found that concept maps pinpoint well understanding of the topic as well as misconceptions regarding it. Scoring is highly reliable. Bartels (1995) found concept mapping a useful tool for "explicitly stressing mathematical connections" (542). Other recent studies have yielded positive results: with university calculus classes (Williams 1998), with monitoring of fourth and sixth grade students' understandings over the course of time (Hasemann and Mansfield 1995), with college algebra and trigonometry classes (Entrekin 1992), with prospective teachers' concept development (Beyerback 1986), and with domain knowledge gain in a professional development setting (Grunow 1999).

Concept mapping, akin to story mapping, scientific cycle maps, and social studies cause/effect diagrams, is finding current use in curriculum mapping and in interpreting the intent of standards statements. Several districts have used concept mapping to "unpack" the standards and then to plan for implementation of specific concepts. Identification of the ideas that go in the circles often helps with understanding the sentence. Then, arrangement of the circles and the cause/effect arrows helps teachers determine the manner in which the concepts should be taught (see Figure D.16).

Formal Assessments

Shafer and Romberg (1999) list end-of-unit assessments and end-of-grade-level assessments as formal assessments.

Listing end-of-unit assessments and end-of-grade-level assessments as formal assessments, Shafer and Romberg (1999) describe assessment activities offered in realistic contexts different from the ones used during instruction. These activities allow students an opportunity to reflect on class experiences to use as a basis for building approaches to the parallel but unfamiliar problems; it allows for creativity. The questions provide opportunity for students to use reproduction, connections, and analysis. The items allow students "to provide ample evidence of their abilities to apply their mathematical knowledge in new situations, to articulate their understanding of the concepts in a domain, and to make mathematics their own as they mathematize the problem situations" (177). The final component of formal assessment, growth over time, suggests accumulation of evidence based on multiple types of assessments. All formal assessments need to be cognizant of mathematical reasoning.

All formal assessments need to be cognizant of mathematical reasoning.

Portfolios

Portfolios are collections of students' work (see Figure D.17). The collection can include assignments, projects, reports, writings, etc. The portfolio showcases student work and shows "progress in, attitudes toward, and understanding of mathematics in a comprehensive way" (Stenmark 1991, 35). Portfolios "show much more than a single test" or collection of tests.

Using portfolios can be extremely valuable in a variety of ways. They can be used as a tool to create discussion between a child and the parents and can be useful in parent conferences.To prepare these portfolios, students select

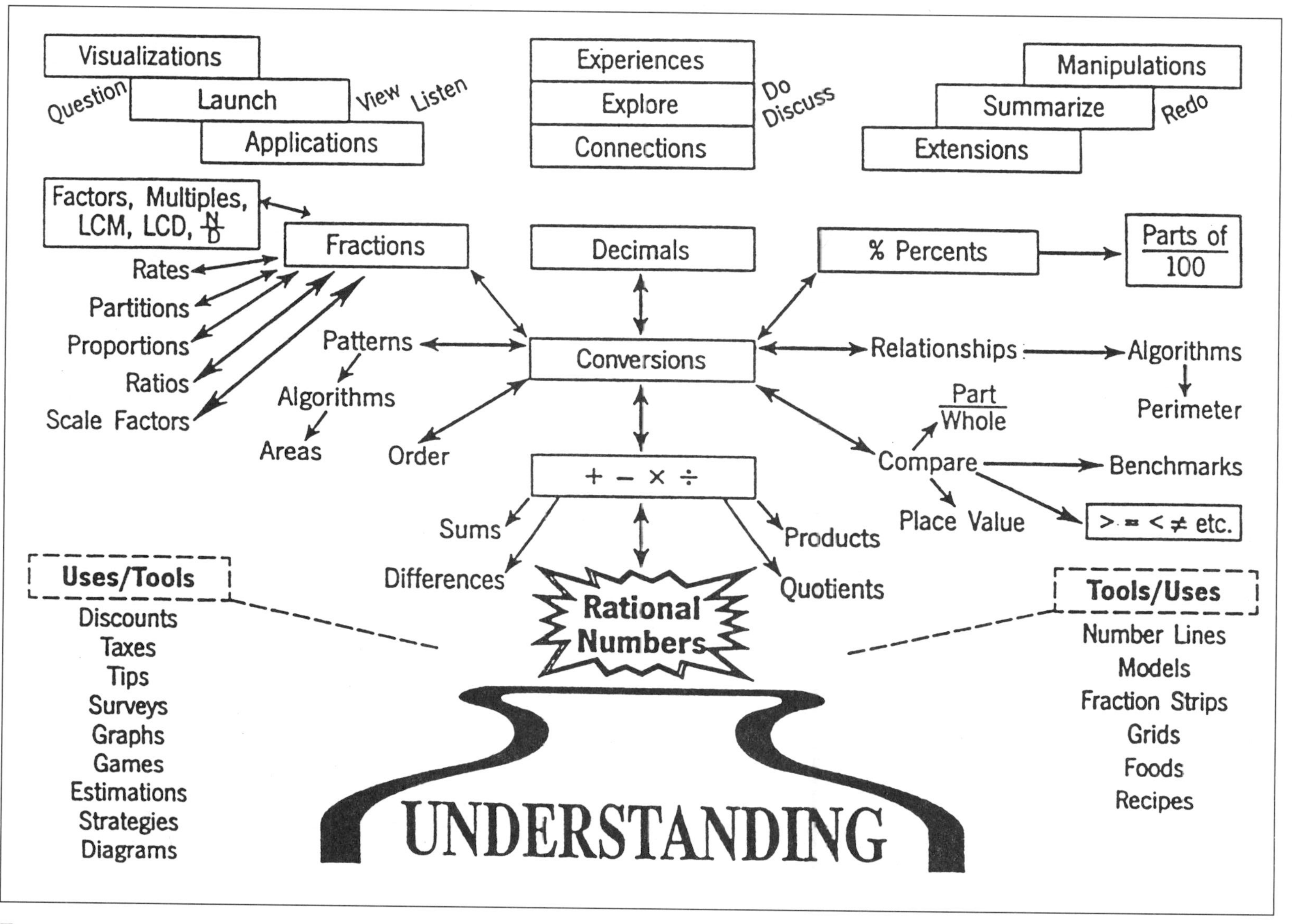

Visualizations
Question
Launch
View
Listen
Applications
Experiences
Explore
Connections
Do
Discuss
Manipulations
Summarize
Redo
Extensions
Factors, Multiples, LCM, LCD, N/D
Fractions
Decimals
% Percents
Parts of 100
Rates
Partitions
Proportions
Ratios
Scale Factors
Patterns
Algorithms
Areas
Order
Conversions
Relationships
Algorithms
Perimeter
Part/Whole
Compare
Benchmarks
Place Value
> = < ≠ etc.
+ − × ÷
Sums
Differences
Products
Quotients
Rational Numbers
Uses/Tools
Discounts
Taxes
Tips
Surveys
Graphs
Games
Estimations
Strategies
Diagrams
Tools/Uses
Number Lines
Models
Fraction Strips
Grids
Foods
Recipes
UNDERSTANDING

FIGURE D.16

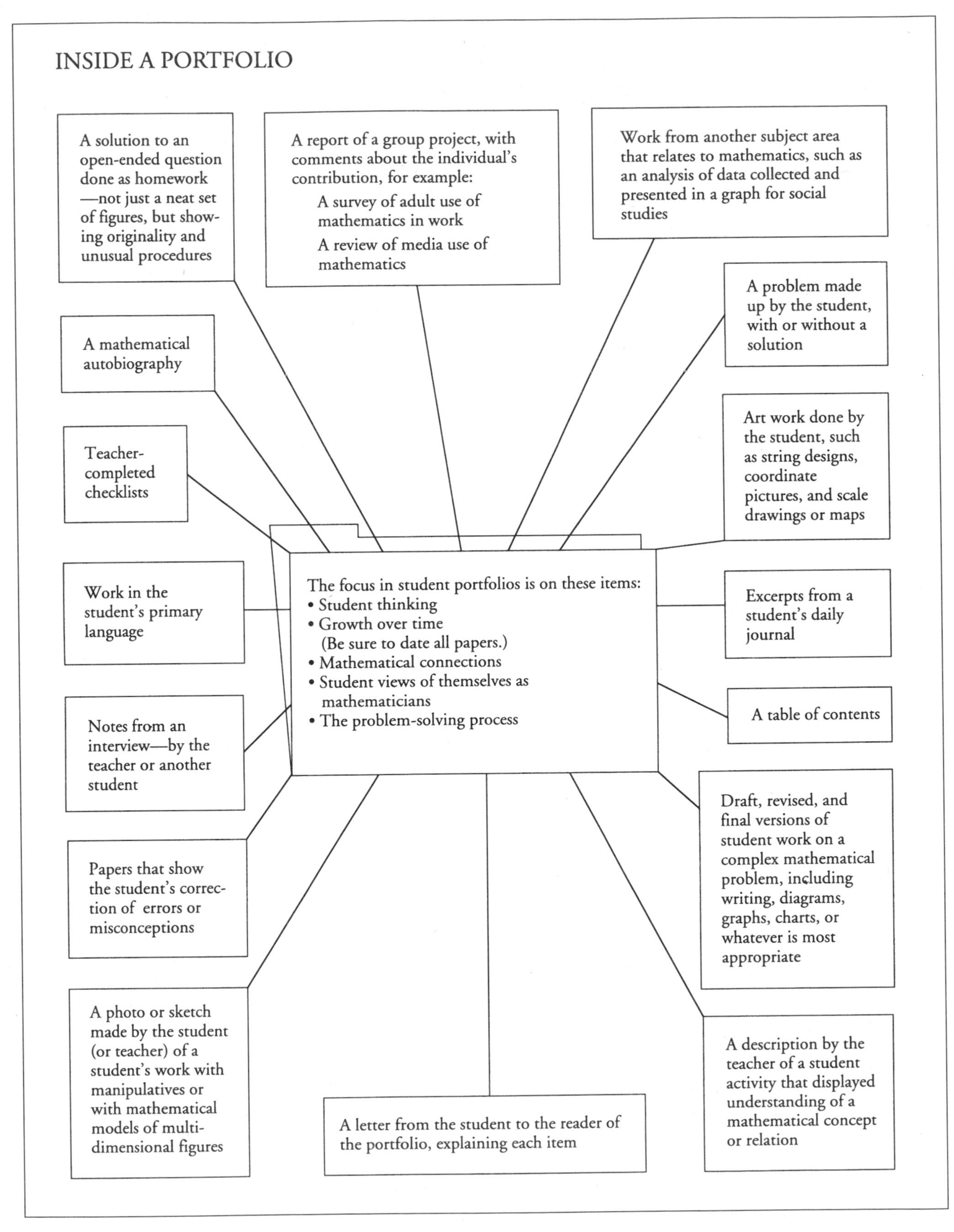

FIGURE D.17 **Inside a Portfolio**

J. K. Stenmark, ed. *Mathematics Assessment: Myths, Models, Good Questions, and Practical Suggestions* (Reston, VA: National Council of Teachers of Mathematics, 1991).

several samples of their work and write comments about why they've chosen those samples.

More extensive portfolios can be used for formal assessment. Students can see their growth over time and can showcase some of their work. Teachers can incorporate a portfolio grade into their grading system for the quarter. Indeed, programs can be evaluated using portfolios (see Figures D.18 and D.19).

One of the greatest values of portfolios is student self-assessment. When the student goes through the pieces he or she has produced, reflection on growth over time is inherent, consistency of performance is abundantly clear, topics covered are reviewed, and the sheer amount of learning that has taken place over the course of time is evident. Selection of tasks representative of his or her work is another reflective, evaluative exercise. The student identifies work that is best, most enlightening, most challenging, most fun, hardest, and so on. Explanation of why the selections were made encourages self-analysis and use of writing skills to demonstrate mathematical learning.

One of the greatest values of portfolios is student self-assessment.

Management of the portfolios is often one of the deterrents to using them. Knowing the worth of the information that portfolios convey, teachers have devised unique systems of organization. One of these methods is to use the portable milk-crate hanging file units. Each class can have its own unit. Color coding can be used effectively to designate classes. Colored file folders can be hung in the crate and labeled with the students' names. Colored file folders labeled also with the students' names can then be placed inside the appropriate hanging file folders. The file folders can then be readily removed to insert "keepers" and replaced easily in the hanging unit. The portable crates allow for portability so teachers can take a class to examine at a time. The crates can also be stacked to conserve space. Other teachers designate a single file cabinet for keeping the materials. Obviously, portfolios are effective only if materials are kept and organized.

Obviously, portfolios are effective only if materials are kept and organized.

Portfolio/Unit Scoring Rubric

Unit Assignments

Neatly assembled ______________

Unit assignment sheet completed ______________

All Papers included in order ______________

Grades recorded ______________

Paper headings complete ______________

Portfolio Assessment

Neatly assembled ______________

Portfolio inclusion sheet completed ______________

All papers included in order ______________

Selections included as requested ______________

Special/unique touches ______________

Total ______________

Total/10 = ______________

FIGURE D.18 **Portfolio/Unit Scoring Rubric**

Jodean Grunow, Dodgeville Middle School, jodean.grunow@dpi.state.wi.us.

Holistic Rubric for Using Portfolios to Evaluate Programs

Here is a sample holistic rubric for evaluating a group of portfolios. As with open-ended questions, many portfolios should be reviewed before your own rubric is finally developed and applied. Collect student work for a while; then take a class set of portfolios and sort them into several stacks.

One of the major benefits of portfolios is the opportunity for students and teachers to present their own best thinking and their most creative work rather than the thinking and work prescribed by others. It is recommended that this rubric be used only as a start in developing your own rubric.

Level 4 (Top Level)
The Level 4 portfolio is exciting to look through. It includes a variety of written and graphic mathematical work, indicating both individual and group work. Projects, investigations, diagrams, graphs, charts, photographs, audiotapes or videotapes, and other work indicate a broad and creative curriculum that leads students to think for themselves. There is evidence of student use of many resources: calculators, computers, reference libraries, and conversations with adults and students. Papers display student organization and analysis of information. Although neatness may not be a primary requisite, clarity of communication is important. Student self-assessment is shown by revisions of drafts, letters that explain why the student chose certain papers, or student-generated assessment lists or reports. Improvement in communication over time is reflected in samples from the beginning, middle, and end of the term. Student work reflects enthusiasm for mathematics.

Level 3
The Level 3 portfolio indicates a solid mathematics program. There is a variety of types of work presented, as in the top level. Students are able to explain fairly well their strategies and problem-solving processes. Some use of resources and group work may be evident, and students indicate good understanding, especially of basic mathematics concepts. Work over a period of time is included. The factors most likely to be missing are indications of student enthusiasm, self-assessment, extensive investigations, and student analysis of information.

Level 2
The Level 2 portfolio indicates an adequate mathematics program, somewhat bound by textbook requirements. There is little evidence of student original thinking as shown by projects, investigation, diagrams, and so on. Student explanations of the process by which they solved problems are minimal. There may be an overconcentration on arithmetic or similar algorithmic topics and a resulting lack of work from other content areas.

Level 1
The Level 1 portfolio includes almost no creative work and may consist mainly of ditto sheets or pages copied from a textbook. There is almost no evidence of student thinking. Papers are likely to be multiple-choice and short-answer and show no evidence that students are discussing mathematical ideas in class. Students do not explain their thinking about mathematical ideas.

FIGURE D.19 **Holistic Rubric for Using Portfolios to Evaluate Programs**
Stenmark, J. K., ed. *Mathematics Assessment: Myths, Models, Good Questions, and Practical Suggestions* (Reston, VA: National Council of Teachers of Mathematics, 1991.

Performance Assessment Rubric Scoring

(See Chapter 3: CIA (Curriculum, Instruction, Assessment)—an Integrated Whole.)

General Proficiency Categories

Advanced	Proficient	Basic	Minimal Performance
The student is distinguished in the content area. Academic achievement is beyond mastery. Test scores provide evidence of in-depth understanding in the academic content area tested.	The student is competent in the content area. Academic achievement includes mastery of the important knowledge and skills. Test scores shows evidence of skills necessary for progress in the academic content area tested.	The student is somewhat competent in the content area. Academic achievement includes mastery of most of the important knowledge and skills. Test score shows evidence of at least one major flaw in understanding the academic content area tested.	There is limited achievement in the content area. Test score shows evidence of major misconceptions or gaps in knowledge and skills in the academic content area tested.

Mathematics Proficiency Descriptors
Elementary Mathematics at Grade 4

Proficiency Level *(Scale Scores)*	**Proficiency Descriptors**
Advanced *(659 and above)*	• Demonstrates all the characteristics of proficient performance; consistently exhibits superior performance, especially in problem solving and mathematical communication. • In data analysis, draws information from multiple sources and infers solutions, providing data-based conclusions. • When solving real-world, non-routine problems, employs multiple strategies, where applicable, as well as shows in-depth reasoning. • Displays a highly developed sense of number and number relations and an understanding of number theory and the properties of numbers and operations.
Proficient *(623 to 658)*	• Consistently exhibits mastery of basic conceptual knowledge, skills, and problem solving. • Applies the four fundamental operations with whole numbers, adds and subtracts decimals and fractions, and determines the reasonableness of answers. • In geometry identifies two- and three-dimensional figures, congruence of figures; uses physical models to examine relationships. • Analyzes data from visual displays and applies it to solve problems. • Uses appropriate tools, understands appropriate units, and measures length to a specified degree of accuracy.
Basic *(581 to 622)*	• Demonstrates a good portion of expected conceptual knowledge and skills but may not be as proficient in applying them to problem solving situations. • Solves simple one-step story problems. • Mathematical computation is limited to addition and subtraction of whole numbers, simple basic multiplication facts, and addition of decimals without regrouping. • Recognizes, fills-in and extends numerical and geometric patterns. Reads a ruler and a thermometer.
Minimal Performance	Limited achievement. Evidence of major misconceptions or gaps in the knowledge and skills tested.

Taken from *Final Summary Report of the Proficiency Score Standards* (Department of Public Instruction, Office of Educational Accountability, 1997).

Middle Level Mathematics at Grade 8

Proficiency Level (*Scale Scores*)	Proficiency Descriptors
Advanced (*750 and above*)	• Consistently demonstrates very high levels of conceptual understanding, numerical, geometric and measurement skills, and problem solving ability. • Accurately applies computational skills with whole numbers, fractions, decimals, percents and integers to the solution of non-routine problems. • Uses knowledge of statistical techniques and theory of probability to establish conclusions and infer future events. • Communicates in a thorough and logical manner about solution strategies, the validity of their own conjectures, and the inferences of others.
Proficient (*718 to 740*)	• Demonstrates mastery of computational and estimation skills with decimals, fractions, and integers and applies these skills to the solution of two-step problems. • Shows ability to work with various kinds of visual displays of data, using them to support conclusions. • Applies measurement skills to determine perimeter and area in both customary and metric units. • Demonstrates competent analysis, solving, and evaluation of solutions to real-world problems by using appropriate symbols, tables, graphs, and algebraic expressions.
Basic (*674 to 717*)	• Demonstrates somewhat competent conceptual knowledge and skills. • Performs basic operations on whole numbers, decimals, and some fractions. In data analysis, works with bar and line graphs and determines possible outcomes of given events. • In geometry and measurement, recognizes most two- and three-dimensional figures, identifies congruence and similarity, and solves simple indirect measurement problems with physical models. • Works backward and uses guess-and-check as techniques to solve problems.
Minimal Performance	Limited achievement. Evidence of major misconceptions or gaps in the knowledge and skills tested.

High School Mathematics at Grade 10

Proficiency Level *(Scale Scores)*	Proficiency Descriptors
Advanced *(782 and above)*	• Consistently demonstrates in-depth understanding of conceptual knowledge, problem solving skills and ability to communicate in a thorough, logical, and articulate fashion. • Utilizes tools of data analysis, probability, and statistics to thoroughly examine data, make inferences, and draw conclusions. • Demonstrates use of a wide variety of high level algebraic, geometric and measurement skills. • Uses direct and indirect reasoning, gives examples while solving problems, makes conjectures, and/or judges the validity of the inferences of other persons.
Proficient *(744 to 781)*	• Consistently demonstrates the ability to apply conceptual knowledge and skills to a variety of problems. • Shows mastery of computation with and without calculators and estimates computations in real-life situations. • Other numerical skills include working with patterns, ratio and proportion, formulas, and translating amongst equivalent forms such as exponents, fractions, decimals, percents, and scientific notation. • Uses data presented in graphical form to rationalize and support arguments, inferences, or conclusions. • Works with probability of simple events, communicating about it with fractions, decimals, and percents. • Competent demonstration of measurement skills, including facility with scale drawings, are well developed.
Basic *(716 to 743)*	• Demonstrates somewhat competent success with most conceptual knowledge and skills, although level of mastery is less than that of proficient performance. • Supports conclusions with some clarity. • Somewhat competent with the basic operations with whole numbers, decimals, fractions, decimals and percents. • Uses appropriate measuring tools to obtain direct measurements and ratio and proportion for indirect measurements. • Algebraic skills include pattern recognition, substitution to solve equations and formulas, interpretation and use of expressions, and solution of one-step equations. • Works backwards and uses guess-and-check to solve problems.
Minimal Performance	Limited achievement. Evidence of major misconceptions or gaps in the knowledge and skills tested.

Taken from *Final Summary Report of the Proficiency Score Standards* (Department of Public Instruction, Office of Educational Accountability, 1997).

Sample Scoring Process

Scoring Process	Sun Prairie Experience
Assemble team of people to score the tasks.	**We used two models:** • Volunteers scored for 4 to 5 days during the summer. • One day during the school year was arranged for those that were interested in only a one-day commitment. This gave an opportunity for more teachers to have first-hand experience in the assessment process. Teachers from a variety of grade levels were invited to participate, even though their grade level was not being assessed.
Work through one task as individuals. Discuss questions and responses. Repeat for each task to be scored.	This is an extremely important step. The teachers who score the task need to actually *do* them to fully understand the task and the potential range of acceptable responses. The discussion is also extremely valuable as varying perspectives were shared.
Review "task-specific" rubric.	In order to score the tasks, teachers need to fully understand the rubric: what the various criteria mean and how to evaluate students' responses.
Look at student work examples and group papers according to rubric criteria.	All teachers first score the same set of student work and then, in small groups, compare results to determine consistent interpretation of the rubric criteria.
Discuss concerns.	As a large group, it is important to look at any possible misinterpretations of the rubric.
Score sets of papers using the rubric. Using teams of two, each person scores independently. When the entire set of papers has been scored, compare results. When the scores agree, record the score on the recording sheet (see below for a sample). When scores do not coincide, discuss reasons. If a mutually agreeable score cannot be reached, a third person will evaluate the task to determine the final score.	The total set of papers should be divided among the number of two-person scoring teams. Every effort should be made to have teachers score papers that are *not* from their own class to avoid "inferring what Johnny or Suzie probably meant."
Record results.	In a district with multiple buildings, it is helpful to have papers alphabetized and scoring sheets printed with student names (see sample below).
Analyze data.	Recording results on a spreadsheet will allow for analyzing data in various charts and graphs.

Used with permission of Sun Prairie School District. Contact: Diana Kasbaum, Eastside Elementary School, e-mail diana_kasbaum@wetn.pbs.org.

General Rubric
Sun Prairie Area School District
K–12 Mathematics Rubric

	0 Not Scorable	1–2 Minimal/Basic	3–4 Proficient (Meets Standard)	5 Advanced (Exceeds Standard)
Reasoning/Approach	No response Irrelevant response Inappropriate response	Partially addresses elements of the problem Uses incomplete or poorly executed strategy Draws faulty or flawed conclusions	Addresses important elements of the problem Uses appropriate strategy(ies) Draws logical conclusions	Uses sophisticated strategy(ies) Makes inferences, generalizations, conjectures
Communication	No response Irrelevant response Inappropriate response	Attempts to represent the solution process Partially organizes response Attempts to use mathematical terminology and symbols	Clearly represents the solution process (words, diagrams/pictures, charts/tables, graphs, etc.) Logically organizes response Uses mathematical terminology and symbols correctly	Uses multiple and detailed and/or sophisticated representations to explain the solution process
Knowledge & Skills	No response Irrelevant response Inappropriate response	Significant errors Inaccurate		

Student Rubric

Criteria	0	Minimal	Basic	Proficient	Advanced
Reasoning/ Approach	I did not try to solve the problem on paper.	I did not show any strategy to solve the problem.	I used a simple strategy. I showed how I solved part of the problem.	I used at least one efficient strategy correctly. I showed how I solved the whole problem.	I used at least one advanced strategy to solve the problem. My solution showed that I used advanced thinking to solve the problem.
Communication	I did not explain how I solved the problem.	I had trouble explaining how I solved the problem.	My explanation was partly organized. I used some math words and symbols.	My explanation was clear and organized. I used math words and symbols correctly.	My explanation was well organized and detailed. I used math words and symbols correctly.
Mathematical Knowledge/Skills	I did not try to solve the problem.	My solution was not correct. I did not understand how to solve the problem.	Part of my solution was correct. I understand how to solve most of the problem.	My solution was correct. I understand how to solve all parts of the problem.	My solution(s) were correct. I understood the problem and showed an advanced way to solve it.

Rubric for Scoring Mathematics Performance Assessment

Milwaukee Public Schools

	Score	Description
	4	**Advanced/Exceeding**
Strategy		An appropriate strategy is used effectively.
Organization		The response is highly organized and well documented.
Communication		Responses to all parts of the prompt are effective and appropriate.
Solution		A correct solution is presented.
Mathematics		Efficient, accurate use of mathematics is evident throughout.
	3	**Proficient**
Strategy		An appropriate strategy is used.
Organization		The response is sufficiently organized and documented.
Communication		Response is communicated and understandable.
Solution		A basically correct solution is presented.
Mathematics		Appropriate mathematics is applied with only minor flaws.
	2	**Basic**
Strategy		A strategy is attempted but is incomplete or poorly carried out.
Organization		The response is poorly organized and insufficiently documented.
Communication		Response is vague or reflects inadequate understanding.
Solution		A partial or incorrect solution is presented.
Mathematics		Some of the mathematics used is inappropriate or frequently flawed.
	1	**Minimal**
Strategy		No strategy is attempted, or it is unclear what the strategy is.
Organization		The response is disorganized.
Communication		Response reflects random thoughts or haphazard restatements of fact.
Solution		Minimal to no response is presented.
Mathematics		Little or no use of appropriate mathematics is presented.
	0	**Not Scorable**

Used with permission of Milwaukee Public Schools. Contact: Dr. DeAnn Huinker, University of Wisconsin–Milwaukee, huinker@uwm.edu.

Piloting Materials

(See Chapter 3: CIA (Curriculum, Instruction, Assessment)—an Integrated Whole.)

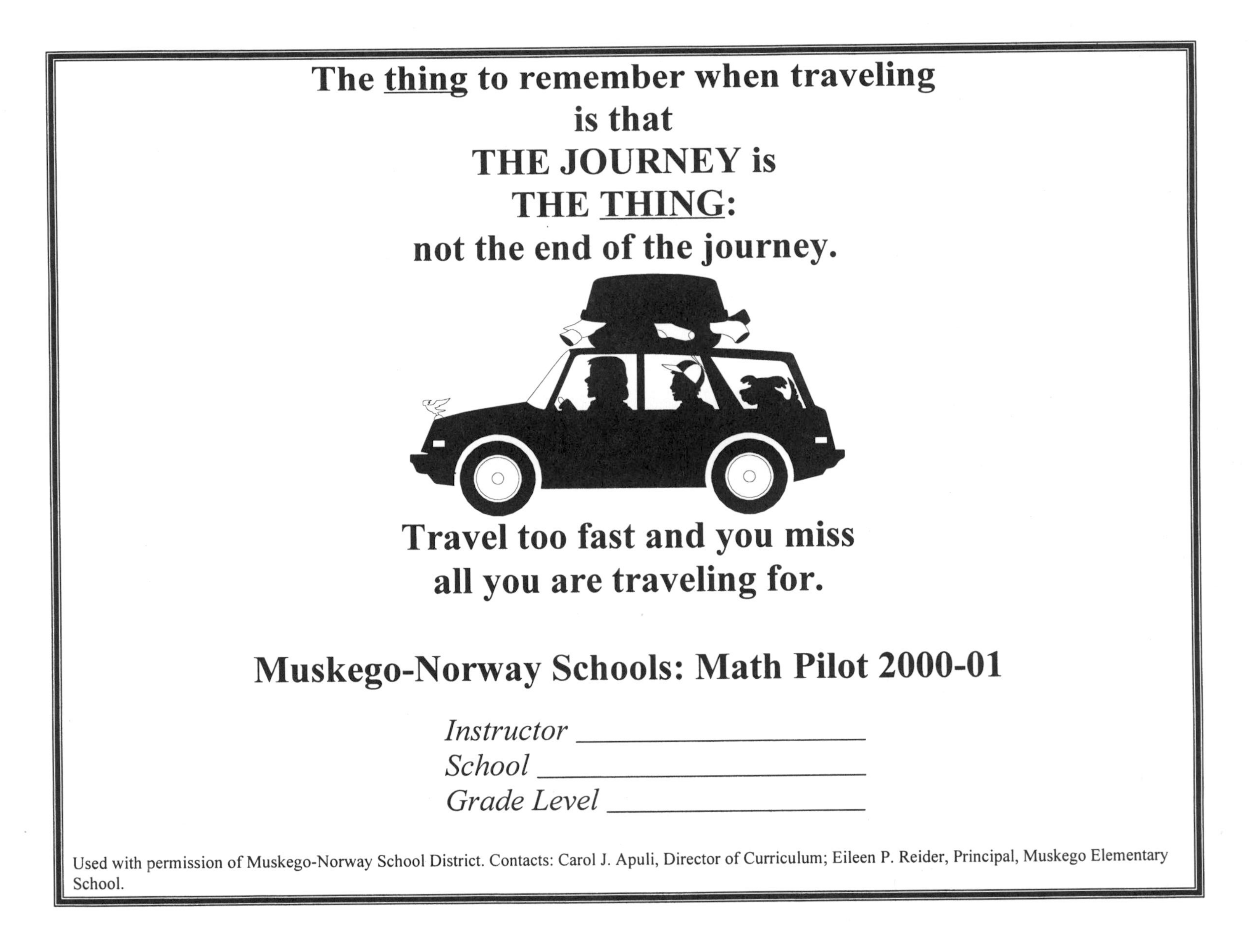
The thing to remember when traveling
is that
THE JOURNEY is
THE THING:
not the end of the journey.
Travel too fast and you miss
all you are traveling for.
Muskego-Norway Schools: Math Pilot 2000-01
Instructor ____________
School ____________
Grade Level ____________
Used with permission of Muskego-Norway School District. Contacts: Carol J. Apuli, Director of Curriculum; Eileen P. Reider, Principal, Muskego Elementary School.

Why Journal?

In a majority of classrooms that are identified as exemplary, we see children engaging in reflective activities. Often this reflection comes by way of the journal.

Over the course of the upcoming school year, you will be asked to respond weekly to a series of focus questions—one question each week. Some of these questions will cause you to examine your teaching, some will ask you to think about student learning. Others will invite sharing parent input or feedback; still others will ask you to collect student work samples. In every case it is our hope that you will take time to reflect on the work you are doing with children and grow as a professional.

What value does a journal have?

- It records important news of the day.
- It provides commentary, experiences, and editorial insights.
- It gathers the thoughts of our neighbors on the important issues of the day.
- With time, it offers us an overview (a bird's-eye view) about the really important concepts of the day.

So, too, this journal will provide both our Muskego-Norway math pilot teachers and other interested readers with insights into

- Support systems for the math piloters to better enable networking
- Dipsticking for areas of strength and concern
- Measurement of the climate of the district regarding math instruction
- Direction and helpful guidelines
- Qualitative data which gives the innermost feelings of our mathematics pilot staff
- Resources for addressing the needs and concerns of the staff

Used with permission of Muskego-Norway School District. Contacts: Carol J. Apuli, Director of Curriculum; Eileen P. Reider, Principal, Muskego Elementary School.

"What we can see easily is only a small percentage of what is possible. Imagination is having the vision to see what is just below the surface; to picture that which is essential, but invisible to the eye."

As you work through each unit, record your thoughts and ideas in *The Parking Lot* section.

What excites you about the prospect of working with this new material and what concerns you? (Explain)

Teacher's Response

Other Teacher Comments

Reply from Eileen and/or Carol

Used with permission of Muskego-Norway School District. Contacts: Carol J. Apuli, Director of Curriculum; Eileen P. Reider, Principal, Muskego Elementary School.

The Parking Lot
(Issues to be pursued later)

Unit ________________

	STUDENT LEARNING	STRATEGIES
LESSON	What are the main math ideas that you want to keep in mind?	What are management strategies you need to deal with?

Used with permission of Muskego-Norway School District. Contacts: Carol J. Apuli, Director of Curriculum; Eileen P. Reider, Principal, Muskego Elementary School.

Deciding on a Program

(See Chapter 3: CIA (Curriculum, Instruction, Assessment)—an Integrated Whole.)

K–5 Mathematics Program Review

Program/Text:

The following rubrics (Mathematical Content, Organization & Structure, Student Experiences, Teacher's Role, Assessment, and Standards) were used by teachers who piloted various elementary mathematics programs which were considered for a districtwide adoption. The adoption committee, some of whom also piloted materials, reviewed these completed rubrics and then used the final rubric when making the final decision.

Category	Description	Does Not Meet	Partially Meets	Fully Meets	Exceeds	Comments
Mathematical Content	Problem solving is built into the program. Mathematics is developed from problem situations. Amenable to individual, small-group, or large-group instruction. Uses open and flexible methods.					
	Communication is built into the program. Students have many opportunities to communicate mathematical ideas. Students are asked to explain, conjecture, and defend ideas orally and in writing. Students are asked to form multiple representations of ideas and relationships and formulate generalizations. Develops mathematical vocabulary.					
	Reasoning is built into the program. Students are asked to explain and justify their thinking. Students do inductive and deductive reasoning.					
	Connections are clear. Mathematics is approached as a whole. Concepts, procedures, and processes are interrelated and connect ideas among math topics, other content areas, and life situations. Program emphasizes both content standards (number & operations, algebra, geometry, measurement, and data analysis and probability) and content standards (problem solving, reasoning and proof, communication, connections, and representation) as defined by NCTM, WI Standards, and district expectations.					

Comments:

K–5 Mathematics Program Review

Program/Text:

Category	Description	Does Not Meet	Partially Meets	Fully Meets	Exceeds	Comments
Organization and Structure	Program is organized so students have sufficient time to explore and investigate in-depth major mathematics ideas.					
	Units include lessons/activities/projects that are multiday and emphasize connections between concepts. Promotes attainment of several (vs. just one) instructional objectives.					
	Students work on worthwhile mathematical tasks. Tasks do not separate mathematical thinking from concepts or skills. Tasks capture student curiosity and invite them to speculate and pursue their hunches. Tasks can be approached in more than one way and may have more than one reasonable solution. Students need to reason about different strategies.					
	Technology is incorporated. Program is designed to expect the use of calculators.					
	Program is appropriate for all students (LD, ED, Gifted & Talented, ESL).					

Comments:

Used with permission of Sun Prairie School District. Contact: Diana Kasbaum, Eastside Elementary School, e-mail diana_kasbaum@wetn.pbs.org.

K–5 Mathematics Program Review **Program/Text:**

Category	Description	Does Not Meet	Partially Meets	Fully Meets	Exceeds	Comments
Student Experiences	Students are active learners. Students are encouraged to explore and investigate mathematical ideas. Students are expected to read, write, and discuss mathematics. Students are asked to reason about strategies, outcomes, alternatives and pursue various paths.					
	Students are expected to work on group and individual projects and assignments.					
	Students are expected to construct their own understanding of mathematics and use prior knowledge when approaching a new task. Activities lead to the development and understanding of algorithms.					
	Students engage in mathematical discourse. Students talk with one another. Students reason and make conjectures. Students communicate and justify their thinking orally and in writing. Students represent their solution process in a variety of ways.					
	Students use manipulatives and technology to explore mathematical ideas, model situations, analyze data, calculate, and solve problems. Students decide what tools are needed and when to use them.					
	Students determine the need to determine exact or approximate answer and choose appropriate procedure (paper/pencil, mental calculation, calculator).					
	Students self-assess performance and disposition.					

Comments:

K–5 Mathematics Program Review

Program/Text:

Category	Description	Does Not Meet	Partially Meets	Fully Meets	Exceeds	Comments
Teacher's Role	Materials provide suggestions to assist teachers in shifting toward NCTM vision. Suggestions are provided to help students make sense of mathematics, reason, form conjectures, solve problems, and connect mathematics to other topics within mathematics and other disciplines.					
	Suggestions to enhance mathematical discourse. Encourages teachers to ask students to explain their thinking and reasoning.					
	Suggestions to engage all students. Students are expected to express themselves in writing, pictures, numbers, and concretely with manipulatives.					
	Suggestions for establishing a learning environment focused on sense making: • Structuring time • Physical space • Collaborative/cooperative activities					
	Teachers are encouraged to reflect on what happens in the classroom. Suggestions on how to observe, listen and gather information.					
	Suggestions on parent involvement. Materials to keep parents informed about the program.					

Comments:

K–5 Mathematics Program Review **Program/Text:**

Category	Description	Does Not Meet	Partially Meets	Fully Meets	Exceeds	Comments
Assessment	Assessment is integrated throughout the program. Assessment activities are similar to learning activities. Assesses applications to situations that require reasoning and creative thinking. Assesses communication and representation of ideas, both in writing and orally.					
	Uses multiple means of assessment. Calculators, computers, and manipulatives are built into assessment activities. Assessment is continuous, dynamic, and often an informal process.					
	All aspects of mathematical knowledge are assessed. Conceptual understanding and procedural knowledge are assessed through tasks. Students are asked to apply information in a new/novel situation.					

Comments:

Used with permission of Sun Prairie School District. Contact: Diana Kasbaum, Eastside Elementary School, e-mail diana_kasbaum@wetn.pbs.org.

K–5 Mathematics Program Review

Program/Text:

Category	Description	Does Not Meet	Partially Meets	Fully Meets	Exceeds	Comments
Standards WI/SPASD	Standard A: Mathematical Processes					
	Standard B: Number Operations and Relationships					
	Standard C: Geometry					
	Standard D: Measurement					
	Standard E: Statistics and Probability					
	Standard F: Algebraic Relationships					

Comments:

Used with permission of Sun Prairie School District. Contact: Diana Kasbaum, Eastside Elementary School, e-mail diana_kasbaum@wetn.pbs.org.

K–5 Mathematics Program Review

Program/Text:

The following rubric was used by the adoption committee to make the final decision. It was used in conjunction with the in-depth rubric completed by pilot teachers.

Category	Description	Does Not Meet	Partially Meets	Fully Meets	Exceeds	Comments
General Criteria	Inquiry based. Incorporates NCTM Standards.					
	Incorporates use of algebraic thinking.					
	Engages student-to-student discourse.					
	Includes open-ended questions.					
	Students are expected to explain their thinking.					
	Evidence of equity—ALL students have access to the lessons.					
	Incorporates the use of technology.					
	Incorporates the use of physical materials to make math meaningful.					
	Students construct their own meaning/ understanding of mathematics.					
	Students apply mathematics to real-world situations.					
	Communicates concepts learned at school to the home.					
	Teaches concepts first, not skills.					
	Concept driven.					
	Introduces mathematical concepts via other curricular areas.					
	Materials easy to use for lesson preparation.					

Comments:

Used with permission of Sun Prairie School District. Contact: Diana Kasbaum, Eastside Elementary School, e-mail diana_kasbaum@wetn.pbs.org.

Standards-Based Model

(See Chapter 3: CIA (Curriculum, Instruction, Assessment)—an Integrated Whole.)

Standards-Based Model in Action

Subject: MATHEMATICS **Example of recording activities used to build understanding of a standard.**

Content Standard: Mathematical Processes

Students in Wisconsin will draw on a broad body of mathematical knowledge and apply a variety of mathematical skills and strategies, including reasoning, oral and written communication, and the use of appropriate technology, when solving mathematical, real-world and non-routine problems.

By the end of Grade 4, students will

A.4.1 Use reasoning abilities to

- Perceive patterns.
- Identify relationships.
- Formulate questions for further exploration.
- Justify strategies.
- Test reasonableness of results.

What?	How?		How do you know?
Content	**Teaching/Learning Strategies**	**Resources**	**Assessment**
What students will learn. Relationships among and between factors and multiples. How knowledge of reverse relationships can be used to solve problems. What is estimation? What is involved in making a good estimate? Number and place value understanding.	*Activities that will be used to develop understanding of this standard.* Patterns in multiplication/division table. Using relationships between multiplication and division as a means for solving division problems. Estimate products. Work with missing digit problems.	*Materials that will be used to develop understanding of this standard.* Blank number chart. Problems that capitalize specifically on application of the inverse. Ask questions re: How much do you think . . . ? If you were to multiply a three digit number by a two digit number, what arrangement of numbers would give you the greatest product?	*Assessments that will be used to measure true understanding of this standard.* Are there more even or odd numbers in this multiplication chart? Explain how your knowledge of multiplication and division helped you in solving these problems. Why did you think your estimate would be . . .? Why did you select that particular sequence?

*Adapted from J. Mortell, *Curriculum Articulation Model*

Grunow/Kniess - 6/15/98

What?	How?		How do you know?
Content	**Teaching/Learning Strategies**	**Resources**	**Assessment**
What students will learn.	*Activities that will be used to develop understanding of this standard.*	*Materials that will be used to develop understanding of this standard.*	*Assessments that will be used to measure true understanding of this standard.*

Integration: Connections made within mathematics.	Connections with other subjects.
Make addition chart. Compare or contrast with multiplication chart. Connection and analysis of the four mathematical operations can then be developed.	The function of estimation in other disciplines. The concept of doing and undoing to solve problems across disciplines.

Subject: MATHEMATICS **Example of using an investigation to build understanding of a standard.**

Content Standard: Number Operations and Relationships

Students in Wisconsin will use numbers effectively for various purposes, such as counting, measuring, estimating, and problem solving.

By the end of Grade 8, students will

B.8.1 Read, represent, and interpret various rational numbers (whole numbers, integers, decimals, fractions, and percents) with verbal descriptions, geometric models, and mathematical notation (e.g., expanded, scientific, exponential).

What?	How?		How do you know?
Content	**Teaching/Learning Strategies**	**Resources**	**Assessment**
What students will learn. The effect of geometric declination—halving ($1/2$, $1/4$, $1/8$, etc.) Addition and subtraction of fractions Fractional equivalence Patterning to develop an algebraic equation Nonlinear graphing Limits Infinite series Exponential notation	*Activities that will be used to develop understanding of this standard.* **Brownie Problem** Use a brownie as a geometric model to study the effects of successive halving.	*Materials that will be used to develop understanding of this standard.* Brownie Napkins Knife Overheads Problem package, including: • Problem • Blank brownie models to record effects of successive halving • Graph paper • Graphing calculator • Read “Anno's Magic Multiplying Jar”	*Assessments that will be used to measure true understanding of this standard.* Can you develop a related extension question? What would happen with successive thirds? Describe how you added fractions with like denominators? With different denominators? List situations in which you use successive halving.

Adapted from J. Mortell (1998). *Curriculum Articulation Model.*
Grunow/Kniess - 6/15/98

<table>
<tr><th>What?</th><th colspan="2">How?</th><th>How do you know?</th></tr>
<tr><th>Content</th><th>Teaching/Learning Strategies</th><th>Resources</th><th>Assessment</th></tr>
<tr><td>What students will learn.</td><td>Activities that will be used to develop understanding of this standard.</td><td>Materials that will be used to develop understanding of this standard.</td><td>Assessments that will be used to measure true understanding of this standard.</td></tr>
<tr><td colspan="2">Integration: Connections made within mathematics.
Decimal and percent equivalences for successive halves
Exponential growth and decay
Infinite series
Limits
Geometric model/numerical pattern/algebraic equation/graphical representation</td><td colspan="2">Connections with other subjects.
Science:
• Successive 1/10ths and metric relations
• Scientific notation
Art:
• Chambered nautilus
• Effect of successive halving on rectangles</td></tr>
</table>

How Could a Brownie Be A Problem?

Each weekend Mom tries to bake a treat. The family especially likes it when she bakes brownies. To make sure that Dad gets in on the treats, Mom always takes the first brownie and sets it aside for him.

Then, Dad likes to make that brownie last as long as he can. So, on the first day, he sneaks into the kitchen and eats half of the brownie. Then he carefully puts what's left in a baggie so it'll stay fresh. The second day he eats half of what remains of the brownie from the first day; on the third day he eats half of what remains from the second day, and so on.

If Dad continues this process over the course of the week (seven days), how much of the brownie has he eaten? How much is left? If the process continues, can he make the brownie last all week until Mom bakes again?

Day	Amount of Brownie Eaten	Sum	Amount of Brownie Not Eaten
Start			
1			
2			
3			
4			
5			
6			
.			
.			
.			
10			
.			
.			
.			
n			

Subject: MATHEMATICS **Example of use of the chart to chronicle building understanding of a concept. Done over the course of the year, this can facilitate the study of curriculum alignment.**

Content Standard: Geometry

Students in Wisconsin will be able to use geometric concepts, relationships, and procedures to interpret, represent, and solve problems.
[Note: Familiar mathematical content dealing with measurement of geometric objects (e.g., length, area, volume) is presented in "D: Measurement."]

By the end of Grade 12, students will
C.12.3 Present convincing arguments by means of demonstration, informal proof, counterexamples, or any other logical means to show the truth of

- Statements (e.g., "These two triangles are not congruent.")
- Generalizations (e.g., "The Pythagorean theorem holds for all right triangles.")

What?	How?		How do you know?
Content	**Teaching/Learning Strategies**	**Resources**	**Assessment**
What students will learn. Attributes of shapes Comparison of attributes Modeling to solve shape/size questions Prediction from space and shape exploration Defense of conjecture through modeling, logical argument, demonstration	*Activities that will be used to develop understanding of this standard.* Chapter 5—Patterns in Space and Visualization Informal study of triangle transformations using Mira, coordinate graph paper, triangle models Informal Pythagorean Theorem investigation. "How can we demonstrate that c^2 does equal $a^2 + b^2$?	*Materials that will be used to develop understanding of this standard.* Text Variety of manipulatives Following investigations, Project Mathematica "Pythagorean Theorem" video "Wheel of Theodorus" Investigation, Connected Mathematics Project (if not experienced)	*Assessments that will be used to measure true understanding of this standard.* Why is it important to justify conjectures? What means can be used to present convincing arguments? When arguments are presented, what are important considerations in developing effective and convincing proofs?

Adapted from J. Mortell (1998). *Curriculum Articulation Model.*
Grunow/Kniess - 6/15/98

<table>
<tr><th>What?</th><th colspan="2">How?</th><th>How do you know?</th></tr>
<tr><th>Content</th><th>Teaching/Learning Strategies</th><th>Resources</th><th>Assessment</th></tr>
<tr><td>What students will learn.</td><td>Activities that will be used to develop understanding of this standard.

Further Study:
Unit 2—Patterns of Location, Shape, and Size
Unit 6—Geometric Form and Its Function</td><td>Materials that will be used to develop understanding of this standard.</td><td>Assessments that will be used to measure true understanding of this standard.</td></tr>
<tr><td colspan="2">Integration: Connections made within mathematics.

Progress to more formal methods of proof.
Connect to trigonometric relationships.
Connect to indirect measurement methods.</td><td colspan="2">Connections with other subjects.

Shape and size transformations in art and design.
Triangle relationships in engineering.</td></tr>
</table>

Standards-Based Model:
Curriculum/Instruction/Assessment

Subject:

Content Standard:

By the end of Grade _____, students will

What?	How?		How do you know?
Content	**Teaching/Learning Strategies**	**Resources**	**Assessment**
What students will learn.	*Activities that will be used to develop understanding of this standard.*	*Materials that will be used to develop understanding of this standard.*	*Assessments that will be used to measure true understanding of this standard.*

Adapted from J. Mortell (1998). *Curriculum Articulation Model.*
Grunow/Kniess - 6/15/98

What?	How?		How do you know?
Content	**Teaching/Learning Strategies**	**Resources**	**Assessment**
What students will learn.	*Activities that will be used to develop understanding of this standard.*	*Materials that will be used to develop understanding of this standard.*	*Assessments that will be used to measure true understanding of this standard.*

Integration: Connections made within ____________________.	Connections with other subjects.

References

Assessment Standards Working Groups. 1995. *Assessment Standards for School Mathematics*. Reston, VA: National Council of Teachers of Mathematics.

Azzolino, A. 1990. "Writing as a Tool for Teaching Mathematics: The Silent Revolution." In *Teaching and Learning Mathematics in the 1990s: 1990 Yearbook*, edited by T. J. Cooney and C. R. Hirsch. Reston, VA: National Council of Teachers of Mathematics.

Bartels, B. 1995. "Promoting Mathematics Connections with Concept Mapping." *Mathematics Teaching in the Middle School, 1* (7), (November–December): 542–549.

Beyerback, B. A. 1986. *Concept Mapping in Assessing Prospective Teacher's Concept Development.* (ERIC Document Reproduction Service No. ED 291 800).

Brown, M. 1988. *The Graded Assessment in Mathematics Project*. Unpublished manuscript, King's College, London, U.K.

Carpenter, T. P., and E. Fennema. 1988. "Research and Cognitively Guided Instruction." In *Integrating Research on Teaching and Learning Mathematics,* edited by E. Fennema, T. P. Carpenter, and S. J. Lamon. Madison: Wisconsin Center for Education Research.

Clarke, D. 1988. *Assessment Alternatives in Mathematics.* Melbourne, Australia: Jenkin Buxton Printers Pty. Ltd.

Commission on Teaching Standards for School Mathematics. 1991. *Professional Standards for Teaching Mathematics*. Reston, VA: National Council of Teachers of Mathematics.

Danielson, C. 1997. *A Collection of Performance Tasks and Rubrics.* Larchmont, NY: Eye On Education.

Entrekin, V. S. 1992. "Sharing Teaching Ideas: Mathematical Mind Mapping." *Mathematics Teacher, 85* (September): 444–445.

Fennema, E. 1998. Grunow Dissertation Discussion. Madison: The University of Wisconsin.

Grunow, J. E. 1999. *Using Concept Maps in a Professional Development Program to Assess and Enhance Teachers' Understanding of Rational Number.* Ann Arbor, MI: UMI (UMI Microform 9910421).

Howell, J. A. 1999. Cray Academy WASDI Teacher Leadership Workshop.

MacDonell, A., and C. Anderson, eds. 1999. *Balanced Assessment for the Mathematics Curriculum: Berkeley, Harvard, Michigan State, Shell Centre*. White Plains, NY: Dale Seymour Publications. (Available for elementary, middle school, and high school, two levels).

Mathematical Sciences Education Board. 1993. *Measuring Up: Prototypes for Mathematics Assessment.* Washington DC: National Academy Press.

New Standards. 1997. *Performance Standards.* Pittsburgh: National Center on Education and the Economy and the University of Pittsburgh.

Peressini, D., and N. Webb. 1999. "Analyzing Mathematical Reasoning in Students' Responses Across Multiple Performance Assessment Tasks." In *Developing Mathematical Reasoning in Grades K–12: 1999 Yearbook*, edited by L. V. Stiff and F. R. Curcio. Reston, VA: National Council of Teachers of Mathematics.

Ruiz-Primo, M. A., and R. J. Shavelson. 1996. "Problems and Issues in the Use of Concept Maps in Science Assessment." *Journal of Research in Science Teaching, 33*(6): 569–600.

Shafer, M. C., and T. A. Romberg. 1999. "Assessment in Classrooms that Promote Understanding." In *Mathematics Classrooms that Promote Understanding*, edited by E. Fennema and T.A. Romberg. Mahwah, NJ: Lawrence Erlbaum Associates.

Shell Centre for Mathematical Education. 1984. *Problems with Patterns and Numbers*. University of Nottingham, U.K.: Joint Matriculation Board.

Silver, E. A., J. Kilpatrick, and B. Schlesinger. 1990. *Thinking Through Mathematics.* New York: College Entrance Examination Board.

Standards 2000 Project. 2000. *Principles and Standards for School Mathematics*. Reston, VA: National Council of Teachers of Mathematics.

Stenmark, J. K., ed. 1991. *Mathematics Assessment: Myths, Models, Good Questions, and Practical Suggestions*. Reston, VA: National Council of Teachers of Mathematics.

Stenmark, J. K., EQUALS Staff and Assessment Committee of the Campaign for Mathematics. 1989. *Assessment Alternatives in Mathematics*. Berkeley, CA: University of California.

Webb, N. L. 1992. "Assessment of Students' Knowledge of Mathematics: Steps toward a Theory." In *Handbook of Research on Mathematics Teaching and Learning*, edited by D. A. Grouws. New York: Macmillan Publishing Co.

Webb, N. L., and D. L. Briars. 1990. "Assessment in Mathematics Classrooms, K–8." In *Teaching and Learning Mathematics in the 1990s: 1990 Yearbook*, edited by T. J. Cooney and C. R. Hirsch. Reston, VA: National Council of Teachers of Mathematics.

Appendix E Readings

E

APPENDIX E

"Never Say Anything a Kid Can Say!" is an article written by one of the guidelines task force members, Steven Reinhart, that appeared in the NCTM journal, *Mathematics Teaching in the Middle School.* It is a wonderful article that provides insight into the development of meaningful classroom discourse congruent with the vision of the various standards pieces and this document. Thanks to Steve for sharing this with us!

"Getting Started at Improving Your Math Education Program: One District's Ongoing Story," was written by Glen Miller, director of curriculum for the Kiel School District, another one of the task force members. It is a reflective piece written to share some of the thoughts, processes, and results involved in focusing district effort on improvement of the mathematics program. Glen has lead his district through an extensive evaluation of the Kiel mathematics program and is now in the process of supporting implementation of three standards-based curricula. The process has been one of the most carefully devised and executed and is a model for other districts to emulate. We appreciate Glen's efforts and his sharing of his district's story.

Never Say Anything a Kid Can Say!
Steven C. Reinhart[1]

After extensive planning, I presented what should have been a masterpiece lesson. I worked several examples on the overhead projector, answered every student's question in great detail, and explained the concept so clearly that surely my students understood. The next day, however, it became obvious that the students were totally confused. In my early years of teaching, this situation happened all too often. Even though observations by my principal clearly pointed out that I was very good at explaining mathematics to my students, knew my subject matter well, and really seemed to be a dedicated and caring teacher, something was wrong. My students were capable of learning much more than they displayed.

Implementing Change over Time

The low levels of achievement of many students caused me to question how I was teaching, and my search for a better approach began. Making a commitment to change 10 percent of my teaching each year, I began to collect and

[1]Steven Reinhart, steven_reinhart@wetn.pbs.org, taught mathematics at Chippewa Falls Middle School, Chippewa Falls, WI 54729. He is interested in the teaching of algebraic thinking at the middle school level and in the professional development of teachers.

use materials and ideas gathered from supplements, workshops, professional journals, and university classes. Each year, my goal was simply to teach a single topic in a better way than I had the year before.

Each year, my goal was simply to teach a single topic in a better way than I had the year before.

Before long, I noticed that the familiar teacher-centered, direct-instruction model often did not fit well with the more in-depth problems and tasks that I was using. The information that I had gathered also suggested teaching in nontraditional ways. It was not enough to teach better mathematics; I also had to teach mathematics better. Making changes in instruction proved difficult because I had to learn to teach in ways that I had never observed or experienced, challenging many of the old teaching paradigms. As I moved from traditional methods of instruction to a more student-centered, problem-based approach, many of my students enjoyed my classes more. They really seemed to like working together, discussing and sharing their ideas and solutions to the interesting, often contextual, problems that I posed. The small changes that I implemented each year began to show results. In five years, I had almost completely changed both *what* and *how* I was teaching.

It was not enough to teach better mathematics; I also had to teach mathematics better.

The Fundamental Flaw

At some point during this metamorphosis, I concluded that a fundamental flaw existed in my teaching methods. When I was in front of the class demonstrating and explaining, I was learning a great deal, but many of my students were not! Eventually, I concluded that if my students were to ever really learn mathematics, *they* would have to do the explaining, and *I*, the listening. My definition of a good teacher has since changed from "one who explains things so well that students understand" to "one who gets students to explain things so well that they can be understood."

Eventually, I concluded that if my students were to ever really learn mathematics, they would have to do the explaining, and I, the listening.

Getting middle school students to explain their thinking and become actively involved in classroom discussions can be a challenge. By nature, these students are self-conscious and insecure. This insecurity and the effects of negative peer pressure tend to discourage involvement. To get beyond these and other roadblocks, I have learned to ask the best possible questions and to apply strategies that require all students to participate. Adopting the goals and implementing the strategies and questioning techniques that follow have helped me develop and improve my questioning skills. At the same time, these goals and strategies help me create a classroom atmosphere in which students are actively engaged in learning mathematics and feel comfortable in sharing and discussing ideas, asking questions, and taking risks.

I have learned to ask the best possible questions and to apply strategies that require all students to participate.

Questioning Strategies That Work for Me

Although good teachers plan detailed lessons that focus on the mathematical content, few take the time to plan to use specific questioning techniques on a regular basis. Improving questioning skills is difficult and takes time, practice, and planning. Strategies that work once will work again and again. Making a list of good ideas and strategies that work, revisiting the list regularly, and planning to practice selected techniques in daily lessons will make a difference.

Create a plan

The following is a list of reminders that I have accumulated from the many outstanding teachers with whom I have worked over several years. I revisit this list often. None of these ideas is new, and I claim none, except the first one, as my own. Although implementing any single suggestion from this list may not result in major change, used together, these suggestions can help transform a classroom. Attempting to change too much too fast may result in frustration and failure. Changing a little at a time by selecting, practicing, and refining one or two strategies or skills before moving on to others can result in continual, incremental growth. Implementing one or two techniques at a time also makes it easier for students to accept and adjust to the new expectations and standards being established.

Used together, these suggestions can help transform a classroom.

Changing a little at a time by selecting, practicing, and refining one or two strategies or skills before moving on to others can result in continual, incremental growth.

1. Never say anything a kid can say! This one goal keeps me focused. Although I do not think that I have ever met this goal completely in any one day or even in a given class period, it has forced me to develop and improve my questioning skills. It also sends a message to students that their participation is essential. Every time I am tempted to tell students something, I try to ask a question instead.
2. Ask good questions. Good questions require more than recalling a fact or reproducing a skill. By asking good questions, I encourage students to think about, and reflect on, the mathematics they are learning. A student should be able to learn from answering my question, and I should be able to learn something about what the student knows or does not know from her or his response. Quite simply, I ask good questions to get students to think and to inform me about what they know. The best questions are open-ended, those for which more than one way to solve the problem or more than one acceptable response may be possible.
3. Use more process questions than product questions. Product questions—those that require short answers or a yes or no response or those that rely almost completely on memory—provide little information about what a student knows. To find out what a student understands, I ask process questions that requires the student to reflect, analyze, and explain his or her thinking and reasoning. Process questions require students to think at much higher levels.
4. Replace lectures with sets of questions. When tempted to present information in the form of a lecture, I remind myself of this definition of a lecture: "The transfer of information from the notes of the lecturer to the notes of the student without passing through the minds of either." If I am still tempted, I ask myself the humbling question "What percent of my students will actually be listening to me?"
5. Be patient. Wait time is very important. Although some students always seem to have their hands raised immediately, most need more time to

I now believe that all students learn more when I pose a high-quality problem and give them the necessary time to investigate, process their thoughts, and reflect on and defend their findings.

process their thoughts. If I always call on one of the first students who volunteers, I am cheating those who need more time to think about, and process a response to, my question. Even very capable students can begin to doubt their abilities, and many eventually stop thinking about my questions altogether. Increasing wait time to five seconds or longer can result in more and better responses.

Good discussions take time; at first, I was uncomfortable in taking so much time to discuss a single question or problem. The urge to simply tell my students and move on for the sake of expedience was considerable. Eventually, I began to see the value in what I now refer to as a "less is more" philosophy. I now believe that all students learn more when I pose a high-quality problem and give them the necessary time to investigate, process their thoughts, and reflect on and defend their findings.

Students should understand that all their statements are valuable to me, even if they are incorrect or show misconceptions.

Share with students reasons for asking questions

Students should understand that all their statements are valuable to me, even if they are incorrect or show misconceptions. I explain that I ask them questions because I am continuously evaluating what the class knows or does not know. Their comments help me make decisions and plan the next activities.

Teach for success

If students are to value my questions and be involved in discussions, I cannot use questions to embarrass or punish. Such questions accomplish little and can make it more difficult to create an atmosphere in which students feel comfortable sharing ideas and taking risks. If a student is struggling to respond, I move on to another student quickly. As I listen to student conversations and observe their work, I also identify those who have good ideas or comments to share. Asking a shy, quiet student a question when I know that he or she has a good response is a great strategy for building confidence and self-esteem. Frequently, I alert the student ahead of time: "That's a great idea. I'd really like you to share that with the class in a few minutes."

Be nonjudgmental about a response or comment

This goal is indispensable in encouraging discourse. Imagine being in a classroom where the teacher makes this comment: "Wow! Brittni, that was a terrific insightful response! Who's next?" Not many middle school students have the confidence to follow a response that has been praised so highly by a teacher. If a student's response reveals a misconception and the teacher replies in a negative way, the student may be discouraged from volunteering again. Instead, encourage more discussion and move on to the next comment. Often, students disagree with one another, discover their own errors, and correct their thinking. Allowing students to listen to fellow classmates is a far more positive way to deal with misconceptions than announcing to the class that an answer is incorrect. If several students remain confused, I might say, "I'm hearing that we do not agree on this issue. Your comments and ideas have given me an idea for an activity that will help you clarify your thinking." I then plan to revisit the concept with another activity as soon as possible.

Try not to repeat students' answers

If students are to listen to one another and value one another's input, I cannot repeat or try to improve on what they say. If students realize that I will repeat or clarify what another student says, they no longer have a reason to listen. I must be patient and let students clarify their own thinking and encourage them to speak to their classmates, not just to me. All students can speak louder—I have heard them in the halls! Yet I must be careful not to embarrass someone with a quiet voice. Because students know that I never accept just one response, they think nothing of my asking another student to paraphrase the soft-spoken comments of a classmate.

If students realize that I will repeat or clarify what another student says, they no longer have a reason to listen.

"Is this the right answer?"

Students frequently ask this question. My usual response to this question might be that "I'm not sure. Can you explain your thinking to me?" As soon as I tell a student that the answer is correct, thinking stops. If students explain their thinking clearly, I ask a "What if?" question to encourage them to extend their thinking.

Participation is not optional!

I remind my students of this expectation regularly. Whether working in small groups or discussing a problem with the whole class, each student is expected to contribute his or her fair share. Because reminding students of this expectation is not enough, I also regularly apply several of the following techniques:

Each student is expected to contribute his or her fair share.

1. Use the think-pair-share strategy. Whole-group discussions are usually improved by using this technique. When I pose a new problem, present a new project, task, or activity, or simply ask a question, all students must think and work independently first. In the past, letting students begin working together on a task always allowed a few students to sit back while others took over. Requiring students to work alone first reduces this problem by placing the responsibility for learning on each student. This independent work time may vary from a few minutes to the entire class period, depending on the task.

   After students have had adequate time to work independently, they are paired with partners or join small groups. In these groups, each student is required to report his or her findings or summarize his or her solution process. When teams have had the chance to share their thoughts in small groups, we come together as a class to share our findings. I do not call for volunteers but simply ask one student to report on a significant point discussed in the group. I might say, "Tanya, will you share with the class one important discovery your group made?" or "James, please summarize for us what Adam shared with you." Students generally feel much more confident in stating ideas when the responsibility for the response is being shared with a partner or group. Using the think-pair-share strategy helps me send the message that participation is not optional.

A modified version of this strategy also works in whole-group discussions. If I do not get the responses that I expect, either in quantity or quality, I give students a chance to discuss the question in small groups. On the basis of the difficulty of the question, they may have as little as fifteen seconds or as long as several minutes to discuss the question with their partners. This strategy has helped improve discussions more than any others that I have adopted.

I explain that it is all right to be confused, but students are responsible for asking questions that might help them understand.

2. If students or groups cannot answer a question or contribute to the discussion in a positive way, they must ask a question of the class. I explain that it is all right to be confused, but students are responsible for asking questions that might help them understand.
3. Always require students to ask a question when they need help. When a student says, “I don't get it,” he or she may really be saying, “Show me an easy way to do this so I don't have to think.” Initially, getting students to ask a question is a big improvement over “I don't get it.” Students soon realize that my standards require them to think about the problem in enough depth to ask a question.

Never accept only one response to a question.

4. Require several responses to the same question. Never accept only one response to a question. Always ask for other comments, additions, clarifications, solutions, or methods. This request is difficult for students at first because they have been conditioned to believe that only one answer is correct and that only one correct way is possible to solve a problem. I explain that for them to become better thinkers, they need to investigate the many possible ways of thinking about a problem. Even if two students use the same method to solve a problem, they rarely explain their thinking in exactly the same way. Multiple explanations help other students understand and clarify their thinking. One goal is to create a student-centered classroom in which students are responsible for the conversation. To accomplish this goal, I try not to comment after each response. I simply pause and wait for the next student to offer comments. If the pause alone does not generate further discussion, I may ask, “Next?” or “What do you think about __________'s idea?”

No one in a group is finished until everyone in the group can explain and defend the solution.

5. No one in a group is finished until everyone in the group can explain and defend the solution. This rule forces students to work together, communicate, and be responsible for the learning of everyone in the group. The learning of any one person is of little value unless it can be communicated to others, and those who would rather work on their own often need encouragement to develop valuable communication skills.
6. Use hand signals often. Using hand signals—thumbs up or thumbs down (a horizontal thumb means “I'm not sure”)—accomplishes two things. First, by requiring all students to respond with hand signals, I ensure that all students are on task. Second, by observing the responses, I can find out how many students are having difficulty or do not understand. Watching students' faces as they think about how to respond is very revealing.

7. Never carry a pencil. If I carry a pencil with me or pick up a student's pencil, I am tempted to do the work for the student. Instead, I must take time to ask thought-provoking questions that will lead to understanding.
8. Avoid answering my own questions. Answering my own questions only confuses students because it requires them to guess which questions I really want them to think about, and I want them to think about all my questions. I also avoid rhetorical questions.
9. Ask questions of the whole group. As soon as I direct a question to an individual, I suggest to the rest of the students that they are no longer required to think.
10. Limit the use of group responses. Group responses lower the level of concern and allow some students to hide and not think about my questions.
11. Do not allow students to blurt out answers. A student's blurted out answer is a signal to the rest of the class to stop thinking. Students who develop this habit must realize that they are cheating other students of the right to think about the question.

As soon as I direct a question to an individual, I suggest to the rest of the students that they are no longer required to think.

Summary

Like most teachers, I entered the teaching profession because I care about children. It is only natural for me to want them to be successful, but by merely telling them answers, doing things for them, or showing them shortcuts, I relieve students of their responsibilities and cheat them of the opportunity to make sense of the mathematics that they are learning. To help students engage in real learning, I must ask good questions, allow students to struggle, and place the responsibility for learning directly on their shoulders. I am convinced that children learn in more ways than I know how to teach. By listening to them, I not only give them the opportunity to develop deep understanding but also am able to develop true insights into what they know and how they think.

By listening to them, I not only give them the opportunity to develop deep understanding but also am able to develop true insights into what they know and how they think.

Making extensive changes in curriculum and instruction is a challenging process. Much can be learned about how children think and learn, from recent publications about learning styles, multiple intelligences, and brain research. Also, several reform curriculum projects funded by the National Science Foundation are now available from publishers. The Connected Mathematics Project, Mathematics in Context, and Math Scape, to name a few, artfully address issues of content and pedagogy.

Bibliography

Burns, Marilyn. *Mathematics: For Middle School.* New Rochelle, NY: Cuisenaire Co. of America, 1989.

Johnson, David R. *Every Minute Counts.* Palo Alto, CA: Dale Seymour Publications, 1982.

National Council of Teachers of Mathematics (NCTM). *Professional Standards for Teaching Mathematics.* Reston, VA: NCTM, 1991.

Getting Started at Improving Your Math Education Program: One District's Ongoing Story

You've come to realize your math program isn't doing the job you'd like of focusing on the Wisconsin math standards—and you have a feeling that there must be better ways of approaching the teaching and learning of math, particularly in view of the call for shifts in mathematical emphases and the need in today's world for much better math literacy on the part of ALL students. Still, it's difficult to know what to do, especially in the face of what former U.S. Secretary of Education Richard W. Riley has referred to as the "polarizing math wars" that have been going on for several years now.

Secretary Riley has further said that "we need to raise [our] standards higher and ensure that all students are learning more challenging concepts in addition to the traditional basics. . . . Those states that have developed challenging standards of learning, aligned their assessments to those standards, and provided substantial professional development for teachers, have demonstrated improvement in student achievement . . . these are issues of real importance—as opposed to politically inspired debates that will serve to sidetrack us from real improvement. We are suffering from an 'either-or' mentality . . . when what we need is a mix of information and styles of providing that information. We need to provide traditional basics, along with more challenging concepts, as well as the ability to problem solve, and to apply concepts in real world settings. The goal must be that all teachers teach better math *and* teach math better."

The goal must be that all teachers teach better math and teach math better.

—Riley 1998

How do you begin to rationally look at something so emotionally and intellectually polarizing yet at the same time so much in need of attention; that is, changing a curriculum and methodology that has been pretty much unchanged for literally hundreds of years (and in a world that is changing radically and rapidly)? Here's how one small community approached the problem beginning in 1996.

The Hometown School District serves about 1,525 K–12 students in a small city of 3,000 and its surrounding rural community. There has always been strong community pride in the local schools, though by the mid-1990s new assessment indicators such as Wisconsin Knowledge and Concepts Examinations (WKCE) (state achievement testing) scores were showing quite average results on subtests . . . and this in spite of local demographics that did *not* correlate well with such modest achievement levels (a natural concern for the school board).

The emphasis was on beginning to move all personnel from habits of too much day-to-day detail management toward more important real educational questions of what was going to be needed to make the schools the best they could be.

By the early 1990s the Hometown Board of Education was becoming sensitive to perceived internal stagnation in parts of the school system. In 1994 the district administrator of more than a decade retired and was replaced by a successor whose assigned mission was to focus on school improvement and giving new impetus to making the Hometown schools the best they could be for the full range of students. During this administrator's initial two years between 1994 and 1996, the emphasis was on beginning to move all personnel from habits of too much day-to-day detail management toward more important real educational questions of what was going to be needed to make the schools the best they could be.

During the 1995–96 school year, there occurred (for the first time) a strategic planning process that involved nearly fifty people from the schools and community, led by an outside facilitator. From the time "the Plan" was adopted by the board of education in spring 1996, it was held to be the district's organizational "driver," which has translated to school improvement issues (e.g., the correlates of effective schools) and more focus on customer satisfaction. The board was clear about its desire to see positive change and growth as well as efforts toward customizing education to the needs of the broad variety of families and individual customers (this being one of the four strategic objectives of the plan). Administrative team members were told they'd need to become the "enzymes acting as catalytic agents" for this renewed focus on school improvement and customer attitudes.

The board was clear about its desire to see positive change and growth as well as efforts toward customizing education to the needs of the broad variety of families and individual customers.

As for curriculum in particular, the district administrator and school board indicated that math was an area they wanted to see given immediate attention. The director of curriculum and assessment was to get this effort in motion on the part of the K–12 Math Committee and was expected to provide ongoing guidance and support for that group's efforts. (All this was taking place during that time when the state of Wisconsin was still encouraging local districts to develop their own academic standards and just prior to the existence of WMAS.) It was expected by the Hometown district administrator that leadership in the work of standards development and curriculum review and alignment would merge from within the various K–12 curriculum committees in the district—and in fact the math staff had been clearly given to understand that foot-dragging in reviewing and upgrading the program (which was perceived in some quarters to have been going on in previous years) would no longer be accepted. If the local staff would not actively involve themselves in the effort to look at NCTM standards and those offered in the Mid-continent Regional Educational Laboratory (McREL) Compendium of Standards for K–12 Education and did not also look at the local math program in light of the literature on the renewal of math education to better serve recommended standards, then the district had determined to have this accomplished "externally," with teachers subsequently being informed what the local math standards would be and what program(s) would be put in place to serve those standards (along with expectations by the board for higher WKCE state test results).

It was expected by the Hometown district administrator that leadership in the work of standards development and curriculum review and alignment would merge from within the various K–12 curriculum committees in the district.

In early September 1996, the curriculum director was directed to search for and bring in an outside math consultant to aid in this process, to help with what was perceived as a bogged-down subject area, disinclined toward self-review and change, and to make a subsequent recommendation to the board by February 1997. By early November the district had settled on two consultants, one from DPI and another working at UW–Madison who had previously been a DPI math consultant for a number of years. During this same time period, the curriculum director spoke informally with individual members of the K–12 Math Committee, telling them about the consultant search and trying to communicate excitement over the board's attitude of empowerment of the committee to truly study math education and arrive at recommendations for the board. The director learned that in fact most members of the committee

1. University of Chicago School Mathematics Project (UCSMP) Everyday Math (one classroom group in each of grades 1 through 4).
2. Connected Math Project (CMP in selected units in grades 5 through 8).
   NOTE: This district has a grades 5–8 middle school, which accounted for the decision to switch from Everyday Math to CMP at Grade 5.
3. Interactive Math Program (IMP) was used in three high school math sections that would otherwise have been algebra and pre-algebra classes—there were also two other regular algebra classes in place.

The board was impressed and excited by the attitudes of the teachers.

The math teachers impressed the members of the Curriculum Committee with their excitement and enthusiasm for the three programs, so they were also asked to make their full presentation to the whole board in early June. The board was impressed and excited by the attitudes of the teachers, and they voted to support the pilot programs for 1997–98, financially and philosophically.

During that summer of 1997 before the pilots were to begin, two middle school members of the math committee attended a two-week CMP math training session at UW–Platteville, both to immerse themselves in the program and to help enable them to subsequently train the other middle-level teachers as the pilot began in fall. Also, the two high school math teachers who were to be teaching the IMP math (Level I) pilot spent a week in training in the Minneapolis schools (followed by several follow-up training days in Minneapolis during the course of the pilot year)—paid for via both local and Eisenhower funds. The elementary pilot team (one teacher at each of grades 1 through 4) did not have the benefit of a specific training workshop prior to the pilot. Instead they worked to familiarize themselves with their new materials on their own during the summer. As the school year approached, everyone had feelings of both excitement and trepidation, feeling that the board and families would be watching as things progressed.

As the 1997–98 year began, the first math meeting to take place was a full-day CMP training session (during a preschool in-service day) for the 11-member middle-level group, led by the two members who had been trained in Platteville as mentioned. Because not all these 11 people had been directly involved on the K–12 math committee, there were varied levels of understanding and commitment for the changes to come. On the other hand, the aura of excitement and the informal leadership positions held by the leaders carried the day and got everyone very involved. At the high school and elementary levels, the teachers involved in the pilot year had in fact all been involved on the K–12 math committee, making for a smoother implementation from the outset.

The group was provided several half-days to meet off campus to train and to discuss their experiences, successes, and concerns. This was very important to the process of developing understanding and ownership on the part of all teachers.

Again on October 10 (and on other occasions during the pilot year), the middle-level group was provided several half-days to meet off campus to train and to discuss their experiences, successes, and concerns. This was very important to the process of developing understanding and ownership on the part of all teachers.

The first full K–12 math committee meeting of 1997–98 took place in mid-October, once the piloting teachers were really into their projects and getting anxious to compare notes across grade levels and buildings. Incidentally, one result of the way this whole project had taken shape as a K–12 "movement" was a truly K–12 point of view by all the teachers as we approached a redevelopment of the math program—the high school was very interested in the experiences of the elementary teachers and so forth. During this pilot year, the formal half-day K–12 math meetings took place in October, November, January, and February. During those months some of the subjects dealt with included

- Success and concerns
- Need for parent education and parent reactions at fall parent conferences
- Plans for "math night" for families
- Collection of evidence
- Timelines for the future
- Mapping of the pilot programs to the by-now-finalized Wisconsin standards (later also provided by the vendors of each program)
- Discussions leading to a unanimous and resounding decision to recommend board adoption of all three programs for 1998–99, and the development of an adoption calendar as follows:
  1. March: Math committee presentation to Curriculum Coordinating Council (CCC) to obtain its support
  2. March: Math committee presentation to board's Curriculum Committee to request its support for adoption
  3. April: Math committee presentation to full board for a first reading of motion to adopt and fund all three programs for 1998–99
  4. May: Second reading of motion by full board
- Level subcommittees (K–4, 5–8, 9–12) were assigned to develop budgets for two possible degrees of adoption/implementation for fall 1998
- At the February math committee meeting, each subcommittee did a practice presentation of its recommendations, data, and budget for reactions and recommendations by the rest of the K–12 math committee.

One result of the way this whole project had taken shape as a K–12 "movement" was a truly K–12 point of view by all the teachers as we approached a redevelopment of the math program.

The CCC meeting went very well, with the group posing helpful questions and suggestions to the math committee. Owing to the outstanding subsequent presentation to the Curriculum Committee, the chairperson (a board member) asked that the math committee do their full (lengthy) presentation to the whole board, which is not the usual way the board deals with issues when they come from a standing board committee. This was meant to be a compliment to the math committee and its work and was taken as such. The board approved all three programs and funding for full adoption on May 12, 1998.

Now our attention shifted to issues of materials procurement and broader staff training, particularly at the elementary level (remember, at this level even the pilot teachers had not had formal training). To begin, we made it a point to obtain all the Everyday Math (EM) materials before the close of the 1997–98 school year so that on the closing in-service day, all the teachers could have a half-day session with the company's area representative, who briefly familiarized them with the materials and methodologies. The teachers

We made it a point to obtain all the materials before the close of the school year so that on the closing in-service day, all the teachers could have a half-day session with the company's area representative.

then took the materials home over the summer. During the August 1998 preschool in-service, the elementary teachers had a full day (in house) with a program developer from the University of Chicago. This full implementation of EM in 1998 was no mean feat on the part of our teachers, especially because they were also implementing the Full Option Science System (FOSS) science program at the same time (our local work in math was not, after all, taking place in isolation—and especially so in view of the general state of urgency throughout Wisconsin regarding the standards movement).

At the middle-level grades, those two math teachers who had attended a two-week training session in CMP math in summer 1997 were having their own locally sponsored 20 hour (each) curriculum workshop in summer, 1998 to work on agendas and materials to do further in-house training of the other nine middle-level math teachers during the coming school year.

As to the high school's IMP program, this training began in summer 1997, with two teachers going to training in IMP Level I in the Minneapolis School District in August, with follow-up on several occasions during the 1997–98 school year. Then in August 1998, these two teachers went back to Minneapolis for IMP Level II training (again with follow-up sessions during the year), while three other teachers now went for IMP I training and so forth ... with Level III training in summer 1999 and Level IV in summer 2000. Bear in mind that IMP was (unlike EM and CMP) implemented one level at a time over four years.

This excitement and general feeling of renewed prospects for the math achievement of all children is a crucial element to large change.

As our programs become fully implemented (from fall 1998 on), there was much internal excitement. In retrospect, this excitement and general feeling of renewed prospects for the math achievement of all children is a crucial element to large change, as this was to be. Perhaps not all involved teachers need to be convinced and committed from the outset, but some level of critical mass (both people and attitudes) is needed to carry the day for a long period of time after the initial euphoria of doing something new and on through those time periods like the "implementation dip" and the second thoughts on the part of true believers and doubters alike.

Perhaps it is true that the pendulum of change and growth must swing from one extreme to the other for it to eventually end up in the middle.

I write this in May 2000, two years later, and where are we? Still the math wars rage on, as Secretary Riley commented over two years ago. Many questions are still being asked about math education, and many convincing answers come from both polarized viewpoints on the battlefield. Sometimes the changes in math education have been referred to as "whole math," an implied pejorative meant to refer to the similar "whole language" versus phonics battles of several year ago. Interestingly, it might be observed that those battles culminated in an eventual meeting point wherein the best of both approaches were blended into a strong hybrid methodology. Perhaps the same thing is now happening in math education. Perhaps it is true that the pendulum of change and growth must swing from one extreme to the other for it to eventually end up in the middle. Or to put it another way, suggestions for modest change often lead to no change at all, as people follow their natural tendency to drift back to the comfortable ground of the way they've always done it.

In May 2000, the Hometown Board of Education is beginning to scrutinize its math programs again, knowing that the programs do have vocal detractors and (less vocal) supporters. Without knowing at this point where this scrutiny will lead, it is still probably safe to say that local Hometown math education can never be the same as it was, with true math literacy for an elite few and most students bailing out of math as early in their high school career as possible. Our teachers have learned a great deal about the potential that exists in both the general student body and in the teaching of mathematics. The process that has gone on in Hometown has reminded teachers and students alike that they can indeed again feel excited about what they can do and learn.

The process that has gone on in Hometown has reminded teachers and students alike that they can indeed again feel excited about what they can do and learn.

References

Covey, S.R. 1989. *The 7 Habits of Highly Effective People: Powerful Lessons in Personal Change.* New York: Fireside.

Riley, R. R. 1998. Presentation "The State of Mathematics Education: Building a Strong Foundation for the Twenty-First Century" at the Conference of the American Mathematical Society and Mathematical Association of America, January 9, 1998, in Baltimore, MD.

Wisconsin Department of Public Instruction. 1998. *Wisconsin's Model Academic Standards for Mathematics.* Madison: Wisconsin Department of Public Instruction.

Appendix F What Are Some Professional Development Programs That Have Reaped Promising Results?

F

The following programs are described as illustrative of professional development that can make a difference.

Wisconsin has been a leader in developing outstanding professional development programs for its mathematics teachers. With professional development receiving much attention and with school districts seeking exemplary models, the following programs are described as illustrative of professional development that can make a difference. The names and addresses of the coordinators of the programs are given so they can be contacted to answer questions regarding the experiences.

Eisenhower Higher Education Partnerships: University of Wisconsin–Platteville

Using Eisenhower federal funding, coupled with university funding, the University of Wisconsin–Platteville offered a middle school mathematics professional development institute for several years that impacted heavily middle school mathematics programs in the state. Participants applied for admittance. Teams were encouraged. The groups that were selected came from all parts of the state and usually numbered around thirty. Selected participants stayed on campus for the two weeks of the institute and received college credit as well as seed funds for classroom supplies.

A unique combination of instructors was used—a college professor and two middle school practicing teachers, one with an elementary background and one with a secondary orientation. The institute, offered for two weeks, focused on current proven middle school curricular materials, instructional approaches designed to help participants augment their knowledge base, and assessment practices designed to be congruent with presented materials and instructional methodologies.

Unique features of the institute were 1) the on-campus, live-in experience; 2) the granting of college credit; 3) the funds for supply purchases for the classroom; 4) the diverse mix of instructors; and 5) the intrastate experience. Outstanding results of this endeavor were 1) stronger middle school mathematics teachers—teachers gained in knowledge of content, of instructional methodologies, of pedagogical content knowledge, and of congruent assessment strategies; 2) collegiality; attending teachers bonded over the two-weeks and continued to interact and support one another as they left the institute; 3) a middle school teacher-leader group—these teachers emerged as workshop leaders, their classrooms served as demonstration sites, and they contributed to regional and state efforts in subsequent years.

Contacts

Jane Howell
313 Tamarac Trace
Platteville, WI 53818
(608) 348-9093
howellja@mhtc.net

Jodean Grunow, Mathematics Consultant
Wisconsin Department of Public Instruction
PO Box 7841 (125 S. Webster St.)
Madison, WI 53707-7841
jodean.grunow@dpi.state.wi.us

Mathematics for the New Millennium—Department of Public Instruction

Responding to felt needs from the field and recognizing that special-needs students were often one of the identified populations in need of improvement, a unique professional development experience was jointly developed by the Title I and Content and Learning teams at the Department of Public Instruction. Schools were encouraged to apply as teams. Title I funds could be used for subscription purposes.

Subscribing schools agreed to participate in a two-day orientation workshop and a week-long summer institute and were then entitled to three mentoring visits in their district to address their identified needs. Participants were treated to sessions with nationally known leaders such as Thomas A. Romberg, University of Wisconsin–Madison, and Glenda Lappan, then president of NCTM. State leaders from the Title I team, the Content and Learning team, and the Assessment team at the Department of Public Instruction met with the groups. Outstanding practicing teachers shared activities and ideas with subscribers. Participants considered mathematics instruction per se, experienced rich mathematical endeavors in meaningful contexts, met with local team members to plan for their district, met with teachers from other districts who taught at specific levels, and had the opportunity to interact with the institute instructors and, in some cases, with the mentors who were to serve their district.

A partnership with institutions of higher learning was a thrust of the mentorships. Districts were able to access either a mentor from the Department of Public Instruction or from a teacher education institution. These mentors worked with districts at district invitation and addressed the specific self-identified needs of the districts.

Unique features of the experience were 1) the Content and Learning and Title I partnership; 2) the focus on systemic assistance; 3) the exposure to a wide variety of expertise; 4) the substantial support provided through the initial learning experience, the interim in-depth augmentation of knowledge, and the continued district-specific support; 5) the relatively sustained timeframe; 6) the collegiality experienced as a by-product of the considerable interaction between and among school districts; and 7) the magnitude of the response (140 Wisconsin school districts participated in the endeavor). Outstanding results of the effort were 1) significant district awareness of mathematics today and of current mathematical endeavors; 2) dialogue between and among deliverers of mathematics regarding efforts to meet the need for challenging and meaningful mathematics for ALL students; 3) development of a cadre of mathematically astute school districts; 4) recognition of similar needs; and 5) formulation of plans for moving mathematics forward in districts and in the state.

Contacts

Myrna Toney, Title I Team Leader
Myrna.toney@dpi.state.wi.us

Jodean Grunow, Mathematics Consultant
jodean.grunow@dpi.state.wi.us

Jennifer Thayer, Eisenhower Coordinator
jennifer.thayer@dpi.state.wi.us

Wisconsin Department of Public Instruction
(800) 441-4563

Wisconsin Academy Staff Development Initiative

The Wisconsin Academy Staff Development Initiative (WASDI) is a coordinated, systematic, statewide dissemination of a documented successful K–12 staff development program to improve mathematics, science, and technology education in Wisconsin. The Wisconsin Academy Staff Development Initiative is a teacher enhancement program, originally funded from 1994 through 1999 by NSF, which has received continuation funding from the U.S. Department of Education under its Fund for the Improvement of Education.

WASDI consists of two major components: a network of summer academies modeled after the Cray Academy in Chippewa Falls, which was initiated and developed with support from Cray Research and its successor SGI, and a Lead Teacher Institute.

The academies offer a series of one-week workshops during the summer. The workshops are led by teachers in most cases and some highly recognized professors, consultants, or other experts. The instructors are identified for their abilities to model effective teaching practices and to provide the teacher-participants with the tools, activities, tasks, and so on to engage their students in mathematics, science, and technology. The integration of these disciplines and networking among participants is emphasized. In addition, during each one-week session, teachers tour local businesses and industries and participate in discussions with business representatives to gain an understanding of the application of science, mathematics, and technology in the workplace. By the summer of 2000, 16 weeks of such academies were held in geographically scattered locations throughout the state.

From 1994 through 2000, 13,255 teachers attended academy workshops and this number included 9,500 unique individual teachers. By the 2000–2001 school year, over 1.6 million students will have been taught by a teacher whose skills were enhanced at a WASDI Academy. Six hundred business and industries provided tours for teacher academy participants, and 75 corporations made financial contributions to various academies. In the six years of the project, 1,000 business representatives participated in structured discussions with teachers about skills needed by the workforce of the present and future.

Evaluations of the academies have documented that teachers have changed their attitude toward math and science and have significantly increased the amount of hands-on activities in their classrooms. Evaluators have found that the academies also provided teachers with the unique opportunity to work in other subject areas, to renew interest in their profession and revitalize their energy, to work with teachers from grade levels other than their own, and to further understand the value of their relationships with people in the business community.

The second major WASDI component, the Lead Teacher Institute, prepared teachers to act as agents of change in their schools and to conduct staff development workshops across the state. The institute consisted of approximately 32 days of training over an 18-month period on topics such as leadership, change process, team building, content (math, science, and technology education), best classroom practice, alternatives for assessment, TIMSS, presentation skills, standards-based education, and strategies of professional development. From 1994 through 2000, 336 WASDI Lead Teachers received training and are now providing leadership in schools, districts, regions, and professional organizations and at WASDI Academies. Evaluation reports noted that Lead Teachers have conducted staff development workshops for teachers in their own schools and in other districts and have received outstanding evaluations; have become involved in national and state standards development; have received numerous professional awards; have written and received program related grants; and have assumed leadership roles in other

NSF-funded programs. The evaluator concluded that the Lead Teacher Institute has provided a cadre of well-trained, dedicated teachers who consistently provide leadership in science, mathematics, and technology education throughout the state.

The WASDI infrastructure, the academies and Lead Teacher cadre, exist in a viable fashion to continue to move Wisconsin ahead in implementing standards-based education. Schools that need staff development in the reform curriculum materials or other current topics can request academy directors to offer the desired topic. Teachers from a variety of districts who need similar training can then benefit from an economy of numbers since each district does not need to duplicate offerings. Approximately 30 percent of the districts in the state have a WASDI Lead Teacher on their staff. These Lead Teachers can also be resources to neighboring districts.

More information is available at http://www.wasdi.org.

Contacts

Dr. Julie Stafford, Project Director
Wisconsin Academy Staff Development Initiative
140 West Elm Street, Math Department
Chippewa Falls, WI 54729
(715) 723-1181
julie_stafford @wetn.pbs.org

Dr. Billie Earl Sparks, Project Codirector
University of Wisconsin–Eau Claire
Eau Claire, WI 54702
(715) 836-3778
sparksbe@uwec.edu

Eisenhower Higher Education/University of Wisconsin–Eau Claire Mentor Program

Under Eisenhower Higher Education funding and university contributions, the University of Wisconsin–Eau Claire has offered a series of programs centered around the concept of talented teachers mentoring other teachers. After refinements over five years of such programs, the model that shows the most promise is one in which outstanding teachers who are experienced in teaching the NSF-sponsored reform curriculum projects have an intern teacher placed in their classroom for the year. The teacher and intern coplan daily, and the intern relieves the teacher of part of their teaching responsibilities. On one or two days a week the intern covers all teaching responsibilities, and the teacher is free to consult with teachers in other districts. The other partners in the project are school districts that are starting implementation of reform curricula in their mathematics programs. Teachers in these schools then receive consultations from the mentor teachers whose classrooms are covered by the interns. These consultations may involve the mentor teacher teaching a demonstration class, observing and critiquing a class, working on lesson plans, developing a plan for coverage of required topics, and answering any questions about detail or philosophy of implementing the materials. This professional development has been seen to be highly effective as it is targeted to teacher need, it is on site and in the classroom, and it utilizes the talents of teachers who are ready to take next steps in the profession. It also has a tremendous advantage in preservice education as the intern works with an outstanding teacher who is a model of professionalism. The model is easily adaptable to other locations and may be used at all levels.

Contact

Dr. Billie Earl Sparks, Department of Mathematics
UW–Eau Claire
Eau Claire, WI 54702

WECB On-line Professional Development Programs

Using a model originally developed by PBS MATHLINE, the Wisconsin Educational Communications Board (ECB) provides unique on-line, professional development programs for K–12 mathematics teachers as well as for higher education faculty who prepare mathematics teachers.

Participants enroll for one year or more. The on-line professional development is lead by teacher-leaders who are trained and supported in the skills and knowledge involved with creating and sustaining on-line learning communities. Called facilitators, each teacher leader is responsible for a Learning Community of 20 to 30 participants.

The ECB partially underwrites these programs. Additional needed revenues come from individual yearly registration fees. Teacher registration fees can be paid by districts using federal and state professional development grant program funds. The ECB solicits business and industry underwriting or grants and uses these funds to offer teacher scholarships. In some programs the participants' fees are paid through NSF, Department of Education (DOE), or Eisenhower federal funding.

The ECB provides and maintains the network server and user licenses for WECB On-line. Participants have multiple ways to access the network: dial-up modem, Internet, or Web access. A toll-free line allows users to access the network from locations throughout the country. A toll-free help line is provided as well as on-line technical support. There are both virtual and face-to-face training opportunities in use of the special on-line conferencing software and in the specific elements of various projects, including the professional development program objectives and how to create and sustain an on-line learning community. The propriety conferencing software that is used by the ECB was specially chosen and is maintained/upgraded because of the results. Facilitators and participants report that it creates an environment that engages teachers in discourse about systemic reform and moving to a vision of mathematics teaching embodied in the state and national standards.

MATHLINE

Wisconsin teachers can enroll in four professional development programs offered by the ECB: the Elementary School Math Project for grades K–5 teachers, the Middle School Math Project for grades 5–8 teachers, the High School Math Project for grades 8–12 teachers, and the Algebraic Thinking Math Project for grades 3–8 teachers. Each project has a set of twelve to twenty-six 15–60 minute videos and accompanying lesson guides for teachers that model the vision embodied in the 1991 NCTM Professional Standards for Teaching Mathematics.

Each year the ECB recruits 10–30 Wisconsin teacher-leaders to become the facilitators of the Learning Communities. They represent the recognized, outstanding teacher leadership in the Wisconsin K–12 mathematics education community.

In collaboration with the facilitators, the ECB prepares an extensive Wisconsin MATHLINE Resource book for each participant. The book contains a variety of professional development materials, including journal articles that are matched to particular videos or tied to broader pedagogical or content issues embodied in the NCTM vision of how classrooms must change in order to meet the need for challenging and meaningful mathematics for ALL students.

Wisconsin MATHLINE is primarily a virtual professional development program. Participants rarely have face-to-face meetings. Yearly at the annual meeting of the NCTM affiliate, the Wisconsin Mathematics Council, Inc., a MATHLINE social gathering takes place. The ECB with the Learning Community facilitators sometimes provide one- to three-day regional workshops at the start of the project year.

In operation for six years, results have included at least a 50 percent return rate by participants in the same program or another program. Some teachers have participated for six years, citing the value of the networking with teacher-leaders across the state and the value of the informational and emotional support received from other on-line participants. Many teachers report that they felt alone in their schools and districts in their efforts to change teaching and learning to embody the vision in the state and national standards. MATHLINE provides support from other teachers around the state who also are making similar changes in teaching and learning, often without support from school colleagues. MATHLINE also provides access to a network of teacher leaders whose daily classroom experiences in implementing reform teaching and learning are shared and discussed. This continuing exposure to the state's teacher leaders and their sharing of daily practice is an invaluable resource for teachers who are at early or middle stages of implementation.

A continuing strong result is teachers' increased comfort in using on-line communications technology for other professional development purposes or for use with students in the classroom. Access to and discussion of the MATHLINE videos continues to be important to teachers. The videos are viewed as electronic field trips into classrooms of teachers like themselves who have struggled to change their practices and continue to move forward because of the positive student results—more students are successful with meaningful and challenging mathematics. The on-line discussion focuses on aspects of the NCTM teaching standards modeled in particular videos as well as on ways teachers change the particular lessons to match their particular students or situations. This sharing about ways to develop exemplary lessons is another important result.

NPRIME

The Networking Project for Improvement in Mathematics Education (NPRIME) is the result of recommendations from the statewide Advisory Council whose vision was to extend the MATHLINE model into higher education in Wisconsin. Using Eisenhower federal funding coupled with ECB and university funding NPRIME connects college and university professors who teach content or curriculum and methods courses to preservice elementary or secondary teachers. NPRIME participants meet primarily in virtual reality. They represent faculty from across the state at the 36 institutions that prepare mathematics teachers. The project leadership team and facilitators meet face-to-face for initial design and planning.

Results include the preparation of a new guidebook that PBS will distribute nationally for preservice use with the PBS MATHLINE video materials. Twelve Wisconsin faculty members contributed to writing sections of this guidebook, which also focuses more generally on recommendations for using videos in courses with preservice students.

This writing project leads to extensive discussion among faculty on-line about the importance and value of using videos that model and reflect the pedagogy of the good instruction. The PBS MATHLINE videos are valuable because they are based upon the 1991 NCTM Professional Standards for Teaching. Faculty reported that using the videos in classrooms resulted in preservice students seeing teachers pose worthwhile tasks and model enhancing discourse through the use of a variety of tools, creating learning environments that support and encourage reasoning and expect and encourage teachers to take intellectual risks. Use of the videos engaged preservice students in discourse about teaching and learning mathematics.

The videos were also valued because they created some disequilibrium for preservice students. The students in the videos challenged the preservice teacher's current assumptions about what mathematics elementary students can do, about the nature of algebra and algebraic thinking, and about what it means for high school students to be successful in communicating mathematically and being successful in understanding mathematics.

The purpose of NPRIME is to encourage collaboration among preservice faculty on 1) timely topics at implementation of the new Wisconsin PI-34 Rules and the 2000 NCTM standards and 2) continuing topics such as recruitment of students as mathematics majors/minors and content and curriculum issues for college-level courses. Results for participants include recognition of similar needs across campuses and sharing of plans for moving preservice courses into alignment with state and national standards and rules.

WASDILine

The Wisconsin Academy Staff Development Initiative was a five-year NSF-funded teacher enhancement project for the development of a Lead Teacher Institute for preparing teachers to become leaders in curriculum reform at the school, district, regional, and state levels. Another goal was the establishment of Summer Academies across the state that would be based on the model of the highly successful Cray Academy in Chippewa Falls. Lead teachers are prepared to provide reform-based workshops at these academies.

WASDILine is in its fifth year as part of the program for the lead teachers as well as the communication network to link project staff, lead teachers, and the directors of the 16 academies that have been established across the state. The NSF grant did not originally have an on-line professional development component for the Lead Teacher Institute. The Department of Public Instruction has participated in funding WASDILine, along with registration fees paid by the district of participating teachers. The WASDI project is currently funded by the Department of Education and the Wisconsin Department for Public Instruction.

WASDILine has been highly successful as a professional development element in the Lead Teacher Institute. Many lead teachers cite it as the most important component of the institute because it is available any time, any place, whether a teacher needs support on a classroom-school-district leadership crisis issue or support on planning for change and professional development in his or her district. The emotional support is as important to lead teachers as the informational support. Each year the WASDI lead teachers spend over four thousand hours on-line networking and supporting one another. The most outstanding benefit has been the increased learning about and use of the national and state standards in their own classrooms. Lead teachers value daily access to one another, the project staff, as well as, access to the Summer Academy directors.

Contact

Marge Wilsman
Director WECB On-Line Professional Development Services
ECB
3319 W. Beltline Hwy.
Madison, WI 53713
(608) 264-9691
mwilsman@ecb.state.wi.us

LLMP Projects Piece

The Linked Learning in Mathematics Project (LLMP) is a professional development project designed to prepare teachers to implement algebra for all in Milwaukee-area schools.

Funded by NSF from 1997 through 2000, LLMP is conducted jointly through Marquette University and Education Development Center. All phases of LLMP are closely tied to the NSF-funded Milwaukee Urban Systemic Initiative. Two cohorts of 55 teachers and principals have participated in the project for each year. Teachers participate in teams of two or more teachers from the same school; principals participate in the project in a limited capacity.

Each cohort of participants takes part in an eight-day summer institute. During the school year, participants are enrolled in a three-credit course entitled "Algebraic Thinking: Framework, Activities, and Implementation." Emphasis in the course is on strategies for understanding student thinking in algebra and for helping students experience success in algebra-related mathematics. Although workshop and class sessions are held on the campus of Marquette University, much of the course work is done inside the teachers' own classrooms. In particular, the teachers are observed by other teachers on six occasions as they implement algebra problem solving activities with their students. Observers record the questions teachers and students ask one another as they work on activities. Additionally, teachers conduct three interviews over the course of the year with small groups of students in which they are given the opportunity to listen to student thinking.

Contact

Dr. John M. Moyer
Department of Mathematics, Statistics, and Computer Science
Marquette University
PO Box 1881 (1313 W. Wisconsin Ave.)
Milwaukee, WI 53201-1881
(414) 288-5299
johnm@mscs.mu.edu

Middle School Trainers' Mathematics Project

The Middle School Trainers' Mathematics Project (MSTMP) operated from 1986 to 1997. Marquette University and Milwaukee Public Schools sponsored the project, which was funded through the Dwight D. Eisenhower Professional Development Program. The project objectives and activities changed over the years. Beginning in 1989, the general objective of the project was to prepare a cadre of middle school trainers who would be available to the various school districts in the greater Milwaukee area to help raise the level of implementation of the NCTM Curriculum and Evaluation Standards.

The project helped the participants themselves implement the NCTM Curriculum and Evaluation Standards, particularly those relating to algebra. The project also included a precourse development and assessment component (utilizing a committee of middle school teachers), monthly postcourse follow-up for all the participants, and a comprehensive evaluation component.

Twenty-eight teachers were enrolled in the project at any given time. The teachers enrolled for a minimum of two years. During the summers, the teachers met each afternoon for three weeks. During the school year they met after school once each month. Graduate credit was granted for participation in the project.

Specific objectives were that the participants:

1. Increase their knowledge of the mathematics related to the curricular changes advocated by the NCTM Curriculum and Evaluation Standards with special connections to algebraic concepts.
2. Engage all students, regardless of background, economic status, ethnicity, national origin, or gender in meaningful, high-order mathematics learning and to prepare their students to enroll and experience success in ninth grade algebra.
3. Improve their classroom practice by implementing the NCTM's Curriculum and Evaluation Standards and Professional Teaching Standards.
4. Reach out to help those teachers in their districts who were not able to participate in MSTMP by offering formal in-service programs in their department, school, or district.

Contact

Dr. John M. Moyer
Department of Mathematics, Statistics, and Computer Science
PO Box 1881 (1313 W. Wisconsin Ave.)
Milwaukee, WI 53201-1881
(414) 288-5299
johnm@mscs.mu.edu

Shape and Dimension Project: Infusing Shape and Dimension into the Middle School Mathematics Curriculum

Through a collaborative effort of the Woodrow Wilson National Fellowship Foundation (WWNFF) in Princeton, New Jersey, and the University of Wisconsin–Milwaukee, the Shape and Dimension Project was a one-year professional development program in mathematics for middle grade teachers (grades 4–8). Through this project, participants examined mathematics topics in new ways, learned strategies and activities that were brought into their own classrooms, and saw that shape and dimension were an integral facet of the curriculum. Some of the specific mathematics topics included factor lattices, Pascal's triangle, polyhedrons, polygons, fractals, topology, number theory, and networks. These topics were explored with tools such as paper folding, cubes, tiles, pentominoes, patty paper, graphing calculators, and dynamic geometry computer programs. The development of children's understanding of geometry was also examined through the Van Hiele model of geometric thought.

A special feature of this project was its link with WWNFF in Princeton, New Jersey. The core of the summer activities was facilitated by a team of master teachers from across the country who had studied with mathematicians and scientists through fellowships at the foundation in Princeton and who were part of its Teacher Outreach program. The major funding for this project came from the University of Wisconsin System Eisenhower Professional Development program. Additional support was provided by WWNFF, the Milwaukee Public Schools, Glendale–River Hills School District, and the University of Wisconsin–Milwaukee.

School-based teams of teachers from southeast Wisconsin met for a seven-day summer institute. The participants then attended monthly follow-up sessions on Saturday morning during the school year. Participants also communicated via an on-line network throughout the project. Each participant received four graduate or undergraduate university credits. Each individual was also given cut paper and other materials so that they could more easily engage their students in paper folding and other project-related activities. The school-year instructional team consisted of one of the WWNFF master teachers who was from Wisconsin, another middle school teacher from the area, and a university professor.

Participants in the Shape and Dimension Project improved their mathematical knowledge, enhanced their teaching strategies, and increased their confidence in infusing shape and dimension into their mathematics curriculum. Each participant conducted two action research projects with their students and compiled a portfolio throughout the school year that demonstrated enhanced student learning. One participant remarked, "I understand much more about geometry than a year ago. I wish that every teacher could have this course. It would change the world of mathematics." Another individual wrote, "I have done more geometry this year than any other year. Next year I plan on starting with geometry, unlike previous years. Shape and Dimension will be woven into activities/lessons weekly. I now realize more than ever before the importance of geometry. There are so many misconceptions that I need to be aware of and desire to change." Another participant summarized, "The shape and dimension project affected how I infuse geometry into activities throughout the year. The Van Hiele levels of thinking have had a dramatic impact on how I teach. I am more willing to let students play and more skillful at pressing their thinking." The school-based teams also developed and carried out implementation plans within their schools or districts in order to further share and disseminate their learning from the course. A participant noted, "This course has created enthusiasm amongst us. We share it with others and now we are advocates for improving and strengthening the teaching of geometry and spatial reasoning."

Contact

Dr. DeAnn Huinker, Mathematics Education Department
University of Wisconsin–Milwaukee
PO Box 413
Milwaukee, WI 53201-0413
(414) 229-5467
huinker@uwm.edu

Leadership for Implementing the Investigations Curriculum

The Leadership for Implementing the Investigations Curriculum (LIIC) program was a collaborative effort of the University of Wisconsin–Milwaukee and the Milwaukee Public Schools. The program developed a cadre of Teacher Leaders that have provided district-wide support for the implementation of the Investigations in Number, Data, and Space (INDS) curriculum. This reform-based K–5 mathematics curriculum was developed by TERC under a grant from NSF. It emphasizes the exploration of meaningful mathematical problems and the deepening of mathematical thinking. The participants in the program 1) developed leadership and facilitation skills for professional development; 2) became more familiar with the philosophy, strategies, components, and organization of INDS; 3) deepened their understanding of children's development of mathematical knowledge; 4) examined effective professional development and change strategies; and 5) learned strategies to support colleagues using INDS. The major funding for this program was from the Wisconsin System Eisenhower Professional Development component. Additional support was provided by the Milwaukee Public Schools and the University of Wisconsin–Milwaukee.

The participants in this leadership program were selected through a recommendation and application process. They had been using INDS in their classrooms, were willing to learn how and work hard to become effective facilitators of professional development sessions at the district and school levels, and made a commitment to conduct staff development sessions for other teachers. The participants attended a summer institute that met for eight half-days in June in which they learned more about the background, organization, and development of mathematical ideas in the curriculum and examined effective professional development and change strategies. Then they participated in the school-year component, meeting about once every other week, for 15 sessions that focused on learning more about the curriculum, strategies to support the implementation of the curriculum, and the planning of professional development sessions. In addition, the participants facilitated professional development sessions for other teachers throughout the district. The participants were also linked electronically through an on-line listserv. During the school year, participants videotaped themselves teaching with the curriculum and discussed ways to use their videos in professional development sessions. Each participant received five graduate or undergraduate university credits.

The participants in this leadership program have become Teacher Leaders in the district. The Teacher Leaders became a united voice of support for implementation of INDS, relaying common messages and successful practices to teachers, administrators, and parents throughout the district. They planned and facilitated numerous districtwide professional development sessions, provided support for teachers in their own schools, and were requested to conduct in-service sessions at other schools. In addition, the Teacher Leaders also provided advice to the district on the implementation of the curriculum, such as developing the pacing chart and setting priorities for professional development structures and topics. Many of the Teacher Leaders were nervous as they joined the cadre, but all grew in their understanding of the curriculum, enhanced their own teaching, developed facilitation skills, and gained confidence in themselves as facilitators of professional development. One participant remarked, "I feel my biggest accomplishment is feeling like a successful facilitator. I made it through two banking day workshops and two cluster sessions (4 two-hour sessions each) unscathed and willing to do more. I wasn't sure how I would do way back when in the beginning of this program, but looking back, I feel proud!" Another individual wrote, "My biggest accomplishment was overcoming my fear of speaking in front of a group. I joined this program because I was terrified at the thought of presenting. I was anticipating being a wreck, but I've found that I actually love facilitating. I may even pursue it further."

Contact

Dr. DeAnn Huinker, Mathematics Education Department
University of Wisconsin–Milwaukee
PO Box 413
Milwaukee, WI 53201-0413
(414) 229-5467
huinker@uwm.edu

Grade-Level Cluster Meetings: Support for Reform-Based Curriculum Implementation

The Milwaukee Public Schools developed a program of grade-level cluster meetings to provide ongoing support for teachers as they began implementing the INDS mathematics curriculum. One goal of the cluster meetings was to engage teachers in professional development sessions that focused directly on the grade-level unit book that was currently being taught according to the district-pacing schedule. Another goal was to increase understanding of the role of the teacher and the role of the students in an INDS classroom that supports effective pedagogical practices.

Cluster meetings were held for each grade level, kindergarten through Grade 5. For each set of cluster meetings, teachers met four times after school, receiving eight hours of support. Each set of grade-level cluster meetings was facilitated by a pair of district Teacher Leaders currently teaching at that grade level. Teachers in their grade-level clusters worked through INDS lessons and discussed student learning, teaching approaches, and implementation strategies. A major portion of each evening was devoted to working through sessions from the curriculum units. As teachers worked through sessions, they had many opportunities to share issues and strategies for implementation.

Teachers attending the cluster meetings were expected to work with other teachers at their grade level, bring in samples of their students' work, and review the assessments from the current unit. Three sets of cluster sessions were held during the first year of implementation of the new curriculum—fall, winter, and spring. Teachers could participate in one, two, or all three sets of clusters. Funds from the district Eisenhower Professional Development program were used to support two teachers for every one teacher supported by individual school budgets for the fall and winter clusters. Individual school budgets were used to compensate teachers for their participation at the spring clusters.

The cluster meetings were a huge success. Multiple sections of each grade level were offered based on the large number of teachers who wanted to attend. About eight hundred teachers attended the fall cluster meetings, about six hundred attended the winter cluster meetings, and about two hundred fifty attended the spring cluster meetings. The participants repeatedly noted their appreciation of the teachers-teaching-teachers model that was used with the cluster meetings. The Teacher Leaders encouraged an exchange of ideas and created a safe atmosphere for learning. One individual wrote, "The strength of this workshop was Teacher Leaders who had taught the program, not corporate people telling us how to use the program." Another remarked, "Great opportunity to dialogue with teachers from our own district who have already implemented this program." Another noted, "Our presenters had taught this before and could bring their experiences to our discussion groups." The teachers valued the opportunity to work through sessions using the manipulatives, materials, and resources they were using in their own classrooms. The cluster meetings created a professional learning community and support network for teachers working together to implement the curriculum. Some of the strengths of the cluster meetings as noted in their written evaluations included: "The sharing of experiences relieved my anxiety," "Thanks for the opportunity to exchange ideas and express concerns," and "I feel more secure—we are all learning together."

Contact

Dr. DeAnn Huinker, Mathematics Education Department
University of Wisconsin–Milwaukee
PO Box 413
Milwaukee, WI 53201-0413
(414) 229-5467
huinker@uwm.edu

Mathematics and Science Beginning Teacher Project

The Mathematics and Science Beginning Teacher Project is a two-year discipline-based teacher induction program for elementary school teachers. Through this project, first-year teachers in the Milwaukee Public Schools (MPS) will receive sustained support for two years as they make the transition from being students in a teacher preparation program to being full-time teachers in a large urban district. The context for support is successful implementation of the district elementary mathematics and science programs, which are standards-based programs that use an inquiry approach to teaching and learning. Participants will become well-versed in the philosophy, components, approaches, content, and assessment of the MPS elementary mathematics and science programs; receive ongoing support in using an inquiry-approach to engage all children in standards-based mathematics and science learning; and form a supportive learning community with other new teachers around the teaching of mathematics and science. The major funding for this program comes from the Wisconsin System Eisenhower Professional Development component. Additional support was provided by MPS and the University of Wisconsin–Milwaukee.

In the first year of the project, participants will meet for four full-day Saturday sessions and 11 evening sessions during the school year. In the second year of the project, they will meet for one week in the summer and approximately twice a month during the school year. Participants will also interact through an on-line conference. To bring these novice teachers into the mathematics and science education professions, each individual will be given memberships in national and state professional organizations in mathematics and science education and be given support to attend state mathematics and science conferences. They will also receive other resources and materials to support their mathematics and science instruction. Participants will earn seven university graduate credits over the two years. The instructional team will consist of university faculty in mathematics and science education and experienced teachers from MPS.

The immediate outcome of this project is a group of beginning teachers who will support each other and receive support from experienced teachers and university faculty in their critical first two years of teaching. They will not experience the isolation of many beginning teachers but will have a sustained support network. Further, they will be more knowledgeable of standards-based mathematics and science programs and more skilled in their implementation resulting in improved student learning. Long-term outcomes of this project are the establishment of new advocates for standards-based mathematics and science and the retention of these participants in the teaching profession, particularly retention within a large urban district.

Contact

Dr. DeAnn Huinker, Mathematics Education Department
University of Wisconsin–Milwaukee
PO Box 413
Milwaukee, WI 53201-0413
(414) 229-5467
huinker@uwm.edu

Teaching Integrated Mathematics in the High School: Implementing the Core-Plus Mathematics Curriculum

This one-year professional development project prepared teachers of mathematics to implement an integrated high school mathematics curriculum with an emphasis on the Course 1 of Contemporary Mathematics in Context (Core-Plus). Participants examined the mathematics content in depth, studied the instructional units with emphasis on learning activities and student responses, experienced and practiced instructional and assessment strategies, and gained knowledge and skills in the use of technology. The project was focused on teacher content and actions necessary to prepare their students to learn and use algebra and functions, data analysis and statistics, geometry, and discrete mathematics. The major funding for this project came from the Wisconsin System Eisenhower Professional Development program and MPS. Additional support was provided by T³—Teachers Teaching with Technology and the University of Wisconsin–Milwaukee.

This project was targeted on the teaching of mathematics to all students, with attention to MPS teachers and students. The project activities were facilitated by nationally recognized teachers who were using the curriculum and graphing calculator technology and by university faculty from education and mathematics departments. The participants met during an intense eight-day summer institute and for eight Saturday-morning sessions during the school year. The summer session focused on learning content and studying mathematics topics from new perspectives. All sessions were very interactive, with hands-on use of concrete materials, computers, and graphing calculators. Curriculum implementation and action plans were developed for implementing the new curriculum into the participants' mathematics classroom and assessing student work in classroom and district assessments. The school-year meetings focused on strategies to implement an integrated mathematics curriculum and provide ongoing support to the participants and their students. Participants also engaged in ongoing discussion and sharing of tasks and activities through an on-line network.

The participants gained an understanding of the rationale for and development of the four-year Core-Plus integrated mathematics curriculum. They developed confidence as they were supported during the implementation of the Core-Plus Course 1 curriculum materials in their classrooms. They received focused instruction on specific mathematics content, instructional strategies, and student assessment. They learned techniques, ideas, and discussions for facilitating collaborative learning and classroom management. They developed skill in effective use of technology embedded in mathematics learning settings. Finally, they developed strategies for gaining support from parents and administrators and practical suggestions for school–home communication.

Contact

Dr. Henry Kepner, Mathematics Education Department
University of Wisconsin–Milwaukee
PO Box 413
Milwaukee, WI 53201-0413
(414) 229-5253
kepner@uwm.edu

Appendix G National Council of Teachers of Mathematics Position Papers

G

These papers are attached for ready reference.

The National Council of Teachers of Mathematics has, through the years, provided outstanding assistance to mathematics teachers. Because their research is so thorough and their input so often sought regarding specific topics, position papers have been developed regarding those issues. Those papers are printed here as a resource. Additional information can be obtained from the NCTM website: http://www.nctm.org.

Calculators and the Education of Youth, July 1998

The National Council of Teachers of Mathematics recommends the integration of calculators into the school mathematics program at all grade levels.

Appropriate instruction that includes calculators can extend students' understanding of mathematics and will allow all students access to rich problem-solving experiences. Such instruction must develop students' ability to know how and when to use a calculator. Skill in estimation, both numerical and graphical, and the ability to determine if a solution is reasonable are essential elements for the effective use of calculators.

Assessment and evaluation must be aligned with classroom uses of calculators. Instruments designed to assess students' mathematical understanding and application must acknowledge students' access to, and use of, calculators.

Research and experience support the potential for appropriate calculator use to enhance the learning and teaching of mathematics. Calculator use has been shown to enhance cognitive gains in areas that include number sense, conceptual development, and visualization. Such gains can empower and motivate all teachers and students to engage in richer problem-solving activities.

Therefore, the National Council of Teachers of Mathematics makes the following recommendations:

- All students should have access to calculators to explore mathematical ideas and experiences, to develop and reinforce skills, to support problem-solving activities, and to perform calculations and manipulations.

- Mathematics teachers at all levels should promote the appropriate use of calculators to enhance instruction by modeling calculator applications, by using calculators in instructional settings, by integrating calculator use in assessment and evaluation, by remaining current with state-of-the-art calculator technology, and by considering new applications of calculators to enhance the study and the learning of mathematics.
- School districts should provide professional development activities that enhance teachers' understanding and application of state-of-the-art calculator technology.
- Teacher education institutions should develop and provide preservice and inservice programs that use a variety of calculator technology.
- Those responsible for the selection of curriculum materials should remain cognizant of how technology—in particular, calculators—affects the curriculum.
- Authors, publishers, and writers of assessment, evaluation, and mathematics competition instruments should integrate calculator applications into their published work.
- Mathematics educators should inform students, parents, administrators, school boards, and others of research results that document the advantages of including the calculator as one of several tools for learning and teaching mathematics.

Early Childhood Mathematics Education, December 1998

Position

The National Council of Teachers of Mathematics believes that early childhood mathematics education for children aged three to eight should take into account children's intellectual, social, emotional, and physical needs.

Rationale

Children's understandings develop as they observe, explore, investigate, and discuss mathematical concepts. The opportunity to interact and play with materials and other individuals enables children to construct, share, modify, and integrate their ideas.

Children's confidence can be developed through varied instructional strategies, meaningful child-related contexts, and opportunities for their active participation in the learning process.

To support children's active development of knowledge, instruction should facilitate learning through explorations and interaction with materials and people. Young children need opportunities to make choices, and classroom practices should be organized around the needs of the child, and should use materials that support developmentally appropriate mathematics education. Materials and rigid time lines should not dictate instruction. Early childhood mathematics education should foster a positive environment, provide equal access for all children, and accommodate cultural and ethnic diversity.

Recommendations

Children's understandings develop as they observe, explore, investigate, and discuss mathematical concepts. The opportunity to interact and play with materials and other individuals enables children to construct, share, modify, and integrate their ideas.

Children's confidence can be developed through varied instructional strategies, meaningful child-related contexts, and opportunities for their active participation in the learning process.

To support children's active development of knowledge, instruction should facilitate learning through exploration and interaction with materials and people. Young children need opportunities to make choices, and classroom practices should be organized around the needs of the child, and should use materials that support developmentally appropriate mathematics education. Materials and rigid time lines should not dictate instruction. Early childhood mathematics education should foster apositive environment, provide equal access for all children, and accommodate cultural and ethnic diversity.

Recommendations

Curriculum and evaluation guidelines for early childhood mathematics instruction should do the following:

- Identify the needs of individual learners at different stages of readiness and plan instruction that builds on children's prior experiences, cultural background, learning styles, and cognitive abilities.
- Incorporate active and interactive learning.
- Provide opportunities for children to develop and expand language acquisition while structuring, restructuring, and connecting mathematical understandings. Concepts should be repeatedly experienced through concrete, visual, verbal, and pictorial formats. Gradually, children should be encouraged to translate and record their experiences in more abstract representations.
- Focus on the development and integration of mathematical thinking, reasoning, understanding, and relationship through concrete problem-solving experiences across the curriculum.
- Encourage children to become capable mathematical thinkers and to have confidence in their mathematical abilities.
- Include ongoing informal and formal assessments that provide evidence of children's thinking processes and their understanding of concepts.

Evaluation of Teacher Performance and Effectiveness, March 2000

Position

It is the position of the National Council of Teachers of Mathematics that the evaluation of teaching performance and effectiveness must be conducted in light of the Council's Professional Standards for Teaching Mathematics and that teaching performance must be connected to the professional development necessary to improve practice.

Rationale

Increasing demands for accountability have led state and provincial legislatures and agencies, school districts, professional organizations, and teacher education institutions to expand their efforts to evaluate a teacher's knowledge, performance, and effectiveness.

Recommendations

The National Council of Teachers of Mathematics believes that the evaluation of teaching performance and effectiveness must:

- include the active participation of the teacher in all phases of the process;
- be conducted by evaluators who possess a knowledge of the vision of teaching mathematics as presented in the Council's Professional Standards for Teaching Mathematics and Principles and Standards for School Mathematics and who us criteria directly related to the content of mathematics, effective instructional strategies, and assessments of students' performance that are aligned with this vision;
- be based on a mutually agreed-on plan for professional growth that arises from clear goals for both the teacher and the evaluator(s);
- include the collection and analysis of a wide range of data, drawn from various sources such as self-assessment, observations, interviews, peers, teacher's portfolios, supervisors, and administrators—as well as from multiple measures of students' performance;
- link the evaluation of teaching with professional development that addresses the needs identified by both the teacher and the evaluator(s);
- balance the identification of strengths and positive aspects of performance with the identification of areas needing improvement.

In addition to resulting in plans for professional improvement, the processes and procedures for the evaluation of teaching performance and effectiveness must also encourage risk taking and personal and collegial reflection on practice. Therefore, the evaluator(s) must work closely with the teacher to assure that the evaluation process is used to enhance the professional development of the teacher, increase the effectiveness of mathematics teaching, and realize the improvement of students' performance.

Guiding Students' Attitudes and Decisions Regarding Their Mathematics Education, February 1994

Educators, including parents, teachers, counselors, and administrators, are positioned to influence the formation of students' attitudes toward, and perceptions about, mathematics. These attitudes and perceptions are formed at an early age and reinforced by educators throughout the students' school years.

The following issues should be addressed at all levels:

- Equal access to the world of mathematics regardless of gender, ethnicity, or race
- The broad range of options available to students of mathematics
- That success in mathematics is more dependent on effort and opportunity to learn than on innate ability.

For students to succeed in their study of mathematics, they must believe that they can "do" mathematics and that it is worth "doing." Therefore, the National Council of Teachers of Mathematics believes the following statements should characterize student programs:

Elementary School Level

Young children enter school having natural interest, curiosity, and an eagerness to learn. As children move through the elementary school level, educators build on these attitudes, which support children's mathematical development. By connecting mathematics with real-life experiences, educators can help children not only understand mathematics but see its value and usefulness.

Middle School Level

During the middle school years, students are experiencing the greatest changes in their development. It is during these years that educators must continue to provide encouragement to students and enable them to develop confidence in their ability to make sense of mathematics. A focal point for this age group is making connections between mathematics and their future options so that their academic and work-related choices are kept as broad as possible.

Secondary School Level

At the secondary school level, students should recognize the importance of the relationships between academic choices and career choices. It is essential that educators reinforce students' belief in the value of mathematics, in their ability to "do" mathematics now and in the future, and in their need to continually learn.

In addition, students should continually be provided reliable, up-to-date information on:

- How what they are learning in mathematics relates to future options;
- The increasing number of career opportunities that are available as a consequence of further mathematical study;
- How careers in fields other than mathematics and the physical sciences are increasingly dependent on mathematical knowledge;
- Their school's graduation requirements in mathematics;
- The entrance and graduation requirements in mathematics for vocational schools, technical schools, colleges, and universities.

Therefore, the NCTM recommends that all educators, including parents, teachers, counselors, and administrators, work together to shape, guide, and inform students' attitudes, perceptions, and decisions regarding their education in mathematics.

The Mathematics Education of Underrepresented Groups, April 1995

The National Council of Teachers of Mathematics is committed to the principle that groups underrepresented in mathematics-based fields of study should be full participants in all aspects of mathematics education. These groups are composed of students who do not take advanced mathematics courses and do not enter mathematics-related vocations and careers in proportion to their representation in the population. The groups include females, African Americans, Hispanics, Native Americans, Alaskan Natives, and Native Pacific Islanders, among others.

Paths to continuing education and employment opportunities are often hindered by powerful social and institutional influences that discourage these students from studying mathematics. In our challenging and complex society, it is paramount that all students be considered valuable resources to be afforded the opportunity to reach their full potential and contribute to society.

Schools and districts should examine programs including assessment and instructional practices that may lead to mathematics avoidance. Mathematics educators must make an individual and collective commitment to eliminate any psychological and institutional barriers to the study of mathematics. Educators must explore and implement innovative ways to convince students and families from these groups of the vital importance of taking mathematics courses.

Teachers, administrators, and counselors at all educational levels should expect that students from all segments of the population will be successful in mathematics. The teacher is in a pivotal position to motivate and encourage all students to continue the study of mathematics.

All students must study a core curriculum in mathematics that reflects the spirit and the content of the NCTM *Standards*. The NCTM will continue to work to increase the participation and achievement of underrepresented groups in mathematics education.

Mathematics for Second-Language Learners, July 1998

The National Council of Teachers of Mathematics is committed to the principle that second-language learners should be full participants in all aspects of mathematics education.

Language and cultural background must not be a barrier to full participation. All students, regardless of their language and cultural background, should study a comprehensive mathematics curriculum with high-quality instruction and coordinated assessment.

The National Council of Teachers of Mathematics believes that:

- Second-language learners should be given appropriate first- and second-language support while learning mathematics;
- Teachers, counselors, and other professionals who have expertise should carefully assess the language and mathematics proficiencies of each student in order to make appropriate curricular decisions and recommendations;
- The importance of mathematics and the nature of the mathematics program should be communicated, with appropriate language support, to both students and parents;
- Mathematics teaching, curriculum, and assessment strategies should build on the prior knowledge and experiences of students and on their cultural heritage.

To verify that barriers have been removed, educators should monitor enrollment and achievement data to determine whether second-language learners have gained access to, and are succeeding in, mathematics courses. Reviews should be conducted at school, district, state or provincial, and national levels.

Mathematics Program Leaders in Elementary, Middle Schools, and High Schools, March 2000*

Position

The National Council of Teachers of Mathematics believes that the designation and the support of school- and district-level mathematics program leaders are essential to the improvement of mathematics instruction and achievement for all students.

*Adapted from "Improving Student Achievement through Designated District and School Mathematics Program Leaders," a position statement of the National Council of Supervisors of Mathematics.

Rationale

The demands on teachers of mathematics have never been greater:

- Society in general and a changing workplace in particular demand broader mathematical literacy for every student.
- Technology forces a reconsideration of what mathematics is essential and how best to teach this mathematics.
- National standards and higher academic expectations have heightened the need for teachers to update their knowledge of, and skills in, mathematics.

Recommendations

The designation of school- and district-level mathematics program leaders must be seen as an indispensable component of providing high-quality mathematics programs for all students. These program leaders should

- be knowledgeable about mathematics, the mathematics curriculum, and mathematics pedagogy;
- coordinate the planning, implementation, and evaluation of mathematics programs;
- promote excellence in mathematics education for all students;
- be visionary agents of positive change, knowledgeable about national standards, aware of current research, and able to translate these standards and this research into classroom practice;
- regularly visit classrooms and maintain an awareness of needs, problems, and issues;
- link with stakeholders in education and enlist their support in improving the quality of the teaching and learning of mathematics.

In addition, these designated program leaders should be able to serve as informed resources in the areas of curriculum design, instructional strategies, student and program assessment, professional development, and the development of partnerships with the broader community. Among the expectations for school- and district-level mathematics program leaders are the following:

In the area of **curriculum design**:

- To coordinate the development and implementation of a sound and coherent grades K–12 mathematics curriculum
- To ensure curricular alignment and coordination between grade levels and courses
- To assist teachers in using rich and challenging problems and activities that integrate mathematics into other disciplines and the content of other disciplines into mathematics
- To guide the ongoing review and revision of the curriculum and ensure alignment with state and local guidelines

In the area of **instructional strategies** and the **use of technology and materials**:

- To recommend programs, instructional technologies and materials, oversee their piloting and adoption, and evaluate their effectiveness
- To acquaint teachers with successful and innovative strategies, including translating research findings into practice
- To assist teachers in the effective use of technology in daily instruction
- To assist teachers by modeling effective instructional strategies
- To encourage reflection on, and the discussion of, what is working, what is not working, and how to make improvements

In the area of **assessment**:

- To ensure the alignment of assessment instruments with the curriculum
- To assist teachers in designing and implementing a broad range of assessment tools
- To collect and analyze data about what is and is not working and to use these data, including results from students' assessment, to improve curriculum and instruction
- To interpret the results of assessment for parents and the community at large

In the area of **professional development**:

- To collaborate with the staff to determine needs and priorities for professional development
- To conduct or facilitate professional development activities and motivate colleagues to engage in ongoing professional growth and evelopment
- To encourage involvement in professional organizations
- To encourage professional communication between and among colleagues
- To promote the mentoring of colleagues and professional visits among teachers

In the area of **forging partnerships**:

- To communicate with committees, school boards, administrators, teachers, parents, and students about the importance of mathematics and the need for high-quality mathematics programs
- To cultivate connections with local businesses, industry, and post-secondary mathematics and mathematics education communities
- To establish and support forums and encourage dialogue among groups that influence the shape and direction of school mathematics programs

Mathematics Leaders in Secondary Schools, April 1988

Strong mathematics leadership in each school building is a significant factor in the improvement of secondary mathematics instruction. Leaders in school mathematics are essential for improving the mathematical knowledge and pedagogical competence of the staff, for providing information about mathematics instruction within the school and between the school and its feeder schools, and for assuring the implementation of a comprehensive, high-quality program. Such leaders give assistance and support to teachers and administrators in promoting excellence in mathematics for all students.

For these reasons, the National Council of Teachers of Mathematics recommends that every secondary school identify a mathematics leader, often called the department head, to provide ongoing leadership and assistance in planning, implementing, and evaluating a comprehensive mathematics program. The mathematics leader must have an in-depth understanding of mathematics and a thorough knowledge of, and competence in, effective teaching methodology, coupled with the ability to work well with others. Such a person would be available at the building level to provide leadership in matters concerning (a) curriculum, (b) methodology and materials, (c) student assessment, (d) professional development, and (e) procedural duties.

Specifically, the mathematics leader should provide leadership and assistance to teachers, principals, supervisors, and other administrators in the following ways:

1. Curriculum
   a. Fostering the development, implementation, evaluation, and updating of departmental goals and objectives
   b. Coordinating the implementation of an effective instructional scope and sequence for mathematics
   c. Contributing toward the development of course offerings to meet diverse student needs
   d. Facilitating the review and revision of curricula to incorporate new developments in mathematics education
   e. Encouraging the integration of mathematics with other content areas
   f. Providing curriculum articulation between schools and among grade levels
   g. Assuring consistent standards among teachers and among courses
2. Methodology and Materials
   a. Acquainting teachers with successful and innovative strategies for classroom instruction through demonstration lessons and conferences
   b. Helping teachers select and implement activities that improve students' motivation and attitude toward mathematics
   c. Facilitating the review and selection of textbooks
   d. Making recommendations for the acquisition of teaching resources such as manipulative materials, calculators, computers, computer software, and other instructional media

3. Student Assessment
    a. Participating in the review and selection of assessment techniques
    b. Assisting teachers in the design and implementation of all forms of assessment
    c. Helping teachers interpret and use assessment data
    d. Working with the guidance staff to facilitate appropriate student placement
4. Professional Development
    a. Encouraging teachers to pursue further study of the mathematical sciences
    b. Providing information about innovative materials and teaching strategies recommended by the mathematics education community
    c. Encouraging teachers to become participating members in professional mathematics organizations
    d. Facilitating the attendance of teachers at professional meetings and conferences
    e. Encouraging discussion of the professional literature in mathematics education and related fields
5. Procedural Duties
    a. Orienting new teachers
    b. Assisting substitute teachers and student teachers
    c. Conducting departmental meetings
    d. Participating in the development of the master schedule
    e. Formulating the departmental budget
    f. Participating in the recruitment and selection of mathematics teachers
    g. Facilitating students' participation in competitions and special programs

For the mathematics leader to accomplish the expected tasks, it is recommended that school systems provide a monetary supplement, released time, sufficient support services, and the opportunity to participate in professional activities in mathematics education at the local, state/provincial, regional, and national levels. The school mathematics leader is a critical component of an effective mathematics instructional program. The need for strong leadership at the building level is not merely desirable, it is essential.

Metrication, March 2000

Position

The National Council of Teachers of Mathematics recommends the use of the metric system as the primary measurement system in mathematics instruction.

Background and Rationale

Measurement is a common daily activity performed throughout the world and in all sectors of society. As a well-conceived, logical system of units, the metric system provides models that reinforce concepts and skills involving numeration, decimal relationships, and estimation, connecting mathematics to the rest of the pre-K–12 curriculum. For example:

- Conversion between metric units is facilitated by the decimal nature and consistent prefixes of the system, which leads to a high degree of accuracy in measurements.
- Units for volume, capacity, and mass are interrelated (for example, one cubic decimeter of water, which is one liter, has a mass of one kilogram).

On an international level in the scientific and industrial worlds, the metric system is the standard system of measurement. Canada adopted the metric system long ago and immediately mandated that all school curricula use the metric system exclusively. To compete in a world that already functions with this system, our students also need to be competent with the metric system.

Recommendations

- Given that the English customary system is well established in the United States and the metric system is used throughout the rest of the world, the pre-K–12 curriculum should include both systems.
- For all teachers to become proficient with the metric system, professional development and support will be necessary and should be provided.
- All mathematics educators should assume responsibility for providing leadership and direction in metrication by integrating it into the entire curriculum at every grade level.

Professional Development of Teachers of Mathematics, July 1998

The National Council of Teachers of Mathematics believes that teachers of mathematics must have access to professional programs and activities that reinforce the instructional and assessment strategies it advocates for students in the Standards and that the primary focus for all professional development programs is helping teachers teach mathematical content to their students.

The professional development of teachers of mathematics is a process of learning: learning mathematics and about mathematics; learning about students and how they learn, individually and in the social setting of school; and learning the craft of teaching. In light of the Standards, professional development must also include learning new ways to develop mathematical power in all students.

Professional development must:

- Actively engage teachers in developing an understanding of mathematics, pedagogy, and students, just as good instruction actively engages students in learning and developing their own understanding;

- Allow adequate time and furnish support for teachers to explore, practice, discuss, refine, apply, and reflect on new ideas, techniques, and practices, just as good instruction allows students to explore, practice, discuss, refine, and apply new understandings;
- Provide feedback and support from colleagues and supervisors, particularly when changes are being implemented, just as good instruction supports students with feedback from teachers and peers;
- Focus on school or department faculties as the primary unit of attention, just as classrooms are the primary unit of focus for students;
- Set students' learning of mathematics as the ultimate outcome of all professional development, just as students' learning of mathematics is the ultimate goal of all classroom mathematics instruction.

NCTM recommends, therefore, that all professional development programs must:

- Be planned and implemented as a regular part of teachers' ongoing work;
- Be guided by a comprehensive framework that emphasizes:
  —Building teachers' knowledge of content, pedagogy, and students;
  —Translating the vision of the Standards into daily teaching practice;
  —Creating vibrant support networks for teachers at the local, state, or provincial, regional, and national levels;
- Attend to teachers' needs as they work to enhance the learning of all students—including students from groups typically underrepresented in mathematics—in their classrooms, schools, and communities;
- Provide teachers with sustained opportunities to collaborate with colleagues, to share and critique ideas, and to draw on the knowledge and perspectives of others as reflective and analytical members of a professional community;
- Recognize that the needs of teachers should evolve as they move from being prospective teachers to new teachers to experienced teachers;
- Involve teachers in the identification of their needs and in the program's design and implementation;
- Be shaped by the most current knowledge of effective practice and ways to achieve it;
- Be supported by adequate long-term financial and structural resources.

Position Statement on Research, February 1999

Position

The National Council of Teachers of Mathematics believes that research must be conducted to study student learning, instructional practices, policies, and programs to improve mathematics education, whether the focus of this research is on individual students, classrooms, schools, school districts or states or provinces. Further, it is the position of the National Council of Teachers of Mathematics that if mathematics education research is to be responsive to questions regarding pedagogy and student learning, then collaboration between teachers and researchers is critical.

Rationale

Educators who have studied student learning, created new programs, proposed new ideas, or expressed new commitments to old ideas should be encouraged to hold their proposals up for critical analysis. One should collect evidence to assess a program or theory whenever possible, even if it increases the likelihood that one's ideas will be discredited. The research literature then serves not as a repository of facts and answers but as a resource that educators can draw on in their question to generate more valid and useful knowledge. When viewed from this dynamic perspective, research becomes a relevant tool in the improvement of educational practice.

Both quantitative and qualitative analyses, on a large as well as a small scale, are required to provide rich and informative pictures of how policies, programs, and theoretical perspectives affect classroom practice and student learning. Students of past efforts make clear that fundamental educational problems cannot be solved by altering or examining just one, or even a few, aspects of a system as complex as schooling. Given the challenge to create appropriate research designs to analyze complex systems, a wide range of design must be employed.

Teachers who are reflecting on their practice can apply those same skills in professional conversations. Teachers should meet with other teachers and with researchers to discuss outcomes and observations, to consider curricular and instructional dilemmas from a variety of perspectives, and to offer classroom data for analysis and interpretation.

Recommendations

Researchers should work together with teachers and administrators to address, examine, and document the initiation, the implementation, and the long-term implications of changes in state or provincial, district, and school-based mathematics programs, in mathematics curricula, and in classroom mathematics instruction and organization, as well as the impact of these changes on students' mathematical knowledge and achievement.

Researchers and teachers should develop effective modes of professional interaction and communication within and between their communities.

Researchers must learn to use the rich sources of information and interpret the information that teachers may share from their mathematics classrooms and schools.

Educational policies, programs, and theories as well as student learning and instructional practices in mathematics education should be examined from an unbiased viewpoint using appropriate research methods and exemplifying differing designs and durations.

Researchers must share their research-based or interpretative insights in ways that address teachers' daily concerns for mathematics classroom practice.

Teaching Mathematics in the Middle Grades, July 1998

The National Council of Teachers of Mathematics advocates that all teachers of mathematics for middle-grades students have extensive preparation in mathematics and its pedagogy as related to the characteristics and needs of students in those grades.

All students have the right to a mathematics education that ensures mathematical literacy and the development of the concepts, skills, and dispositions necessary for a meaningful and productive life. Middle-grades students in particular are at a crucial stage when the attitudes they develop toward mathematics can have a significant impact on their changes for success in high school mathematics and on their life choices. It is therefore essential that mathematics teaching in the middle grades help students experience mathematics as a personally meaningful and worthwhile endeavor.

Therefore, middle-grades mathematics teachers should:

- Have high-quality preparation in mathematics and mathematical pedagogy as outlined in NCTM's *Professional Standards for Teaching Mathematics;*
- Create a learning environment that fosters students' confidence in their own abilities and skills in problem solving, reasoning, making connections, and communicating with and about mathematics;
- Work with other teachers as a team to develop meaningful mathematics programs;
- Modify mathematics instruction as needed to motivate and nurture students;
- Interact sensitively with students and parents from diverse racial and cultural backgrounds.

Schools, school districts, colleges, universities, and provincial or state licensing agencies are encouraged to collaborate in designing and implementing undergraduate, in-service, and graduate programs that:

- Include preparation for the middle-grades teacher's license or certification;
- Enable middle-grades teachers of mathematics to teach for the present and future demands of society in accordance with the developmental needs of the students.

The Use of Technology in the Learning and Teaching of Mathematics, July 1998

The appropriate use of instructional technology tools is integral to the learning and teaching of mathematics and to the assessment of mathematics learning at all levels.

Technology has changed the ways in which mathematics is used and has led to the creation of both new and expanded fields of mathematical study. Thus, technology is driving change in the content of mathematics programs, in methods for mathematics instruction, and in the ways that mathematics is learned and assessed. A vital aspect of such change is a teacher's ability to select and use appropriate instructional technology to develop, enhance, and extend students' understanding and application of mathematics. It is essential that teachers continue to explore the impact of instructional technology and the perspectives it provides on an expanding array of mathematics concepts, skills, and applications.

Therefore, the National Council of Teachers of Mathematics makes the following recommendations:

- Every student should have access to an appropriate calculator.
- Every mathematics teacher should have access to a computer with appropriate software and network connections for instructional and non-instructional tasks.
- Every mathematics classroom should have computers with Internet connections available at all times for demonstrations and students' use.
- Every school mathematics program should provide students and teachers access to computers and other appropriate technology for individual, small-group, and whole-class use, as needed, on a daily basis.
- Curriculum development, evaluation, and revision must take into account the mathematical opportunities provided by instructional technology. When a curriculum is implemented, time and emphasis must be given to the use of technology to teach mathematics concepts, skills, and applications in the ways they are encountered in an age of ever increasing access to more-powerful technology.
- Professional development for preservice and in-service teachers must include opportunities to learn mathematics in technology-rich environments.

Interdisciplinary Learning, Pre-K–Grade 4, September 1994

Recent calls for educational reform focus on the need for curricula emphasizing conceptual learning that is integrated across traditional subject areas. Responding to this need, the major national subject-matter organizations—the National Council of Teachers of Mathematics, the National Council of Teachers of English, the International Reading Association, the National Science Teachers Association, the National Council of the Social Studies, the Speech

Communication Association, and the Council for Elementary Science International—met to discuss and develop guidelines for integrating the curriculum from Pre-K–Grade 4. A result of their discussions is this position statement, which outlines the principles that should guide the implementation of an integrated curriculum.

Basic to this effort is the belief that educational experiences are more authentic and of greater value to students when the curricula reflect real lie, which is multifaceted rather than being compartmentalized into neat subject-matter packages. Interdisciplinary instruction capitalizes on natural and logical connections that cut across content areas and is organized around questions, themes, problems, or projects rather than along traditional subject-matter boundaries. Such instruction is likely to be responsive to children's curiosity and questions about real life and to result in productive learning and positive attitudes toward school and teachers.

The participating organizations believe that educational experiences should help develop children's natural curiosity and their inclination to construct meaning. A focus on relationships across disciplines should encourage creative problem solving and decision making because it makes available to student the perspectives, knowledge, and data-gathering skills of all the disciplines. Such an instructional process should also encourage children to interact with others in a learning community where diversity of thought and culture is valued.

With the above statements in mind, the participating organizations recommend the following guiding principles:

Interdisciplinary Pre-K–Grade 4 curricula should:

1. Maintain the integrity of content drawn from the disciplines by using meaningful connections to sustain students' inquiry between and among those disciplines. Interdisciplinary instruction should be authentic and worthwhile. It is important for students to develop familiarity with the knowledge, assumptions, and methods of inquiry used in many subject-matter areas in order to be able to select that which is most appropriate for any given situation. Major concepts and methods from the various disciplines should be taught as part of integrated units and at times that are appropriate to students' interests and cognitive and social development.
2. Foster a learning community in which students and teachers determine together the issues, questions, and strategies for investigation. An appropriate balance should be maintained between student-initiated and teacher-initiated learning experiences.
3. Develop democratic classrooms. Select curricula and organize classrooms that will cultivate a learning community in which students develop both independence as investigators and the ability to collaborate with each other and with teachers to raise questions, investigate issues, and solve problems. Students should be encouraged to assume increasing responsibility for their learning sot that they can gain confidence in their abilities to find information, understand and articulate ideas, and make decisions.

4. Provide a variety of opportunities for interaction among diverse learners: for example, discussion, investigation, product development, drama, and telecommunications. Collaborative interaction among students who differ in abilities, perspectives, experiences, ethnicity, and interests promotes learning for all students and fosters positive attitudes towards others and toward learning.
5. Respect diversity of thought and culture. Students should learn by employing a variety of learning strategies, engaging in a wide range of learning experiences, and examining many and varied perspectives.
6. Teach students to use a wide variety of sources, including primary sources, oral communication, direct observation, and experimentation. The use of multiple and diverse sources accommodates various learning styles, interests, and abilities; teaches the importance of cross-checking for accuracy and bias; and develops students' ability to choose the most appropriate and productive sources for investigating specific questions or problems.
7. Use multiple symbol systems as tools to learn and present knowledge. These can include symbols used in language, mathematics, music, and art, as well as those that translate knowledge into tables, charts, and graphs.
8. Use wide-ranging assessments to evaluate both the processes and outcomes of student learning. Ongoing assessment during the inquiry process should lead students and teachers to determine what criteria can be used to identify quality work. Decisions about instruction should be based on a variety of formal and informal assessment strategies that move beyond the exclusive use of objective measures to include observation, portfolios, and performance assessments.

An interdisciplinary education which draws from the knowledge and processes of multiple disciplines should encourage students to become active learners equipped with the analytical, interpretative, and evaluative skills needed to solve real life problems. Eliminating artificial barriers among subject areas gives students a broader context for solving real-life problems.

Teacher Time, March 2000

Position

The National Council of Teachers of Mathematics believes that schools must restructure the use of teacher time to enable teachers to implement a variety of activities aimed at enhancing students' learning. These activities should provide opportunities for teachers to reflect on their practice, support teachers' ongoing professional development, and support a culture of professionalism.

Background and Rationale

High standards for mathematics curricular content and the increased diversity of the student population require teachers to have an in-depth understanding of mathematics content and effective methods for instruction and assessment. Internationally, mathematics achievement is higher among those students whose teachers have more opportunities within the school day to reflect on their teaching and discuss instructional and related issues with their colleagues.

The professional literature indicates that teachers frequently cite professional isolation as a serious impediment to their ability to improve curriculum, instruction, and assessment practices. Teachers improve their practice when time is provided to collaborate with their colleagues by sharing ideas and materials, discussing mathematics content, comparing instructional methods, learning about different forms of student assessment, and exploring an ever-expanding array of technological tools and their appropriate uses.

Moreover, current research indicates that teachers need time to reflect on their current teaching practices and to increase their own mathematical knowledge. In order for teachers to increase the mathematics learning of their students and to grow professionally, they must be afforded adequate opportunity during the regular school day for reflection, personal planning, and collaboration with colleagues. This vitally important time must be provided for in addition to structured, on-going professional development opportunities.

Recommendations

School schedules should be restructured to ensure the implementation of two types of strategies:

- Strategies that provide for the well-planned allocation of personal reflective time for all teachings as well as time for collaborative professional interaction with their peers. This allocated time should include opportunities for
  —deepening understanding of mathematical content;
  —common planning among teachers;
  —the mentoring of new teachers;
  —reading, sharing, and discussing current research and educational literature.
- Strategies for long-term, coordinated staff development that is based on a long-range plan that include support for both the pedagogical and the mathematical needs of teachers.

High-Stakes Testing, November 2000

Position

The National Council of Teachers of Mathematics believes that far-reaching and critical educational decisions should be made only on the basis of multiple measures. A well-conceived system of assessment and accountability must consist of a number of assessment components at various levels.

Rationale

High-stakes tests are tests that are used to make significant educational decisions about children, teachers, schools, or school districts. To use a single objective test in the determination of such things as graduation, course credit, grade placement, promotion to the next grade, or placement in special groups is a serious misuse of such tests. This misuse of tests is unacceptable. The movement toward high-stakes testing marks a major retreat from fairness, accuracy, and educational equity. When test use is inappropriate, especially in making high-stakes decisions about a child's future, it undermines the quality of education and equality of opportunity.

Just as disturbing as the serious misuse of these tests is the manner in which the content and format of these high-stakes tests tends to narrow the curriculum and limit instructional approaches. Test results may also be invalidated by teaching so narrowly to the objectives of a particular test that scores are raised without actually improving the broader, often more important, set of academic skills that the test is intended to measure.

Assessment should be a means of fostering growth toward high expectations and should support high levels of student learning. When assessments are used in thoughtful and meaningful ways, students' scores provide important information that, when combined with information from other sources, can lead to decisions that promote student learning and equality of opportunity. The misuse of tests for high-stakes purposes has subverted the benefits these tests can bring if they are used appropriately.

Recommendations

- Multiple sources of assessment information should be used when making high-stakes decisions. No single high-stakes test should be used for making decisions about the tracking, promotion, or graduation of individual children.
- Assessment methods must be appropriate for their purposes.
- All aspects of mathematical knowledge and its connections should be assessed.
- Instruction and curriculum should be considered equally in judging the quality of a program.
- Assessment should advance students' learning and inform teachers as they make instructional decisions.
- Assessment should be an open process with everyone knowing what is expected, what will be measured, and what the results imply for what should be done next.
- If tests are used as one of multiple measures in making high-stakes decisions about students, those tests must be valid and reliable for the purposes for which they are used; they must measure what the student was taught; they must provide students with multiple opportunities to demonstrate proficiency; and they must provide appropriate accommodations for student with special needs or limited English proficiency.
- All standardized assessments of mathematical understanding at the national, state, province, district, or classroom level should be aligned with the NCTM standards.